Stochastic Geometric Analysis with Applications

Other World Scientific Titles by the Author

An Informal Introduction to Stochastic Calculus with Applications
ISBN: 978-981-4678-93-3
ISBN: 978-981-4689-91-5 (pbk)

Deterministic and Stochastic Topics in Computational Finance
ISBN: 978-981-320-307-5
ISBN: 978-981-320-308-2 (pbk)

An Informal Introduction to Stochastic Calculus with Applications
Second Edition
ISBN: 978-981-12-4709-5
ISBN: 978-1-944660-27-7 (pbk)

Stochastic Geometric Analysis with Applications

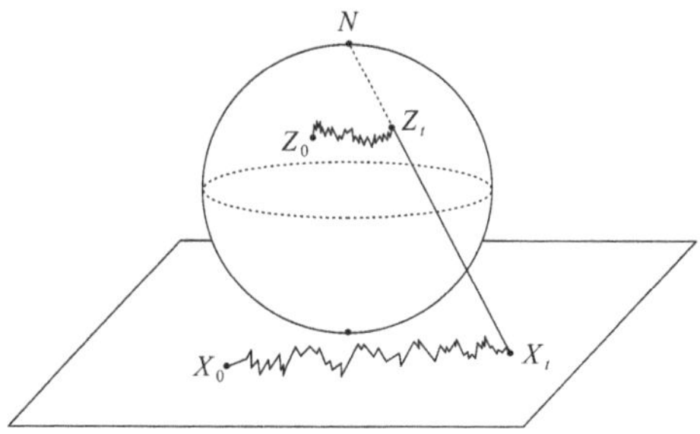

Ovidiu Calin
Eastern Michigan University, USA

World Scientific

NEW JERSEY • LONDON • SINGAPORE • BEIJING • SHANGHAI • HONG KONG • TAIPEI • CHENNAI • TOKYO

Published by

World Scientific Publishing Co. Pte. Ltd.
5 Toh Tuck Link, Singapore 596224
USA office: 27 Warren Street, Suite 401-402, Hackensack, NJ 07601
UK office: 57 Shelton Street, Covent Garden, London WC2H 9HE

Library of Congress Control Number: 2023049613

British Library Cataloguing-in-Publication Data
A catalogue record for this book is available from the British Library.

STOCHASTIC GEOMETRIC ANALYSIS WITH APPLICATIONS

ISBN 978-981-12-8327-7 (hardcover)
ISBN 978-981-12-8328-4 (ebook for institutions)
ISBN 978-981-12-8329-1 (ebook for individuals)

For any available supplementary material, please visit
https://www.worldscientific.com/worldscibooks/10.1142/13592#t=suppl

To the memory of Professor Peter Charles Greiner
who has inspired us through his teachings, guidance and mentorship support.

Preface

This book is a thorough exploration of the intersection of Stochastic Analysis, Geometry and Partial Differential Equations (PDEs). Its main focus is on the study of diffusions that are induced by an underlying geometry, which can be either Riemannian or sub-Riemannian. The exposition will delve into the various ways in which geometry shapes and influences the behavior of these diffusions, and how a deeper understanding of this relationship can be used to solve a wide range of problems in PDEs, mathematical finance and other fields. The book will also examine the ways in which curvature influences the diffusion of Brownian movement along a given curve or surface, and how these diffusions can be intrinsic or extrinsic to the given manifold. The book will cover both theory and applications, providing readers with a comprehensive understanding of the topic.

This book aims to use stochastic methods to not only offer alternative proofs for results in geometric analysis, but also to uncover new questions and properties that arise from this approach. From this point of view, the book endeavors to unify the relationship between PDEs, nonholonomic geometry and stochastic processes. They key is a specific condition imposed on some vector fields, which appears in both PDEs and nonholonomic geometry and goes by different names such as the bracket-generating condition, Chow's condition, the full rank condition, or Hörmander's condition. In PDEs, this condition implies the hypoellipticity of a sum of squares of vector fields operator, while in geometry, it represents a global connectedness condition by curves tangent to the distribution generated by the vector fields that make up the aforementioned operator. The question that this book seeks to answer is how these two results can be unified. The author proposes that this can be achieved by the use of diffusions.

The PDEs results imply that such a diffusion has a smooth transition probability, while the geometric considerations state that the diffusion can spread from any initial point to all other points. This type of stochastic process is called a sub-elliptic diffusion or a non-holonomic diffusion, since it is a diffusion with constraints on the directions it is allowed to diffuse. The book presents several examples of such diffusions and computes their transition probability, which in turn, provide the heat kernels of the associated differential characteristic operator.

The objective was not to present the cutting-edge advancements in the field, but rather to explore various aspects related to the computation of heat kernels in closed form. Additionally, we examined their connections with other fields such as stochastic calculus, geometry and equations, as well as explore the physical interpretations of the obtained formulas.

This book is intended for researchers and practitioners in the fields of mathematics, physics, and engineering who are interested in the application of stochastic techniques in geometry and PDEs and how it relates to mathematical finance and electrical circuits. It aims to provide a comprehensive understanding of the topic and the ways in which PDEs, nonholonomic geometry, and stochastic processes can be unified.

Overview

Chapter 1 provides a comprehensive introduction to the fundamental concepts and prerequisites of Stochastic Calculus, which will be essential for understanding the subsequent chapters of the book. It includes important results such as Cameron-Martin's formula, Lévy's area, Kolmogorov's equation, and Feynman-Kac's theorem, among others. These concepts and results serve as the foundation for the rest of the book.

In **Chapter 2**, the author investigates the relationship between Brownian motion and geometry by studying Brownian motions on various shapes such as lines, circles, and curves. The chapter also examines the area of Brownian triangles, and diffusions near a curve. This chapter provides the challenge of understanding of how Brownian motion is influenced by underlying geometric structures and shapes. Thus, we consider intrinsic diffusions, which are constrained to live on the manifold, such as an electron in a wire or metal sheet that spreads stochastically without leaving the manifold, and how their behavior depends on the geometry of the manifold. On the other hand, the chapter also examines extrinsic diffusions, which are exterior to the surface, such as wind molecules passing by an airplane wing, which are also influenced by the underlying geometry.

Chapter 3 introduces the basic concepts of elliptic and sub-elliptic operators, including Hörmander's theorem, and explore the relationship between these operators and underlying Riemannian and sub-Riemannian geometries. The Chow-Rashevski connectivity theorem is also covered, making the relation to connectivity property of distribution. The chapter provides examples of distributions and non-holonomic systems which will be studied in more detail in later chapters of the book. This chapter lays the foundation for the understanding of the geometric and analytic properties of the operators and systems that are studied throughout the book.

Chapter 4 is dedicated to the computation of heat kernels for a variety of operators, which are sums of squares of vector fields, using the stochastic method. Since the heat kernel is the transition probability of the associated diffusion, stochastic techniques can be used to find the transition density for elliptic or sub-elliptic operators, such as the Kolmogorov operator, Grushin operator, Heisenberg operator, and Casimir operator among others. The smoothness of the transition density is a direct consequence of the Hörmander theorem.

Chapter 5 focuses on the explicit computation of fundamental solutions for a selection of important operators such as the Euclidean Laplacian, Laplacian on the hyperbolic space, Heisenberg, Grushin or saddle laplacians. This chapter builds on the results of Chapter 3. The computation of these solutions is done in explicit forms and is a key step in the analysis of these operators.

Chapter 6 approaches elliptical diffusions. These are diffusions that can propagate in any direction of the space and their generator operators are elliptic. The chapter is focused on elliptic diffusions on manifolds that become Brownian motions on the associated Riemannian manifolds; these includes the familiar cases of diffusions on spheres, hyperbolic spaces, etc. This chapter also includes a financial interpretation of the hyperbolic diffusion. The study of elliptical diffusions is a crucial step in understanding the dynamics of diffusions on different geometric structures and their applications in mathematical finance.

In **Chapter 7**, the author examines sub-elliptic diffusions. These are diffusions that are constrained to move only in certain allowed directions and can be considered as diffusions on non-integrable distributions. The chapter covers the study of Heisenberg diffusion, Grushin diffusion, Engel diffusion as well as some nonholonomic diffusions which come from nonholonomic geometry (such as the rolling coin or knife edge). This chapter provides a deeper understanding of the dynamics of sub-elliptic diffusions and their applications in nonholonomic geometry. The study of these diffusions is an important step in understanding the behavior of diffusions on non-integrable distributions and their geometric and analytic properties.

Chapter 8 presents a sub-Riemannian version of the Poincaré lemma. This provides equivalent closeness and exactness conditions for a sub-elliptic differential system of equations, which means a system with a number of equations less than the dimension of the space. This result is important as it allows to study the properties of sub-elliptic differential systems in a similar way as it is done for Riemannian differential systems. This chapter provide a deeper understanding of the geometric and analytic properties of sub-elliptic differential systems and their relationship with the underlying sub-Riemannian geometry. The chapter covers the cases of Heisenberg, Grushin, and Engel groups in detail.

Chapter 9 contains a variational approach to the study of LC-circuits, which are oscillating circuits formed by an inductor and a capacitor. Both Lagrangian and Hamiltonian methods are used to analyze the dynamics of these circuits. The chapter also explores the relationship between the heat kernel and LC-circuits and provides a stochastic interpretation of the dynamics of electrons flowing through the circuit. This leads to a variant of the Cameron-Martin formula, which is a key result in the study of stochastic processes. This chapter provides a deeper understanding of the dynamics of LC-circuits and their connection to stochastic analysis and geometric analysis.

Bibliographical Remarks

It is important to note that there are already several books on Stochastic Geometry available in the literature, but the content and approach of the present book sets it apart from the others.

In general, the field of Stochastic Geometry deals with the mathematical foundations of random geometric shapes and their properties, without specifically focusing on the connection to partial differential equations or diffusions on manifolds. It usually involves the study of geometric objects that are defined by a probability measure, such as random polygons, random points in space, and random tessellations. There are many books available that cover topics of Stochastic Geometry from both theoretical and practical perspectives, such as [124], [54], [71], [127], [116], [20], [13], [76], and [8], among others.

However, this book is unique in its approach as it focuses on the use of diffusions to study partial differential equations. This concept is not novel, as it was first explored in the 1930s by Kolmogorov [90], who calculated the heat kernel of a degenerate operator using this method. The idea was further developed in 1970s by Hulanicki [79] and Gaveau [60], who expanded the computation to include the case of the Heisenberg group. The technique was shortly expanded to more general hypoelliptic operators in Gaveau [61].

This book is focused on the study of diffusion processes that are associated with a Riemannian or sub-Riemannian geometry. These diffusions are characterized by a generator, such as the Laplacian or sub-Laplacian, which is constructed from a sum of squares of smooth vector fields. If these vector fields satisfy a nonholonomy condition, which means that their iterated Lie brackets span the entire tangent space at each point, then it has been shown by Chow [46] and Rachevskii [119] in late 1930s that any two points in the space can be connected by a piece-wise smooth curve that is tangent to the distribution generated by the aforementioned vector fields. These spaces, known as Carnot spaces, have applications in several applied fields such as robotics, thermodynamics, image reconstruction, and control theory.

The connection between PDEs and nonholonomic geometry was first established in the 1960s by Hörmander [77], who proved that a sum of squares operator satisfying the nonholonomy condition, known also as Hörmander's condition, is hypoelliptic. This means that the heat kernel of the corresponding differential operator has a smooth transition probability, and as a result, the associated diffusion process will also be smooth.

There is a wealth of research on deriving explicit expressions for the heat kernels of hypoelliptic operators in the literature. A few books dealing with the subject are Widder [138], Schulman [125], Avramidi [6], and Calin et al. [29]. One notable approach to this problem has been to utilize the complex Hamiltonian method developed by Beals, Gaveau and Greiner. This method is based on the idea that heat primarily propagates along geodesic flows, with geodesics determined by the underlying geometry imposed by the given vector fields. In Riemannian geometry, the geodesic between two points is unique if they are close enough. However, in sub-Riemannian geometry, the geodesic is not unique and the methods used to find geodesics, and hence heat kernels, must be adapted accordingly. This approach has been applied in several notable publications, such as: [10], [30], [35], [38], [43], [44], [62], [63], [65], and [66], among others.

While previous approaches, such as the complex Hamiltonian method, have been successful for operators of step $k \leq 2$, they have limitations for operators of higher step. This book aims to investigate whether stochastic methods can be used to overcome these limitations and provide new insights into the problem.

To achieve this goal, we have drawn upon results related to Brownian motions collected from the books of Borodin and Salmilen [15], Karatzas and Shreve [85], and Revuz and Yor [120]. We used the concept of Brownian motion on Riemannian manifolds, which was introduced and used in the work of McKean [82], [107], Strook [128], Meyer [109], Oksendal [114] and others. In particular, we focus on the use of hypoelliptic diffusions, which evolve along a horizontal nonholonomic distribution, an approach that bridge the connection between sub-Riemannian geometry and the study of sub-elliptic operators, as it was introduced in Calin [21] and Boscain et al. [16], [17]. It is worth noting that the last two papers deal with applications of hypoellitic diffusions to image reconstruction and human vision.

The section of the book that covers the study of integrability in sub-elliptic systems draws heavily from the research of Calin et al. [27, 28, 31, 32, 33], Isagurova [81], and Kristály [91].

The chapter that deals with the application of Lagrangians in electrical circuits draws inspiration from the papers of Werne [137] and Jettsema [84], and incorporates theoretical concepts from the book of Calin and Chang [36].

It is certainly possible that there are many more publications in literature that we have overlooked due to negligence or ignorance, and we apologize for that. We have strictly limited ourselves mainly to the material that was used in the present book.

Acknowledgements

The monograph is the result of a collaborative effort and the author would like to take this opportunity to acknowledge the support of several institutions and individuals. The research for this book was partially supported by the Summer Research Award 2022, Dean's Faculty Professional Development Award 2022, as well as a Research Teaching Release at Eastern Michigan University.

The author wishes to express his deep appreciation to the late Professor Peter Greiner from the University of Toronto for his invaluable guidance and support throughout the research process.

The author would also like to acknowledge the contributions of colleagues and friends who have provided valuable feedback, suggestions and support during the preparation of the manuscript. The book is the result of many years of research and the author is grateful to all the institutions where he has been able to present this work and receive feedback, including Princeton University, Hong Kong University of Science and Technology, Academia Sinica (Taiwan) and Kuwait University.

Finally, the author would like to thank the World Scientific Press team for their encouragement and support in making this publication possible. Their efforts have been invaluable in bringing this book to fruition.

August, 2023
Michigan, USA
O.C.

Interdependence chart of the chapters

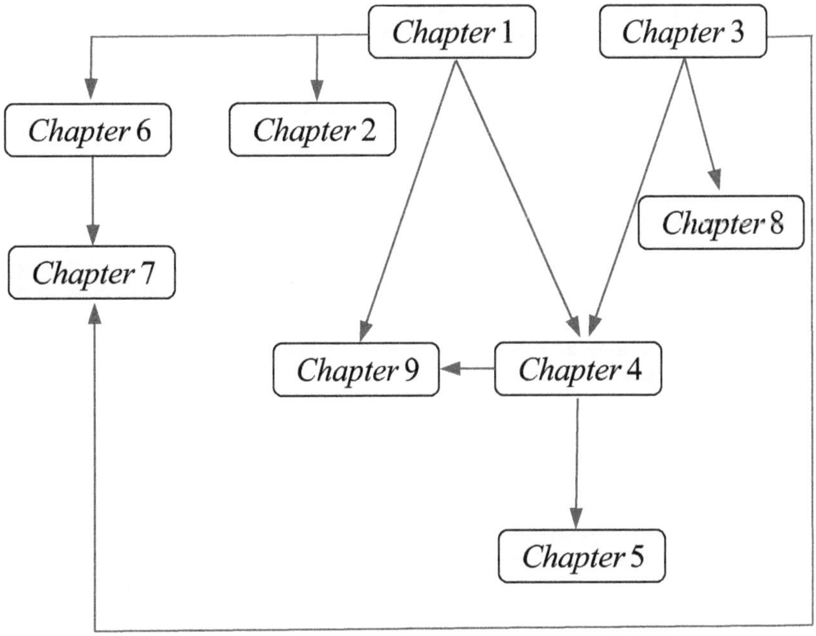

The chart illustrates the relationship between each chapter and its preceding chapters (with a few peripheral references excluded).

List of Notations and Symbols

The following notations have been frequently used in the text.

Linear Algebra

$\mathbb{R}$	Set of real numbers
e_i	Vector $(0, \cdots, 0, 1, 0, \cdots, 0)$
e_i	Vector $(0, \cdots, 0, 1, 0, \cdots, 0)$
$\mathbb{R}^n$	n-dimensional Euclidean space
$\det A$	Determinant of matrix A
A^{-1}	Inverse of matrix A
$\mathbb{I}$	Unitary matrix
A^T	The transpose of matrix A
$\mathcal{R}_\theta(\mathbf{x})$	The rotation of agle θ of the vector $\mathbf{x}$
$\langle \cdot \rangle$	The Euclidean inner product

Calculus

$\Gamma(n)$	Euler's Gamma function
$I_\nu(x)$	The modified Bessel function of the first kind
$K_\nu(x)$	The modified Bessel function of the second kind
$\nabla f(x)$	The gradient of $f(x)$
$\theta_3(z\|\tau)$	The theta 3 function
L_X	The Lie derivative
$H(x,p)$	The Hamiltonian function
$\langle \cdot \rangle$	The Euclidean inner product
∂_t	Partial derivative in t
d/dx	derivative with respect to x
$\sim$	Asymptotic correspondence
$\mathcal{F}_x$	The Fourier transform in x
$o(x)$	The Landau symbol

Stochastic Processes

$(\Omega, \mathcal{F}, P)$	Probability space
Ω	Sample space
ω	Element of the sample space
X	Random variable
X_t	Stochastic process
$\mathcal{F}_t$	Filtration
W_t	Brownian motion
$\mathbb{E}(X)$	The mean of X
$\mathbb{E}[X_t \mid X_0 = x]$	Expectation of X_t, given $X_0 = x$
$\mathbb{E}(X \mid \mathcal{F})$	Conditional expectation of X, given $\mathcal{F}$
$\mathbb{E}_x(X_t)$	Expectation of X_t, given $X_0 = x$
$\mathbb{E}_x(X_t \mid W[0,t])$	Conditional expectation of X_t, given the history $\{W_s; 0 \le s \le t\}$
$Var(X)$	Variation of X
$\mathbb{E}^P[\cdot]$	Expectation operator in the distribution P
$Var(X)$	Variance of the random variable X
$cov(X, Y)$	Covariance of X and Y
$corr(X, Y)$	Correlation of X and Y
$\mathcal{N}(\mu, \sigma^2)$	Normally distributed with mean μ and variance σ^2
1_A	The characteristic function of set A
$\phi_X(t)$	The characteristic function of the random variable X
$p_X(t)$	The probability density function of X
$p_t(x_0, x)$	The transition density of X_t
$\delta(x - a)$	The Dirac delta function sitted at a
$\mathcal{L}(f)$	The Laplace transform of f
$\mathcal{L}^{-1}(F)$	The inverse Laplace transform of F
$\langle M \rangle_t$	The quadratic variation of M_t
$I_t^{(n)}$	The inner product of two indepdendent n-dim Brownian motions
H_t	The Heisenberg diffusion
G_t	The Grushin diffusion
S_t	The Lévi area process
$R_t^{(n)}$	The n-dim Bessel process

Differential Geometry

U, V	Vector fields
g_{ij}	Coefficients of the first fundamental form
$\Gamma_{ijk}, \Gamma_{ij}^k$	Christoffel coefficients of the first and second kind
$D_U V, \nabla_U V$	Derivation of vector field V in direction U
$\dot{c}(t)$	Velocity vector along curve $c(t)$
$\ddot{c}(t)$	Acceleration vector along curve $c(t)$
$T(s)$	The tangent vector field
$N(s)$	The normal vector field
$\kappa(s)$	The curvature at s
$\nabla_g f$	Gradient of function f in metric g
∇^0	Levi-Civita connection
$\text{div}X$	Divergence of vector field X
$\text{curl}X$	Curl of vector field X
$\|\dot{\gamma}(t)\|_g$	The length of velocity $\dot{\gamma}(t)$ in metric g
Δ_g	Laplace-Beltrami operator
$S_{cls}(t, x, y)$	Riemannian action between x and y within time t
$d_g(x, y)$	Riemannian distance between x and y within respect to g

Contents

Chapter 1

Topics of Stochastic Calculus

This chapter is designed to give readers a comprehensive understanding of the key concepts and properties of stochastic calculus and prepare them for the more advanced material covered in the subsequent chapters. It serves as an essential foundation for the rest of the book and will be an important reference throughout the reading.

1.1 Prerequisites of Stochastic Calculus

This book's first section provides a comprehensive review of the fundamental concepts and properties of Stochastic Calculus, which will be used throughout the text. The section sets the notation used throughout the book and covers a wide range of topics including Brownian motion's properties, Ito's lemma, Kolmogorov's equation, Feynman-Kac formula, Levy's characterization theorem of Brownian motion, Girsanov's theorem, and other material. These concepts and properties are essential for understanding the later chapters of the book, which delve deeper into the application of stochastic calculus in geometry and PDEs.

Stochastic processes

A *probability space* will be denoted by $(\Omega, \mathcal{F}, P)$, where Ω stands for the sample space. An arbitrary element of Ω will be denoted by ω; $\mathcal{F}$ denotes the *sigma-algebra of events*, while P is a *probability measure* on $\mathcal{F}$. We assume the reader is familiar with the concept of a random variable on a probability space.

A *stochastic process* is a family of random variables, either discrete or continuous. It is denoted customarily by X_t or $X(t)$, where t denotes the time parameter, which in the continuous case belongs to the interval $[0, T]$, with $T \leq \infty$. For any fixed $\omega \in \Omega$, the application $t \to X_t(\omega)$ is a realization or a sample path of the process X_t corresponding to the state ω.

The statistical distribution of $X_t(\omega)$ is measured by the *transition density function* $p_t(x, y) = p(x, y; 0, t)$ given by

$$P(X_t \in dy | X_0 = x) = p_t(x, y)dy.$$

The associated *distribution function* is defined by

$$F_{X_t}(u) = P(X_t \leq u), \qquad u \in \mathbb{R}.$$

The stochastic process which plays a central role in this book is the one-dimensional *Brownian motion*, which is customarily denoted by W_t. Occasionally, it will be denoted by B_t, β_t or γ_t. Its definition is recalled as follows:

(1) The process starts at zero: $W_0 = 0$.

(2) For $0 \leq s < t$ the random variable $W_t - W_s$ is normally distributed with mean zero and variance $t - s$.

(3) W_t has independent increments, i.e., if $0 \leq s < t < u$, then the differences $W_u - W_t$ and $W_t - W_s$ are independent random variables.

(4) The sample paths $t \to W_t(\omega)$ are continuous.

The filtration defined by W_t is the sigma-algebra generated by the family of random variables $\{W_s; 0 \leq s \leq t\}$ and is denoted by $\mathcal{F}_t$, or in some cases by $W[0, t]$. We also called it the history of the Brownian motion until time t.

Sometimes we shall use the term of *standard Brownian motion* to refer to a one-dimensional Brownian motion starting at zero and having no drift.

There is also an n-dimensional version of the Brownian motion defined by the vector values process $W_t = \big(W_1(t), \dots, W_n(t)\big)$, with $W_j(t)$ one-dimensional independent Brownian motions.

There are several processes that can be derived from Brownian motions and will be used in this book. One of them is the *integrated Brownian motion*, $Z_t = \int_0^t W_s ds$, which is normally distributed with mean zero and variance $t^3/3$.

Another one, is the *Bessel process* of dimension n, which is obtained by taking the norm of an n-dimensional Brownian motion, $R_t^{(n)} = \sqrt{\sum_{k=1}^n W_j(t)^2}$. The Brownian motion is used in the definition of more complex stochastic processes. For example, an *Ito diffusion* X_t starting at $x \in \mathbb{R}^n$ is a stochastic process given by

$$dX_t = b(X_t)dt + \sigma(X_t)dW_t, \qquad X_0 = x, \tag{1.1.1}$$

where W_t is an m-dimensional Brownian motion with continuous drift given by $b : \mathbb{R}^n \to \mathbb{R}^n$ and dispersion $\sigma : \mathbb{R}^n \to \mathbb{R}^n \times \mathbb{R}^m$. The solutions of (1.1.1) are understood in the *strong sense*, i.e.,

(i) The process X_t lives in the space $(\Omega, \mathcal{F}, P)$;

(*ii*) X_t has continuous sample paths;
(*iii*) X_t is $W[0,t]$-adapted;
(*iv*) The following integral identity holds almost surely

$$X_t = x + \int_0^t b(X_s)ds + \int_0^t \sigma(X_s)dW_s$$

(*v*) For any $t \geq 0$ and any i, j we have

$$\int_0^t \left(|b_i(X_s)| + \sigma_{ij}^2(X_s)\right) ds < \infty$$

almost surely.

A few obvious examples of Ito diffusions are:
1. The *n-dimensional Brownian motion with drift*: $dX_t = \mu dt + dW_t$, with $\mu \in \mathbb{R}^n$.
2. The *geometric Brownian motion*: $dX_t = \mu X_t dt + X_t dW_t$, with $\mu \in \mathbb{R}$.
3. The *Ornstein-Uhlenbeck process*: $dX_t = -X_t dt + \sigma dW_t$, with $\sigma \in \mathbb{R}$.

For other processes, such as the Bessel process, it is not obvious that they are Ito diffusions. To accomplish this, we need to apply some more advanced results, see page 9. Similarly with the Bessel process case, other processes in this book are shown to be Ito diffusions. We are mostly interested in those diffusions which are induced by an underlying geometry, which can be either Riemannian (elliptic diffusions) or sub-Riemannian (sub-elliptic diffusions), as shall be seen in Chapters 6 and 7.

Ito's Formula

This is an analog of the second order Taylor approximation of a function from the classical calculus to the stochastic environment. It is used to find the stochastic differential equations of processes derived from more simple processes.

The autonomous case Let X_t be a process given by

$$dX_t = b_t dt + \sigma_t dW_t,$$

where W_t is a one-dimensional Brownian motion and $b_t(\omega)$ and $\sigma_t(\omega)$ are measurable real-valued functions. If $F_t = f(X_t)$, with f twice continuously differentiable real-valued function, then

$$dF_t = f'(X_t)dX_t + \frac{1}{2}f''(X_t)(dX_t)^2.$$

Informally, using formulas $dt^2 = 0$, $dW_t^2 = dt$ and $dt dW_t = 0$, we obtain Ito's formula

$$dF_t = \Big[b_t f'(X_t) + \frac{1}{2}\sigma_t^2 f''(X_t)\Big] dt + \sigma_t f'(X_t)dW_t. \tag{1.1.2}$$

The equivalent integral form of Ito's formula is

$$F_t = f(X_0) + \int_0^t \big(b_s f'(X_s) + \frac{1}{2}\sigma_s^2 f''(X_s)\big)ds + \int_0^t \sigma_s f'(X_s)\,dW_s.$$

For instance, if $F_t = W_t^n$, with $n \geq 1$ integer, then Ito's formula provides

$$dF_t = \frac{1}{2}n(n-1)W_t^{n-2}dt + nW_t^{n-1}dW_t.$$

The non-autonomous case Ito's formula can be extended for the non-autonomous case $F_t = f(t, X_t)$, with $f : \mathbb{R}_+ \times \mathbb{R} \to \mathbb{R}$ be $\mathcal{C}^{(1,2)}$-differentiable as

$$dF_t = \partial_t f(t, X_t)dt + \partial_x f(t, X_t)dX_t + \frac{1}{2}\partial_x^2 f(t, X_t)(dX_t)^2,$$

which leads to the formula

$$dF_t = \Big(\partial_t f(t, X_t) + b_t \partial_x f(t, X_t) + \frac{1}{2}\sigma_t^2 \partial_x^2 f(t, X_t)\Big)dt + \sigma_t \partial_x f(t, X_t)dW_t.$$

For instance, if $F_t = tW_t^2$, then choosing $b_t = 0$, $\sigma_t = 1$ and $f(t, x) = tx^2$, we obtain

$$dF_t = (1 + W_t^2)dt + 2tW_t dW_t.$$

The multi-dimensional case Let $W_1(t), \dots, W_m(t)$ denote m Brownian motions and consider n Ito processes

$$dX_j(t) = b_j(t, X_t)dt + \sum_{k=1}^m \sigma_{j,k}(t, X_t)dW_k(t), \quad 1 \leq j \leq n.$$

Let $F(t, x) = \big(F_1(t, x), \dots, F_p(t, x)\big)$ be a twice differentiable map from $\mathbb{R}_+ \times \mathbb{R}^n$ to $\mathbb{R}$, and denote $X_t = \big(X_1(t), \dots, X_n(t)\big)$ and $F_t = F(t, X_t)$. Then

$$dF_t = \sum_k \partial_{x_k} F(t, X_t)dX_k(t) + \partial_t F(t, X_t)dt + \frac{1}{2}\sum_{i,j} \partial_{x_i}\partial_{x_j} F(t, X_t)dX_i(t)dX_j(t), \tag{1.1.3}$$

with the conventions $dW_i(t)dW_j(t) = \delta_{ij}dt$ and $dtdW_i(t) = 0$.

For instance, if $F_t = \sqrt{W_1(t)^2 + W_2(t)^2}$, then choosing $X_j(t) = W_j(t)$, $j = 1, 2$, and $f(t, x_1, x_2) = \sqrt{x_1^2 + x_2^2}$, we obtain

$$dF_t = \frac{1}{2F_t}dt + \frac{W_1(t)}{F_t}dW_1(t) + \frac{W_2(t)}{F_t}dW_2(t).$$

The conditional expectation

Let $(X_t)_{t\geq 0}$ be a stochastic process in $\mathbb{R}^n$. In the following the notation $\mathbb{E}^x$ stands for the expectation operator given the initial condition $X_0 = x$. This means that for any bounded function f we have

$$\mathbb{E}^x[f(X_t)] = \mathbb{E}[f(X_t)|X_0 = x] = \int_{\mathbb{R}^n} f(y)p_t(x,y)dy,$$

where $p_t(x,y) = p(x,y;0,t)$ is the transition density of X_t, given that $X_0 = x$.

Quadratic variation

For a continuous process X_t on the probability space $(\Omega, \mathcal{F}, P)$ the *quadratic variation* is defined by the following limit in probability

$$\langle X\rangle_t = \lim_{\max_i |t_{i+1}-t_i|\to 0} \sum_{i=0}^{n-1} |X_{t_{i+1}} - X_{t_i}|^2,$$

where $0 = t_1 < t_2 < \cdots < t_n = t$. For the purposes of this book we shall compute the quadratic variation by the stochastic integral

$$\langle X\rangle_t = \int_0^t (dX_t)^2.$$

It can be shown that if X_t is a continuous square integrable martingale, then $\langle X\rangle_t$ is increasing such that $\langle X\rangle_0 = 0$ and $X_t^2 - \langle X\rangle_t$ is a martingale. Here are a few examples involving Ito, Wiener and Riemann integrals:

1. Let W_t be a one-dimensional Brownian motion. If $X_t = \int_0^t f(W_s)\,dW_s$, then $dX_t = f(W_t)\,dW_t$, and hence

$$\langle X\rangle_t = \int_0^t (dX_t)^2 = \int_0^t f(W_s)^2\,ds,$$

where we used that $dW_t^2 = dt$.

2. Set $X_t = \int_0^t f(s)\,dW_s$. Then $dX_t = f(t)\,dW_t$, and hence

$$\langle X\rangle_t = \int_0^t (dX_t)^2 = \int_0^t f(s)^2\,ds.$$

3. If $X_t = \int_0^t f(W_s)\,ds$, then $dX_t = f(W_t)\,dt$. Using $dt^2 = 0$ we obtain

$$\langle X\rangle_t = \int_0^t (dX_t)^2 = 0.$$

I particular, if $X_t = W_t$ we obtain $\langle W_t \rangle_t = \int_0^t (dW_s)^2 = \int_0^t ds = t$. Furthermore, $W_t^2 - t$ is a martingale.

The *quadratic covariance* of two continuous square integrable martingales, X_t and Y_t is defined by the polarization formula

$$\langle X, Y \rangle_t = \frac{1}{4}\Big(\langle X+Y, X+Y \rangle_t - \langle X-Y, X-Y \rangle_t \Big).$$

This can be also computed by the integral formula

$$\langle X, Y \rangle_t = \int_0^t dX_t dY_t.$$

We also note that $\langle X, X \rangle_t = \langle X \rangle_t$.

The infinitesimal generator

The relation with partial differential equations is done by associating a second order differential operator with each Ito process as follows. Let $(X_t)_{t\geq 0}$ be a stochastic process starting at $X_0 = x$. The *infinitesimal generator* of X_t is the differential operator $\mathcal{A}$ defined by

$$\mathcal{A}f(x) = \lim_{t\searrow 0} \frac{\mathbb{E}^x[f(X_t)] - f(x)}{t} = \lim_{t\searrow 0} \mathbb{E}^x\Big[\frac{f(X_t) - f(X_0)}{t}\Big],$$

for any smooth compact supported function $f \in C_0^2(\mathbb{R}^n)$.

If the process X_t is an Ito diffusion given by

$$dX_t = b(X_t)dt + \sigma(X_t)dW_t, \qquad X_0 = x$$

where W_t is an m-dimensional Brownian motion with continuous drift $b : \mathbb{R}^n \to \mathbb{R}^n$ and dispersion $\sigma : \mathbb{R}^n \to \mathbb{R}^n \times \mathbb{R}^m$, then the infinitesimal generator of X_t becomes the following second order differential operator

$$\mathcal{A} = \frac{1}{2}\sum_{i,j} (\sigma\sigma^T)_{ij} \frac{\partial^2}{\partial x_i \partial x_j} + \sum_k b_k(x) \frac{\partial}{\partial x_k}. \tag{1.1.4}$$

The relation between the expectation and the infinitesimal generator is given by the next result. This will be useful in the study of properties of sub-elliptic diffusions in Chapter 7.

Theorem 1.1.1 (Dynkin's formula) *Let $f \in C_0^2(\mathbb{R}^n)$ and X_t be an Ito diffusion starting at x. If T is a stopping time with $\mathbb{E}[T] < \infty$, then*

$$\mathbb{E}^x[f(X_T)] = f(x) + \mathbb{E}^x\Big[\int_0^T \mathcal{A}f(X_s)\, ds\Big],$$

where $\mathcal{A}$ is the infinitesimal generator of X_t.

The following result is a consequence of Dynkin's formula and makes the relation with partial differential equations.

Theorem 1.1.2 (Kolmogorov's backward equation) *For any $f \in C_0^2(\mathbb{R}^n)$ the function $v(t,x) = \mathbb{E}^x[f(X_t)]$ satisfies the following Cauchy problem*

$$\begin{aligned} \frac{\partial v}{\partial t} &= \mathcal{A}v, \quad t > 0 \\ v(0,x) &= f(x). \end{aligned}$$

The previous formula had been extended by Richard Feynman and Mark Kac in 1940s to represent the solution of a heat equation with external cooling terms using an expectation involving a Brownian motion. We start with its simplest version.

Theorem 1.1.3 (one-dim Feynman-Kac's formula) *Let $U(x)$ be a non-negative, continuous function and $f(x)$ be a bounded continuous function. We assume that $u(t,x)$ is a bounded function that satisfies the equation*

$$\frac{\partial u}{\partial t} = \frac{1}{2}\frac{\partial^2 u}{\partial x^2} - U(x)u$$

with the initial condition $u(0,x) = f(x)$. Then we can represent it as

$$u(t,x) = \mathbb{E}^x[e^{-\int_0^t U(B_s)ds} f(B_t)],$$

where B_s is a one-dimensional Brownian motion starting at $B_0 = x$.

The multidimensional version of the previous statement takes the following form:

Theorem 1.1.4 (n-dim Feynman-Kac's formula) *Let $U : \mathbb{R}^n \to [0, +\infty)$ be a continuous function and $f : \mathbb{R}^n \to \mathbb{R}$ be a continuous bounded function. We assume that $u(t,x)$ is a bounded function that satisfies the equation*

$$\frac{\partial u}{\partial t} = \frac{1}{2}\Delta u - U(x)u$$

with the initial condition $u(0,x) = f(x)$. Then we can represent it as

$$u(t,x) = \mathbb{E}^x[e^{-\int_0^t U(B_s)ds} f(B_t)],$$

where B_s is an n-dimensional Brownian motion starting at $B_0 = x$.

We have used the notation $\Delta = \sum_{j=1}^n \partial_{x_j}^2$ to denote the Laplacian on $\mathbb{R}^n$.

The following general form of the aforementioned result can be found in Durrett [53], page 130 and Karatzas and Shreve [85], page 268.

Theorem 1.1.5 *For $T > 0$ let $F : \mathbb{R} \to \mathbb{R}$, $U : [0,T] \times \mathbb{R} \to \mathbb{R}$ and $G : [0,T] \times \mathbb{R} \to \mathbb{R}$ be given functions such that G is bounded and F is bounded and Hölder continuous. If G and U are Hölder continuous locally in t, i.e. for any $N < \infty$ there exist $C_i > 0$ and $\alpha_i > 0$ such that*

$$\begin{aligned}|G(t,x) - G(t,y)| &< C_1|x-y|^{\alpha_1}\\ |U(t,x) - U(t,y)| &< C_2|x-y|^{\alpha_2}\end{aligned}$$

for all $t < N$, then the function

$$u(t,x) = \mathbb{E}^x\left[F(B_t)e^{-\int_0^t U(t-s,B_s)ds} + \int_0^t G(t-s,B_s)e^{-\int_0^s U(t-v,B_v)dv}ds\right]$$

is the unique solution of the problem

$$\begin{aligned}\frac{\partial u}{\partial t} &= \frac{1}{2}\frac{\partial^2 u}{\partial x^2} - U(t,x)u + G(t,x), \quad t > 0, x \in \mathbb{R}\\ u(0,x) &= F(x), \quad x \in \mathbb{R}.\end{aligned}$$

Martingale characterization of Brownian motions

Many processes are actually hidden Brownian motions, as the next result shows. For a proof the reader is referred to Revus and Yor [120], page 150, or Øksendal [114], page 154.

Theorem 1.1.6 (Lévy's characterization theorem) *Let X_t be an $\mathcal{F}_t$-adapted continuous d-dimensional martingale vanishing at zero such that*

$$\langle X^i, X^j\rangle_t = \delta_{ij}t$$

for every $1 \le i,j \le d$. Then X_t is an $\mathcal{F}_t$-Brownian motion.

The previous result is often used in the following one-dimensional variant:

Corollary 1.1.7 *The one-dimensional Brownian motion is the only continuous martingale with t as its quadratic variation process.*

Another variant of Lévy's martingale characterization of a Brownian motion is given by:

Theorem 1.1.8 *Let X_t be a continuous real-valued process in the filtered probability space $(\Omega, \mathcal{F}, \mathcal{F}_t, P)$. Then X_t is an $\mathcal{F}_t$-Brownian motion if and only if both X_t and $M_t = X_t^2 - t$ are $\mathcal{F}_t$-martingales.*

As an application, we can associate a one-dimensional Brownian motion β_t to each n-dimensional Brownian motion W_t as follows.

Proposition 1.1.9 *Let $W_t = (W_1(t), \dots, W_n(t))$ be an n-dimensional Brownian motion. Then*

$$\beta_t = \sum_{i=1}^{n} \int_0^t \frac{W_i(s)}{\|W_s\|} dW_i(s)$$

is a one-dimensional Brownian motion, where $\|W_t\| = \sqrt{\sum_{i=1}^n W_i(t)^2}$.

The previous proposition is used in the construction of the stochastic equation of the n-dimensional Bessel process $R_t^{(n)} = \|W_t\|$. In this direction, Ito's lemma provides

$$\begin{aligned} dR_t^{(n)} &= \sum_{i=1}^{n} \frac{W_i(t)}{R_t^{(n)}} dW_i(t) + \frac{n-1}{2R_t^{(n)}} dt \\ &= d\beta_t + \frac{n-1}{2R_t^{(n)}} dt. \end{aligned} \tag{1.1.5}$$

Therefore, the generator of $R_t^{(n)}$ is given by

$$\mathcal{A} = \frac{1}{2}\frac{d^2}{dx^2} + \frac{n-1}{2x}\frac{d}{dx}, \qquad x > 0.$$

We note that $R_t^{(1)} = |W_t|$ is a reflected Brownian motion.

Another application is for the case of a two-dimensional Brownian motion as follows:

Proposition 1.1.10 *Let $(W_1(t), W_2(t))$ be a two-dimensional Brownian motion. Then*

$$\beta_t = \int_0^t \frac{W_1(s)dW_1(s) + W_2(s)dW_2(s)}{\sqrt{W_1(s)^2 + W_2(s)^2}}$$

$$\gamma_t = \int_0^t \frac{W_1(s)dW_2(s) - W_1(s)dW_2(s)}{\sqrt{W_1(s)^2 + W_2(s)^2}}$$

are two independent Brownian motions.

Time change for Brownian motions

The next result can be considered as a natural generalization of Lévy's theorem and provides conditions under which a martingale becomes a Brownian motion running at a modified time clock.

Theorem 1.1.11 (Dambis, Dubins, Schwartz) *Let M_t be a continuous, square integrable $\mathcal{F}_t$-martingale satisfying $\lim_{t\to\infty}\langle M, M\rangle_t = +\infty$, a.s. Then M_t can be written as a time-transformed Brownian motion as*

$$M_t = B_{\langle M,M\rangle_t},$$

where B_t is a one-dimensional Brownian motion. Moreover, if we set

$$T_s = \inf\{t; \langle M, M\rangle_t > s\},$$

then the time-changed process $B_s = M_{T_s}$ is a $\mathcal{F}_{T_s}$-Brownian motion.

The Brownian motion B_t will be referred to as the *DDS Brownian motion* of the martingale M_t. The proof can be found in Revuz and Yor [120], page 181.

The next result is a consequence of Dambis, Dubins and Schwartz's theorem.

Theorem 1.1.12 (Gaussian martingales) *If M_t is a continuous martingale with $M_0 = 0$, and $\langle M\rangle_t = f(t)$, a.e., for a deterministic non-negative increasing function f, then M_t has a Gaussian distribution,*

$$M_t \sim \mathcal{N}(0, f(t)).$$

It is worth noting that we have $M_t = B_{f(t)} = \int_0^t \sqrt{f'(s)}dW_s$, provided f is differentiable.

The multidimensional version of the Dambis, Dubins-Schwartz theorem takes the following form, see [120], page 183.

Theorem 1.1.13 (Knight) *Let $M_t = (M_t^1, \cdots, M_t^d)$ be a continuous vector valued martingale such that $M_0 = 0$, $\lim_{t\to\infty}\langle M^k, M^k\rangle_t = \infty$ for any $k \in \{1, \dots d\}$ and $\langle M^k, M^j\rangle_t = 0$ for $k \neq j$. If we set*

$$T_s^k = \inf\{t; \langle M^k, M^k\rangle_t > s\},$$

and $B_s^k = M_{T_s}^k$, then the process $B_t = (B_t^1, \dots, B_t^d)$ is a d-dimensional Brownian motion.

The next result, see Øksendal [114], page 148, answers the following question: Under which conditions an Ito process becomes an Ito diffusion under a stochastic time change?

Consider the non-decreasing time change $\beta_t(\omega) = \int_0^t c(s, \omega)ds$, with an $\mathcal{F}_t$-adapted rate $c(t, \omega) \geq 0$. Define $\alpha_t(\omega) = \inf\{s; \beta_s(\omega) > t\}$. It can be shown that α_t is right-continuous in t and that $\beta_{\alpha_t} = t$. In the case $c(t, \omega) > 0$, then α_t is continuous in t and we also have $\alpha_{\beta_t} = t$.

Theorem 1.1.14 *Let Y_t be an Ito process given by*

$$dY_t = u(t,\omega)dt + v(t,\omega)dW_t, \quad u \in \mathbb{R}^n, \quad v \in \mathbb{R}^{n\times m},\ Y_0 = x$$

and X_t an Ito diffusion given by

$$dX_t = b(X_t)dt + \sigma(X_t)dW_t, \quad b \in \mathbb{R}^n, \quad \sigma \in \mathbb{R}^{n\times m},\ X_0 = x$$

Assume that

$$u(t,\omega) = c(t,\omega)b(Y_t), \qquad vv^T(t,\omega) = c(t,\omega)\,\sigma\sigma^T(Y_t)$$

Then

$$Y_{\alpha_t} = X_t, \qquad X_{\beta_t} = Y_t.$$

Girsanov's Theorem

This result of stochastic calculus does not have an analog in the classical calculus. It will be used to reduce the computation of expectations of diffusions with drift to the no drift case, which can be handled easier.

Let $(Y_t)_{0\le t\le T}$ be a one-dimensional $\mathcal{F}_t$-adapted stochastic process on the probability space $(\Omega, \mathcal{F}, \mathcal{F}_t, P)$, satisfying *Novikov's condition*

$$\mathbb{E}\Big[e^{\frac{1}{2}\int_0^T Y_s^2 dt}\Big] < \infty. \tag{1.1.6}$$

If $(W_t)_{0\le t\le T}$ is an $\mathcal{F}_t$-Brownian motion, then the exponential process

$$M_t = e^{\int_0^t Y_s dW_s - \frac{1}{2}\int_0^t Y_s^2 ds}, \quad 0 \le t \le T \tag{1.1.7}$$

is a martingale under the probability P and the filtration $\mathcal{F}_t$. Consequently, $\mathbb{E}[M_t] = M_0 = 1$.

Another measure Q can be defined by $dQ_{|\mathcal{F}_t} = M_t\, dP_{|\mathcal{F}_t}$, namely

$$Q(A) = \int_A dQ = \int_A M_t\, dP = \mathbb{E}^P[1_A M_t], \qquad \forall A \in \mathcal{F}_t.$$

Q is a probability measure since

$$Q(\Omega) = \int_\Omega dQ = \int_\Omega M_t\, dP = \mathbb{E}^P[M_t] = 1.$$

Theorem 1.1.15 (Girsanov) *Under the previous assumptions, the process*

$$X_t = W_t - \int_0^t Y_s\, ds, \qquad 0 \le t \le T$$

becomes a Brownian motion under the measure Q.

For some particular processes Y_s we obtain the following versions of Girsanov's theorem:

Corollary 1.1.16 *Let $u \in L^2[0,T]$ be a deterministic function. Then the process*

$$X_t = W_t - \int_0^t u(s)ds, \qquad 0 \le t \le T$$

is a Brownian motion with respect to the probability measure Q defined by

$$dQ = e^{\int_0^T u(s)dW_s - \frac{1}{2}\int_0^T u(s)^2} dP.$$

Corollary 1.1.17 *The process*

$$X_t = W_t + \lambda t, \qquad 0 \le t \le T$$

is a Brownian motion with respect to the probability measure Q defined by

$$dQ = e^{-\frac{1}{2}\lambda^2 T - \lambda W_T} dP.$$

These results can be used for drift reduction purposes, as shown next.

Proposition 1.1.18 (Drift reduction formula) *Let B_s be a one-dimensional Brownian motion and consider the process $A_t^{(\nu)} = \int_0^t e^{2(B_s+\nu s)}ds$. Then for any Borel function $f : \mathbb{R}_+ \to \mathbb{R}$ we have*

$$\mathbb{E}[f(A_t^{(\nu)})] = \mathbb{E}[f(A_t^{(0)})e^{\nu B_t - \frac{1}{2}\nu^2 t}].$$

Proof: If we change measures from P to Q and apply Corollary 1.1.17, we obtain

$$\begin{aligned}
\mathbb{E}[f(A_t^{(\nu)})] &= \mathbb{E}^P[f(A_t^{(\nu)})] = \mathbb{E}^Q[f(A_t^{(\nu)})M_t^{-1}] = \mathbb{E}^Q[f(A_t^{(\nu)})e^{\frac{\nu^2 t}{2}+\nu B_t}] \\
&= \mathbb{E}^Q[f(A_t^{(\nu)})e^{-\frac{\nu^2 t}{2}} e^{\nu(B_t+\nu t)}] \\
&= \mathbb{E}^Q\Big[f\Big(\int_0^t e^{2(B_s+\nu s)}ds\Big)e^{-\frac{\nu^2 t}{2}} e^{\nu(B_t+\nu t)}\Big] \\
&= \mathbb{E}^Q\Big[f\Big(\int_0^t e^{2W_s}ds\Big)e^{-\frac{\nu^2 t}{2}} e^{\nu W_t}\Big] \\
&= \mathbb{E}^P\Big[f(A_t^{(0)})e^{-\frac{\nu^2 t}{2}} e^{\nu B_t}\Big],
\end{aligned}$$

where we used that $W_t = B_t + \nu t$ is a Brownian motion with respect to the measure Q.

■

In a similar way we can prove the following two drift reduction formulas:

Proposition 1.1.19 *Let W_t be a Brownian motion and f a Borel function. Then*
(i) $\mathbb{E}[f(\lambda t + W_t)] = e^{-\frac{\lambda^2 t}{2}} \mathbb{E}[f(W_t)e^{\lambda W_t}]$;
(ii) $\mathbb{E}[f(\lambda t + W_t)e^{-\lambda W_t}] = e^{\frac{\lambda^2 t}{2}} \mathbb{E}[f(W_t)]$.

A proof of Girsanov's theorem can be found, for instance, in Øksendal [114].

1.2 Exponentially Stopped Brownian Motion

In this section, we explore the properties of a Brownian motion process that is stopped at an exponential time and its connection to Laplace and exponential distributions. Additionally, we introduce an integral operator T that plays a significant role in representing solutions of certain PDEs in later sections of the book.

We consider an exponentially distributed random variable S_θ, with the law

$$S_\theta \sim \theta e^{-\theta t}, \qquad t \geq 0, \quad \theta > 0,$$

and we shall investigate a one-dimensional Brownian motion B_t stopped at S_θ, with S_θ independent of B_t. For this purpose we consider $X = B_{S_\theta}$ and study the distribution law of X. It is worth noting that both random variables, B_t and S_θ, are defined on the same probability space. The relation with the exponential distribution is given by the next result.

Proposition 1.2.1 *The random variable $X = B_{S_\theta}$ has the same distribution law as the difference $Z_1 - Z_2$ of two independent exponentially distributed random variables with $Z_i \sim \sqrt{2\theta}e^{-\sqrt{2\theta}t}, t \geq 0$, $i = 1, 2$.*

Proof: It suffices to show that the random variables X and $Z_1 - Z_2$ have the same characteristic function. We start by recalling the characteristic functions of S_θ and B_τ:

$$\phi_{S_\theta}(t) = \frac{\theta}{\theta - it}, \qquad \phi_{B_\tau}(t) = e^{-\frac{1}{2}t^2\tau}.$$

Then using the independence property, the characteristic function of the difference $Z_1 - Z_2$ becomes

$$\begin{aligned} \phi_{Z_1-Z_2}(t) &= \mathbb{E}[e^{it(Z_1-Z_2)}] = \mathbb{E}[e^{itZ_1}]\mathbb{E}[e^{-itZ_2}] \\ &= \frac{\sqrt{2\theta}}{\sqrt{2\theta} - it} \cdot \frac{\sqrt{2\theta}}{\sqrt{2\theta} + it} = \frac{2\theta}{2\theta + t^2}. \end{aligned} \tag{1.2.8}$$

Next we compute the characteristic function $\phi_X(t)$ by integrating over the distribution of S_θ

$$\begin{aligned}\phi_X(t) &= \mathbb{E}[e^{itX}] = \mathbb{E}[e^{itB_{S_\theta}}] = \int_0^\infty \mathbb{E}[e^{itB_{S_\theta}}|S_\theta = u]f_{S_\theta}(u)\,du \\ &= \int_0^\infty \mathbb{E}[e^{itB_u}]f_{S_\theta}(u)\,du = \int_0^\infty e^{-\frac{1}{2}t^2u}\theta e^{-\theta u}\,du \\ &= \theta\int_0^\infty e^{-(\frac{t^2}{2}+\theta)u}\,du = \frac{2\theta}{2\theta+t^2}. \end{aligned} \tag{1.2.9}$$

Relations (1.2.8) and (1.2.9) yield $\phi_X(t) = \phi_{Z_1-Z_2}(t)$, which implies that $X \sim Z_1 - Z_2$. We used the fact that if the characteristic functions of two random variables X and Y are the same, then X and Y have the same distribution. This is a consequence of the injectivity of the Fourier transform. ■

The law of the exponentially stopped Brownian motion is given as follows.

Proposition 1.2.2 *The variable $X = B_{S_\theta}$ has the Laplace distribution law*

$$p_X(x) = \frac{c}{2}e^{-c|x|}, \quad x \in \mathbb{R},\ c = \sqrt{2\theta}.$$

Proof: For any bounded Borel function F, we shall compute the expectation $\mathbb{E}[F(X)]$ in two ways. First, using the expectation definition, we have

$$\mathbb{E}[F(X)] = \int_{\mathbb{R}} F(x)p_X(x)\,dx. \tag{1.2.10}$$

On the other hand, by conditioning over S_θ and integrating out the condition, after changing the order of integration by Fubini's theorem, we obtain

$$\begin{aligned}\mathbb{E}[F(X)] &= \mathbb{E}[F(B_{S_\theta})] = \int_0^\infty \mathbb{E}[F(B_{S_\theta})|S_\theta = u]f_{S_\theta}(u)\,du \\ &= \int_0^\infty \mathbb{E}[F(B_u)]f_{S_\theta}(u)\,du = \int_0^\infty\int_{\mathbb{R}} F(x)f_{B_u}(x)\,dx\,f_{S_\theta}(u)\,du \\ &= \int_{\mathbb{R}} F(x)\Big(\int_0^\infty f_{B_u}(x)f_{S_\theta}(u)\,du\Big)\,dx. \end{aligned} \tag{1.2.11}$$

Comparing (1.2.10) and (1.2.11) yields the following expression for the probability density of X in terms of the densities of B_u and S_θ

$$p_X(x) = \int_0^\infty f_{B_u}(x)f_{S_\theta}(u)\,du. \tag{1.2.12}$$

In order to find a closed form expression for the integral (1.2.12) we shall consider the integral as a function of x and show that it satisfies a differential equation with a given initial condition. Then solving the equation by a standard technique yields the expression for $p_X(x)$.

We let

$$p(x) = p_X(x) = \int_0^\infty f_{B_u}(x) f_{S_\theta}(u)\,du = \theta \int_0^\infty \frac{1}{\sqrt{2\pi u}} e^{-(\frac{x^2}{2u}+\theta u)}\,du. \quad (1.2.13)$$

The initial value of $p(x)$ is computed at $x = 0$. For this we use the substitution $t = \theta u$ and the gamma function value $\Gamma(1/2) = \sqrt{\pi}$ as follows

$$\begin{aligned} p(0) &= \theta \int_0^\infty \frac{1}{\sqrt{2\pi u}} e^{-\theta u}\,du = \frac{\sqrt{\theta}}{\sqrt{2\pi}} \int_0^\infty \frac{1}{\sqrt{t}} e^{-t}\,dt = \frac{\sqrt{\theta}}{\sqrt{2\pi}} \cdot \Gamma\Big(\frac{1}{2}\Big) \\ &= \frac{\sqrt{\theta}}{\sqrt{2\pi}} \cdot \sqrt{\pi} = \sqrt{\frac{\theta}{2}}. \end{aligned}$$

Since $p(x)$ is an even function, it suffices to assume $x > 0$. Differentiating with respect to x in (1.2.13) yields

$$p'(x) = \theta \int_0^\infty \frac{1}{\sqrt{2\pi u}} e^{-(\frac{x^2}{2u}+\theta u)} \Big(-\frac{x}{u}\Big)\,du.$$

Employing the substitution

$$\theta v = \frac{x^2}{2u}$$

we obtain

$$\begin{aligned} e^{-(\frac{x^2}{2u}+\theta u)} &= e^{-(\frac{x^2}{2v}+\theta v)} \\ du &= -\frac{x^2}{2\theta v^2}\,dv \\ \frac{1}{\sqrt{u}}\Big(-\frac{x}{u}\Big)du &= \frac{\sqrt{2\theta}}{\sqrt{v}}\,dv. \end{aligned}$$

This transforms the previous integral into an integral with respect to v

$$p'(x) = -\sqrt{2\theta} \cdot \theta \int_0^\infty \frac{1}{\sqrt{2\pi v}} e^{-(\frac{x^2}{2v}+\theta v)}\,dv = -\sqrt{2\theta}\,p(x).$$

Therefore, $p(x)$ satisfies the following Cauchy problem

$$\begin{aligned} p'(x) &= -\sqrt{2\theta}\,p(x), \qquad x \geq 0 \\ p(0) &= \sqrt{\frac{\theta}{2}}. \end{aligned}$$

This has the unique solution $p(x) = \sqrt{\frac{\theta}{2}} e^{-\sqrt{2\theta} x}$, $x \geq 0$. Using that $p(-x) = p(x)$, then we have

$$p(x) = p(|x|) = \sqrt{\frac{\theta}{2}} e^{-\sqrt{2\theta}|x|}, \qquad \forall x \in \mathbb{R}, \quad (1.2.14)$$

which proves the desired result. ∎

1.3 The operator T

Inspired by the proof of the previous result we introduce the integral transform

$$T(f)(x) = \int_0^\infty \phi_u(x) f(u)\, du, \qquad (1.3.15)$$

where $\phi_u(x) = \dfrac{1}{\sqrt{2\pi u}} e^{-\frac{x^2}{2u}}$ is the standard normal kernel and $f(u)$ is a probability density over $(0, \infty)$. The function $T(f)$ represents the probability density of a Brownian motion stopped at a random variable distributed by the law $f(u)$, $u \geq 0$.

We will investigate first the circumstances that ensure the operator T is well-defined. This analysis will be carried out in two cases, namely when x is not equal to zero and when x equals zero.

(*i*) Let $x \neq 0$ be fixed. Since the function $u \to \phi_u(x)$ has a maximum at $u = x^2$, we have

$$\max_{u>0} \phi_u(x) = \phi_{x^2}(x) = \frac{1}{\sqrt{2\pi e}} \frac{1}{|x|}.$$

Using $\int_0^\infty f(u)\, du = 1$, an estimation provides

$$0 \leq T(f)(x) \leq \int_0^\infty \max_u \phi_u(x)\, f(u)\, du = \frac{1}{\sqrt{2\pi e}} \frac{1}{|x|} < \infty.$$

(*ii*) Let $x = 0$. Then

$$T(f)(0) = \int_0^\infty \frac{1}{\sqrt{2\pi u}} f(u)\, du$$

is finite if the function $f(u)$ has a behavior about $u = 0$ no worst than the singularity $u^{-1/2}$, i.e.

$$f(u) \sim u^\alpha, \quad \alpha > -\frac{1}{2}.$$

This is equivalent to asking the density $f(u)$ to satisfy the limit condition

$$\lim_{u \searrow 0} \sqrt{u} f(u) = 0. \qquad (1.3.16)$$

Condition (1.3.16) is satisfied by all the densities discussed in this section, including exponential, uniform, Erlang, and others.

The main properties of the transform T are given as follows.

Proposition 1.3.1 *Let T be the transform given by* (1.3.15). *Then*

1. $T(cf) = cT(f)$ *for any* $c \in \mathbb{R}$*;*
2. $T(f+g) = T(f) + T(g)$*;*
3. $T(f)(-x) = T(f)(x)$*;*
4. $\int_{-\infty}^{\infty} T(f)(x)\,dx = 1$*;*
5. $T(f')(x) = -f(0)\delta(x) - \frac{1}{2}\partial_x^2 T(f)(x)$ *if* $f(0) < \infty$*;*
6. $T(f')(x) = -\frac{1}{2}\partial_x^2 T(f)(x)$ *if* $f(0) = 0$*;*
7. $T(\delta_u)(x) = \phi_u(x)$, $a > 0$

where $\delta(x)$ *and* $\delta_a(x) = \delta(x-a)$ *are the Dirac distributions centered at* $x = 0$ *and* $x = a$*, respectively.*

Proof: Parts 1 and 2 follow from the linearity of the integral operator. Part 3 follows from the fact that the normal kernel is even with respect to x. Part 4 is a consequence of Fubini's theorem of interchanging the order of integration and the fact that both $\phi_u(x)$ and $f(u)$ are probability densities.
For part 5 we first we note that $\lim_{u\to\infty} \phi_u(x)f(u) = 0$ and $\lim_{u\to 0} \phi_u(x)f(u) = f(0)\delta(x)$, where $\delta(x)$ denotes the Dirac distribution centered at zero. Then applying integration by parts and the fact that $\phi_u(x)$ satisfies the heat equation $\partial_u \phi_u(x) = \frac{1}{2}\partial_x^2 \phi_u(x)$, we obtain

$$\begin{aligned} T(f')(x) &= \int_0^\infty \phi_u(x) f'(u)\,du = \phi_u(x)f(x)\Big|_{u=0}^{u=\infty} - \int_0^\infty \partial_u \phi_u(x) f(u)\,du \\ &= -f(0)\delta(x) - \int_0^\infty \frac{1}{2}\partial_x^2 \phi_u(x) f(u)\,du \\ &= -f(0)\delta(x) - \frac{1}{2}\partial_x^2 T(f)(x). \end{aligned}$$

Part 6 follows from part 5. For 7 we use the property of Dirac distribution

$$T(\delta_a)(x) = \int_0^\infty \phi_u(x)\delta(u-a)\,du = \int_{-\infty}^\infty \phi_u(x)\delta(u-a)\,du = \phi_a(x).$$

■

In the following the operator T and its primary properties will be illustrated through several immediate applications.

1.4 Uniformly Stopped Brownian Motion

Let $0 < a < b$ and consider a Brownian motion B_t stopped uniformly in the interval $[a, b]$. The density of a uniformly distributed random variable, U, on the interval $[a, b]$ is given by

$$f(u) = \begin{cases} \dfrac{1}{b-a}, & \text{if } a \le u \le b \\ 0, & \text{otherwise.} \end{cases}$$

We shall find the distribution law of B_U. A straightforward application of relation (1.3.15) provides the density of B_U in the integral form

$$p(x) = T(f)(x) = \frac{1}{b-a}\int_a^b \frac{1}{\sqrt{2\pi u}} e^{-\frac{x^2}{2u}}\, du.$$

This represents the integral average over the interval $[a, b]$ of a normal distribution with a variable variance. In order to obtain a closed form expression we shall use Proposition 1.3.1. First we write the uniform density in terms of Heaviside functions as

$$f(u) = \frac{1}{b-a}\Big(H(u-a) - H(u-b)\Big)$$

and take the derivative (in the generalized sense)

$$f'(u) = \frac{1}{b-a}\Big(\delta(u-a) - \delta(u-b)\Big).$$

Then using the properties of the transform T we obtain

$$T(f')(x) = \frac{1}{b-a}\Big(T(\delta_a)(x) - T(\delta_b)(x)\Big) = \frac{1}{b-a}\Big(\phi_a(x) - \phi_b(x)\Big).$$

On the other side, we have

$$T(f')(x) = -\frac{1}{2}\partial_x^2 T(f)(x).$$

If let $p(x) = T(f)(x)$, then $p(x)$ satisfies the ordinary differential equation

$$p''(x) = \frac{2}{b-a}\Big(\phi_b(x) - \phi_a(x)\Big)$$

with vanishing boundary conditions at infinity $\lim_{x\to\infty} p(x) = \lim_{x\to\infty} p'(x) = 0$. We shall express the solution $p(x)$ in terms of the cumulative normal distribution function $\Phi(x) = \frac{1}{\sqrt{2\pi}}\int_0^x e^{-\frac{x^2}{2}}\, dx$. Integrating once yields

$$p'(x) = \frac{2}{b-a}\Big(\Phi(x/\sqrt{b}) - \Phi(x/\sqrt{a})\Big),$$

which further implies

$$p(x) = \frac{2}{b-a}\Big(\int_{-\infty}^x \Phi(y/\sqrt{b}) - \Phi(y/\sqrt{a})\Big)\, dy.$$

This is the probability density of the uniformly stopped Brownian motion B_U.

1.5 Erlang Stopped Brownian Motion

The Erlang distribution is a particular type of gamma distribution with the shape parameter being a positive integer. Its density is given by

$$f_k(u) = \frac{\theta^k u^{k-1} e^{-u\theta}}{(k-1)!}, \quad u > 0,\ \theta > 0,\ k = 1, 2, 3, \ldots \tag{1.5.17}$$

The fact that the Erlang distribution models the waiting time until the kth jump of a Poisson process with parameter θ made it useful in the study of queueing systems and stochastic processes.

Here we are interested in stopping a Brownian motion at the kth jump of a Poisson process. Its density is provided by the transform $T(f_k)$, which can be computed recursively over k.

We denote $U_k(x) = T(f_k)(x)$. We note that for $k = 1$ the Erlang distribution becomes an exponential distribution with parameter θ, so by Proposition 1.2.2 we have

$$U_1(x) = \sqrt{\frac{\theta}{2}} e^{-\sqrt{2\theta}|x|}, \qquad \forall x \in \mathbb{R}. \tag{1.5.18}$$

Logarithmic differentiation applied to (1.5.17) implies

$$f_k'(u) = \theta f_{k-1}(u) - \theta f_k(u).$$

Using Proposition 1.3.1 we have

$$T(f_k')(x) = \theta T(f_{k-1})(x) - \theta T(f_k)(x) = \theta U_{k-1}(x) - \theta U_k(x)$$

and

$$T(f_k')(x) = -\frac{1}{2}\partial_x^2 T(f_k)(x) = -\frac{1}{2}U_k''(x).$$

From the last two relations we infer the following differential recursive formula

$$U_k''(x) = 2\theta U_k(x) - 2\theta U_{k-1}(x), \qquad x \in \mathbb{R},\ k = 1, 2, 3, \cdots \tag{1.5.19}$$

This can be solved as a second order differential nonhomogeneous equation as

$$U_k(x) = U_k^0(x) + U_k^p(x),$$

where

$$U_k^0(x) = A_1 e^{\sqrt{2\theta}x} + A_2 e^{-\sqrt{2\theta}x}, \quad A_1, A_2 \in \mathbb{R}$$

is the solution of the associated homogeneous equation

$$U_k''(x) = 2\theta U_k(x)$$

and $U_k^p(x)$ is a particular solution of the nonhomogeneous equation (1.5.19).

Let $k = 2$. We shall find the particular solution $U_2^p(x)$ for $x > 0$. Using the expression of $U_1(x)$ provided by (1.5.18), we obtain that $U_2^p(x)$ satisfies the equation

$$U''(x) = 2\theta U(x) - 2\theta\sqrt{\frac{\theta}{2}}e^{-\sqrt{2\theta}x}, \quad x > 0.$$

Using the method of undetermined coefficients, we look for a particular solution in the form

$$U_2^p(x) = Cxe^{-\sqrt{2\theta}x}, \quad x > 0$$

and we obtain $C = \theta/2$. Therefore

$$U_2(x) = U_2^0(x) + U_2^p(x) = A_1e^{\sqrt{2\theta}x} + A_2e^{-\sqrt{2\theta}x} + \frac{\theta}{2}xe^{-\sqrt{2\theta}x}, \qquad x > 0.$$

Since $\lim_{x\to\infty} U_2(x) = 0$, we obtain $A_1 = 0$. Now, using that $U_2(x)$ is an even function of x, we can extended it symmetrically over all real numbers as

$$U_2(x) = A_2e^{-\sqrt{2\theta}|x|} + \frac{\theta}{2}|x|e^{-\sqrt{2\theta}|x|}, \qquad x \in \mathbb{R}.$$

The constant A_2 is obtained from the condition $\int_{\mathbb{R}} U_2(x)\,dx = 1$ as follows

$$\begin{aligned} 1 &= 2A_2\int_0^\infty e^{-\sqrt{2\theta}x}\,dx + \theta\int_0^\infty xe^{-\sqrt{2\theta}x}\,dx \\ &= \frac{2A_2}{\sqrt{2\theta}} + \frac{1}{2}, \end{aligned}$$

which implies $A_2 = \frac{1}{2}\sqrt{\frac{\theta}{2}}$. Hence, the probability density of a Brownian motion stopped at the second arrival time of a Poisson process is given by

$$U_2(x) = \frac{1}{2}\Big(\sqrt{\frac{\theta}{2}} + \theta|x|\Big)e^{-\sqrt{2\theta}|x|}, \qquad x \in \mathbb{R}. \tag{1.5.20}$$

This is a symmetric, single peaked distribution. The larger the parameter θ the more peaked the density is.

1.6 Inversion Formula for T

Until now we have computed the density of a stopped Brownian motion at a random time X defined on the semi-axis $[0, \infty)$. Namely, given the density f of X, we found the density of the stopped Brownian motion B_X by formula of $T(f)$ given by (1.3.15).

In this section we are interested in inverting formula (1.3.15). This is to find the density f that satisfies $T(f)(x) = U(x)$, in terms of $U(x)$. This answers the following question: *Given the density of the stopped Brownian motion B_X, what is the law of the stopping variable X?* This goal will be accomplished by using the inverse Laplace transform as follows:

Given a symmetric density function $U(x)$ on $\mathbb{R}$, we consider the integral equation

$$\int_0^\infty \frac{1}{\sqrt{2\pi u}} e^{-\frac{x^2}{2u}} f(u)\, du = U(x).$$

For the time being we assume $x > 0$ and substitute $u = 1/v$ to obtain

$$\int_0^\infty e^{-\frac{x^2}{2}v}\, v^{-3/2} f\Big(\frac{1}{v}\Big)\, dv = \sqrt{2\pi}\, U(x).$$

Let $g(v) = v^{-3/2} f\big(\frac{1}{v}\big)$ and $s = x^2/2$. Then the previous integral can be written as a Laplace transform

$$\mathcal{L}(g(v))(s) = \sqrt{2\pi}\, U(\sqrt{2s}). \tag{1.6.21}$$

If we assume $x \le 0$ and perform the same transformations as before, we arrive at

$$\mathcal{L}(g(v))(s) = \sqrt{2\pi}\, U(-\sqrt{2s}), \tag{1.6.22}$$

which is the same as (1.6.21) due to the symmetry $U(\sqrt{2s}) = U(-\sqrt{2s})$.

Applying the inverse Laplace transform yields

$$g(v) = \sqrt{2\pi}\, \mathcal{L}^{-1}(U(\sqrt{2s}))(v). \tag{1.6.23}$$

Therefore, the desired density function is

$$f(u) = u^{-3/2} g\Big(\frac{1}{u}\Big), \qquad u > 0. \tag{1.6.24}$$

We will now discuss several cases where the inverse Laplace transform (1.6.23) can be computed successfully.

The case $U(x) = \rho(x^2)$ We assume there is a smooth function $\rho(\cdot)$ such that $U(x) = \rho(x^2)$, for all $x \in \mathbb{R}$. Then (1.6.23) becomes

$$g(v) = \sqrt{2\pi}\, \mathcal{L}^{-1}(U(\sqrt{2s}))(v) = \sqrt{2\pi}\, \mathcal{L}^{-1}(\rho(2s))(v) = \sqrt{\pi/2}\, \mathcal{L}^{-1}(\rho(s))(v/2).$$

Therefore, as long as the inverse Laplace transform of ρ is known, we can retrieve the density $f(u)$. We shall provide an example.

Example 1.6.1 We assume a Brownian motion stopped at a random time X is Cauchy-distributed with parameter $a > 0$ as follows

$$B_X \sim \frac{a}{\pi(x^2 + a^2)}, \qquad x \in \mathbb{R}.$$

In this case $U(x) = \rho(x^2)$ with $\rho(x) = \dfrac{a}{\pi(x+a^2)}$. Its inverse Laplace transform is $\mathcal{L}^{-1}(\rho)(v) = \dfrac{a}{\pi}e^{-a^2 v}$, which implies

$$g(v) = \frac{a}{\sqrt{2\pi}}e^{-a^2 v/2}.$$

Then the density of the random time X is given by (1.6.24) as

$$f(u) = \frac{a}{\sqrt{2\pi}}\frac{1}{u^{3/2}}e^{-\frac{a^2}{2u}}, \qquad u > 0. \tag{1.6.25}$$

This is known as the Pearson 5 distribution. It can be shown that this is the density of the first passage of time when a Brownian motion hits the level $a > 0$ for the first time, see for instance Calin [22], page 81. Therefore, the density of a Brownian motion that is stopped when it reaches the level $a > 0$ is given by a Cauchy distribution with parameter a.

Another situation where equation (1.6.23) can be expressed using a closed-form solution is when the inverse Laplace transform of $U(x)$ is known. To accomplish this, we require preliminary information, which is supplied by the following two technical lemmas.

Lemma 1.6.2 *We have*

$$\frac{y}{2\sqrt{\pi}}\int_0^\infty e^{-sx}x^{-3/2}e^{-\frac{y^2}{4x}}\,dx = e^{-\sqrt{s}y}, \quad \forall s \geq 0. \tag{1.6.26}$$

Proof: We shall consider both sides as functions of the variable s and show that they satisfy the same initial condition ODE. We denote $\phi(s) = e^{-\sqrt{s}y}$, with y fixed parameter. It is easy to see that $\phi(s)$ is a solution of the Cauchy problem

$$Y'(s) = -\frac{y}{2\sqrt{s}}Y(s) \tag{1.6.27}$$

$$Y(0) = 1. \tag{1.6.28}$$

With substitution $z = \frac{1}{2}yx^{-1/2}$, the expression on the left side of identity (1.6.26) becomes

$$\begin{aligned}\psi(s) &= \frac{y}{2\sqrt{\pi}}\int_0^\infty x^{-3/2}e^{-(\frac{y^2}{4x}+sx)}\,dx\\ &= \frac{2}{\sqrt{\pi}}\int_0^\infty e^{-(z^2+\frac{y^2 s}{4z^2})}\,dz.\end{aligned}$$

Evaluating at $s = 0$ we get

$$\psi(0) = \frac{2}{\sqrt{\pi}} \int_0^\infty e^{-z^2}\, dz = 1.$$

Differentiating in s and using the substitution $u = \frac{y}{2z}\sqrt{s}$, we obtain

$$\begin{aligned}
\psi'(s) &= \frac{2}{\sqrt{\pi}}\partial_s \int_0^\infty e^{-(z^2+\frac{y^2 s}{4z^2})}\, dz = \frac{2}{\sqrt{\pi}} \int_0^\infty e^{-(z^2+\frac{y^2 s}{4z^2})}\Big(-\frac{y^2}{4z^2}\Big)\, dz \\
&= -\frac{y}{2\sqrt{s}} \cdot \frac{2}{\sqrt{\pi}} \int_0^\infty e^{-(u^2+\frac{y^2 s}{4u^2})}\, du = -\frac{y}{2\sqrt{s}}\psi(s).
\end{aligned}$$

Therefore, $\psi(s)$ is also a solution of the system (1.6.27)-(1.6.28). From the solution uniqueness theorem it follows that $\psi(s) = \phi(s)$, for all $s \geq 0$, which proves the desired identity. ■

Lemma 1.6.3 *Let $F(s) = \mathcal{L}(f)(s) = \int_0^\infty e^{-sx} f(x)\, dx$. Then*

$$\mathcal{L}^{-1}(F(\sqrt{s}))(y) = \frac{1}{2\sqrt{\pi} y^{3/2}} \int_0^\infty x e^{-\frac{x^2}{4y}} f(x)\, dx.$$

Proof: Denote

$$g(y) = \frac{1}{2\sqrt{\pi} y^{3/2}} \int_0^\infty x e^{-\frac{x^2}{4y}} f(x)\, dx.$$

We need to show $\mathcal{L}(g(y))(s) = F(\sqrt{s})$. We have

$$\begin{aligned}
\mathcal{L}(g(y))(s) &= \int_0^\infty e^{-sy} g(y)\, dy \\
&= \int_0^\infty e^{-sy}\Big(\frac{1}{2\sqrt{\pi} y^{3/2}} \int_0^\infty x e^{-\frac{x^2}{4y}} f(x)\, dx\Big)\, dy \\
&= \int_0^\infty x\Big(\int_0^\infty e^{-sy} \frac{1}{2\sqrt{\pi} y^{3/2}} e^{-\frac{x^2}{4y}}\, dy\Big) f(x)\, dx \\
&= \int_0^\infty y\Big(\int_0^\infty e^{-sx} \frac{1}{2\sqrt{\pi} x^{3/2}} e^{-\frac{y^2}{4x}}\, dx\Big) f(y)\, dy \\
&= \int_0^\infty e^{-\sqrt{s}y} f(y)\, dy = \mathcal{L}(f)(\sqrt{s}) = F(\sqrt{s}),
\end{aligned}$$

where we used Fubini's theorem and Lemma 1.6.2. ■

The next proposition will be used to invert the integral operator T.

Proposition 1.6.4 *Let $\mathcal{L}(\varphi)(s) = U(s)$ and assume that $\varphi(y)$ increases to infinity sub-exponentially. Then the integral equation*

$$\int_0^\infty \frac{1}{\sqrt{2\pi u}} e^{-\frac{x^2}{2u}} f(u)\, du = U(x)$$

has a unique solution, which is given by

$$f(u) = \int_0^\infty y e^{-y^2 u/2} \varphi(y)\, dy. \tag{1.6.29}$$

Proof: Let $F(s) = U(\sqrt{2}s)$. If consider the inverse Laplace transforms $\phi = \mathcal{L}^{-1}(F)$ and $\varphi = \mathcal{L}^{-1}(U)$, then

$$\phi(x) = \mathcal{L}^{-1}(F(s))(x) = \mathcal{L}^{-1}(U(\sqrt{2}s))(x) = \frac{1}{\sqrt{2}}\varphi\Big(\frac{x}{\sqrt{2}}\Big).$$

Next we shall compute $g(v)$ given by (1.6.23) using Lemma 1.6.3

$$\begin{aligned} g(v) &= \sqrt{2\pi}\,\mathcal{L}^{-1}(U(\sqrt{2s}))(v) = \sqrt{2\pi}\,\mathcal{L}^{-1}(F(\sqrt{s}))(v) \\ &= \sqrt{2\pi}\frac{1}{2\sqrt{\pi}}\frac{1}{v^{3/2}}\int_0^\infty x e^{-\frac{x^2}{4v}}\phi(x)\, dx \\ &= \frac{1}{2}\frac{1}{v^{3/2}}\int_0^\infty x e^{-\frac{x^2}{4v}}\varphi\Big(\frac{x}{\sqrt{2}}\Big)\, dx = \frac{1}{v^{3/2}}\int_0^\infty y e^{-\frac{y^2}{2v}}\varphi(y)\, dy. \end{aligned}$$

Then

$$g\Big(\frac{1}{u}\Big) = u^{3/2}\int_0^\infty y e^{-\frac{y^2 u}{2}}\varphi(y)\, dy.$$

Substituting in (1.6.24) we obtain the solution

$$f(u) = u^{-3/2} g\Big(\frac{1}{u}\Big) = \int_0^\infty y e^{-\frac{y^2 u}{2}}\varphi(y)\, dy.$$

The uniqueness follows from the way the solution was derived. ■

Remark 1.6.5 The inverse of the operator T defined by (1.3.15) has the property

$$T^{-1}\Big(\mathcal{L}(\varphi)(x)\Big) = \int_0^\infty y e^{-\frac{y^2 u}{2}}\varphi(y)\, dy.$$

Remark 1.6.6 The solution (1.6.29) has the following probabilistic interpretation. First we recall the Rayleigh distribution with parameter σ

$$\rho(y;\sigma) = \frac{y}{\sigma^2}e^{-\frac{y^2}{2\sigma^2}}, \quad y \geq 0.$$

Then the solution $f(u)$ can be represented as an expectation of a function of a Rayleigh distributed variable

$$\begin{aligned} f(u) &= \int_0^\infty y e^{-\frac{y^2 u}{2}}\varphi(y)\, dy = \frac{1}{u}\int_0^\infty \rho(u; u^{-1/2})\varphi(y)\, dy \\ &= \frac{1}{u}\mathbb{E}[\varphi(R)], \end{aligned}$$

where R is a Rayleigh distributed random variable with parameter $\sigma = u^{-1/2}$.

Now, let's consider $R = |W_t| = \sqrt{W_1(t)^2 + W_2(t)^2}$ be the radial part of a 2-dimensional Brownian motion $(W_1(t), W_2(t))$. Then R is Rayleigh distributed with parameter $\sigma = \sqrt{t}$. Substituting $u = 1/t$ yields

$$f(1/t) = t\,\mathbb{E}[\varphi(|W_t|)], \qquad t > 0.$$

Example 1.6.7 Similarly with Example 1.6.1, we assume the stopped Brownian motion B_X has a Cauchy distribution with parameter a

$$U(x) = \frac{a}{\pi(x^2 + a^2)}, \qquad x \in \mathbb{R}.$$

This time, in order to find the density $f(u)$ of the stopping time X we shall use Proposition 1.6.4. Since

$$\phi(y) = \mathcal{L}^{-1}(U)(y) = \frac{1}{\pi}\sin(ay),$$

then relation (1.6.29) provides

$$f(u) = \int_0^\infty ye^{-y^2u/2}\varphi(y)\,dy = \frac{1}{\pi}\int_0^\infty ye^{-y^2u/2}\sin(ay)\,dy.$$

We shall show that this integral is in fact equal to the Pearson 5 distribution (1.6.25). Since the integrand is even, we can extend the integral over the entire real axis as

$$f(u) = \frac{1}{2\pi}\int_{-\infty}^\infty ye^{-y^2u/2}\sin(ay)\,dy.$$

With the use of the vanishing integral

$$\frac{1}{2\pi}\int_{-\infty}^\infty ye^{-y^2u/2}\cos(ay)\,dy = 0$$

and Euler's formula, we obtain

$$if(u) = \frac{1}{2\pi}\int_{-\infty}^\infty ye^{-\frac{y^2u}{2}+iay}\,dy.$$

Substituting $\alpha = u/2$ and $\beta = ai$ in the known improper integral formula

$$\int_{-\infty}^\infty ye^{-\alpha y^2+\beta y}\,dy = \sqrt{\frac{\pi}{\alpha}}\Big(\frac{\beta}{2\alpha}\Big)e^{\beta^2/(4\alpha)}$$

yields

$$if(u) = i\frac{a}{\sqrt{2\pi}}u^{-3/2}e^{-\frac{a^2}{2u}},$$

which produces after simplification a Pearson 5 distribution for $f(u)$ as given by (1.6.25).

1.7 Cameron-Martin's Formula

In this section several famous mathematical relations will come into action. They involve the use of the heat kernel for the Hermite operator and Feynman-Kac's formula to compute the moment generating function of the stochastic process $X_t = \int_0^t (W_s + x)^2\, ds$.

The idea is to solve an initial value problem for the Hermite operator using two different methods: PDE techniques and the Feynman-Kac's formula. By comparing these two solutions, we can derive Cameron-Martin's formula.

We start with a result regarding a solution for the Cauchy problem of the Hermite operator $\frac{1}{2}\frac{d^2}{dx^2} - \frac{1}{2}\gamma^2 x^2$, where $\gamma \in \mathbb{R}$ is a constant.

Lemma 1.7.1 *The function*

$$u(t,x) = \frac{1}{\sqrt{\cosh(\gamma t)}} e^{-\frac{1}{2}\gamma x^2 \tanh(\gamma t)}, \quad x \in \mathbb{R}, t > 0 \tag{1.7.30}$$

solves the following initial value problem

$$\frac{\partial u}{\partial t} = \frac{1}{2}\frac{\partial^2 u}{\partial x^2} - \frac{1}{2}\gamma^2 x^2 u \tag{1.7.31}$$
$$u(0,x) = 1. \tag{1.7.32}$$

Proof: Instead of just checking by a straightforward computation that the function given by the expression (1.7.30) is a solution, we shall provide two ways to get constructively to this expression.

The first proof variant. We shall provide first a proof based on an educated guess. Neglecting the second order derivative $\frac{1}{2}\frac{\partial^2 u}{\partial x^2}$, the initial value problem becomes

$$\begin{aligned} \frac{\partial v}{\partial t} &= -\frac{1}{2}\gamma^2 x^2 v \\ v(0,x) &= 1. \end{aligned}$$

with the solution given by $v(t,x) = e^{-\frac{1}{2}\gamma^2 x^2 t}$. Inspired by this expression, we shall look for a solution of the problem (1.7.31)-(1.7.32) under the form

$$u(t,x) = e^{\alpha(t) - \frac{1}{2}x^2 \beta(t)}, \tag{1.7.33}$$

with $\alpha(t)$ and $\beta(t)$ smooth functions, subject to be found later. To accommodate for the initial condition (1.7.32) we shall require $\alpha(0) = 0$ and $\beta(0) = 0$.

Substituting $z(t,x) = \ln u(t,x)$, we obtain that $z(t,x)$ satisfies the following initial value problem

$$\begin{aligned} \frac{\partial z}{\partial t} &= \frac{1}{2}\Big[\Big(\frac{\partial z}{\partial x}\Big)^2 + \frac{\partial^2 z}{\partial x^2}\Big] - \frac{1}{2}\gamma^2 x^2 && (1.7.34)\\ z(0,x) &= 0. && (1.7.35) \end{aligned}$$

On the other side, by (1.7.33), we have $z(t,x) = \alpha(t) - \frac{1}{2}x^2\beta(t)$. Substituting in equation (1.7.34) we obtain a vanishing quadratic polynomial in x

$$\frac{1}{2}x^2\big(\beta'(t) + \beta^2(t) - \gamma^2\big) - \big(\alpha'(t) + \frac{1}{2}\beta(t)\big) = 0, \quad \forall x \in \mathbb{R}.$$

Equating the coefficients to zero we arrive at the following initial value problems

$$\begin{aligned} \beta'(t) &= \gamma^2 - \beta^2(t)\\ \beta(0) &= 0 \end{aligned}$$

and

$$\begin{aligned} \alpha'(t) &= -\frac{1}{2}\beta(t)\\ \alpha(0) &= 0. \end{aligned}$$

Integrating yields $\beta(t) = \gamma\tanh(\gamma t)$ and $\alpha(t) = -\frac{\gamma}{2}\int_0^t \tanh(\gamma s)ds = -\frac{1}{2}\ln\cosh(\gamma t)$. Substituting back into the expression of $u(t,x)$ and using the logarithm properties we obtain

$$u(t,x) = e^{\alpha(t) - \frac{1}{2}x^2\beta(t)} = \frac{1}{\sqrt{\cosh(\gamma t)}} e^{-\frac{1}{2}\gamma x^2\tanh(\gamma t)},$$

which is the desired solution.

The second proof variant. The second proof is based on the following expression of the heat kernel of the Hermite operator, see Calin and Chang [36]

$$K(x,y;t) = \frac{\sqrt{\gamma t}}{\sqrt{2\pi t}\sqrt{\sinh(\gamma t)}} e^{-\frac{1}{2t}\frac{\gamma t}{\sinh(\gamma t)}[(x^2+y^2)\cosh(\gamma t) - 2xy]}. \qquad (1.7.36)$$

This means that K satisfies

$$\begin{aligned} \frac{\partial}{\partial t}K(x,y;t) &= \Big(\frac{1}{2}\frac{\partial^2}{\partial x^2} - \frac{1}{2}\gamma^2 x^2\Big)K(x,y;t) && (1.7.37)\\ \lim_{t\to 0} K(x,y,t) &= \delta(x-y). && (1.7.38) \end{aligned}$$

The solution can be computed integrating the initial condition against the heat kernel as follows

$$\begin{aligned} u(t,x) &= \int_{\mathbb{R}} K(x,y;t)u(0,y)\,dy = \int_{\mathbb{R}} K(x,y;t)\,dy \\ &= \frac{\sqrt{\gamma t}}{\sqrt{2\pi t}\sqrt{\sinh(\gamma t)}} e^{-\frac{\gamma}{2}x^2\coth(\gamma t)} \int_{\mathbb{R}} e^{-\frac{\gamma}{2}\coth(\gamma t)y^2 + \frac{\gamma x}{\sinh(\gamma t)}y}\,dy. \end{aligned}$$

The integral can be simplified by substituting $a = \frac{\gamma}{2}\coth(\gamma t)$ and $b = \frac{\gamma x}{\sinh(\gamma t)}$ into the formula

$$\int_{\mathbb{R}} e^{-ay^2+by}\,dy = \sqrt{\frac{\pi}{a}}e^{\frac{b^2}{4a}}.$$

We obtain

$$\begin{aligned} u(t,x) &= \frac{\sqrt{\gamma t}}{\sqrt{2\pi t}\sqrt{\sinh(\gamma t)}}\, e^{-\frac{\gamma}{2}x^2\coth(\gamma t)} \sqrt{\frac{2\pi}{\gamma}}\frac{\sqrt{\sinh(\gamma t)}}{\sqrt{\cosh(\gamma t)}}\, e^{\frac{\gamma x^2}{2\sinh(\gamma t)\cosh(\gamma t)}} \\ &= \frac{1}{\sqrt{\cosh(\gamma t)}}\, e^{\frac{\gamma}{2}x^2\frac{1}{\sinh(\gamma t)}\left(\frac{1}{\cosh(\gamma t)} - \cosh(\gamma t)\right)} \\ &= \frac{1}{\sqrt{\cosh(\gamma t)}} e^{-\frac{1}{2}\gamma x^2\tanh(\gamma t)}, \end{aligned}$$

where we used the relation $\cosh^2 z - \sinh^2 z = 1$.

■

The next result contains the celebrated Cameron-Martin formula.

Theorem 1.7.2 (Cameron-Martin) *Let W_t be a standard Brownian motion starting at zero. Then for any $\gamma \in \mathbb{R}$ we have*

$$\mathbb{E}[e^{-\frac{\gamma^2}{2}\int_0^t (W_s+x)^2\,ds}] = \frac{1}{\sqrt{\cosh(\gamma t)}} e^{-\frac{1}{2}\gamma x^2\tanh(\gamma t)}, \quad \forall x \in \mathbb{R},\ t > 0. \quad (1.7.39)$$

Proof: Since the infinitesimal generator of the Brownian motion W_t is $\frac{1}{2}\frac{d^2}{dx^2}$, then applying Feynman-Kac's formula with the quadratic potential $U(x) = \frac{\gamma^2}{2}x^2$ implies that

$$u(t,x) = \mathbb{E}^x[e^{-\frac{\gamma^2}{2}\int_0^t W_s^2\,ds}] = \mathbb{E}[e^{-\frac{\gamma^2}{2}\int_0^t (W_s+x)^2\,ds}]$$

is a solution for the initial value problem (1.7.31)-(1.7.32). The conclusion follows now by applying Lemma 1.7.1. ■

Remark 1.7.3 In the case $x = 0$ we obtain the following useful formula

$$\mathbb{E}[e^{-\frac{\gamma^2}{2}\int_0^t W_s^2\, ds}] = \frac{1}{\sqrt{\cosh(\gamma t)}}, \qquad t > 0. \tag{1.7.40}$$

This can be interpreted as a closed form expression for the moment generating function of the integrated squared Brownian motion, $X_t = \int_0^t W_s^2\, ds$. This can be written as the Laplace transform of its density $p_t(x)$

$$\mathcal{L}(p_t(x))\Big(\frac{\gamma^2}{2}\Big) = \frac{1}{\sqrt{\cosh(\gamma t)}}.$$

Substituting $s = \frac{\gamma^2}{2}$ and using that cosh is even, we can write

$$\mathcal{L}(p_t(x))(s) = \frac{1}{\sqrt{\cosh(\sqrt{2s}\, t)}}.$$

Therefore, the density of X_t can be expressed as an inverse Laplace transform

$$p_t(x) = \mathcal{L}^{-1}\Big(\sqrt{\mathrm{sech}(\sqrt{2s}t)}\Big)(x).$$

Remark 1.7.4 Let $W_t = (W_1(t), W_2(t))$ be a two dimensional Brownian motion, with $W_i(t)$ independent. Let $R_t = \|W_t\| = \sqrt{W_1(t)^2 + W_2(t)^2}$ be the norm of W_t. Then using the independence of $W_1(t)$ and $W_2(t)$ as well as formula (1.7.40) we obtain

$$\begin{aligned}
\mathbb{E}[e^{-\frac{\gamma^2}{2}\int_0^t R_s^2\, ds}] &= \mathbb{E}[e^{-\frac{\gamma^2}{2}\int_0^t W_1(s)^2\, ds}]\mathbb{E}[e^{-\frac{\gamma^2}{2}\int_0^t W_2(s)^2\, ds}] \\
&= \frac{1}{\cosh(\gamma t)}, \qquad t > 0.
\end{aligned}$$

1.8 Interpretation of Cameron-Martin's Formula

This section will provide a variational interpretation of the Cameron-Martin formula in terms of the action along a classical path associated with a classical Lagrangian.

We shall start from the expression of the integral of the heat kernel for the Hermite operator (1.7.36), which after integration in y becomes

$$\int_{\mathbb{R}} K(x, y; t)\, dy = \frac{1}{\sqrt{\cosh(\gamma t)}} e^{-\frac{1}{2}\gamma x^2 \tanh(\gamma t)}, \tag{1.8.41}$$

see the computation done at page 27. Then we recall the famous van Vleck formula, see Calin et al. [29], page 133, which expresses the heat kernel as a product

$$K(x, y; t) = V(t) e^{-S_{cl}(x,y;t)},$$

where

$$V(t) = \sqrt{\det\Big(-\frac{1}{2\pi}\frac{\partial^2 S_{cl}(x,y;t)}{\partial x \partial y}\Big)} = \frac{\sqrt{\gamma t}}{\sqrt{2\pi t}\sqrt{\sinh(\gamma t)}} \tag{1.8.42}$$

is the van Vleck determinant] and $S_{cl}(x,y;t)$ stands for the classical action between x and y within time t

$$S_{cl}(x,y;t) = \frac{1}{2t}\frac{\gamma t}{\sinh(\gamma t)}[(x^2+y^2)\cosh(\gamma t) - 2xy]. \tag{1.8.43}$$

The classical action is the time integral of the Lagrangian

$$L(x,\dot{x}) = \frac{1}{2}\dot{x}^2 + \frac{1}{2}\gamma^2 x^2 \tag{1.8.44}$$

along a minimizing path

$$S_{cl}(x,y;t) = \min_{\phi(0)=x,\phi(t)=y} \int_0^t L(\phi(s),\dot{\phi}(s))\,ds.$$

The minimizing path $\phi(s)$ satisfies the Euler-Lagrange variational equations

$$\frac{d}{dt}\frac{\partial L}{\partial \dot{\phi}} = \frac{\partial L}{\partial \phi}$$

with fixed boundaries, which writes as

$$\ddot{\phi}(t) = \gamma^2\phi(t), \quad \phi(0) = x, \quad \phi(t) = y.$$

Solving the previous boundary value ODE we obtain the minimal path

$$\phi(s) = [y - x\cosh(\gamma t)]\frac{\sinh(\gamma s)}{\sinh(\gamma t)} + x\cosh(\gamma s), \quad 0 \le s \le t.$$

Integrating the Lagrangian (1.8.44) over $\phi(s)$ yields the expression of the aforementioned classical action (1.8.43). Since the Lagrangian contains a quadratic potential, $U(x) = \frac{1}{2}\gamma^2 x^2$, it follows that $\phi(s)$ represents the trajectory of a particle under the effect of an elastic force $F(x) = U'(x) = \gamma x$.

Substituting the van Vleck formula into relation (1.8.41) yields

$$\int_{\mathbb{R}} e^{-S_{cl}(x,y;t)}\,dy = \sqrt{\frac{2\pi}{\gamma}\tanh(\gamma t)}e^{-\frac{1}{2}\gamma x^2\tanh(\gamma t)}, \quad t > 0. \tag{1.8.45}$$

It is easy to see that this relation is equivalent to Cameron-Martin's formula. It is worth noting how this depends on time through the $\tanh(\gamma t)$ function.

The exponential $e^{-S_{cl}(x,y;t)}$ is reminiscent from the concept of probability amplitude from Quantum Mechanics,[1] see Feynman [56]. Thus, the integral on the left side of (1.8.45) represents the total cumulative contributions of amplitudes $e^{-S_{cl}(x,y;t)}$ along all paths starting at x and ending at an arbitrary y at time t.

A similar application regarding LC-circuits is given in Chapter 9, where we show how to obtain the expression of the action from both the Lagrangian and Hamiltonian points of view.

1.9 The Law of B_{A_t}

Consider the integrated quadratic Brownian motion, $A_t = \int_0^t W_s^2\, ds$, with W_t standard Brownian motion. Then we choose another standard Brownian motion, B_t, which is independent of W_t and consider the time changed Brownian motion $X_t = B_{A_t}$. We shall compute the characteristic function of X_t.

Using the tower property of conditional expectations and the Cameron-Martin formula, we obtain an explicit expression for the characteristic function of X_t

$$\begin{aligned}\mathbb{E}[e^{ixX_t}] &= \mathbb{E}[e^{ixB_{A_t}}] = \mathbb{E}[\mathbb{E}[e^{ixB_{A_t}}|W[0,t]]] \\ &= \mathbb{E}[e^{-\frac{1}{2}x^2 A_t}] = \mathbb{E}[e^{-\frac{1}{2}x^2 \int_0^t W_s^2 ds}] \\ &= [\cosh(tx)]^{-1/2}, \qquad (1.9.46)\end{aligned}$$

where $W[0,t]$ is the the sigma-algebra generated by W_s, with $0 \le s \le t$ and is called the *history of the Brownian motion* until time t.

Proposition 1.9.1 *The probability density of the process $X_t = B_{A_t}$ is given by*

$$p_{X_t}(\xi) = \frac{1}{2\pi}\int_{\mathbb{R}} \frac{e^{-ix\xi}}{\sqrt{\cosh(tx)}}\, dx.$$

Proof: We use the fact that the characteristic function of X_t is the Fourier transform of the probability density p_{X_t}

$$\mathcal{F}(p_{X_t})(x) = \int_{\mathbb{R}} e^{ix\xi} p_{X_t}(\xi)\, d\xi = \mathbb{E}[e^{ixX_t}] = [\cosh(tx)]^{-1/2},$$

[1] In Quantum Mechanics this is actually $e^{-\frac{i}{\hbar}S_{cl}(x,y;t)}$, fact that provides to S_{cl} the interpretation of phase.

where the last identity is validated by formula (1.9.46). Then inverting the Fourier transform we obtain the desired result

$$p_{X_t}(\xi) = \mathcal{F}^{-1}([\cosh(tx)]^{-1/2})(\xi) = \frac{1}{2\pi}\int_{\mathbb{R}} \frac{e^{-ix\xi}}{\sqrt{\cosh(tx)}}\,dx.$$

∎

Remark 1.9.2 Since $\cosh(x)$ is an even function, after the variable change $y = -x$ we have

$$p_{X_t}(\xi) = \frac{1}{2\pi}\int_{\mathbb{R}} \frac{e^{-ix\xi}}{\sqrt{\cosh(tx)}}\,dx = \frac{1}{2\pi}\int_{\mathbb{R}} \frac{e^{iy\xi}}{\sqrt{\cosh(ty)}}\,dy = \overline{p}_{X_t}(\xi).$$

Therefore, we can write it as a cosine transform

$$p_{X_t}(\xi) = \frac{1}{2}(p_{X_t}(\xi) + \overline{p}_{X_t}(\xi)) = \frac{1}{2\pi}\int_{\mathbb{R}} \frac{\cos(x\xi)}{\sqrt{\cosh(tx)}}\,dx. \tag{1.9.47}$$

Remark 1.9.3 If f_{A_t} denotes the density of $A_t = \int_0^t W_s^2 ds$, then by a computation similar with the one done at page 14 for obtaining formula (1.2.12) we obtain the density of X_t in terms of f_{A_t}

$$p_{X_t}(\xi) = \int_0^\infty \frac{1}{\sqrt{2\pi u}} e^{-\frac{\xi^2}{2u}} f_{A_t}(u)\,du, \tag{1.9.48}$$

with p_{X_t} given by (1.9.47). Using Proposition 1.6.4 we can retrieve the density of A_t from the inverse Laplace transform of p_{X_t}, namely

$$f_{A_t}(u) = \int_0^\infty y e^{-y^2 u/2} \varphi(y)\,dy, \tag{1.9.49}$$

where $\mathcal{L}(\varphi)(s) = \frac{1}{2\pi}\int_{\mathbb{R}} \frac{\cos(x\xi)}{\sqrt{\cosh(tx)}}\,dx$. The fact that the aforementioned Laplace transform cannot be easily inverted makes f_{A_t} to have a very complex expression. As a matter of fact, we tried to avoid using an explicit formula for f_{A_t} while computing the moment generating function of A_t.

Remark 1.9.4 Now we consider the more general case when A_t is a functional of the history of the Brownian motion W_s until time t

$$A_t = F(W_s; 0 \le s \le t).$$

We shall consider the particular case when $A_t = \int_0^t g^2(W_s)\,ds$, with g continuous function. Then conditioning over the history of the Brownian motion $W[0,t] = \{W_s; 0 \le s \le t\}$ and using the tower property, we obtain an identity between the characteristic function of $X_t = B_{A_t}$ and the moment generating function of A_t

$$\begin{aligned}
\mathbb{E}[e^{ixX_t}] &= \mathbb{E}[e^{ixB_{A_t}}] = \mathbb{E}[\mathbb{E}[e^{ixB_{A_t}}|W[0,t]]] \\
&= \mathbb{E}[e^{-\frac{1}{2}x^2A_t}]. \qquad (1.9.50)
\end{aligned}$$

The inner conditional expectation has been computed as if A_t would be deterministic, using formula $\mathbb{E}[e^{ixB_u}] = e^{-\frac{1}{2}x^2u}$. The expectation (1.9.50) can be further computed using Feynman-Kac's formula, provided the solution of the initial value problem

$$\begin{aligned}
\partial_t u(t,y) &= \frac{1}{2}\partial_y^2 u(t,y) - \frac{1}{2}x^2 g^2(y)u(t,y) \\
u(0,y) &= 1
\end{aligned}$$

is known. Explicit solutions can been found, for instance, in the cases $g(x) = x$ and $g(x) = e^x$.

Remark 1.9.5 In the following we shall present a direct proof of Proposition 1.9.1. First we write the heat kernel as the expectation of a Dirac function and then use the tower property to obtain

$$\begin{aligned}
p_{X_t}(\xi) &= \int_{\mathbb{R}} \delta(y-\xi)p_{X_t}(y)\,dy = \mathbb{E}[\delta(X_t-\xi)] = \mathbb{E}[\delta(B_{A_t}-\xi)] \\
&= \mathbb{E}[\,\mathbb{E}[\delta(B_{A_t}-\xi)|W[0,t]\,]\,] = \mathbb{E}\Big[\mathbb{E}\Big[\frac{1}{2\pi}\int_{\mathbb{R}} e^{ix(B_{A_t}-\xi)}\,dx|W[0,t]\Big]\Big] \\
&= \mathbb{E}\Big[\frac{1}{2\pi}\int_{\mathbb{R}} e^{-ix\xi}\mathbb{E}\Big[e^{ixB_{A_t}}|W[0,t]\Big]\,dx\Big] \\
&= \mathbb{E}\Big[\frac{1}{2\pi}\int_{\mathbb{R}} e^{-ix\xi}e^{-\frac{1}{2}x^2A_t}\,dx\Big] \qquad (1.9.51) \\
&= \frac{1}{2\pi}\int_{\mathbb{R}} e^{-ix\xi}\mathbb{E}\big[e^{-\frac{1}{2}x^2A_t}\big]\,dx \\
&= \frac{1}{2\pi}\int_{\mathbb{R}} e^{-ix\xi}\frac{1}{\sqrt{\cosh(tx)}}\,dx.
\end{aligned}$$

We commuted the integral with the expectation, as an application of Fubini's theorem and wrote the delta function as the inverse Fourier transform of 1. In the last relation we have applied Cameron-Martin's formula.

Remark 1.9.6 Using the Gaussian integral

$$\int_{\mathbb{R}} e^{-ax^2-i\xi x}\,dx = \sqrt{\frac{\pi}{a}}e^{-\frac{\xi^2}{4a}}, \qquad a \in \mathbb{R}^*,$$

we compute the integral in (1.9.51) before taking the expectation

$$\begin{aligned} p_{X_t}(\xi) &= \mathbb{E}\Big[\frac{1}{2\pi}\int_{\mathbb{R}} e^{-ix\xi} e^{-\frac{1}{2}x^2 A_t}\, dx\Big] = \mathbb{E}\Big[\frac{1}{\sqrt{2\pi A_t}} e^{-\frac{\xi^2}{2A_t}}\Big] \\ &= \int_0^\infty \frac{1}{\sqrt{2\pi u}} e^{-\frac{\xi^2}{2u}} f_{A_t}(u)\, du \\ &= T(f_{A_t})(\xi), \end{aligned}$$

where T is the operator introduced in section 1.3. Therefore, we obtained $p_{X_t} = T(f_{A_t})(x)$, which is equivalent to formula (1.9.48).

1.10 Ornstein-Uhlenbeck Process

The Langevin equation, introduced by Langevin [94], describes the velocity of a particle in a fluid undergoing random collisions with the fluid molecules. Its solution is known as the Ornstein-Uhlenbeck process. In this section, we focus on the one- and two-dimensional processes and establish their connection with the Bessel process. This connection is later used in the proofs of both Cameron-Martin's and Lévy's area formulas.

The one-dimensional case

The Ornstein-Uhlenbeck process is the solution of the Langevin's equation

$$dX_t = -\lambda X_t dt + dW_t \qquad (1.10.52)$$

$$X_0 = x_0, \qquad (1.10.53)$$

where W_t is a one-dimensional Brownian motion and λ a positive constant. Multiplying by the integrating factor $e^{\lambda t}$ we obtain the exact equation

$$d(e^{\lambda t} X_t) = e^{\lambda t} dW_t,$$

which after integration leads to the unique strong solution

$$X_t = x_0 e^{-\lambda t} + e^{-\lambda t}\int_0^t e^{\lambda s}\, dW_s. \qquad (1.10.54)$$

It follows that X_t is normally distributed, $X_t \sim N(\mu, \sigma^2)$, with mean $\mu = x_0 e^{-\lambda t}$ and variance $\sigma^2 = e^{-2\lambda t}\int_0^t e^{2\lambda s}\, ds = \frac{1-e^{-2\lambda t}}{2\lambda}$. Therefore, for all $t > 0$

and $x, x_0 \in \mathbb{R}$, the transition probability density of X_t becomes

$$
\begin{aligned}
p_t(x_0, x) &= \frac{1}{\sqrt{2\pi}\sigma} e^{-\frac{(x-\mu)^2}{2\sigma^2}} = \sqrt{\frac{\lambda}{\pi(1-e^{-2\lambda t})}} e^{-\frac{\lambda}{1-e^{-2\lambda t}}(x-x_0 e^{-\lambda t})^2} \\
&= \sqrt{\frac{\lambda}{2\pi \sinh(\lambda t)}} e^{\frac{\lambda t}{2}} e^{-\frac{\lambda}{2\sinh(\lambda t)}(x^2 e^{\lambda t} + x_0^2 e^{-\lambda t} - 2x x_0)} \\
&= \sqrt{\frac{\lambda}{2\pi \sinh(\lambda t)}} e^{\frac{\lambda t}{2}} e^{-\frac{\lambda}{2\sinh(\lambda t)}((x^2+x_0^2)\cosh(\lambda t) - 2x x_0 + (x^2 - x_0^2)\sinh(\lambda t))} \\
&= \sqrt{\frac{\lambda}{2\pi \sinh(\lambda t)}} e^{\frac{\lambda}{2}(t - x^2 + x_0^2)} e^{-\frac{\lambda}{2\sinh(\lambda t)}((x^2+x_0^2)\cosh(\lambda t) - 2x x_0)}.
\end{aligned}
$$

Using the classical action between x_0 and x within time t denoted by $S_{cl}(x_0, x; t)$ and given by formula (1.8.43) and the van Vleck's determinant $V(t)$ provided by (1.8.42), we can further write the transition density as

$$
p_t(x_0, x) = e^{\frac{\lambda}{2}(t - x^2 + x_0^2)} V(t) e^{-S_{cl}(x_0, x; t)}, \tag{1.10.55}
$$

where we considered $\gamma = \lambda$.

The two-dimensional case

We consider a two-dimensional Ornstein-Uhlenbeck process satisfying the system of stochastic differential equations

$$
\begin{aligned}
dY_1(t) &= -\lambda Y_1(t)dt + dW_1(t) && (1.10.56) \\
dY_2(t) &= -\lambda Y_2(t)dt + dW_2(t), && (1.10.57)
\end{aligned}
$$

with $W_1(t)$ and $W_2(t)$ independent standard Brownian motions. The solution can be written in the vector form as

$$
Y(t) = y_0 e^{-\lambda t} + e^{-\lambda t} \int_0^t e^{\lambda s}\, dW(s),
$$

where $Y(t) = \big(Y_1(t), Y_2(t)\big)^T$, and $W(s) = \big(W_1(s), W_2(s)\big)^T$. The process is Gaussian, with mean $y_0 e^{-\lambda t}$ and covariance matrix $\frac{1-e^{-2\lambda t}}{2\lambda}\mathbb{I}_2$, so the transition density is given by

$$
p_t(y_0, y) = \frac{\lambda}{\pi(1-e^{-2\lambda t})} e^{-\frac{\lambda}{1-e^{-2\lambda t}}\|y - y_0 e^{-\lambda t}\|^2}, \quad t > 0,\ y, y_0 \in \mathbb{R}^2. \tag{1.10.58}
$$

It is worth noting that the infinitesimal generator of the process $Y(t)$ is the operator

$$
L = \frac{1}{2}(\partial_{y_1}^2 + \partial_{y_2}^2) - \lambda(y_1 \partial_{y_1} + y_2 \partial_{y_2}). \tag{1.10.59}
$$

Using polar coordinates, $y_1 = r\cos\theta$, $y_2 = r\sin\theta$, we have

$$\partial_{y_1}^2 + \partial_{y_2}^2 = \partial_r^2 + \frac{1}{r}\partial_r + \frac{1}{r^2}\partial_\theta^2, \quad y_1\partial_{y_1} + y_2\partial_{y_2} = r\partial_r,$$

so the generator in polar coordinates is given by

$$L = \frac{1}{2}\partial_r^2 + \Big(\frac{1}{2r} - \lambda r\Big)\partial_r + \frac{1}{2r^2}\partial_\theta^2.$$

The radial part of L, which is

$$L_{rad} = \frac{1}{2}\partial_r^2 + \Big(\frac{1}{2r} - \lambda r\Big)\partial_r$$

should be the infinitesimal generator of the radial part of $Y(t)$, which is $V_t = \|Y(t)\|$. Therefore, the associated stochastic differential equation of V_t is

$$dV_t = \Big(\frac{1}{2V_t} - \lambda V_t\Big)dt + d\beta_t,$$

with β_t standard Brownian motion. In the following we shall show this relation directly using Ito's lemma.

The norm of a 2D Ornstein-Uhlenbeck process Let $V_t = \|Y_t\| = \sqrt{Y_1^2 + Y_2^2}$ be the radial part of the two-dimensional Ornstein-Uhlenbeck process (1.10.56)-(1.10.57). First, we notice that the martingale

$$M_t = \int_0^t \frac{Y_1(s)dW_1(s) + Y_2(s)dW_2(s)}{V_s}$$

has the quadratic variation $\langle M\rangle_t = \int_0^t \frac{Y_1^2+Y_2^2}{V_s^2}ds = t$. Then by Lévy's Theorem there is a Brownian motion β_t such that $M_t = \beta_t$. In differential form this becomes

$$\frac{Y_1(t)dW_1(t) + Y_2(t)dW_2(t)}{V_t} = d\beta_t.$$

Applying Ito's formula we obtain

$$\begin{aligned}
dV_t &= \frac{Y_1}{V_t}dY_1 + \frac{Y_2}{V_t}dY_2 + \frac{1}{2}\frac{Y_1^2 + Y_2^2}{V_t^3}dt \\
&= \frac{1}{V_t}(Y_1\, dY_1 + Y_2\, dY_2) + \frac{1}{2V_t}dt \\
&= \frac{1}{V_t}[Y_1(-\lambda Y_1 dt + dW_1) + Y_2(-\lambda Y_2 dt + dW_2)] + \frac{1}{2V_t}dt \\
&= -\frac{\lambda(Y_1^2 + Y_2^2)}{V_t}dt + \frac{Y_1 dW_1 + Y_2 dW_2}{V_t} + \frac{1}{2V_t}dt \\
&= -\lambda V_t\, dt + d\beta_t + \frac{1}{2V_t}dt \\
&= \Big(\frac{1}{2V_t} - \lambda V_t\Big)dt + d\beta_t.
\end{aligned}$$

Therefore, the norm $V_t = \|Y_t\|$ of a two-dimensional Ornstein-Uhlenbeck process, $Y_t = (Y_1, Y_2)$, satisfies

$$dV_t = \Big(\frac{1}{2V_t} - \lambda V_t\Big)dt + d\beta_t. \tag{1.10.60}$$

Remark 1.10.1 It is worth noting that taking $\lambda = 0$ the process Y_t becomes the two-dimensional Bessel process $R_t = \sqrt{W_1(t)^2 + W_2(t)^2}$ and (1.10.60) becomes

$$dR_t = \frac{1}{2R_t}dt + d\beta_t. \tag{1.10.61}$$

The law of V_t In the following we compute the law of the norm $V_t = \|Y_t\|$, provided the Ornstein-Uhlenbeck process starts at the origin, $V_0 = 0$. Employing polar coordinates and using Fubini's theorem we compute the cumulative distribution function of V_t as

$$\begin{aligned} F(\rho) &= P(V_t \le \rho) = \iint_{y_1^2+y_2^2\le\rho^2} p_t(0,y)\,dy_1dy_2 \\ &= 2\pi\frac{\lambda}{\pi(1-e^{-2\lambda t})}\int_0^\rho r\exp\Big(-\frac{\lambda}{(1-e^{-2\lambda t})}r^2\Big)\,dr \end{aligned}$$

and hence the density of V_t is given by

$$p_{V_t}(\rho) = F'(\rho) = \frac{2\lambda}{(1-e^{-2\lambda t})}\rho\exp\Big(-\frac{\lambda}{(1-e^{-2\lambda t})}\rho^2\Big), \quad \rho > 0. \tag{1.10.62}$$

The expectation of V_t is

$$\begin{aligned} \mathbb{E}[V_t] &= \int_0^\infty \rho p_{V_t}(\rho)\,d\rho = \sqrt{\frac{1-e^{-2\lambda t}}{\lambda}}\int_0^\infty y^{1/2}e^{-y}\,dy \\ &= \sqrt{\frac{\pi}{4\lambda}(1-e^{-2\lambda t})}, \end{aligned}$$

where we used $\int_0^\infty y^{1/2}e^{-y}\,dy = \Gamma(3/2) = \sqrt{\pi}/2$. Since $\lim_{t\to\infty}\mathbb{E}[V_t] = \frac{1}{2}\sqrt{\frac{\pi}{\lambda}}$, it follows that at any time the Ornstein-Uhlenbeck process is expected to be inside the disk of radius $r = \frac{1}{2}\sqrt{\frac{\pi}{\lambda}}$; in other words the process is expected to stay in a neighborhood of the origin. Similarly, we can compute the second moment

$$\mathbb{E}[V_t^2] = \int_0^\infty \rho^2 p_{V_t}(\rho)\,d\rho = \frac{1-e^{-2\lambda t}}{\lambda},$$

so the variance of the norm becomes

$$Var(V_t) = \frac{1-e^{-2\lambda t}}{\lambda}\Big(1-\frac{\pi}{4}\Big).$$

It is worth noting that the variance stays bounded for all $t \ge 0$.

The relation with the 2D Bessel process

We shall show that the norm process $V_t = \|Y_t\|$ given by (1.10.60) becomes a two2-dimensional Bessel process R_t with respect to a change of the probability measure. First, we consider the process $\gamma_s = \beta_s + \lambda \int_0^s R_u\, du$, $0 \leq s \leq t$, where β_s is the Brownian motion which appears in the equation $dR_s = \frac{1}{2R_s} ds + d\beta_s$.

By Girsanov's theorem, Theorem 1.1.15, γ_t becomes a Brownian motion with respect to the probability measure $Q = M_t P$, where M_t is the exponential martingale given by

$$M_t = e^{-\lambda \int_0^t R_s d\beta_s - \frac{\lambda^2}{2} \int_0^t R_s^2 ds}.$$

Since $d\gamma_t = d\beta_t + \lambda R_t\, dt$ then

$$\begin{aligned} dR_t &= \frac{1}{2R_t} dt + d\beta_t = \frac{1}{2R_t} dt + d\gamma_t - \lambda R_t\, dt \\ &= \Big(\frac{1}{2R_t} - \lambda R_t\Big) dt + d\gamma_t, \end{aligned}$$

which retrieves equation (1.10.60). Hence, the two-dimensional Bessel process R_t becomes the norm of a two-dimensional Ornstein-Uhlenbeck process under the probability measure Q. Consequently,

$$\mathbb{E}^Q[R_t] = \sqrt{\frac{\pi}{4\lambda}(1 - e^{-2\lambda t})}, \quad Var^Q(R_t) = \frac{1 - e^{-2\lambda t}}{\lambda}\Big(1 - \frac{\pi}{4}\Big).$$

The subsequent proposition furnishes an Ito integral that can be computed explicitly and will be utilized shortly.

Lemma 1.10.2 *If R_t is a two-dimensional Bessel process and β_t is its driving Brownian motion, then*

$$\int_0^t R_s d\beta_s = \frac{1}{2} R_t^2 - t. \tag{1.10.63}$$

Proof: Multiplying by R_t in

$$dR_t = \frac{1}{R_t} dt + d\beta_t, \tag{1.10.64}$$

and then integrating yields

$$\int_0^t R_s dR_s = \frac{1}{2} t + \int_0^t R_s d\beta_s. \tag{1.10.65}$$

To obtain the integral on the left side we apply Ito's formula

$$d(R_t^2) = 2R_t dR_t + dt$$

and then integrate to obtain

$$\int_0^t R_s dR_s = \frac{1}{2}R_t^2 - \frac{1}{2}t. \tag{1.10.66}$$

Substituting equation (1.10.66) into (1.10.65) and (1.10.66) and solving for $\int_0^t R_s d\beta_s$ leads to relation (1.10.63). ■

In the following proposition both expectations are taken with respect to the probability measure P.

Proposition 1.10.3 *Let f be a bounded Borel function. Then*

$$\mathbb{E}[f(R_t)e^{-\frac{\lambda^2}{2}\int_0^t R_s^2 ds}] = e^{-\lambda t}\mathbb{E}[f(V_t)e^{\frac{\lambda}{2}V_t^2}], \tag{1.10.67}$$

where R_t denotes the two-dimensional Bessel process starting at $R_0 = 0$ and V_t is the norm of a two-dimensional Ornstein-Uhlenbeck process starting at $(0, 0)$.

Proof: Changing measures from P to Q, using the definition of the martingale M_t and Lemma 1.10.2, we obtain

$$\begin{aligned}
\mathbb{E}^P[f(R_t)e^{-\frac{\lambda^2}{2}\int_0^t R_s^2 ds}] &= \mathbb{E}^Q[M_t^{-1} f(R_t)e^{-\frac{\lambda^2}{2}\int_0^t R_s^2 ds}] \\
&= \mathbb{E}^Q[e^{\lambda\int_0^t R_s d\beta_s + \frac{\lambda^2}{2}\int_0^t R_s^2 ds} f(R_t)e^{-\frac{\lambda^2}{2}\int_0^t R_s^2 ds}] \\
&= \mathbb{E}^Q[f(R_t)e^{\lambda\int_0^t R_s d\beta_s}] = e^{-\lambda t}\mathbb{E}^Q[f(R_t)e^{\frac{\lambda}{2}R_t^2}] \\
&= e^{-\lambda t}\mathbb{E}^P[f(V_t)e^{\frac{\lambda}{2}V_t^2}],
\end{aligned}$$

where in the last identity we used that R_t in the measure Q has the law of V_t with respect to measure P.

■

Since the law of V_t is known, see formula (1.10.62), then the right side of (1.10.67) can be explicitly computed for several functions f, as shall be seen next.

Cameron-Martin's formula and its generalizations

For instance, choosing $f(x)=1$ in Proposition 1.10.3, we obtain another proof for Cameron-Martin's formula

$$\begin{aligned}
\mathbb{E}[e^{-\frac{\lambda^2}{2}\int_0^t R_s^2 ds}] &= e^{-\lambda t}\mathbb{E}[e^{\frac{\lambda}{2}V_t^2}] = e^{-\lambda t}\int_0^\infty e^{\frac{\lambda}{2}\rho^2} p_{V_t}(\rho)d\rho \\
&= e^{-\lambda t}\int_0^\infty e^{\frac{\lambda}{2}\rho^2}\frac{2\lambda}{1-e^{-2\lambda t}}\rho\exp\Big(-\frac{\lambda}{1-e^{-2\lambda t}}\rho^2\Big)d\rho d\rho \\
&= \frac{\lambda e^{-\lambda t}}{1-e^{-2\lambda t}}\int_0^\infty e^{\left(\frac{\lambda}{2}-\frac{\lambda}{1-e^{-2\lambda t}}\right)v}dv \\
&= \frac{\lambda}{2\sinh(\lambda t)}\int_0^\infty e^{-\frac{\lambda}{2}\coth(\lambda t)\,v}dv \\
&= \frac{1}{\sinh(\lambda t)}\tanh(\lambda t) = \frac{1}{\cosh(\lambda t)}.
\end{aligned}$$

Using the independence of $W_1(s)$ and $W_2(s)$ we have

$$\mathbb{E}[e^{-\frac{\lambda^2}{2}\int_0^t R_s^2 ds}] = \mathbb{E}[e^{-\frac{\lambda^2}{2}\int_0^t W_1(s)^2 ds}]\mathbb{E}[e^{-\frac{\lambda^2}{2}\int_0^t W_2(s)^2 ds}] = \mathbb{E}[e^{-\frac{\lambda^2}{2}\int_0^t W_1(s)^2 ds}]^2,$$

so

$$\mathbb{E}[e^{-\frac{\lambda^2}{2}\int_0^t W_1(s)^2 ds}] = \frac{1}{\sqrt{\cosh(\lambda t)}},$$

where W_1 is a Brownian motion starting at the origin.

By performing some minor computation adjustments, it is possible to establish a more general version of Cameron-Martin's formula, which is presented below.

Proposition 1.10.4 *For $\xi,\lambda\in\mathbb{R}$ we have*

$$\mathbb{E}[e^{\xi R_t^2-\frac{\lambda^2}{2}\int_0^t R_s^2 ds}] = \frac{1}{\cosh(\lambda t)}\cdot\frac{1}{1-\dfrac{2\xi}{\lambda}\tanh(\lambda t)}, \qquad t\geq 0. \tag{1.10.68}$$

Proof: Let $f(x)=e^{\xi x^2}$ in Proposition 1.10.3. Then a similar computation provides

$$\begin{aligned}
\mathbb{E}[e^{\xi R_t^2-\frac{\lambda^2}{2}\int_0^t R_s^2 ds}] &= e^{-\lambda t}\mathbb{E}[e^{(\xi+\frac{\lambda}{2})V_t^2}] = e^{-\lambda t}\int_0^\infty e^{(\xi+\frac{\lambda}{2})\rho^2}p_{V_t}(\rho)d\rho \\
&= \frac{\lambda}{2\sinh(\lambda t)}\int_0^\infty e^{\left(\xi+\frac{\lambda}{2}-\frac{\lambda}{1-e^{-2\lambda t}}\right)v}dv \\
&= \frac{\lambda}{2\sinh(\lambda t)}\int_0^\infty e^{\left(\xi-\frac{\lambda}{2}\coth(\lambda t)\right)v}dv \\
&= \frac{\lambda}{2\sinh(\lambda t)}\cdot\frac{1}{\frac{\lambda}{2}\coth(\lambda t)-\xi} \\
&= \frac{1}{\cosh(\lambda t)}\cdot\frac{1}{1-\dfrac{2\xi}{\lambda}\tanh(\lambda t)}.
\end{aligned}$$

■

The next result can be also regarded as a generalization of the Cameron-Martin formula, which is retrieved for $k = 0$.

Proposition 1.10.5 *For $k \geq 0$ integer, we have*

$$\mathbb{E}[R_t^{2k}e^{-\frac{\lambda^2}{2}\int_0^t R_s^2 ds}] = k!\Big(\frac{2}{\lambda}\tanh(\lambda t)\Big)^k\operatorname{sech}(\lambda t), \quad t\geq 0. \tag{1.10.69}$$

Proof: Substituting $f(x) = x^{2k}$ in Proposition 1.10.3 and using the Euler integral

$$\int_0^\infty v^k e^{-av}dv = \frac{k!}{a^{k+1}}, \qquad a>0,$$

we have

$$\begin{aligned}
\mathbb{E}[R_t^{2k}e^{-\frac{\lambda^2}{2}\int_0^t R_s^2 ds}] &= e^{-\lambda t}\mathbb{E}[V_t^{2k}e^{\frac{\lambda}{2}V_t^2}] = e^{-\lambda t}\int_0^\infty \rho^{2k}e^{\frac{\lambda}{2}\rho^2}p_{V_t}(\rho)d\rho \\
&= \frac{\lambda}{2\sinh(\lambda t)}\int_0^\infty v^k e^{-\frac{\lambda}{2}\coth(\lambda t)v}dv \\
&= \frac{2^k}{\lambda^k}k!\frac{\tanh(\lambda t)^{k+1}}{\sinh(\lambda t)} = k!\Big(\frac{2}{\lambda}\tanh(\lambda t)\Big)^k\operatorname{sech}(\lambda t).
\end{aligned}$$

■

Using the same proof idea, we shall provide next an even more general statement. In the following $\mathcal{L}(f)$ will denote the Laplace transform of f.

Proposition 1.10.6 *Let f be a continuous function with a sub-exponential increase. Then*

$$\mathbb{E}[f(R_t^2)e^{-\frac{\lambda^2}{2}\int_0^t R_s^2 ds}] = \frac{\lambda}{2\sinh(\lambda t)}\mathcal{L}(f)\big(\frac{\lambda}{2}\coth(\lambda t)\big), \quad t\geq 0. \tag{1.10.70}$$

Proof: By Proposition 1.10.3 and the definition of Laplace transform we have

$$\begin{aligned}\mathbb{E}[f(R_t^2)e^{-\frac{\lambda^2}{2}\int_0^t R_s^2 ds}] &= e^{-\lambda t}\mathbb{E}[f(V_t^2)e^{\frac{\lambda}{2}V_t^2}] \\ &= \frac{\lambda}{2\sinh(\lambda t)}\int_0^\infty f(v)e^{-\frac{\lambda}{2}\coth(\lambda t)v}dv \\ &= \frac{\lambda}{2\sinh(\lambda t)}\mathcal{L}(f)\Big(\frac{\lambda}{2}\coth(\lambda t)\Big).\end{aligned}$$

■

We provide some examples where the aforementioned Laplace transform can be evaluated in closed form.

Example 1.10.1 Let $f(x) = \delta(x - c)$ be the Dirac delta function sited at c. Then

$$\mathbb{E}[\delta(R_t^2 - c)e^{-\frac{\lambda^2}{2}\int_0^t R_s^2 ds}] = \frac{\lambda}{2\sinh(\lambda t)}e^{-\frac{c\lambda}{2}\coth(\lambda t)}, \quad t \geq 0.$$

Example 1.10.2 Let $f(x) = \sinh(ax)$. Then

$$\mathbb{E}[\sinh(aR_t^2)e^{-\frac{\lambda^2}{2}\int_0^t R_s^2 ds}] = \frac{\lambda}{2\sinh(\lambda t)} \cdot \frac{a}{\frac{\lambda^2}{4}\coth^2(\lambda t) - a^2}.$$

Choosing $a = \lambda/2$, we obtain the simple expression

$$\mathbb{E}\Big[\sinh\Big(\frac{\lambda}{2}R_t^2\Big)e^{-\frac{\lambda^2}{2}\int_0^t R_s^2 ds}\Big] = \sinh(\lambda t).$$

A similar computation provides the formula

$$\mathbb{E}\Big[\cosh\Big(\frac{\lambda}{2}R_t^2\Big)e^{-\frac{\lambda^2}{2}\int_0^t R_s^2 ds}\Big] = \cosh(\lambda t).$$

The following section applies the formulas and results derived in the previous sections to prove Lévy's area formula.

1.11 Lévy's Area

Lévy's area is defined by

$$S_t = \int_0^t W_1(s)dW_2(s) - W_2(s)dW_1(s), \tag{1.11.71}$$

where $W_t = \big(W_1(t), W_2(t)\big)$ is a two-dimensional standard Brownian motion. This "stochastic area" is the area enclosed in the plane by the trajectory of

$(W_s)_{s\le t}$ and its chord. It has been introduced by Lévy [99] and then studied in the subsequent papers [100], [101], [102], [103], where he found the characteristic function of S_t and its conditional characteristic function, given the position of the Brownian motion at time t. Lévy's original proof of the conditional characteristic function of S_t is based on the series expansion of a Brownian motion as

$$W_t = \frac{t}{\sqrt{\pi}}Y_0 + \sqrt{\frac{2}{\pi}}\sum_{m=1}^{\infty}\frac{\sin(mt)}{m}Y_m, \qquad 0\le t\le \pi, \tag{1.11.72}$$

where $Y_0, Y_1, \dots,$ are independent normally distributed random variables with mean zero and variance 1, see [103].

The proof of Theorem 1.11.3 presented here is due to Marc Yor [141] and employs a result on Bessel and Orstein-Uhlenbeck processes, see section 1.10. The reader is also referred to the online notes of Fabrice Baudoin.[2]

We shall start with a few preliminary results. In the following δ will denote the Dirac delta function.

Lemma 1.11.1 *Let R_t be the two-dimensional Bessel process and $x\in\mathbb{R}^2$. Then*

$$\mathbb{E}[\delta(R_t^2 - \|x\|^2)] = \frac{1}{2t}e^{-\frac{\|x\|^2}{2t}}, \qquad t>0.$$

Proof: The relation between the distribution functions of R_t^2 and R_t is given by

$$F_{R_t^2}(\rho^2) = P(R_t^2\le\rho^2) = P(R_t\le\rho) = F_{R_t}(\rho).$$

Differentiating in ρ we obtain

$$2\rho p_{R_t^2}(\rho^2) = p_{R_t}(\rho) = \frac{1}{t}\rho e^{-\frac{\rho^2}{2t}},$$

where the last identity is the Wald's distribution of the 2-dimensional Bessel process. This implies

$$p_{R_t^2}(\rho^2) = \frac{1}{2t}e^{-\frac{\rho^2}{2t}}. \tag{1.11.73}$$

Using the definition of expectation and a property of the Delta function, we have

$$\begin{aligned}\mathbb{E}[\delta(R_t^2 - \|x\|^2)] &= \int_0^\infty p_{R_t^2}(\rho)\,\delta(\rho - \|x\|^2)d\rho\\ &= p_{R_t^2}(\|x\|^2) = \frac{1}{2t}e^{-\frac{\rho^2}{2t}},\end{aligned}$$

[2]https://fabricebaudoin.wordpress.com/2016/06/30/lecture-1-the-paul-levys-stochastic-area-formula/

where in the last identity we used (1.11.73).

■

Lemma 1.11.2 *If X and Y are two continuous random variables then*

$$\mathbb{E}[X|Y=y] = \frac{\mathbb{E}[X\,\delta(Y-y)]}{\mathbb{E}[\delta(Y-y)]}.$$

Proof: The formula can be derived using the properties of conditional expectations and delta functions, as shown below

$$\begin{aligned}\mathbb{E}[X|Y=y] &= \int x f_{X|Y}(x|y)\,dx = \int x\frac{f(x,y)}{f_Y(y)}dx\\ &= \frac{\int xf(x,y)dx}{f_Y(y)} = \frac{\int\int x\delta(u-y)f(x,u)dudx}{\int f_Y(u)\delta(u-y)du}\\ &= \frac{\mathbb{E}[X\,\delta(Y-y)]}{\mathbb{E}[\delta(Y-y)]}.\end{aligned}$$

■

The next result provides the conditional characteristic function of S_t, given the position of the Brownian motion at time t.

Theorem 1.11.3 (Lévy's area formula) *Let S_t be the Lévy area process (1.11.71) and $x \in \mathbb{R}^2$. Then for any $\lambda > 0$ we have*

$$\mathbb{E}[e^{i\lambda S_t}|W_t = x] = \frac{\lambda t}{\sinh(\lambda t)}e^{-\frac{\|x\|^2}{2t}(\lambda t\cosh(\lambda t)-1)}, \qquad t>0. \tag{1.11.74}$$

Proof: To prove the Lévy area formula, we will go through the following steps. First, we will represent the Lévy area using a Brownian motion with a stochastic clock. Then, we will use the rotational invariance of the Brownian motion to simplify the conditional expectation. Finally, we will complete the computation to obtain the desired result.
Step 1: We show that there is a Brownian motion β_t such that

$$S_t = \beta_{\int_0^t R_u^2 du}.$$

where $R_t = \|W_t\|$ is the two-dimensional Bessel process associated with the plane Brownian motion W_t.

By Lévy's theorem, the continuous martingale

$$\gamma_t = \int_0^t \frac{W_1(s)dW_2(s) - W_2(s)dW_1(s)}{\sqrt{W_1(s)^2+W_2(s)^2}}$$

has the quadratic variation $\langle \gamma \rangle_t = t$, and hence it is a Brownian motion. It follows that the Lévy area satisfies $dS_t = R_t d\gamma_t$. Since S_t is a continuous martingale and $\langle S \rangle_t = \int_0^t (dS_u)^2 = \int_0^t R_u^2 du$, then by Theorem 1.1.11 there is a DDS Brownian motion β_t such that $S_t = \beta_{\int_0^t R_u^2 du}$.

Step 2: We show that

$$\mathbb{E}[e^{i\lambda S_t} | W_t = x] = \mathbb{E}[e^{-\frac{\lambda^2}{2}\int_0^t R_u^2 du} | R_t = \|x\|].$$

Since the Brownian motion W_t and the Lévy area S_t are rotational invariant processes, we have

$$\begin{aligned}\mathbb{E}[e^{i\lambda S_t} | W_t = x] &= \mathbb{E}[e^{i\lambda S_t} | \|W_t\| = \|x\|] \\ &= \mathbb{E}[e^{i\lambda S_t} | R_t = \|x\|].\end{aligned}$$

Since the last conditional expectation is $W[0,t]$-measurable, using properties of conditional expectations, we write

$$\begin{aligned}\mathbb{E}[e^{i\lambda S_t} | R_t = \|x\|] &= \mathbb{E}[\,\mathbb{E}[e^{i\lambda S_t} | R_t = \|x\|] \,|\, W[0,t]\,] \\ &= \mathbb{E}[\,\mathbb{E}[e^{i\lambda S_t} |\, W[0,t]] \,|\, R_t = \|x\|\,] \\ &= \mathbb{E}[\,\mathbb{E}[e^{i\lambda \beta_{\int_0^t R_u^2 du}} |\, W[0,t]] \,|\, R_t = \|x\|\,] \\ &= \mathbb{E}[e^{-\frac{\lambda^2}{2}\beta_{\int_0^t R_u^2 du}} \,|\, R_t = \|x\|],\end{aligned}$$

where we used Step 1 and the fact that $\int_0^t R_u^2 du$ is determined by the history $W[0,t]$.

Step 3: We show that

$$\mathbb{E}[e^{-\frac{\lambda^2}{2}\int_0^t R_u^2 du} | R_t = \|x\|] = \frac{\lambda t}{\sinh(\lambda t)} e^{-\frac{\|x\|^2}{2t}(\lambda t \cosh(\lambda t) - 1)}.$$

To this end, we substitute $c = \|x\|^2$ in Example 1.10.1 to obtain

$$\mathbb{E}[\delta(R_t^2 - \|x\|^2) e^{-\frac{\lambda^2}{2}\int_0^t R_s^2 ds}] = \frac{\lambda}{2\sinh(\lambda t)} e^{-\frac{\|x\|^2 \lambda}{2}\coth(\lambda t)}. \tag{1.11.75}$$

Using Lemmas 1.11.1 and 1.11.2 together with relation (1.11.75) we have

$$\begin{aligned}\mathbb{E}[e^{-\frac{\lambda^2}{2}\int_0^t R_u^2 du} | R_t = \|x\|] &= \mathbb{E}[e^{-\frac{\lambda^2}{2}\int_0^t R_u^2 du} | R_t^2 = \|x\|^2] \\ &= \frac{\mathbb{E}[\delta(R_t^2 - \|x\|^2) e^{-\frac{\lambda^2}{2}\int_0^t R_s^2 ds}]}{\mathbb{E}[\delta(R_t^2 - \|x\|^2))} \\ &= \frac{\frac{\lambda}{2\sinh(\lambda t)} e^{-\frac{\|x\|^2 \lambda}{2}\coth(\lambda t)}}{\frac{1}{2t} e^{-\frac{\|x\|^2}{2t}}} \\ &= \frac{\lambda t}{\sinh(\lambda t)} e^{-\frac{\|x\|^2}{2t}(\lambda t \cosh(\lambda t) - 1)}.\end{aligned}$$

Finally, Lévy's area formula follows by combining the relations proved at steps 2 and 3. ■

For a variant of Lévy's stochastic area formula in higher dimensions we refer the reader to Helmes and Schwane [74]. As noted by Gaveau, [60] Lévy's area formula can be used to find closed form expressions for the heat kernels of generators of diffusion (W_t, S_t), which form a class of hypoelliptic operators. These types of applications will be done in subsequent chapters of this book.

Remark 1.11.4 Gaveau's generalization for the Lévy's area formula is as follows:
Let A be an $n \times n$ skew-symmetric matrix and $W_t = (W_1(t), \dots, W_n(t))$ be a Brownian motion on $\mathbb{R}^n$. If let $\langle AW_t, dW_t\rangle = \sum_{i,j=1}^n a_{ij} W_i(t) dW_j(t)$, then for any $t > 0$ and $\mathbf{x} \in \mathbb{R}^n$, we have

$$\mathbb{E}[e^{i\int_0^t \langle AW_s, dW_s\rangle\, ds} | W_t = \mathbf{x}] = \det\Big(\frac{tA}{\sin(tA)}\Big)^{1/2} \exp\Big\langle \frac{I_n - tA\cot(tA)}{2t}\mathbf{x}, \mathbf{x}\Big\rangle.$$

The classical Lévy's area formula is retrieved for the case $A = \begin{pmatrix} 0 & -1 \\ 1 & 0 \end{pmatrix}$.

1.12 Characteristic Function for the Lévi Area

We recall the notation for the Lévi area

$$S_t = \int_0^t W_1(s)dW_2(s) - \int_0^t W_2(s)dW_1(s),$$

where $W_1(t)$ and $W_2(t)$ are two independent standard Brownian motions. We shall show that the characteristic function of S_t is given by

$$\mathbb{E}[e^{i\lambda S_t}] = \frac{1}{\cosh(\lambda t)}. \tag{1.12.76}$$

The proof of this formula will be accomplished in two ways, as being a consequence of either Cameron-Martin's formula or Lévy's area formula, as follows.

1. Proof using Cameron-Martin's formula
Since S_t is a martingale with the quadratic variation

$$\langle S\rangle_t = \int_0^t (dS_s)^2 ds = \int_0^t (W_1(s)^2 + W_2(s)^2)\, du = A_t,$$

Theorem 1.1.11 implies the existence of a DDS Brownian motion B_t such that $S_t = B_{A_t}$, i.e., the Lévi area can be represented as the Brownian motion

B_t under the stochastic clock A_t. Using the tower property of conditional expectations and the Cameron-Martin's formula (1.7.39) we have

$$\begin{aligned}
\mathbb{E}[e^{i\lambda S_t}] &= \mathbb{E}[e^{i\lambda B_{A_t}}] = \mathbb{E}[\mathbb{E}[e^{i\lambda B_{A_t}} | W_1[0,t], W_2[0,t]]] \\
&= \mathbb{E}[e^{-\frac{1}{2}\lambda^2 A_t}] = \mathbb{E}[e^{-\frac{1}{2}\lambda^2 \int_0^t W_1^2 ds}]\mathbb{E}[e^{-\frac{1}{2}\lambda^2 \int_0^t W_2^2 ds}] \\
&= \Big(\frac{1}{\sqrt{\cosh(\lambda t)}}\Big)^2 = \frac{1}{\cosh(\lambda t)}.
\end{aligned}$$

We also used that $\mathbb{E}[e^{i\lambda B_u}] = e^{-\frac{1}{2}\lambda^2 u}$ and the independence of Brownian motions W_1 and W_2.

2. Proof using Lévi's area formula
Lévi's area formula provides the following conditional expectation

$$\mathbb{E}[e^{i\lambda S_t} | W_t = x] = \frac{\lambda t}{\sinh \lambda t} e^{-\frac{\|x\|^2}{2t}(\lambda t \coth(\lambda t) - 1)},$$

where $W_t = (W_t^1, W_t^2)$. Using tower property and integrating over the condition $W_t = x$, we obtain

$$\begin{aligned}
\mathbb{E}[e^{i\lambda S_t}] &= \mathbb{E}[\mathbb{E}[e^{i\lambda S_t} | W_t = x]] \\
&= \iint_{\mathbb{R}^2} \mathbb{E}[e^{i\lambda S_t} | W_t = x] \cdot \frac{1}{2\pi t} e^{-\frac{\|x\|^2}{2t}}\, dx_1 dx_2 \\
&= \frac{1}{2\pi t} \iint_{\mathbb{R}^2} \frac{\lambda t}{\sinh \lambda t} e^{-\frac{\|x\|^2}{2t}\lambda t \coth \lambda t}\, dx_1 dx_2 \\
&= \frac{\lambda t}{2\pi t} \cdot \frac{1}{\sinh \lambda t} \Big(\int_{\mathbb{R}} e^{-\frac{x_1^2}{2t}\lambda \coth \lambda t}\Big)\Big(\int_{\mathbb{R}} e^{-\frac{x_2^2}{2t}\lambda \coth \lambda t}\Big) \\
&= \frac{\lambda t}{2\pi t} \cdot \frac{1}{\sinh \lambda t} \cdot \frac{2\pi}{\lambda \coth \lambda t} = \frac{1}{\coth \lambda t},
\end{aligned}$$

where we used the Gaussian integral $\int_{\mathbb{R}} e^{-ax^2}\, dx = \sqrt{\frac{\pi}{a}}$, with $a = \frac{\lambda t \coth \lambda t}{2t}$.

An consequence of formula (1.12.76) is to obtain the density of S_t using an inverse Fourier transform

$$f_{S_t}(x) = \frac{1}{2\pi} \int_{\mathbb{R}} \frac{e^{-i\lambda x}}{\cosh(\lambda t)}\, d\lambda = \frac{1}{2t} \text{sech}\Big(\frac{\pi x}{2t}\Big), \qquad x \in \mathbb{R}, \tag{1.12.77}$$

as we shall see in Proposition 2.11.2, page 92. It is worth noting that S_t is equal in law to the product of two independent one-dimensional Brownian motions, $S_t \sim B_t W_t$, see Proposition 2.11.1. The geometrical interpretation of this result is that the 3rd component of a Heisenberg diffusion has the same law as the product of two independent Brownian motions.

1.13 Asymmetric Area Process

The calculation of the following *asymmetric area* is needed in the case of the non-symmetrical Heisenberg group or the Grushin operator with multiple missing directions:

$$\tilde{S}_t = \int_0^t W_1(s)dW_2(s), \tag{1.13.78}$$

where $W_t = (W_1(t), W_2(t))$ is a two-dimensional Brownian motion starting at the origin. The process $\tilde{S}_t$ integrates a Brownian motion with respect to another Brownian motion, which is independent of the former. The process $\tilde{S}_t$ can be regarded as the mechanical work done by a Brownian force acting along the trajectory of another independent Brownian motion. Its quadratic variation is given by

$$\langle \tilde{S} \rangle_t = \int_0^t W_1(s)^2 ds = A_t,$$

and since $\tilde{S}_t$ is a continuous martingale, by Theorem 1.1.11 there is a DDS Brownian motion, B_t, independent of $W_1(t)$ such that $\tilde{S}_t = B_{A_t}$. The use of tower property and Cameron-Martin's formula yields

$$\begin{aligned} \mathbb{E}[e^{i\lambda \tilde{S}_t}] &= \mathbb{E}[e^{i\lambda B_{A_t}}] = \mathbb{E}[\mathbb{E}[e^{i\lambda B_{A_t}} | W_1[0,t]] \\ &= \mathbb{E}[e^{-\frac{1}{2}\lambda^2 A_t}] = \frac{1}{\sqrt{\cosh \lambda t}}. \end{aligned} \tag{1.13.79}$$

Applying the inverse Fourier transform provides the density of $\tilde{S}_t$

$$f_{\tilde{S}_t}(x) = \frac{1}{2\pi} \int_{\mathbb{R}} e^{-iyx} \frac{1}{\sqrt{\cosh yt}}\, dy. \tag{1.13.80}$$

The following result presents a formula analogous to Lévy's area formula, but for the process $\tilde{S}_t$.

Proposition 1.13.1 *Let* $\tilde{S}_t = \int_0^t W_1(s)dW_2(s)$. *Then*

$$\mathbb{E}[e^{i\lambda \tilde{S}_t} | W_1(t) = w_1, W_2(t) = w_2] = \frac{\lambda t}{2\sinh(\lambda t/2)} e^{\frac{i\lambda}{2} w_1 w_2 - \frac{\|w\|^2}{2t}\left(\frac{\lambda t}{2}\coth(\frac{\lambda t}{2}) - 1\right)}. \tag{1.13.81}$$

Proof: Even if $\tilde{S}_t$ is not rotational invariant, we shall manage to reduce it to the usual Levy's area formula. We start by considering a two-dimensional Brownian motion, $(B_1(t), B_2(t))$, which is obtained from $(W_1(t), W_2(t))$ by a rotation of angle $\theta = -\pi/4$

$$B_1(t) = \frac{1}{\sqrt{2}}(W_1(t) - W_2(t)) \tag{1.13.82}$$

$$B_2(t) = \frac{1}{\sqrt{2}}(W_1(t) + W_2(t)). \tag{1.13.83}$$

Since this can be inverted as

$$W_1(t) = \frac{1}{\sqrt{2}}(B_1(t) + B_2(t))$$
$$W_2(t) = \frac{1}{\sqrt{2}}(B_2(t) - B_1(t)),$$

we shall compute $W_1(t)dW_2(t)$ as follows

$$\begin{aligned} W_1 dW_2 &= \frac{1}{2}(B_1 + B_2)(dB_2 - dB_1) \\ &= \frac{1}{2}\big(B_1 dB_2 - B_2 dB_1\big) + \frac{1}{2}\big(B_2 dB_2 - B_1 dB_1\big). \end{aligned}$$

Integrating, we obtain

$$\begin{aligned} \tilde{S}_t &= \int_0^t W_1(s)dW_2(s) \\ &= \frac{1}{2}\int_0^t \big(B_1 dB_2 - B_2 dB_1\big) + \frac{1}{2}\Big(\int_0^t B_2 dB_2 - \int_0^t B_1 dB_1\Big) \\ &= \frac{1}{2}S_t + \frac{1}{4}\Big(B_2(t)^2 - B_1(t)^2\Big) \\ &= \frac{1}{2}S_t + \frac{1}{4}\big(B_2(t) - B_1(t)\big)\big(B_2(t) + B_1(t)\big) \\ &= \frac{1}{2}S_t + \frac{1}{2}W_1(t)W_2(t), \end{aligned} \tag{1.13.84}$$

where we used the stochastic integral

$$\int_0^t B(s)dB(s) = \frac{B(t)^2}{2} - \frac{t}{2},$$

for any Brownian motion $B(t)$. Therefore, the asymmetric area $\tilde{S}_t$ can be written in terms of the symmetric area S_t as

$$\tilde{S}_t = \frac{1}{2}S_t + \frac{1}{2}W_1(t)W_2(t).$$

Let $b_1 = (w_1 - w_2)/\sqrt{2}$ and $b_2 = (w_1 + w_2)/\sqrt{2}$. Then the conditional expectation becomes

$$\begin{aligned} \mathbb{E}[e^{i\lambda \tilde{S}_t} | W_1(t) = w_1, W_2(t) = w_2] &= \mathbb{E}[e^{\frac{i\lambda}{2}S_t} e^{\frac{i\lambda}{2}W_1(t)W_2(t)} | W_1(t) = w_1, W_2(t) = w_2] \\ &= e^{\frac{i\lambda}{2}w_1 w_2}\mathbb{E}[e^{\frac{i\lambda}{2}S_t} | W_1(t) = w_1, W_2(t) = w_2] \\ &= e^{\frac{i\lambda}{2}w_1 w_2}\mathbb{E}[e^{\frac{i\lambda}{2}S_t} | B_1(t) = b_1, B_2(t) = b_2] \\ &= e^{\frac{i\lambda}{2}w_1 w_2}\frac{\lambda t}{2\sinh(\lambda t/2)}e^{-\frac{\|b\|^2}{2t}\left(\frac{\lambda t}{2}\coth(\frac{\lambda t}{2}) - 1\right)}, \end{aligned}$$

where the last identity is obtained by applying Levy's area formula (1.11.74). Since the rotational symmetry implies $\|b\| = \|w\|$, we obtain the desired formula.

■

1.14 Integrated Geometric Brownian Motion

We have seen that Cameron-Martin's formula provides a closed form expression for the moment generating function associated with the stochastic process $X_t = \int_0^t W_s^2\,ds$. We would like to obtain a similar formula in the case when the quadratic potential is replaced by an exponential one, so that the process becomes the integrated geometric Brownian motion $X_t = \int_0^t e^{2W_s}\,ds$.

We shall use a similar method based on the comparison of the expressions of two solutions of the same initial value problem and then use the solution uniqueness property.

The next result deals with a solution for the Cauchy problem for the heat problem of the operator $\frac{1}{2}\frac{d^2}{dx^2} - \frac{1}{2}\gamma^2 e^{2x}$, where $\gamma \in \mathbb{R}$ is a constant.

Lemma 1.14.1 *Let* $\phi_t(y) = \frac{1}{\sqrt{2\pi t}} e^{-\frac{y^2}{2t}}$, *for* $t > 0$ *and* $y \in \mathbb{R}$. *Then the function*

$$u(t,x) = \int_{\mathbb{R}} e^{i\gamma e^x \sinh(y)} \phi_t(y)\,dy \tag{1.14.85}$$

solves the following initial value problem

$$\frac{\partial u}{\partial t} = \frac{1}{2}\frac{\partial^2 u}{\partial x^2} - \frac{1}{2}\gamma^2 e^{2x} u \tag{1.14.86}$$

$$u(0,x) = 1. \tag{1.14.87}$$

Proof: We start by looking for a solution of the form

$$u(t,x) = \int_{\mathbb{R}} e^{i\gamma a(x)b(y)} \phi_t(y)\,dy,$$

with $a(x)$ and $b(y)$ convenient chosen. For the clarity sake we split the proof into several steps.

Step 1. Requiring $b(0) = 0$, we show that condition (1.14.87) is satisfied

$$u(0,x) = \lim_{t\to 0} \int_{\mathbb{R}} e^{i\gamma a(x)b(y)} \phi_t(y)\,dy = \int_{\mathbb{R}} e^{i\gamma a(x)b(y)} \delta(y)\,dy = e^{i\gamma a(x)b(0)} = 1.$$

Step 2. We require

$$\frac{\partial^2}{\partial y^2} e^{i\gamma a(x)b(y)} = \frac{\partial^2}{\partial x^2} e^{i\gamma a(x)b(y)} - \gamma^2 e^{2x} e^{i\gamma a(x)b(y)} \tag{1.14.88}$$

and determine $a(x)$ and $b(y)$ satisfying this condition. Substituting

$$\begin{aligned}\frac{\partial^2}{\partial x^2}e^{i\gamma a(x)b(y)} &= \big(-\gamma^2 a'(x)^2 b(y)^2 + i\gamma a''(x)b(y)\big)e^{i\gamma a(x)b(y)} \\ \frac{\partial^2}{\partial y^2}e^{i\gamma a(x)b(y)} &= \big(-\gamma^2 a(x)^2 b'(y)^2 + i\gamma a(x)b''(y)\big)e^{i\gamma a(x)b(y)}\end{aligned}$$

in (1.14.88) and identifying real and imaginary parts we obtain

$$a(x)b''(y) = a''(x)b(y) \tag{1.14.89}$$
$$a(x)^2 b'(y)^2 = a'(x)^2 b(y)^2 + e^{2x}. \tag{1.14.90}$$

Separating variables in equation (1.14.89) provides

$$a''(x) = \kappa a(x), \qquad b''(y) = \kappa b(y),\ b(0) = 0,$$

where κ is a separation constant. We note that $\kappa > 0$, because otherwise $a(x)$ would be either oscillatory or linear, fact that does not match the exponential term e^{2x} from the right side of (1.14.90). Therefore

$$\begin{aligned}a(x) &= A_1 e^{\sqrt{\kappa}x} + A_2 e^{-\sqrt{\kappa}x} \\ b(y) &= C\sinh(\sqrt{\kappa}y),\end{aligned}$$

with C, A_1, A_2 real constants. Substituting $b(0) = 0$ and $b'(0) = C\sqrt{\kappa}$ in equation (1.14.90) we get

$$(A_1^2 e^{2\sqrt{\kappa}x} + A_2 e^{-2\sqrt{\kappa}x} + 2A_1A_2)C^2\kappa = e^{2x}.$$

By identification we obtain $\kappa = 1, A_1 = \pm 1, A_2 = 0$ and $C = \pm 1$. Therefore, relation (1.14.88) holds for $a(x) = \pm e^x$ and $b(y) = \pm\sinh(y)$.

Step 3. Using that $\phi_t(y)$ satisfies the heat equation $\dfrac{\partial \phi_t(y)}{\partial t} = \dfrac{1}{2}\dfrac{\partial^2 \phi_t(y)}{\partial y^2}$ and applying condition (1.14.88) we obtain

$$\begin{aligned}\frac{\partial u}{\partial t} &= \int_{\mathbb{R}} e^{i\gamma a(x)b(y)}\frac{\partial}{\partial t}\phi_t(y)\,dy = \frac{1}{2}\int_{\mathbb{R}} e^{i\gamma a(x)b(y)}\frac{\partial^2}{\partial y^2}\phi_t(y)\,dy \\ &= \frac{1}{2}\int_{\mathbb{R}}\frac{\partial^2}{\partial y^2}e^{i\gamma a(x)b(y)}\phi_t(y)\,dy \\ &= \frac{1}{2}\int_{\mathbb{R}}\Big(\frac{\partial^2}{\partial x^2} - \gamma^2 e^{2x}\Big)e^{i\gamma a(x)b(y)}\phi_t(y)\,dy \\ &= \frac{1}{2}\Big(\frac{\partial^2}{\partial x^2} - \gamma^2 e^{2x}\Big)u,\end{aligned}$$

i.e., relation (1.14.86) is satisfied.

We have arrived at two possible solutions

$$u(t,x) = \int_{\mathbb{R}} e^{i\gamma e^x \sinh(y)} \phi_t(y)\,dy, \quad \overline{u}(t,x) = \int_{\mathbb{R}} e^{-i\gamma e^x \sinh(y)} \phi_t(y)\,dy.$$

The fact that $\phi_t(y)$ is even and $\sinh(y)$ is odd implies $u(t,x) = \overline{u}(t,x)$.

■

Proposition 1.14.2 *Let $\phi_t(y)$ be the probability density of the Brownian motion W_t. Then we have*

$$\mathbb{E}[e^{-\frac{1}{2}\gamma^2 \int_0^t e^{2W_s}\,ds}] = \int_{\mathbb{R}} e^{i\gamma \sinh(y)} \phi_t(y)\,dy, \quad \forall \gamma \in \mathbb{R}. \tag{1.14.91}$$

Proof: Since the standard Brownian motion has the infinitesimal generator $\frac{1}{2}\frac{\partial^2}{dx^2}$, the Feynman-Kac formula with potential $U(x) = \frac{1}{2}\gamma^2 e^{2x}$ provides a solution for the system (1.14.86)-(1.14.87) in the expectation form

$$u(t,x) = \mathbb{E}^x[1 \cdot e^{-\int_0^t U(W_s)ds}] = \mathbb{E}[e^{-\int_0^t U(W_s+x)\,ds}] = \mathbb{E}[e^{-\frac{1}{2}e^{2x}\gamma^2 \int_0^t e^{2W_s}}].$$

Comparing with the solution (1.14.85) provided by Lemma 1.14.1 and using the uniqueness of solution we obtain

$$\mathbb{E}[e^{-\frac{1}{2}e^{2x}\gamma^2 \int_0^t e^{2W_s}ds}] = \int_{\mathbb{R}} e^{i\gamma e^x \sinh(y)} \phi_t(y)\,dy. \tag{1.14.92}$$

Substituting $\lambda = e^x$ the relation simplifies to

$$\mathbb{E}[e^{-\frac{1}{2}\lambda^2\gamma^2 \int_0^t e^{2W_s}ds}] = \int_{\mathbb{R}} e^{i\gamma\lambda \sinh(y)} \phi_t(y)\,dy. \tag{1.14.93}$$

Taking $\lambda = 1$ we arrive to the desired solution. ■

Remark 1.14.3 Looking for a solution of the integral type (1.14.85) seems an appealing idea. In the following we shall investigate to what extent we can generalize this method for the case of an arbitrary non-negative potential function $U(x)$.

To this end, we look for a solution for the initial value problem

$$\frac{\partial u}{\partial t} = \frac{1}{2}\frac{\partial^2 u}{\partial x^2} - U(x)u \tag{1.14.94}$$

$$u(0,x) = 1. \tag{1.14.95}$$

under the form

$$u(t,x) = \int_{\mathbb{R}} e^{i\varphi(x,y)} \phi_t(y)\,dy, \tag{1.14.96}$$

where $\varphi(x, y)$ is a phase function subject to be determined. Following a similar idea of proof as in Lemma 1.14.1 we require the phase function $\varphi(x, y)$ to satisfy

$$\partial_y^2 e^{i\varphi(x,y)} = \partial_x^2 e^{i\varphi(x,y)} - 2U(x)e^{i\varphi(x,y)}.$$

Equating the real and imaginary parts in the previous relation yields the system

$$\begin{aligned} \partial_x^2\varphi - \partial_y^2\varphi &= 0 && (1.14.97)\\ (\partial_y\varphi)^2 - (\partial_x\varphi)^2 &= 2U(x). && (1.14.98) \end{aligned}$$

The first equation shows that $\phi(x, y)$ satisfies a one-dimensional wave equation. Using D'Alembert's formula we obtain

$$\varphi(x, y) = F(x - y) - G(x + y),$$

with F and G continuously twice differentiable functions. In order to have satisfied the initial condition $u(0, x) = 1$, we require $\phi(x, 0) = 0$. This implies $F = G$. Hence

$$\varphi(x, y) = F(x - y) - F(x + y).$$

Substituting in equation (1.14.98) and denoting $H = F'$ we arrive at

$$2H(x - y)H(x + y) = U(x).$$

Taking $y = 0$, we obtain $2H^2(x) = U(x)$. Substituting back in the previous equation we obtain the following functional equation for $U(x)$

$$U(x - y)U(x + y) = U(x)^2.$$

To solve this equation, we apply a logarithmic function. With notation $R(x) = \ln U(x)$, we obtain

$$\frac{R(x - y) + R(x + y)}{2} = R(x).$$

This means that the middle of any chord of the graph of $R(x)$ belongs to the graph. Therefore, the function must be linear, i.e., $R(x) = \alpha x + \beta$. This yields the potential $U(x) = e^{\alpha x + \beta}$, with $\alpha, \beta \in \mathbb{R}$.

The rest of the proof is similar with the proof of Lemma 1.14.1

$$\begin{aligned} \partial_t u &= \int_{\mathbb{R}} e^{i\varphi(x,y)} \partial_t \phi_t(y)\, dy = \frac{1}{2}\int_{\mathbb{R}} e^{i\varphi(x,y)} \partial_y^2 \phi_t(y)\, dy \\ &= \frac{1}{2}\int_{\mathbb{R}} \partial_y^2 e^{i\varphi(x,y)} \phi_t(y)\, dy = \frac{1}{2}\int_{\mathbb{R}} \big(\partial_x^2 - 2U(x)\big) e^{i\varphi(x,y)} \phi_t(y)\, dy \\ &= \frac{1}{2}u(t, x) - U(x)u(t, x) \end{aligned}$$

$$u(0, x) = \int_{\mathbb{R}} e^{i\varphi(x,y)} \delta(y)\, dy = e^{i\varphi(x,0)} = 1.$$

Denoting $a = \alpha/2$ and $e^{\beta/2} = \gamma/\sqrt{2}$, we have $H^2(x) = \frac{1}{4}\gamma^2 e^{2ax}$ and then $F(x) = \frac{1}{2a}\gamma e^{ax}$. Therefore

$$\begin{aligned}\varphi(x,y) &= \frac{1}{2a}\gamma e^{a(x-y)} - \frac{1}{2a}\gamma e^{a(x+y)} \\ &= -\frac{\gamma}{a}e^{ax}\sinh(ax).\end{aligned}$$

It follows that

$$u(t,x) = \int_{\mathbb{R}} e^{-i\frac{\gamma}{a}e^{ax}\sinh(ay)}\phi_t(y)\,dy$$

verifies the initial value problem

$$\begin{aligned}\partial_t u &= \frac{1}{2}\partial_x^2 u - \frac{1}{2}\gamma^2 e^{2ax}u \\ u(0,x) &= 1.\end{aligned}$$

The conclusion is that solutions of the integral form (1.14.96) occur just for the case of the exponential potential. An application of Feynman-Kac's formula like in Proposition 1.14.2 provides in this case

$$\mathbb{E}[e^{-\frac{1}{2}\gamma^2\int_0^t e^{2aW_s}ds}] = \int_{\mathbb{R}} e^{-i\frac{\gamma}{a}\sinh(ay)}\phi_t(y)\,dy. \tag{1.14.99}$$

which generalizes relation (1.14.91). The expression is valid for all values of a. The validity at $a = 0$ is checked by taking the limit

$$e^{-\frac{1}{2}\gamma^2 t} = \int_{\mathbb{R}} e^{-i\gamma y}\phi_t(y)\,dy,$$

which retrieves the expression of the Fourier transform of $\phi_t(y)$ with respect to y.

The next result can be found in Bougerol [18]. A slightly different variant of proof can be found in Yor [142]. Another proof variant is adopted in Gulisashvili [69], page 78.

Proposition 1.14.4 (Bougerol, 1983) *Let W_t be a standard Brownian motion. Then the stochastic process $X_t = \sinh(W_t)$ is a time-changed Brownian motion with the stochastic clock given by $A_t = \int_0^t e^{2W_s}ds$, namely there is a Brownian motion B_t independent of W_t such that*

$$\sinh(W_t) = B_{A_t} \tag{1.14.100}$$

holds in law.

Proof: It suffices to show that processes X_t and B_{A_t} have the same characteristic functions.
First we recall the characteristic function of a standard Brownian motion

$$\mathbb{E}[e^{ixB_t}] = e^{-\frac{x^2}{2}t}.$$

The characteristic function of B_{A_t} is computed using the tower property conditioning over the history $W[0,t] = \{W_s; 0 \le s \le t\}$ of the Brownian motion W_s until time t and applying Proposition 1.14.2

$$\begin{aligned}
\mathbb{E}[e^{ixB_{A_t}}] &= \mathbb{E}[\mathbb{E}[e^{ixB_{A_t}}|W[0,t]]] = \mathbb{E}[e^{-\frac{x^2}{2}A_t}] \\
&= \mathbb{E}[e^{-\frac{x^2}{2}\int_0^t e^{2W_s}ds}] = \int_{\mathbb{R}} e^{ix\sinh(y)}\phi_t(y)\,dy \\
&= \mathbb{E}[e^{ix\sinh(W_t)}].
\end{aligned}$$

Since this relation holds for any real x, then the characteristic functions are equal, fact that ends the proof. ∎

Corollary 1.14.5 *Let W_t and B_t be two independent standard Brownian motions. Then $\int_0^t e^{W_s}\,dB_s$ and $\sinh(W_t)$ have the same probability law.*

Proof: Let $X_t = \int_0^t e^{W_s}\,dB_s$. Then X_t is a continuous martingale starting at zero satisfying $dX_t = e^{W_t}dB_t$, so its quadratic variation is $\langle X\rangle_t = \int_0^t e^{2W_s}ds$. By Theorem 1.1.11, there is a DDS Brownian motion β_t, independent of W_t, such that X_t is a time-changed Brownian motion with the stochastic clock given by $A_t = \int_0^t e^{2W_s}ds$, namely

$$X_t = \beta_{A_t}.$$

By a computation similar with the one employed in Proposition 1.14.4 we obtain

$$\mathbb{E}[e^{ixX_t}] = \mathbb{E}[e^{ix\beta_{A_t}}] = \mathbb{E}[e^{ix\sinh(W_t)}], \qquad \forall x \in \mathbb{R}.$$

Since the characteristic functions are equal, then X_t and $\sinh(W_t)$ have the same law. ∎

The law of $A_t = \int_0^t e^{2W_s}\,ds$
As an application of Bougerol's identity, we shall find the law of $A_t = \int_0^t e^{2W_s}\,ds$, or more precisely, a characterization of the function $a_t(x)$ satisfying

$$P(A_t \in dx) = a_t(x)\,dx.$$

We start by applying an arbitrary bounded Borel function f on both sides of the Bougerol's identity (1.14.100) and then take the expectation to obtain

$$\mathbb{E}[f(\sinh(W_t))] = \mathbb{E}[f(B_{A_t})]. \tag{1.14.101}$$

We compute the left side expectation using the substitution $u = \sinh x$

$$\begin{aligned}\mathbb{E}[f(\sinh(W_t))] &= \int_{\mathbb{R}} f\big(\sinh(x)\big)\,\frac{1}{\sqrt{2\pi t}}e^{-\frac{x^2}{2t}}\,dx\\ &= \frac{1}{\sqrt{2\pi t}}\int_{\mathbb{R}} f(u)\,e^{-\frac{(\sinh^{-1}u)^2}{2t}}\,\frac{du}{\sqrt{1+u^2}}.\end{aligned} \tag{1.14.102}$$

The right side of (1.14.101) is computed using the tower property with a condition over the history $W[0,t]$ as follows

$$\begin{aligned}\mathbb{E}[f(B_{A_t})] &= \mathbb{E}[\mathbb{E}[f(B_{A_t})\,|\,W[0,t]]] = \mathbb{E}\Big[\int_0^\infty f(B_u)\,a_t(u)\,du\Big]\\ &= \int_0^\infty \mathbb{E}[f(B_u)]\,a_t(u)\,du = \int_0^\infty\int_{\mathbb{R}} f(x)\phi_u(x)\,dx\;a_t(u)\,du\\ &= \int_{\mathbb{R}} f(x)\Big(\int_0^\infty \phi_u(x)a_t(u)\,du\Big)\,dx = \int_{\mathbb{R}} f(x)\,T(a_t)(x)\,dx,\end{aligned} \tag{1.14.103}$$

where $\phi_u(x) = \frac{1}{\sqrt{2\pi t}}e^{-\frac{x^2}{2t}}$ is the standard normal density and T is the integral operator defined by (1.3.15).

Comparing the integral expressions (1.14.102) and (1.14.103) and using that f was an arbitrary Borel function, we obtain

$$T(a_t)(x) = \frac{1}{\sqrt{2\pi t}}\frac{1}{\sqrt{1+x^2}}e^{-\frac{(\sinh^{-1}x)^2}{2t}}.$$

The density $a_t(x)$ is obtained by inverting the operator T. Using Proposition 1.6.4 we obtain an identity which characterizes the law of A_t

$$a_t(u) = \int_0^\infty ye^{-y^2u/2}\varphi(y)\,dy,$$

where

$$\varphi(y) = \mathcal{L}_x^{-1}\Big(\frac{1}{\sqrt{1+x^2}}e^{-\frac{(\sinh^{-1}x)^2}{2t}}\Big).$$

Unfortunately, we do not have access to a closed form expression for the last Laplace inverse transform.

1.15 Lamperti-Type Properties

There are several examples of striking identities in law between a function of a Brownian motion B_t and a well-known process under a stochastic clock A_t, where A_t is a functional integral of the path $\{B_s,\, 0 \le s \le t\}$. For instance, in the case of Bougerol's identity we have $\sinh B_t = W_{A_t}$, where W_t is another standard Brownian motion and $A_t = \int_0^t e^{2B_s}\, ds$. In the case of Lamperti's identity [93] we have $e^{B_t} = R_{A_t}$, where R_t is the two-dimensional Bessel process and A_t is defined as in Bougerol's identity. This section deals with several examples of similar identities.

Let B_t be a one-dimensional Brownian motion and consider a twice continuously differentiable function φ and denote $f(x) = (\varphi'(x))^2$. Consider the integral functional

$$A_t = \int_0^t f(B_s)\, ds, \qquad t \ge 0,$$

which will play the role of stochastic clock. The process A_t is continuous and strictly increasing in t, with $A_0 = 0$ and $\lim_{t\to\infty} A_t = \infty$. Then there is an inverse process, T_u, such that

$$\int_0^{T_u} f(B_s)\, ds = u, \qquad u \ge 0,$$

or equivalently, $A_{T_u} = u$. By chain rule

$$\begin{aligned} 1 = \frac{du}{du} = \frac{dA_{T_u}}{du} = \frac{dA_{T_u}}{dT_u}\frac{dT_u}{du} = \frac{dA_s}{ds}_{\,|s=T_u}\frac{dT_u}{du} \\ = f(B_s)_{|s=T_u}\frac{dT_u}{du} = f(B_{T_u})\frac{dT_u}{du}, \end{aligned}$$

so we have

$$dT_u = \frac{1}{f(B_{T_u})}du. \tag{1.15.104}$$

We note that T_u depends on its history $\{T_s; 0 \le s \le u\}$ via formula

$$T_u = \int_0^u \frac{1}{f(B_{T_s})}\, ds,$$

which is obtained by integrating the previous equation and using $T_0 = 0$.

Let $Y_t = \varphi(B_t)$. Applying Ito's formula we have

$$dY_t = \varphi'(B_t)\, dB_t + \frac{1}{2}\varphi''(B_t)\, dt.$$

Integrating yields

$$Y_t = \varphi(0) + M_t + \frac{1}{2}\int_0^t \varphi''(B_s)\,ds, \tag{1.15.105}$$

where

$$M_t = \int_0^t \varphi'(B_u)\,dB_u$$

is a continuous martingale. Since $(dM_t)^2 = (\varphi'(B_t))^2 dt$, the quadratic variation of M_t is

$$\langle M\rangle_t = \int_0^t (\varphi'(B_u))^2 du = \int_0^t f(B_u)\,du = A_t,$$

so $\langle M\rangle_t \to \infty$ as $t \to \infty$. Applying Theorem 1.1.11, there is a DDS Brownian motion W_t, independent of B_t, such that M_t is the Brownian motion W_t under a time change with the stochastic clock A_t

$$M_t = W_{A_t}, \qquad t \geq 0.$$

Substituting back into (1.15.105) yields

$$Y_t = \varphi(0) + W_{A_t} + \frac{1}{2}\int_0^t \varphi''(B_s)\,ds.$$

Then replace t by T_u and use $A_{T_u} = u$ to get

$$Y_{T_u} = \varphi(0) + W_u + \frac{1}{2}\int_0^{T_u} \varphi''(B_s)\,ds. \tag{1.15.106}$$

Taking the differential in (1.15.106) and using (1.15.104) implies

$$\begin{aligned} dY_{T_u} &= dW_u + \frac{1}{2}\varphi''(B_{T_u})\,dT_u \\ &= dW_u + \frac{1}{2}\varphi''(B_{T_u})\,\frac{1}{f(B_{T_u})}\,du \\ &= dW_u + \frac{1}{2}\varphi''(B_{T_u})\,\frac{1}{(\phi'(B_{T_u}))^2}\,du. \end{aligned}$$

We shall assume there is a function F such that

$$\frac{\varphi''(x)}{(\varphi'(x))^2} = F\big(\varphi(x)\big). \tag{1.15.107}$$

Under this assumption the previous stochastic equation becomes

$$dY_{T_u} = dW_u + \frac{1}{2}F(\varphi(B_{T_u}))\,du.$$

Denote $X_u = Y_{T_u} = \varphi(B_{T_u})$. Then the process X_u satisfies

$$dX_u = dW_u + \frac{1}{2}F(X_u)\,du, \qquad (1.15.108)$$

with the initial condition $X_0 = \varphi(0)$. The relation $\varphi(B_{T_u}) = X_u$ can be written equivalently as $\varphi(B_t) = X_{A_t}$. The equation (1.15.108) has a strong solution provided the function F satisfies the Lipschitz condition and does not increase faster than a linear function. We conclude the previous computation with the following generic result:

Proposition 1.15.1 *If B_t is a Brownian motion, then there is a stochastic process X_t, which is given by (1.15.108), such that $\varphi(B_t) = X_{A_t}$, namely $\varphi(B_t)$ is the stochastic process X_t under a time change with stochastic clock $A_t = \int_0^t (\varphi'(B_s))^2\,ds$ and F given by (1.15.107).*

In the following we shall consider a few distinguished particular cases of Proposition 1.15.1.

1. Let $\varphi(x) = e^x$. Since $\dfrac{\varphi''(x)}{\varphi'(x)^2} = \dfrac{1}{\varphi(x)}$, then $F(x) = \dfrac{1}{x}$, and (1.15.108) becomes

$$dX_u = dW_u + \frac{1}{2X_u}\,du, \qquad X_0 = 1.$$

Its solution is a two-dimensional Bessel process, $X_t = R_t$ starting from 1. We obtained the following result:

Proposition 1.15.2 (Lamperti) *The geometric Brownian motion e^{B_t} is a time-transformed Bessel process in the plane with stochastic clock $A_t = \int_0^t e^{2B_s}\,ds$, namely*

$$e^{B_t} = R_{A_t}.$$

2. Let $\varphi(x) = x^2/2$. In this case $\dfrac{\varphi''(x)}{\varphi'(x)^2} = \dfrac{1}{x^2} = \dfrac{1}{2\varphi(x)}$, so $F(x) = \dfrac{1}{2x}$. Then (1.15.108) becomes

$$dX_u = dW_u + \frac{1/2}{2X_u}\,du, \qquad X_0 = 0.$$

Comparing with the equation satisfied by the Bessel process of index ν

$$dR_u = dW_u + \frac{2\nu + 1}{2R_u}\,du, \qquad (1.15.109)$$

we obtain that X_u is the Bessel process of index $\nu = -1/4$, i.e., $X_u = R_u^{(-1/4)}$. We obtained the following result:

Proposition 1.15.3 *Half the square of a Brownian motion,* $\frac{1}{2}B_t^2$, *is a time-transformed Bessel process of index* $\nu = -1/4$ *with stochastic clock* $A_t = \int_0^t B_s^2\, ds$, *namely*

$$\frac{1}{2}B_t^2 = R_{A_t}^{(-1/4)}.$$

3. Consider $\varphi(x) = x^k/k$, with $k = 2p$. We have $\dfrac{\varphi''(x)}{\varphi'(x)^2} = \dfrac{k-1}{k\varphi(x)}$, so $F(x) = \dfrac{k-1}{kx}$ and X_u satisfies

$$dX_u = dW_u + \frac{1}{2}\frac{1-\frac{1}{k}}{X_u}\, du.$$

Comparing with (1.15.109), we obtain the solution $X_u = R_t^{(-\frac{1}{2k})}$. Using Proposition 1.15.1 we obtain:

Proposition 1.15.4 *If* B_t *is a Brownian motion, and* k *an even integer, then*

$$\frac{1}{k}B_t^k = R_{A_t}^{(-\frac{1}{2k})},$$

where $A_t = \int_0^t B_s^{2(k-1)}\, ds$ *and* $R_t^{(-\frac{1}{2k})}$ *is a Bessel process of index* $\nu = -\frac{1}{2k}$.

4. Let $\varphi(x) = \sinh x$. In this case $\dfrac{\varphi''(x)}{\varphi'(x)^2} = \dfrac{\sinh x}{1+\sinh^2 x}$, which implies $F(x) = \dfrac{x}{1+x^2}$. Then the process X_u satisfies the equation

$$dX_u = dW_u + \frac{X_u}{2(1+X_u^2)}\, du, \quad X_0 = 0. \tag{1.15.110}$$

We have:

Proposition 1.15.5 *If* B_t *is a Brownian motion and denote* $A_t = \int_0^t \cosh^2 B_s\, ds$ *then*

$$\sinh B_t = X_{A_t}, \quad t \geq 0,$$

where X_u *is the solution of (1.15.110).*

It is worth noting the similarity with Bougerol's identity.

5. Consider the **erf** function

$$\epsilon(x) = \frac{2}{\sqrt{\pi}}\int_0^x e^{-s^2}\, ds, \qquad x \geq 0,$$

which is continuous and increasing, with $\epsilon(0) = 0$ and $\lim_{x\to\infty} \epsilon(x) = \infty$. Let $\eta(u) = \epsilon^{-1}(u)$ be its inverse, namely a function with $\eta(0) = 0$ and satisfying

$$\frac{2}{\sqrt{\pi}} \int_0^{\eta(u)} e^{-s^2}\, ds = u, \qquad u \geq 0.$$

We recall that Langevin's equation

$$dX_t = X_t dt + dW_t, \qquad X_0 = 0 \tag{1.15.111}$$

has the strong solution given by the Ornstein-Uhlenbeck process

$$X_t = e^t \int_0^t e^{-s}\, dW_s. \tag{1.15.112}$$

Now, we look for a function $\varphi(x)$ such that the equation (1.15.108) becomes the Langevin equation (1.15.111), while X_t becomes the process (1.15.112). Asking $F(x) = 2x$, then $\varphi(x)$ satisfies the ordinary differential equation

$$\frac{\varphi''(x)}{(\varphi'(x))^2} = 2\varphi(x), \quad x \geq 0.$$

Writing the equation as $\dfrac{\varphi''(x)}{\varphi'(x)} = 2\varphi(x)\varphi'(x)$ we integrate to get

$$\ln|\varphi'(x)| = c + \varphi(x)^2.$$

Taking an exponential yields $\varphi'(x) = Ke^{\varphi(x)^2}$, with K constant. Integrating again we have

$$\int_0^x e^{-\varphi(u)^2} \varphi'(u)\, du = Kx.$$

We look for a particular solution with $\varphi(0) = 0$ that corresponds to $K = \sqrt{\pi}/2$. This satisfies

$$\frac{2}{\sqrt{\pi}} \int_0^{\varphi(x)} e^{-s^2}\, ds = x,$$

which implies that $\varphi(x) = \eta(x)$, the inverse of the **erf** function. Then

$$A_t = \int_0^t \varphi'(B_s)^2\, ds = \frac{\pi}{4} \int_0^t e^{2\varphi(B_s)^2}\, ds = \frac{\pi}{4} \int_0^t e^{2\eta(B_s)^2}\, ds.$$

The relation $\varphi(B_t) = X_{A_t}$ becomes $\eta(B_t) = X_{A_t}$, where X_t is the process (1.15.112). We conclude with the following result:

Proposition 1.15.6 *Let B_t be a Brownian motion and $\eta = \epsilon^{-1}$. Then $\eta(B_t)$ is a time-changed Ornstein-Uhlenbeck process (1.15.112)*

$$\eta(B_t) = X_{A_t}, \quad t \geq 0,$$

where $A_t = \frac{\pi}{4}\int_0^t e^{2\eta(B_s)^2}\,ds$.

Corollary 1.15.7 *Let B_t be a Brownian motion and $\eta = \epsilon^{-1}$. Then there is a Brownian motion γ_t, such that*

$$\eta(B_t)e^{-A_t} = \gamma_{_{(1-e^{-2A_t})/2}}, \quad t \geq 0,$$

where $A_t = \frac{\pi}{4}\int_0^t e^{2\eta(B_s)^2}\,ds$.

Proof: Let $Y_t = \int_0^t e^{-s}\,dW_s$. Since Y_t is a continuous martingale with the quadratic variation $\langle Y\rangle_t = \int_0^t e^{-2s}\,ds = \frac{1}{2}(1-e^{-2t})$, there is a Brownian motion γ_t such that $Y_t = \gamma_{_{\langle Y\rangle_t}}$. Therefore, the Ornstein-Uhlenbeck process (1.15.112) can be written in terms of the Brownian motion γ_t as

$$X_t = e^t Y_t = e^t \gamma_{_{(1-e^{-2t})/2}}.$$

The result follows now by applying Proposition 1.15.6. ■

For a similar approach for the case of a Brownian motion with drift the reader can consult Salminen and Yor [123]. For the case of general diffusions we refer the reader to Khoshnevishan et al. [88].

1.16 Reflected Brownian Motion as Bessel Process

Let B_t be a one-dimensional Brownian motion and denote by R_t a two-dimensional Bessel process. The laws of these processes are given by

$$P(|B_t| \in dx) = \sqrt{\frac{2}{\pi t}}e^{-\frac{x^2}{2t}}\,dx$$
$$P(R_t \in d\rho) = \frac{1}{t}\rho e^{-\frac{\rho^2}{2t}}\,d\rho, \qquad t > 0.$$

We assume there is a process A_t such that the reflected Brownian motion is obtained by changing the clock of the Bessel process as

$$|B_t| = R_{A_t}.$$

In this section we shall find the law of A_t. To this end, we equate the expectations for any Borel function $f : [0, +\infty) \to \mathbb{R}$ as

$$\mathbb{E}[f(|B_t|)] = \mathbb{E}[f(R_{A_t})]. \tag{1.16.113}$$

Let $P(A_t \in dx) = a_t(x)dx$. Evaluating each expectation as

$$\mathbb{E}[f(|B_t|)] = \int_0^\infty f(x)\sqrt{\frac{2}{\pi t}}e^{-\frac{x^2}{2t}}\,dx$$
$$\mathbb{E}[f(R_{A_t})] = \int_0^\infty f(x)\int_0^\infty \frac{1}{u}xe^{-\frac{x^2}{2u}}a_t(u)\,du\,dx,$$

and then equating yields

$$\sqrt{\frac{2}{\pi t}}e^{-\frac{x^2}{2t}} = \int_0^\infty \frac{1}{u}xe^{-\frac{x^2}{2u}}a_t(u)\,du.$$

We shall solve this integral equation for $a_t(u)$. To accomplish this, we first substitute $v = 1/u$ in the right side integral to obtain

$$\sqrt{\frac{2}{\pi t}}\frac{1}{x}e^{-\frac{x^2}{2t}} = \int_0^\infty e^{-\frac{x^2}{2}v}b_t(v)\,dv,$$

where $b_t(v) = \dfrac{1}{v}a_t(1/v)$. If let $s = x^2/2$ we obtain

$$\sqrt{\frac{2}{\pi t}}\frac{1}{\sqrt{2s}}e^{-\frac{s}{t}} = \int_0^\infty e^{-sv}b_t(v)\,dv.$$

Applying an inverse Laplace transform, this equation can be solved exactly as

$$b_t(v) = \mathcal{L}^{-1}\left(\frac{1}{\sqrt{\pi t s}}e^{-\frac{s}{t}}\right)(v) = \frac{1}{\pi}\frac{H(vt-1)}{\sqrt{vt-1}},$$

where $H(\cdot)$ stands for the Heaviside function. Substituting $x = 1/v$ we solve for $a_t(x)$ and obtain

$$a_t(x) = \frac{1}{\pi}\frac{H(t-x)}{\sqrt{x(t-x)}} = \begin{cases} \dfrac{1}{\pi}\dfrac{1}{\sqrt{x(t-x)}}, & \text{if } 0 < x < t \\ 0, & \text{otherwise.} \end{cases}$$

This shows that the law of A_t is a beta distribution [3] with $\alpha = \beta = 1/2$.

[3]This is given in general by $f(x) = \frac{1}{B(\alpha,\beta)}\frac{x^{\alpha-1}(t-x)^{\beta-1}}{t^{\alpha+\beta-1}}$, $0 < x < t$, $\alpha, \beta > 0$.

1.17 The Heat Semigroup

The idea of the next few sections is to exploit the relationship between the evolution along a vector field and the associated heat semigroup.

Consider the one-dimensional heat equation on $\mathbb{R}$ given by

$$\begin{aligned} \partial_t u(t,x) &= \frac{1}{2}\partial_x^2 u(t,x), \quad t>0 \\ u(0,x) &= f(x), \end{aligned}$$

with $f \in C_0^\infty(\mathbb{R})$. The classical solution to this equation is given by $u(t,x) = P_t f(x)$, where $P_t = e^{\frac{1}{2}t\partial_x^2}$ is the *heat semigroup* on $\mathbb{R}$. The operator P_t is defined by

$$P_t f(x) = \int_{\mathbb{R}} \phi_t(x-y) f(y)\,dy,$$

where $\phi_t(x)$ is the Gaussian kernel $\phi_t(x) = \frac{1}{\sqrt{2\pi t}} e^{-\frac{x^2}{2t}}$. The heat semigroup $(P_t)_{t\ge 0}$ satisfies the semigroup composition property $P_s \circ P_t = P_{s+t}$ for all $s, t \ge 0$, which also justifies its name. We shall present next a stochastic interpretation of P_t.

Let W_t be a one-dimensional Brownian motion on $\mathbb{R}$ starting at $x_0 = 0$, and define the operator

$$e^{W_t\partial_x} = \sum_{n\ge 0} \frac{1}{n!} W_t^n \partial_x^n$$

acting on infinite differentiable functions with compact support. Taking the expectation and using Exercise 1.21.9 (a) we have

$$\begin{aligned} \mathbb{E}[e^{W_t\partial_x} f(x)] &= \sum_{n\ge 0} \frac{1}{n!} \mathbb{E}[W_t^n] \partial_x^n f \\ &= \sum_{k\ge 0} \frac{1}{k!} \Big(\frac{t}{2}\Big)^k (\partial_x^2)^k f(x) = e^{\frac{1}{2}t\partial_x^2} f(x) = P_t f(x). \end{aligned} \tag{1.17.114}$$

This provides the interpretation of the heat semigroup as an expectation

$$P_t f(x) = \mathbb{E}[e^{W_t\partial_x} f(x)] \approx \frac{1}{N} \sum_{j=1}^{N} e^{W_t(\omega_j)\partial_x} f(x), \quad N \sim \infty. \tag{1.17.115}$$

We can extend this procedure to the heat equation in several dimensions. Consider the heat equation on $\mathbb{R}^n$ given by

$$\begin{aligned} \partial_t u(t,x) &= \Delta_n u(t,x), \quad t>0 \\ u(0,x) &= f(x). \end{aligned}$$

Here, Δ_n is the Euclidean Laplacian operator defined as $\Delta_n = \frac{1}{2}\sum_{j=1}^{n} \partial_{x_j}^2$. The solution to this equation is given by $u(t,x) = e^{t\Delta_n} f(x)$.

We consider an n-dimensional Brownian motion $W_t = (W_t^1, \cdots, W_t^n)$ starting at the origin. Using the independence of the components W_t^j and the commutation of the vector fields ∂_{x_i}, we can reduce the problem to a one-dimensional case. Thus we obtain

$$\mathbb{E}[e^{\sum_{j=1}^{n} W_t^j \partial_{x_j}} f(x)] = \prod_{j=1}^{n} \mathbb{E}[e^{W_t^j \partial_{x_j}} f(x)] \quad = \prod_{j=1}^{n} e^{\frac{1}{2} t \partial_{x_j}^2} f(x) = e^{t\Delta_n} f(x),$$

where $\mathbb{E}$ denotes the expected value. In the next section we will investigate the case of general vector fields that satisfy this type of relation.

1.18 The Linear Flow

Let $X = \sum_{j=1}^{n} X_j \partial_{x_j}$ be a smooth vector field on $\mathbb{R}^n$ and consider the initial value problem

$$\partial_t u(t,x) = -Xu(t,x) \tag{1.18.116}$$
$$u(0,x) = f(x). \tag{1.18.117}$$

The solution $u(t,x) = e^{-tX} f(x)$ represents the propagation of the initial value function f along the integral lines of the vector fields X. To show this, we consider $x(t) = \big(x_1(t), \cdots, x_n(t)\big)$ satisfying the system

$$x_j'(t) = X_j(x(t)), \quad x_j(0) = x_j^0,$$

where $x_j'(t) = dx_j(t)/dt$. The trajectory $x(t)$ is a curve in $\mathbb{R}^n$ called the integral line of the vector field X starting at x_0. By using the chain rule, we can rewrite the derivative of $u(t,x(t))$ as

$$\begin{aligned}\frac{d}{dt} u\big(t, x(t)\big) &= \partial_t u + \sum_{j=1}^{n} x_j'(t) \partial_{x_j} u = \partial_t u + \sum_{j=1}^{n} X_j(x(t)) \partial_{x_j} u \\ &= \partial_t u(t,x) + Xu(t,x) = 0.\end{aligned}$$

As a result, the solution of the system (1.18.116)-(1.18.117) remains constant along the integral lines of the vector field X. Therefore,

$$u(t, x(t)) = u(0, x(0)) = f(x_0). \tag{1.18.118}$$

This explains why the solution $u(t,x)$ represents the propagation of the value $f(x_0)$ along the integral line that connects x_0 and x within time t.

Although there is no closed form formula known for e^{-tX} for a general given vector field X, there are some cases where it is possible to explicitly compute this quantity. Formulas will be presented just for the cases where the vector field X has constant coefficients and where X has linear coefficients.

Example 1.18.1 Let $X = -\sum_{j=1}^{n} c_j \partial_{x_j}$, where $c_j \in \mathbb{R}$. The system of integral lines can be described as $x_j'(t) = -c_j$ with initial conditions $x_j(0) = x_j^0$. Thus, we can derive $x_j(t) = x_j^0 - c_j t$. Consequently, the condition $u(t, x(t)) = f(x_0)$ is satisfied when

$$u(t, x) = f(x_1 + c_1 t, \dots, x_n + c_n t).$$

This further implies that

$$e^{t\sum_{j=1}^{n} c_j \partial_{x_j}} f(x) = f(x_1 + c_1 t, \dots, x_n + c_n t).$$

This means the operator $e^{t\sum_{j=1}^{n} c_j \partial_{x_j}}$ can be represented as a shift operator by the vector $t(c_1, \dots, c_n)$.

Example 1.18.2 Consider $X = \sum_{j=1}^{n} a_j(x)\partial_{x_j}$, where the coefficients are given by the linear functions $a_j(x) = \sum_{k=1}^{n} a_{jk} x_k$ and $a_{jk} \in \mathbb{R}$. The integral lines of X satisfy $x'(t) = Ax(t)$, where $A = (a_{jk})$. The solution is $x(t) = e^{At} x_0$. If we impose the condition $u(t, x(t)) = f(x_0)$, then we have $u(t, x) = f(e^{-At} x)$. Therefore, we can conclude that

$$e^{-t\sum_{j=1}^{n} a_j(x)\partial_{x_j}} f(x) = f(e^{-At} x).$$

(i) For instance, if $X = x_1 \partial_{x_2} - x_2 \partial_{x_1}$ is the angular momentum vector field, then using that $e^{-At} = \begin{pmatrix} \cos t & \sin t \\ -\sin t & \cos t \end{pmatrix}$, it follows that e^{-tX} represents the composition of f by a rotation of angle t.

(ii) If $X = x_1 \partial_{x_1} + x_2 \partial_{x_2}$ is the radial vector field, using $e^{-tI_2} = e^{-t} I_2$, then it can be shown that

$$e^{-t(x_1\partial_{x_1} + x_2\partial_{x_2})} f(x) = f(e^{-t} x),$$

which is an exponential dilation of the argument.

We make the remark that the solution $u(t, x) = e^{-tX} f(x)$ of the initial value problem (1.18.116)-(1.18.117) is unique. To this end, we consider two solutions $u_1(t, x)$ and $u_2(t, x)$ and notice that their difference $u(t, x) = u_1(t, x) - u_2(t, x)$ satisfies the homogeneous initial value problem

$$\partial_t u(t, x) = -Xu(t, x) \tag{1.18.119}$$

$$u(0, x) = 0. \tag{1.18.120}$$

Applying equation (1.18.118) we obtain $u(t, x(t)) = 0$. Therefore, the solution vanishes along all integral curves of the vector X, and hence, vanishes everywhere. Therefore, $u_1 = u_2$, which proves the uniqueness of solution.

1.19 The Stochastic Flow

Let us begin by using linear flow concepts to derive formula (1.17.115). Firstly, we observe that the quantity $e^{\tau}\partial_x f(x)$ can be obtained from Example 1.18.1 by setting $n=1$ and $t=-\tau$, which yields $e^{\tau}\partial_x f(x) = f(x-\tau)$. Next, using the definition of expectation, we can express the expected value of $e^{W_t\partial_x} f(x)$ as

$$\mathbb{E}[e^{W_t\partial_x} f(x)] = \int_{\mathbb{R}} e^{\tau}\partial_x f(x)\frac{1}{\sqrt{2\pi t}}e^{-\frac{\tau^2}{2t}}, d\tau \quad = \int_{\mathbb{R}} f(x-\tau)\frac{1}{\sqrt{2\pi t}}e^{-\frac{\tau^2}{2t}}, d\tau.$$

By changing the variable of integration to $y = x - \tau$, we obtain

$$\mathbb{E}[e^{W_t\partial_x} f(x)] = \int_{\mathbb{R}} f(y)\frac{1}{\sqrt{2\pi t}}e^{-\frac{(x-y)^2}{2t}}\, d\tau = P_t f(x),$$

which retrieves $u(t,x) = P_t f(x)$, i.e., the temperature at time t on the real line $\mathbb{R}$, given the initial temperature $f(x)$.

We will generalize this relationship to accommodate a smooth vector field X on $\mathbb{R}^n$, replacing the one-dimensional vector field ∂_x.

To this end, we consider the evolution equation

$$\partial_t u(t,x) = Xu(t,x), \qquad u(0,x) = f(x) \tag{1.19.121}$$

with the solution given by $u(t,x) = e^{tX}f(x)$. Then we replace t by a one-dimensional Brownian motion, W_t, and compute the expectation by a procedure similar with the one used in formula (1.17.114) as follows

$$\begin{aligned}\mathbb{E}[e^{W_tX} f(x)] &= \sum_{n\geq 0}\frac{1}{n!}\mathbb{E}[W_t^n]X^n f \\ &= \sum_{k\geq 0}\frac{1}{k!}\Big(\frac{t}{2}\Big)^k (X^2)^k f(x) = e^{\frac{1}{2}tX^2} f(x).\end{aligned} \tag{1.19.122}$$

We note that $v(t,x) = e^{\frac{1}{2}tX^2}f(x)$ verifies the parabolic equation

$$\partial_t v(t,x) = \frac{1}{2}X^2 v(t,x) \tag{1.19.123}$$

$$v(0,x) = f(x). \tag{1.19.124}$$

The process $e^{W_tX}f(x)$ is called the *stochastic flow* along the vector field X. Relation (1.19.122) shows how the expectation of the stochastic flow is related to the solution $v(t,x)$ of the associated heat equation.

Building upon the previous calculation, we can expand the problem to incorporate n commuting vector fields and offer an interpretation of temperature on an n-dimensional manifold as an average value obtained by moving along each integral line by an independent Brownian motion.

Let $X_1, \cdots, X_n$ be n independent vector fields on $\mathbb{R}^n$, such that $[X_i, X_j] = 0$. We define the function $G : \mathbb{R}^n \times \mathbb{R}^n \to \mathbb{R}^n$ by

$$G(t_1, \ldots, t_n, x) = e^{t_1 X_1} \circ e^{t_2 X_2} \circ \cdots \circ e^{t_n X_n} x.$$

Then the point $y = G(t_1, \ldots, t_n, x)$ is obtained by concatenating integral lines along the vector fields X_j, of parameter t_j, starting from the initial point $x \in \mathbb{R}^n$. Since the vector fields commute, any order permutation will not affect the output y.

We also consider $\Delta_X = \frac{1}{2}\sum_{j=1}^n X_j^2$ be the sum of squares operator defined by the previous vector fields.

Proposition 1.19.1 *Let $W_t = (W_t^1, \ldots, W_t^n)$ be an n-dimensional Brownian motion on $\mathbb{R}^n$. Then*

$$v(t, x) = \mathbb{E}[G(W_t^1, \ldots, W_t^n, f(x)]$$

satisfies the heat equation

$$\begin{aligned} \partial_t u(t, x) &= \Delta_X u(t, x), \qquad t > 0 \\ u(0, x) &= f(x). \end{aligned}$$

Proof: Using equation (1.19.122) and the properties of the expectation with respect to independence, we have

$$\begin{aligned} e^{\frac{1}{2}t\Delta_X} &= e^{\frac{1}{2}t\sum_{j=1}^n X_j^2} = e^{\frac{1}{2}tX_1^2} \cdots e^{\frac{1}{2}tX_n^2} \\ &= \mathbb{E}[e^{W_t^1 X_1}] \cdots \mathbb{E}[e^{W_t^n X_n}] = \mathbb{E}[e^{W_t^1 X_1} \cdots e^{W_t^n X_n}]. \end{aligned}$$

Applying to $f(x)$ yields

$$e^{\frac{1}{2}t\Delta_X} f(x) = \mathbb{E}[e^{W_t^1 X_1} \cdots e^{W_t^n X_n} f(x)] = \mathbb{E}[G(W_t^1, \ldots, W_t^n, f(x)].$$

■

The previous result provides an explanation of how the temperature changes over time, starting from the initial distribution $f(x)$ on a Riemannian manifold $(\mathbb{R}^n, g)$ with metric g defined by the commuting vector fields X_j such that $g(X_i, X_j) = \delta_{ij}$.

The process $Z_t(x) = G(W_t^1, \ldots, W_t^n, x)$ is the stochastic flow starting at x with respect to the system of commuting vector fields $\{X_1, \ldots, X_n\}$. According to this interpretation, the temperature $u(t, x)$ at time $t > 0$ and point $x \in M$ is the average of values obtained by propagating the initial temperature $f(x)$ along the aforementioned stochastic flow. The problem of stochastic flows associated with commuting vector fields has been studied for the first time by Doss [51] and Sussmann [129] in late 1970s.

Remark 1.19.2 We note that the equation

$$e^{\frac{1}{2}t\Delta_X} = \mathbb{E}[e^{\sum_{j=1}^n W_t^j X_j}]$$

is valid only for vector fields that commute. If the vector fields do not commute, and are defined on a Lie group, the reader is referred to Baudoin [9] for further information.

The previous computation also leads to the following outcome:

Proposition 1.19.3 *Let $u(t,x)$ be the solution of the initial value problem (1.19.121). Then $v(t,x) = \mathbb{E}[u(W_t,x)]$ is a solution of the parabolic equation (1.19.123) with initial condition (1.19.124).*

Proof: Replacing t by W_t in the expression of the solution of problem (1.19.121) yields $u(W_t,x) = e^{W_t X} f(x)$. Then substituting $v(t,x) = e^{\frac{1}{2}tX^2} f(x)$ into the formula

$$\mathbb{E}[e^{W_t X} f(x)] = e^{\frac{1}{2}tX^2} f(x)$$

yields the desired result. ■

Next, we will provide several examples of how the previous result can be applied in various contexts.

Example 1.19.1 Consider the vector field $X = -\sum_{j=1}^n c_j \partial_{x_j}$, $c_i \in \mathbb{R}$. Using Example 1.18.1, the solution of

$$\partial_t u = \sum_{j=1}^n c_j \partial_{x_j} u$$
$$u(0,x) = f(x)$$

is given by $u(t,x) = e^{-tX} f(x) = f(x_1 + c_1 t, \ldots, x_n + c_n t)$. Applying Proposition 1.19.3 we obtain that a solution of

$$\partial_t u = \frac{1}{2} \sum_{i,j=1}^n c_i c_j \partial_{x_i} \partial_{x_j} u$$
$$u(0,x) = f(x)$$

can be expressed using expectation as

$$\begin{aligned} v(t,x) &= \mathbb{E}[f(x_1 + c_1 W_t, \ldots, x_n + c_n W_t)] \\ &= \frac{1}{\sqrt{2\pi t}} \int_{\mathbb{R}} f(x_1 + c_1\theta, \ldots, x_n + c_n\theta) e^{-\frac{\theta^2}{2t}}\, d\theta. \end{aligned}$$

We can see that the differential operator $\sum_{i,j=1}^n c_i c_j \partial_{x_i} \partial_{x_j} u = \langle cc^T \partial_x, \partial_x \rangle$ is degenerate. This is because the matrix cc^T has a rank that is less than or equal to the rank of the vector c, which is 1. As a result, the determinant of cc^T is zero, indicating that the operator is degenerate.

Example 1.19.2 Let $X = x_1\partial_{x_1} + x_2\partial_{x_2}$. Using Example 1.18.2, we can solve the partial differential equation

$$\partial_t u = x_1\partial_{x_1}u + x_2\partial_{x_2}u$$
$$u(0,x) = f(x)$$

and obtain the solution $u(t,x) = f(e^{-t}x)$. By applying Proposition 1.19.3, we can use expectation to express a solution of the partial differential equation

$$\partial_t u = \frac{1}{2}(x_1^2\partial_{x_1}^2 + x_2^2\partial_{x_2}^2 + 2x_1x_2\partial_{x_1}\partial_{x_2} + x_1\partial_{x_1} + x_2\partial_{x_2})u$$
$$u(0,x) = f(x)$$

as

$$v(t,x) = \mathbb{E}[f(e^{-W_t}x)] = \frac{1}{\sqrt{2\pi t}}\int_{\mathbb{R}} f(e^{-\theta}x)e^{-\frac{\theta^2}{2t}}\,d\theta.$$

1.20 Summary

This chapter covers various advanced concepts of Stochastic Calculus, including Ito's formula, Dynkin's formula, Kolmogorov's backward equation, different variants of Feynman Kac's formula, Lévi's characterization theorem of a Brownian motion, Dambis, Dubins, and Schwartz characterization of a martingale as a time-changed Brownian motion, Knight's theorem, and Girsanov's theorem.

It then focuses on stopped Brownian motions at exponentially, uniformly, and Erlang distributed stochastic times. The common relationship between these processes are studied and encapsulated in the properties of an integral operator T. In certain cases, the operator mentioned earlier can be inverted, providing a solution for determining the density of certain stochastic times.

Cameron-Martin's formula is then derived using Feynman Kac's formula, and a variational interpretation of it is provided. A study of the norm of a 2-dimensional Ornstein-Uhlenbeck process leads to a new proof of Cameron-Martin's formula as well as a proof of Lévy's area, which is important in the study of the Heisenberg diffusions. The characteristic function of Lévy's area is computed and inverted to obtain its probability law, which is the same as the law of a product of two independent standard Brownian options.

The chapter also includes a study of asymmetric area variants, the integral of an exponential Brownian motion, and Bougerol's identity in law. Finally, the section concludes with other striking identities in law, such as the Lamperti property.

The last section presents the heat semigroup and its connections to linear and stochastic flow.

1.21 Exercises

Exercise 1.21.1 Let X be a real random variable. Show that X is symmetric if and only if its characteristic function $\phi_X(t)$ is real.

Exercise 1.21.2 Let X and Y be two real-valued random variables. Show that if

$$\mathbb{E}[\cos(\lambda X - \xi)] = \mathbb{E}[\cos(\lambda Y - \xi)], \qquad \forall \lambda, \xi \in \mathbb{R}$$

then X and Y have the same distribution law.

Exercise 1.21.3 (a) Let X be a random variable that is uniformly distributed on $\big(-\frac{\pi}{2}, \frac{\pi}{2}\big)$. Show that $Y = \tan X$ is Cauchy distributed, i.e.

$$Y \sim \frac{1}{\pi(1+x^2)}, \qquad x \in \mathbb{R}.$$

(b) Let Y be a Cauchy distributed random variable. Show that $X = \tan^{-1}(Y)$ is uniformly distributed on $\big(-\frac{\pi}{2}, \frac{\pi}{2}\big)$.

Exercise 1.21.4 (a) Let X and Y be two independent standard normally distributed random variables. Denote $Z = X^2+Y^2$. Show that Z is exponentially distributed.
(b) Let $(W_1(t), W_2(t))$ be a two-dimensional Brownian motion and consider $X_t = W_1(t)/\sqrt{t}$ and $Y_t = W_2(t)/\sqrt{t}$. Apply part (a) to find the probability density of $Z_t = W_1(t)^2 + W_2(t)^2$ and $R_t = Z_t^{1/2}$.

Exercise 1.21.5 Let X and Y be two independent standard normally distributed random variables and consider $R = \sqrt{X^2+Y^2}$ and $W = \tan^{-1}(Y/X)$.
(a) Are the processes R and W independent?
(b) What are the distributions of R and W?

Exercise 1.21.6 Let S_t denote the Lévy area. Show that

$$\mathbb{E}[e^{i\lambda S_t} | W_t = (\sqrt{t}, \sqrt{t})] = \frac{\lambda t}{\sinh(\lambda t)} e^{1-\lambda t \coth(\lambda t)}.$$

Exercise 1.21.7 Let $W_1(t)$ and $W_2(t)$ be two independent Brownian motions and define

$$S_t^{\alpha,\beta} = \alpha^2 \int_0^t W_1(s)dW_2(s) - \beta^2 \int_0^t W_2(s)dW_1(s)$$

for $\alpha, \beta \in \mathbb{R}$. Show that

$$\mathbb{E}[e^{i\lambda S_t^{\alpha,\beta}}] = \sqrt{\mathrm{sech}(\lambda\alpha t)\mathrm{sech}(\lambda\beta t)}.$$

Exercise 1.21.8 Let R_t denote the two-dimensional Bessel process starting at the origin. Compute the expectation $\mathbb{E}[e^{i\xi-\frac{\lambda^2}{2}\int_0^t R_s^2 ds}]$ for $\xi, \lambda \in \mathbb{R}$.

Exercise 1.21.9 Let W_t be a standard Brownian motion.
(a) Using Ito's lemma show that

$$\mathbb{E}[W_t^{2n+1}] = 0, \qquad \mathbb{E}[W_t^{2n}] = \frac{(2n)!}{2^n n!}t^n, \quad n \geq 0.$$

(b) We consider the real-analytic function $\varphi(u) = \sum_{n\geq 0} \frac{1}{n!(2n)!}u^n$. Show that

$$\mathbb{E}[\varphi(W_t^2)] = I_0(t),$$

where I_0 denotes the modified Bessel function of order zero.

Exercise 1.21.10 (a) Use the Feynman-Kac's formula informally to find a solution for the Cauchy's problem

$$\frac{\partial u}{\partial t} = \frac{1}{2}\frac{\partial^2 u}{\partial x^2} - \alpha x u$$
$$u(0,x) = 1,$$

where $\alpha \in \mathbb{R}$.
(b) Check by a direct computation that the solution obtained in part (a) satisfies the aforementioned Cauchy problem.

Exercise 1.21.11 Let W_t be a standard Brownian motion and $\mathcal{F}_t$ be the sigma algebra generated by $\{W_s; 0 \leq s \leq t\}$.
(a) Starting from the $\mathcal{F}_t$-continuous martingale $X_t = W_t^2 - t$ and using Theorem 1.1.11, show that there is a Brownian motion β_t and a stochastic time α_t such that $W_t^2 = t + \beta_{\alpha_t}$.
(b) Show that $\alpha_t = \frac{1}{4}A_t$, and use formula (1.9.46) to find the characteristic function $\phi_{W_t^2}(x) = \mathbb{E}[e^{ixW_t^2}]$.

Exercise 1.21.12 Prove Proposition 1.1.10.

Exercise 1.21.13 Let $Y_t = W_t$ be a standard Brownian motion. Use Cameron-Martin's formula to show Novikov's condition (1.1.6) for $0 < T < \pi/2$.

Exercise 1.21.14 Use Girsanov's theorem to prove the drift reduction formulas provided by Proposition 1.1.19.

Exercise 1.21.15 Find the law of a Brownian motion stopped at the third arrival time of a Poisson process with parameter θ.

Exercise 1.21.16 (*a*) Let $K(x, y; t) = \frac{1}{\sqrt{2\pi t}} e^{-\frac{(x-y)^2}{2t}} e^{-\frac{1}{2}\gamma(x+y)t + \frac{1}{24}\gamma^2 t^3}$, $t > 0$ and $x, y \in \mathbb{R}$. Verify that $K(x, y; t)$ satisfies

$$\begin{aligned} \frac{\partial}{\partial t} K(x, y; t) &= \Big(\frac{1}{2}\frac{\partial^2}{\partial x^2} - \gamma x\Big) K(x, y; t) \\ \lim_{t \to 0} K(x, y, t) &= \delta(x - y). \end{aligned}$$

(*b*) Using the second proof variant of Cameron-Martin's formula presented at page 27, find a similar formula for the expectation $\mathbb{E}[e^{-\gamma \int_0^t (W_s + x)\, ds}]$.
(*c*) Redo for this case the computations suggested at page 30 to find the corresponding classical action $S_{cl}(x, y; t)$ by integrating over the solutions of the Euler-Lagrange equations of a Lagrangian with a linear potential.

Exercise 1.21.17 Consider the process W_t defined by the series expression (1.11.72).
(*a*) Find the characteristic function $\mathbb{E}[e^{i\lambda W_t}]$, $\lambda \in \mathbb{R}$;
(*b*) Prove that W_t is a Brownian motion.
(*c*) Let $Z_t = \int_0^t W_s\, ds$. Show that Z_t is normally distributed with mean zero and variance $t^3/3$.

Exercise 1.21.18 Let W_t be a one-dimensional Brownian motion and consider the process $Z_t(x) = e^{W_t \partial_x} f(x)$, where $f(x)$ is a smooth bounded function on $\mathbb{R}$.
(*a*) Using Ito's lemma find $dZ_t(x)$;
(*b*) Show that

$$Z_t(x) = f(x) + \int_0^t \partial_x Z_t(x)\, dW_s + \frac{1}{2}\int_0^t \partial_x^2 Z_s(x)\, ds.$$

(*c*) Take the expectation in part (*b*) and verify that $\mathbb{E}[Z_t(x)] = v(t, x)$, where $v(t, x)$ satisfies the heat equation

$$\partial_t v = \frac{1}{2}\partial_x^2 v, \qquad v(0, x) = f(x).$$

Exercise 1.21.19 Formulate and solve a similar problem with Exercise 1.21.18 in the case of the process $Z_t(x) = e^{W_t X} f(x)$, where X is a smooth vector field on $\mathbb{R}^n$.

Chapter 2

Stochastic Geometry in Euclidean Space

This chapter focuses on the geometric properties of Brownian motion, including its behavior on various geometric structures like lines, circles, curves and planes, as well as transformations like rotations, reflections, projections, and inner products. Specifically, the area of Brownian triangles is examined, and the chapter explores how the behavior of a Brownian motion can be expressed in terms of curvature when it moves around a plane curve.

2.1 Brownian Motion on a Line

Consider a line in the plane given by the vector equation $r(t) = r_0 + vt$, where $v \in \mathbb{R}^2$ denotes the direction vector and t is the parameter on the line. If the parameter t is replaced by a one-dimensional Brownian motion, β_t, then we obtain a *Brownian motion on the line*. This is given by the two-dimensional process

$$r_t = r_0 + v\beta_t, \qquad t \geq 0.$$

If $r_t = (x_t, y_t)$ and $v = (v_1, v_2)$, then a Brownian motion on the line with direction v that passes through the point $r_0 = (x_0, y_0)$ has the parametric equations

$$\begin{aligned} x_t &= x_0 + v_1\beta_t \\ y_t &= y_0 + v_2\beta_t, \qquad t \geq 0. \end{aligned}$$

An equivalent way of describing a Brownian motion on a line is to think of a point M_t on the line, whose coordinate on the line is stochastic and given by the Brownian motion β_t.

2.2 Directional Brownian Motion

Let $W_t = \big(W_1(t), W_2(t)\big)$ be a two-dimensional Brownian motion and consider the unit vector $\eta \in \mathbb{R}^2$, $\eta = (\cos\theta, \sin\theta)$, with $\theta \in [0, 2\pi)$. The process

$$X_t = \langle W_t, \eta\rangle = W_1(t)\cos\theta + W_2(t)\sin\theta, \qquad t \geq 0$$

is called the *directional Brownian motion* in direction η. The fact that the process X_t is a Brownian motion follows from Lévy's theorem, Theorem 1.1.6. Its hypotheses are verified as follows. The process X_t is a continuous martingale, which is square integrable since $\mathbb{E}[X_t^2] = t$. On the other side, the quadratic variation of X_t is given by $\langle X\rangle_t = \int_0^t (\cos^2\theta + \sin^2\theta)\, ds = t$.

The next result shows the usefulness of the directional Brownian motion. It states that the only way a plane Brownian motion can be applied on a one-dimensional Brownian motion is in a directional manner.

Proposition 2.2.1 *Let $W_t = \big(W_1(t), W_2(t)\big)$ be a two-dimensional Brownian motion and $f : \mathbb{R}^2 \to \mathbb{R}$ a continuous, twice differentiable function, with $f(0,0) = 0$. If the process $X_t = f(W_t)$ is a Brownian motion, then X_t is a directional Brownian motion.*

Proof: We need to show that there is an angle θ such that $X_t = W_1(t)\cos\theta + W_2(t)\sin\theta$. Equivalently stated, this means that f is a linear function, with $f(x_1, x_2) = c_1x_1 + c_2x_2$, where $c_1 = \cos\theta$ and $c_2 = \sin\theta$.

To show this we start by applying Ito's formula

$$dX_t = \frac{\partial f}{\partial x_1} dW_1(t) + \frac{\partial f}{\partial x_2} dW_2(t) + \frac{1}{2}\Big(\frac{\partial^2 f}{\partial x_1^2} + \frac{\partial^2 f}{\partial x_2^2}\Big)\, dt. \tag{2.2.1}$$

This implies

$$X_t = X_0 + \int_0^t \frac{\partial f}{\partial x_1} dW_1(s) + \int_0^t \frac{\partial f}{\partial x_2} dW_2(s) + \frac{1}{2}\int_0^t \Big(\frac{\partial^2 f}{\partial x_1^2} + \frac{\partial^2 f}{\partial x_2^2}\Big)\, ds.$$

Because X_t is a Brownian motion, then it is a martingale, and since the first two integrals on the right side are also martingales, it follows that the drift term must vanish for any $t \geq 0$. This implies

$$\frac{\partial^2 f}{\partial x_1^2} + \frac{\partial^2 f}{\partial x_2^2} = 0, \tag{2.2.2}$$

namely f is a harmonic function. A computation of the quadratic variation of X_t using relation (2.2.1) yields

$$\langle X\rangle_t = \int_0^t (dX_s)^2 = \int_0^t |\nabla f|^2 ds,$$

where ∇f stands for the gradient of f. On the other side, since X_t is a Brownian motion, we have $\langle X \rangle_t = t$, for all $t \geq 0$. Therefore, $|\nabla f|^2 = 1$, or

$$\Big(\frac{\partial f}{\partial x_1}\Big)^2 + \Big(\frac{\partial f}{\partial x_2}\Big)^2 = 1, \tag{2.2.3}$$

which shows that f satisfies an eiconal equation. It follows that there is a continuous function $\theta = \theta(x_1, x_2)$ such that

$$\frac{\partial f}{\partial x_1} = \cos\theta, \qquad \frac{\partial f}{\partial x_2} = \sin\theta. \tag{2.2.4}$$

Differentiating in the same variable yields

$$\frac{\partial^2 f}{\partial x_1^2} = -\sin\theta\frac{\partial\theta}{\partial x_1}, \quad \frac{\partial^2 f}{\partial x_2^2} = \cos\theta\frac{\partial\theta}{\partial x_2}$$

and then substituting into (2.2.3) we obtain

$$\cos\theta\frac{\partial\theta}{\partial x_2} - \sin\theta\frac{\partial\theta}{\partial x_1} = 0. \tag{2.2.5}$$

On the other side, cross-differentiating in (2.2.4) provides

$$\frac{\partial(\cos\theta)}{\partial x_2} = \frac{\partial(\sin\theta)}{\partial x_1},$$

which is

$$\cos\theta\frac{\partial\theta}{\partial x_1} + \sin\theta\frac{\partial\theta}{\partial x_2} = 0. \tag{2.2.6}$$

Relations (2.2.5) and (2.2.6) can be written in the vector form $\mathcal{R}_\theta\nabla\theta = 0$, where $\nabla\theta = (\frac{\partial\theta}{\partial x_1}, \frac{\partial\theta}{\partial x_2})^T$ and $\mathcal{R}_\theta$ denotes the matrix of a rotation of angle θ. It follows that $\nabla\theta = 0$, and hence θ is a constant function. Integrating in (2.2.4) and using that $f(0,0) = 0$ implies that $f(x_1, x_2) = x_1\cos\theta + x_2\sin\theta$, which leads to the desired conclusion. ■

2.3 Rotations

A *rotation* of angle θ is an application $\mathcal{R}_\theta : \mathbb{R}^2 \to \mathbb{R}^2$ given by

$$\mathcal{R}_\theta(\mathbf{x}) = \begin{pmatrix} \cos\theta & \sin\theta \\ -\sin\theta & \cos\theta \end{pmatrix}\begin{pmatrix} x_1 \\ x_2 \end{pmatrix},$$

where $\mathbf{x} = (x_1, x_2)^T$. The next result states that two-dimensional Brownian motions are rotational invariant.

Proposition 2.3.1 *Let $W_t = \big(W_1(t), W_2(t)\big)$ be a two-dimensional Brownian motion and $\mathcal{R}_\theta$ a rotation. Then $B_t = \mathcal{R}_\theta(W_t)$ is also a two-dimensional Brownian motion.*

Proof: If we denote $B_t = \big(B_1(t), B_2(t)\big)$, then

$$\begin{aligned}B_1(t) &= \cos\theta\, W_1(t) + \sin\theta\, W_2(t)\\ B_2(t) &= -\sin\theta\, W_1(t) + \cos\theta\, W_2(t).\end{aligned}$$

We shall perform the proof following the following three steps.

Step 1. We show first that $B_1(t)$ and $B_2(t)$ are Brownian motions.
Since the Brownian motions $W_1(t)$ and $W_2(t)$ are continuous martingales, then $B_1(t)$ and $B_2(t)$ are also continuous martingales, as their linear combinations. Since

$$\begin{aligned}\mathbb{E}[B_i(t)^2] &= (\cos\theta)^2\mathbb{E}[W_1(t)] + (\sin\theta)^2\mathbb{E}[W_2(t)] + \sin 2\theta\, \mathbb{E}[W_1(t)]\mathbb{E}[W_2(t)]\\ &= t < \infty,\end{aligned}$$

then $B_i(t)$ is a square integrable martingale. Since $dB_1(t)^2 = dB_2(t)^2 = dt$, then both $B_1(t)$ and $B_2(t)$ have the quadratic variation equal to t. By Lévy's theorem, Theorem 1.1.6, it follows that $B_1(t)$ and $B_2(t)$ are Brownian motions.

Step 2. We show that $B_1(t)$ and $B_2(t)$ are uncorrelated. To this end it suffices to prove that the covariance of $B_1(t)$ and $B_2(t)$ vanishes. By a direct computation we have

$$\begin{aligned}Cov\big(B_1(t), B_2(t)\big) &= \mathbb{E}[B_1(t)B_2(t)] - \mathbb{E}[B_1(t)]\mathbb{E}[B_2(t)] = \mathbb{E}[B_1(t)B_2(t)]\\ &= \sin\theta\cos\theta\mathbb{E}[W_2(t)^2 - W_1(t)^2]\\ &\quad + (\cos\theta^2 - \sin\theta^2)\mathbb{E}[W_1(t)W_2(t)]\\ &= 0,\end{aligned}$$

where we used $\mathbb{E}[W_i(t)] = 0$ and $\mathbb{E}[W_i(t)^2] = t$.

Step 3. We show that $B_1(t)$ and $B_2(t)$ are independent.

We note that the process $\big(B_1(t), B_2(t)\big)$ has a bivariate normal distribution. This follows from the fact that for any $a, b \in \mathbb{R}$ there are $c_1, c_2 \in \mathbb{R}$ such that the linear combination $aB_1(t) + bB_2(t)$ can be written as $c_1W_1(t) + c_2W_2(t)$, which is normally distributed. Since $\big(B_1(t), B_2(t)\big)$ has a bivariate normal distribution and $B_1(t)$, $B_2(t)$ are uncorrelated (by the step 2), it follows that $B_1(t)$, $B_2(t)$ are independent.

It follows that $B_t = \big(B_1(t), B_2(t)\big)$ is a two-dimensional Brownian motion.

■

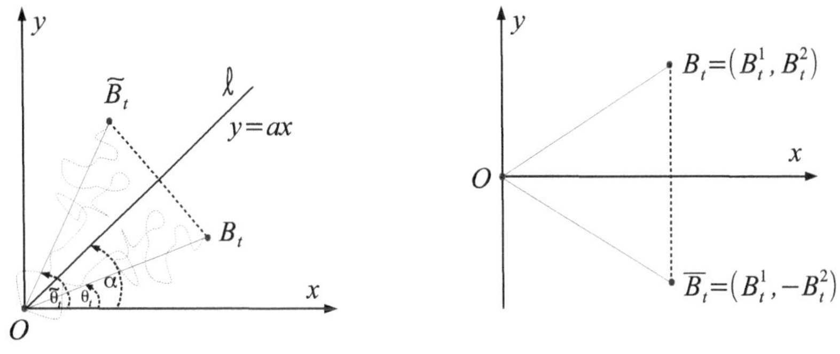

a b

Figure 2.1: (a) *Reflection of a Brownian motion in the line $y = ax$.* (b) *Reflection of a Brownian motion in the x-axis.*

2.4 Reflections in a Line

A *reflection* in a line ℓ is a transformation of the plane in which each point of the original process has an image that is situated at the same Euclidean distance from the line as the original point and belongs to the opposite side of the line. Some stochasticity attributes are expected not to change under a reflection, as the next result shows.

Proposition 2.4.1 *The reflection of a two-dimensional Brownian motion $B_t = (B_1^1, B_t^2)$ in the line $\{y = ax\}$ is also a two-dimensional Brownian motion.*

Proof: Let $\widetilde{B}_t$ be the reflected Brownian motion B_t in the line $\{y = ax\}$, see Fig. 2.1 a. Employing polar coordinates we shall write $B_t = |B_t|e^{i\theta_t}$ and $\widetilde{B}_t = |\widetilde{B}_t|e^{i\widetilde{\theta}_t}$. Reflection properties imply $|B_t| = |\widetilde{B}_t|$ and $\widetilde{\theta}_t = 2\alpha - \theta_t$, where $\alpha = \tan^{-1} a$. Therefore

$$\begin{aligned}\widetilde{B}_t = |\widetilde{B}_t|e^{i\widetilde{\theta}_t} &= e^{2\alpha i}|B_t|e^{-i\theta_t} \\ &= e^{2\alpha i}\bar{B}_t,\end{aligned}$$

where $\bar{B}_t = (B_t^1, -B_t^2)$ is the reflected Brownian motion B_t in the x-axis, see Fig. 2.1 b. Since $-B_t^2$ is also a Brownian motion, and $e^{2\alpha i}$ represents a rotation of angle 2α, applying Proposition 2.3.1 it follows that $\widetilde{B}_t$ is also a Brownian motion in the plane. ■

Remark 2.4.2 If the reflection line does not pass through the origin, then the reflected Brownian motion is a translated Brownian motion, namely it does not start at the origin.

2.5 Projections

A Brownian motion is preserved by projections in certain cases. We shall investigate a few of these in the following.

Proposition 2.5.1 *Let $W_t = \big(W_1(t), W_2(t)\big)$ be a two-dimensional Brownian motion and ℓ a line in the same plane passing through the origin. Then the orthogonal projection of W_t onto ℓ is a Brownian motion on ℓ.*

Proof: Let θ be the angle made by the line ℓ and the x-axis. Consider the rotation $\mathcal{R}_{-\theta}$, which maps the line ℓ into the x-axis and the Brownian motion W_t into $B_t = \mathcal{R}_{-\theta}(W_t)$, which is also a Brownian motion by Proposition 2.3.1. The projection of B_t onto the x-axis is $B_1(t)$, which is a Brownian motion. The orthogonal projection of W_t onto ℓ is obtained by rotating the x-axis onto ℓ by $\mathcal{R}_\theta$, obtaining $\mathcal{R}_\theta(B_1(t))$, which is still a Brownian motion on ℓ. ■

Remark 2.5.2 In fact the orthogonal projection of $W_t = \big(W_1(t), W_2(t)\big)$ onto ℓ is a directional Brownian motion. If $\eta = (\cos\theta, \sin\theta)$ denotes the direction vector of the line ℓ, then the orthogonal projection is given by

$$X_t = \langle W_t, \eta\rangle = W_1(t)\cos\theta + W_2(t)\sin\theta,$$

which shows that X_t is a directional Brownian motion in direction η.

Remark 2.5.3 If in Proposition 2.5.1 the line ℓ does not pass through the origin, then let P be the orthogonal projection of the origin onto ℓ. Under this weaker hypothesis the orthogonal projection of W_t onto ℓ is a Brownian motion on ℓ starting at the point P.

The following result describes the geometric representation of the correlation between two Brownian motions as the cosine of an angle.

Proposition 2.5.4 *Let $W_t = \big(W_1(t), W_2(t)\big)$ be a two-dimensional Brownian motion in the plane and ℓ_1 and ℓ_2 be two lines in the same plane passing through the origin making an acute angle θ. Then the orthogonal projections of W_t on ℓ_1 and ℓ_2 are two Brownian motions with correlation coefficient* $\cos\theta$.

Proof: The fact that the orthogonal projections of W_t on ℓ_1 and ℓ_2 are Brownian motions follows from Proposition 2.5.1. Since a Brownian motion is invariant by rotations, see Proposition 2.3.1, we may assume without restricting generality that the line ℓ_1 is the x-axis. Then the orthogonal projection of $W_t = \big(W_1(t), W_2(t)\big)$ on ℓ_1 is $W_1(t)$, while the orthogonal projection on ℓ_2 is a point with coordinate measured along ℓ_2 given by

$$\beta_t = W_1(t)\cos\theta + W_2(t)\sin\theta,$$

which is a directional Brownian motion. The covariance of the two projections is given by

$$Cov(W_1(t), \beta_t) = t\cos\theta + \sin\theta\, Cov(W_1(t), W_2(t)) = t\cos\theta,$$

while the correlation is

$$Corr(W_1(t), \beta_t) = \frac{Cov(W_1(t), \beta_t)}{\sqrt{Var(W_1(t))Var(\beta_t)}} = \cos\theta.$$

■

Corollary 2.5.5 *Under the hypotheses of Proposition 2.5.4, we denote the orthogonal projections of W_t on ℓ_1 and ℓ_2 by M_t and P_t, respectively, and let $u_t = |M_tP_t|^2$. Then*

$$u_t = \alpha_t^2 + \beta_t^2 - 2\alpha_t\beta_t Corr(\alpha_t, \beta_t),$$

where α_t and β_t are Brownian motions describing the coordinates of M_t and P_t along the two lines.

Proof: It is a consequence of the Cosine Theorem and Proposition 2.5.4. ■

Remark 2.5.6 If in Proposition 2.5.4 the lines ℓ_1 and ℓ_2 do not pass through the origin, then the orthogonal projections of W_t onto these lines are translated Brownian motions whose correlation coefficient is still $\cos\theta$. In particular, if the projections are independent processes, then ℓ_1 and ℓ_2 are perpendicular lines.

2.6 Distance from a Brownian Motion to a Line

In the following we consider a fixed line in the Euclidean plane and a Brownian motion in the same plane.

Proposition 2.6.1 *The distance from a two-dimensional Brownian motion to a line, which passes though the origin, has the law of a reflected Brownian motion.*

Proof: Let $W_t = \big(W_1(t), W_2(t)\big)$ be a two-dimensional Brownian motion and ℓ a line with the equation $Ax + By = 0$. Denote by M_t the point in the plane having coordinates $(W_1(t), W_2(t))$. Then the distance from M_t to ℓ is given by

$$\begin{aligned} d_t = dist(M_t, \ell) &= \frac{1}{\sqrt{A^2 + B^2}} |AW_1(t) + BW_2(t)| \\ &= |\cos\theta W_1(t) + \sin\theta W_2(t)|. \end{aligned}$$

Since the process $\beta_t = \cos\theta W_1(t) + \sin\theta W_2(t)$ is a directional Brownian motion, then $d_t = |\beta_t|$ is a reflected Brownian motion. ∎

Remark 2.6.2 A variant of proof involves Proposition 2.3.1 as follows. Since the line ℓ passes through the origin, there is a plane rotation, which is centered at the origin, that will map the line ℓ into the x-axis. Under this rotation the two-dimensional Brownian motion W_t becomes the two-dimensional Brownian motion $\widehat{W}_t = (\widehat{W}_t^1, \widehat{W}_t^2)$. Since rotations preserve distances, we have $\text{dist}_{\text{Eu}}(W_t, \ell) = \text{dist}_{\text{Eu}}(\widehat{W}_t, 0x) = |\widehat{W}_t^1|$, which is a reflected one-dimensional Brownian motion.

Remark 2.6.3 Since d_t is a reflected Brownian motion, its probability density is given by

$$p_{d_t}(x) = \frac{2}{\sqrt{2\pi t}} e^{-\frac{x^2}{2t}}, \qquad x \geq 0. \tag{2.6.7}$$

We also note that d_t can be seen as a Bessel process under a stochastic clock, see Section 1.16.

Application 2.6.4 We shall provide now an application to finding the area of a triangle with one stochastic vertex and two fixed vertices. Consider a triangle ABC, with vertices B and C fixed in the Euclidean plane and the vertex A following a two-dimensional Brownian motion $W_t = (W_t^1, W_t^2)$. Let $a = \text{dist}_{\text{Eu}}(B, C)$ denote the Euclidean distance between points B and C. Assume the line $\ell = BC$ does pass through the origin and consider the distance $d_t = \text{dist}_{\text{Eu}}(W_t, \ell)$. Then the area of the triangle is given by $\mathcal{A}_t = \frac{1}{2} a d_t$, with d_t distributed as in (2.6.7). It follows that the probability density of the area $\mathcal{A}_t$ is

$$p_{\mathcal{A}_t}(x) = \frac{4}{a\sqrt{2\pi t}} e^{-\frac{2}{t}(\frac{x}{a})^2}, \qquad x \geq 0.$$

The more challenging issue of calculating the area of a triangle formed by three Brownian motions will be postponned until Section 2.17.1.

2.7 Central Projection

Let ℓ be a line of direction $v = (v_1, v_2)$, with $|v| = 1$, which does not pass through the origin O, and $B_t = (B_t^1, B_t^2)$ be a Brownian motion in the plane.

The *central projection* of B_t onto the line ℓ is the intersection between the lines OB_t and ℓ. This intersection is a point M_t, see Fig. 2.2. We shall find the law of the process τ_t that describes the coordinate of M_t along the line ℓ.

We start by writing the vector equation of the line ℓ as $r = r_0 + tv$, where $t \in \mathbb{R}$ and r_0 was chosen such that it is perpendicular on v, i.e., $\langle r_0, v\rangle = 0$. Then the intersection point M_t has the coordinate vector $r_0 + v\tau_t$.
Consider the following two processes

$$\alpha_t = v_1 B_t^1 + v_2 B_t^2$$
$$\beta_t = -v_2 B_t^1 + v_1 B_t^2.$$

If $(v_1, v_2) = (\cos\theta, \sin\theta)$, then (α_t, β_t) is the rotation of (B_t^1, B_t^2) by an angle of $-\theta$. By Proposition 2.3.1, the process (α_t, β_t) is a plane Brownian motion, which will be useful shortly.
Since M_t is collinear with O and B_t, there is a one-dimensional process λ_t such that scaling B_t by this stochastic factor we obtain the point M_t, i.e.,

$$r_0 + v\tau_t = \lambda_t B_t, \qquad t \geq 0.$$

On components, this becomes

$$r_0^1 + v_1\tau_t = \lambda_t B_t^1$$
$$r_0^2 + v_2\tau_t = \lambda_t B_t^2.$$

Multiplying the first equation by v_1 and the second by v_2 and then adding yields $\tau_t = \lambda_t\alpha_t$. On the other side, multiplying the first equation by v_2 and the second by v_1 and then subtracting provides $|r_0| = \lambda_t\beta_t$, where we used that

$$-r_0^1 v_2 + r_0^2 v_1 = \langle r_0, v^\perp\rangle = |r_0|\,|v^\perp|\cos 0 = |r_0|,$$

where $v^\perp = (-v_2, v_1)$ is a unit vector perpendicular on v. Eliminating λ_t from the relations

$$\tau_t = \lambda_t\alpha_t, \quad |r_0| = \lambda_t\beta_t$$

yields

$$\tau_t = |r_0|\frac{\alpha_t}{\beta_t}, \tag{2.7.8}$$

where $|r_0| = dist(O, \ell)$. Therefore, the coordinate τ_t is proportional to the quotient of two independent Brownian motions.

Proposition 2.7.1 *The probability density of the random variable τ_t is given by the Cauchy distribution*

$$f_{\tau_t}(u) = \frac{|r_0|}{\pi}\frac{1}{|r_0|^2 + u^2}, \qquad u \in \mathbb{R}. \tag{2.7.9}$$

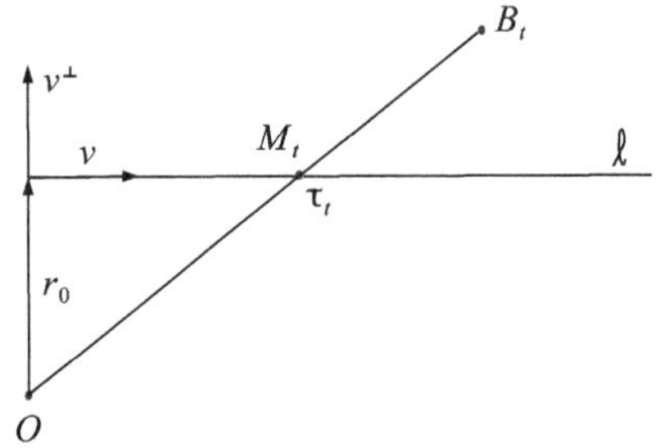

Figure 2.2: *The central projection of the Brownian motion B_t from the origin onto the line ℓ.*

Proof: Since α_t and β_t are independent Brownian motions, there are two independent standard normal distributed random variables, $X, Y \sim N(0,1)$, such that in law we have $\alpha_t = \sqrt{t}X$ and $\beta_t = \sqrt{t}Y$. Therefore, the quotient $\dfrac{\alpha_t}{\beta_t}$ has the same law as $\dfrac{X}{Y}$. By Exercise 2.22.1 the quotient of two independent standard normal variables is the standard Cauchy distribution

$$p_{X/Y}(u) = \frac{1}{\pi(1+u^2)}, \qquad u \in \mathbb{R}.$$

Then using relation (2.7.8) a straightforward computation leads to formula (2.7.9). ■

2.8 Distance Between Brownian Motions in $\mathbb{R}^n$

1. Let W_t and B_t be two independent one-dimensional Brownian motions and consider the process $X_t = \frac{1}{\sqrt{2}}(B_t - W_t)$. Since X_t is a continuous square integrable continuous martingale and $\langle X, X\rangle_t = \frac{1}{2}\int_0^t [(dB_s)^2 + (dW_s)^2] = t$, by Levy's theorem, Theorem 1.1.6, there is a Brownian motion β_t such that $X_t = \beta_t$, $\forall t \geq 0$. Equivalently, $B_t - W_t = \sqrt{2}\beta_t$, namely the difference between two independent Brownian motions is proportional with another Brownian motion.

2. Consider now two independent Brownian motions in $\mathbb{R}^n$ denoted by $W_t = \big(W_1(t), \dots, W_n(t)\big)$ and $B_t = \big(B_1(t), \dots, B_n(t)\big)$. The difference between them is the vector process

$$D_t = W_t - B_t = \big(W_1(t) - B_1(t), \dots, W_n(t) - B_n(t)\big).$$

Applying part 1 on each component we obtain n independent one-dimensional Brownian motions, $\beta_j(t)$, such that $W_j(t)-B_j(t)=\sqrt{2}\beta_j(t)$, $j=1,\dots,n$. Denoting $\beta_t=\big(\beta_1(t),\dots,\beta_n(t)\big)$, we obtain $D_t=\sqrt{2}\beta_t$. Therefore, the distance between the n-dimensional Brownian motions W_t and B_t is given by

$$\|D_t\|=\sqrt{2}\|\beta_t\|=\sqrt{2}R_t^{(n)},$$

where $R_t^{(n)}$ is a Bessel process of dimension n. We conclude by stating that the distance between two Brownian motions in $\mathbb{R}^n$ is proportional to a Bessel process.

2.9 Euclidean Invariance of Brownian Motion

In the context of stochastic processes in the Euclidean plane, only those processes that remain invariant under a change of coordinates are significant for the theory. For example, a two-dimensional Brownian motion retains its essential properties even after being subjected to a rotation or reflection with respect to a line. In this section, we aim to determine all continuous differentiable invertible transformations of the plane that maintain the stochastic nature of the Brownian motion.

Proposition 2.9.1 *Let $F:\mathbb{R}^2\to\mathbb{R}^2$ be a continuous, twice differentiable invertible transformation of the plane, with $F(0,0)=(0,0)$, such that if $W_t=(W_t^1,W_t^2)$ is a plane Brownian motion, then $B_t=F(W_t)$ is also a Brownian motion in the plane. Then F must be either a rotation centered at the origin, a flip about a line passing through the origin, or a composition of these two transformations.*

Proof: We denote $F(x_1,x_2)=\big(F_1(x_1,x_2),F_2(x_1,x_2)\big)$. Then $B_t^1=F_1(W_t^1,W_t^2)$ and $B_t^2=F_2(W_t^1,W_t^2)$ are Brownian motions starting at 0. Applying Proposition 2.2.1 there are two constants $\theta_1,\theta_2\in\mathbb{R}$ such that

$$\begin{aligned}F_1(x_1,x_2)&=x_1\cos\theta_1+x_2\sin\theta_1\\ F_2(x_1,x_2)&=x_1\cos\theta_2+x_2\sin\theta_2,\end{aligned}$$

which shows that F is a linear function. Since F is invertible, its Jacobian is nonsingular

$$\begin{vmatrix}\cos\theta_1 & \sin\theta_1\\ \cos\theta_2 & \sin\theta_2\end{vmatrix}=\sin(\theta_2-\theta_1)\neq 0,$$

which implies $\theta_2\notin\{\theta_1+k\pi,\ k=0,\pm1,\pm2,\dots\}$. We shall compute a tighter relation between θ_1 and θ_2.

Since B_t is a two-dimensional Brownian motion, B_t^1 and B_t^2 are independent processes, which infers $Var(B_t^1, B_t^2) = 0$. On the other side, a direct computation of the variance provides

$$\begin{aligned} Var(B_t^1, B_t^2) &= Var\Big(\cos\theta_1 W_t^1 + \sin\theta_1 W_t^2,\ \cos\theta_2 W_t^1 + \sin\theta_2 W_t^2 \Big) \\ &= \cos\theta_1 \cos\theta_2 Var(W_t^1, W_t^1) + \sin\theta_1 \sin\theta_2 Var(W_t^1, W_t^1) \\ &= t\cos(\theta_1 - \theta_2), \end{aligned}$$

where we used the independence of W_t^1 and W_t^2. The vanishing variance condition implies now

$$\theta_1 \in \Big\{ \theta_2 + \frac{k\pi}{2}, k = 0, \pm 1, \pm 2, \dots \Big\}.$$

We let $\theta = \theta_2$ and note that the coefficients of the linear function F can be written in terms of θ as follows

$$\cos\theta_1 = (-1)^k \sin\theta, \qquad \sin\theta_1 = (-1)^{k+1}\cos\theta$$

and hence

$$F(x_1, x_2) = \begin{pmatrix} (-1)^k \sin\theta & (-1)^{k+1}\cos\theta \\ \cos\theta & \sin\theta \end{pmatrix} \begin{pmatrix} x_1 \\ x_2 \end{pmatrix}.$$

We have two cases: (i) If k is even, then the matrix associated to F is a rotation of angle $\frac{\pi}{2} - \theta$, i.e.,

$$\begin{pmatrix} \sin\theta & -\cos\theta \\ \cos\theta & \sin\theta \end{pmatrix} = \mathcal{R}_{\frac{\pi}{2}-\theta}.$$

(ii) If k is odd, then the matrix associated to F is a rotation of angle θ composed with a flip with respect to the $\{y = x\}$ line, namely

$$\begin{pmatrix} -\sin\theta & \cos\theta \\ \cos\theta & \sin\theta \end{pmatrix} = \mathcal{R}_\theta \circ \mathcal{F},$$

where $\mathcal{F}(x, y) = (y, x)$ and $\mathcal{R}_\theta = \begin{pmatrix} \cos\theta & -\sin\theta \\ \sin\theta & \cos\theta \end{pmatrix}$.

Composing transformations (i) and (ii) generates the desired result. ■

2.10 Some Hyperbolic Integrals

This section presents technical results that will be useful in the upcoming sections. We start by recaling the definitions of some hyperbolic functions

$$\operatorname{sech} x = \frac{1}{\cosh x} = \frac{2}{e^x + e^{-x}}$$

$$\operatorname{csch} x = \frac{1}{\sinh x} = \frac{2}{e^x - e^{-x}}$$

and of the Riemann's zeta function

$$\zeta(s) = \sum_{k\geq 1} \frac{1}{k^s}, \qquad s > 1.$$

The following improper integrals will be useful in evaluating the moments of several density functions.

Proposition 2.10.1 *Let $p > 0$ and $a > 0$. Then*

$$(i) \qquad \int_0^\infty x^p \operatorname{csch}(x)\, dx = p!\Big(2 - \frac{1}{2^p}\Big)\zeta(p+1) \tag{2.10.10}$$

$$(ii) \qquad \int_0^\infty x^p \operatorname{sech}(x)\, dx = 2p! \sum_{k\geq 0} (-1)^k \frac{1}{(2k+1)^{p+1}}. \tag{2.10.11}$$

Proof: First we note that if $p, a > 0$, then using the properties of the Gamma function we have

$$\begin{aligned} \int_0^\infty x^p e^{-ax}\, dx &= \frac{1}{a^{p+1}} \int_0^\infty u^p e^{-u}\, du = \frac{1}{a^{p+1}} \Gamma(p+1) \\ &= \frac{p!}{a^{p+1}}. \end{aligned} \tag{2.10.12}$$

(i) Using a geometric series expansion we write

$$x^p \operatorname{csch}(x) = \frac{x^p}{\sinh x} = \frac{2x^p e^{-x}}{1 - e^{-2x}} = 2\sum_{k\geq 0} x^p e^{-(2k+1)x}.$$

Integrating the above series term by term and using (2.10.12) yields

$$\int_0^\infty x^p \operatorname{csch}(x)\, dx = 2\sum_{k\geq 0} \int_0^\infty x^p e^{-(2k+1)x}\, dx = 2p! \sum_{k\geq 0} \frac{1}{(2k+1)^{p+1}}. \tag{2.10.13}$$

Since the last series can be written in terms of the zeta function as

$$\sum_{k\geq 0}\frac{1}{(2k+1)^{p+1}} = \zeta(p+1) - \Big(\frac{1}{2^{p+1}} + \frac{1}{4^{p+1}} + \frac{1}{6^{p+1}} + \frac{1}{8^{p+1}} + \cdots\Big)$$

$$= \zeta(p+1) - \frac{1}{2^{p+1}}\zeta(p+1) = \Big(1 - \frac{1}{2^{p+1}}\Big)\zeta(p+1),$$

then substituting in (2.10.13) yields relation (2.10.10).

(*ii*) Expanding using an alternating geometric series we write

$$x^p \operatorname{sech}(x) = \frac{2x^p e^{-x}}{1+e^{-2x}} = 2\sum_{k\geq 0}(-1)^k x^p e^{-(2k+1)x}.$$

Integrating the series term by term using (2.10.12) we have

$$\int_0^\infty x^p \operatorname{sech}(x)\,dx = 2\sum_{k\geq 0}(-1)^k \int_0^\infty x^p e^{-(2k+1)x}\,dx$$

$$= 2p!\sum_{k\geq 0}\frac{(-1)^k}{(2k+1)^{p+1}}.$$

■

Remark 2.10.2 1. Let $p = 1$ in (2.10.10). Using the identity $\sum_{n\geq 1}\frac{1}{n^2} = \frac{\pi^2}{6}$ we obtain

$$\int_0^\infty x \operatorname{csch}(x)\,dx = \frac{\pi^2}{4}.$$

2. If let $p = 2$ in relation (2.10.10) we obtain

$$\int_0^\infty x^2 \operatorname{csch}(x)\,dx = \frac{7}{2}\zeta(3). \tag{2.10.14}$$

3. The sum of the series (2.10.11) can sometimes be written in closed form. Here are a few particular cases:

(*i*) Let $C = \sum_{k=0}^{\infty}(-1)^k\frac{1}{(2k+1)^2} \approx 0.915\ldots$ denote the Catalan constant. Then if we let $p = 1$ in (2.10.11) we obtain

$$\int_0^\infty x \operatorname{sech}(x)\,dx = 2C. \tag{2.10.15}$$

(*ii*) Taking the limit $p \searrow 0$ into (2.10.11) and using Leibnitz' formula

$$1 - \frac{1}{3} + \frac{1}{7} - \frac{1}{9} + \cdots = \frac{\pi}{4},$$

we obtain

$$\int_0^\infty \operatorname{sech}(x)\,dx = \frac{\pi}{2}.$$

2.11 The Product of 2 Brownian Motions

Let W_t and B_t be two independent, one-dimensional Brownian motions and consider the product process $X_t = W_t B_t$, $t \geq 0$. By Ito's lemma we have

$$dX_t = W_t\, dB_t + B_t\, dW_t,$$

which implies that the quadratic variation is

$$\langle X \rangle_t = \int_0^t (W_s^2 + B_s^2)\, ds = \int_0^t R_s^2\, ds,$$

where R_t denotes a Bessel process of dimension 2. Since we can write

$$X_t = \int_0^t W_s\, dB_s + \int_0^t B_s\, dW_s,$$

then X_t is a continuous martingale, with

$$\mathbb{E}[X_t^2] = \mathbb{E}[W_t^2]\mathbb{E}[B_t^2] = t^2 < \infty.$$

By Theorem 1.1.11 the process X_t can be written as a Brownian motion, β_t, under a stochastic time as

$$X_t = \beta_{\int_0^t R_s^2\, ds}, \quad t \geq 0.$$

Using the tower property of conditional expectations and Cameron-Martin's formula we compute the characteristic function of X_t as

$$\begin{aligned}
\mathbb{E}[e^{i\lambda X_t}] &= \mathbb{E}[e^{i\lambda\beta_{\int_0^t R_s^2\, ds}}] \\
&= \mathbb{E}[\,\mathbb{E}[e^{i\lambda\beta_{\int_0^t R_s^2\, ds}}|W[0,t], B[0,t]]\,] \\
&= \mathbb{E}[e^{-\frac{\lambda^2}{2}\int_0^t R_s^2\, ds}] = \mathbb{E}[e^{-\frac{\lambda^2}{2}\int_0^t W_s^2\, ds}]\,\mathbb{E}[e^{-\frac{\lambda^2}{2}\int_0^t B_s^2\, ds}] \\
&= \left(\frac{1}{\sqrt{\cosh(\lambda t)}}\right)^2 = \frac{1}{\cosh(\lambda t)}, \qquad \forall \lambda \in \mathbb{R}.
\end{aligned} \tag{2.11.16}$$

If $p_{X_t}(x)$ denotes the probability density of X_t, the previous computation can be written in the following integral form

$$\int_{\mathbb{R}} e^{i\lambda x} p_{X_t}(x)\, dx = \frac{1}{\cosh(\lambda t)}.$$

Then the probability density of X_t can be retrieved as an inverse Fourier transform as

$$p_{X_t}(x) = \frac{1}{2\pi}\int_{\mathbb{R}} e^{-i\lambda x}\frac{1}{\cosh(\lambda t)}\, d\lambda, \qquad x \in \mathbb{R}. \tag{2.11.17}$$

The integral on the right side of (2.11.17) can be computed using the Residues Theorem. First, using symmetry reasons we have that $p_{X_t}(x)$ is even, i.e.,

$$p_{X_t}(-x) = p_{X_t}(x), \quad \forall x \in \mathbb{R}.$$

This follows from an application of the Euler's formula, $e^{-i\lambda x} = \cos(\lambda x) - i\sin(\lambda x)$, and the fact that $\sin x$ is an odd function, while $\cos x$ and $\cosh x$ are even functions, as well as from properties of integrals of even and odd functions. Therefore, it suffices to assume $x \geq 0$. In order to apply the Residues Theorem, we notice that the poles of $\dfrac{1}{\cosh(\lambda t)}$ are the zeros of $\cos(i\lambda t)$, i.e.,

$$\lambda \in \Big\{ \pm i\frac{\pi}{2t}, \pm i\frac{3\pi}{2t}, \pm i\frac{5\pi}{2t}, \dots \Big\}.$$

To find the residues we use the following result: If $f(z) = \frac{g(z)}{h(z)}$ is a complex function with $h(c) = 0$ and $h'(c) \neq 0$ and h and g holomorphic about c, then the residue of $f(z)$ at c is given by

$$Res(f, c) = \frac{g(c)}{h'(c)}.$$

In our case we let $g(\lambda) = e^{-i\lambda x}$ and $h(\lambda) = \cosh(\lambda t)$, so

$$Res(f, c) = \frac{e^{-icx}}{t\sinh(ct)}. \tag{2.11.18}$$

We shall compute the residues of $f(\lambda) = \frac{e^{-i\lambda x}}{\cosh(\lambda x)}$ only at the poles with negative coordinate. An elementary evaluation of (2.11.18) at those poles yields

$$\begin{aligned}
Res\Big(f, -i\frac{\pi}{2t}\Big) &= \frac{i}{t}e^{-\frac{\pi}{2t}x} \\
Res\Big(f, -i\frac{3\pi}{2t}\Big) &= -\frac{i}{t}e^{-\frac{3\pi}{2t}x} \\
Res\Big(f, -i\frac{5\pi}{2t}\Big) &= \frac{i}{t}e^{-\frac{5\pi}{2t}x} \\
&\dots\dots\dots\dots\dots\dots\dots \\
Res\Big(f, -i\frac{(2k+1)\pi}{2t}\Big) &= (-1)^k\frac{i}{t}e^{-\frac{(2k+1)\pi}{2t}x}, \quad k \geq 0.
\end{aligned}$$

Now we compute the improper integral as a limit

$$\int_{\mathbb{R}} f(\lambda)\, d\lambda = \lim_{a\to\infty} \int_{-a}^{a} f(\lambda)\, d\lambda, \tag{2.11.19}$$

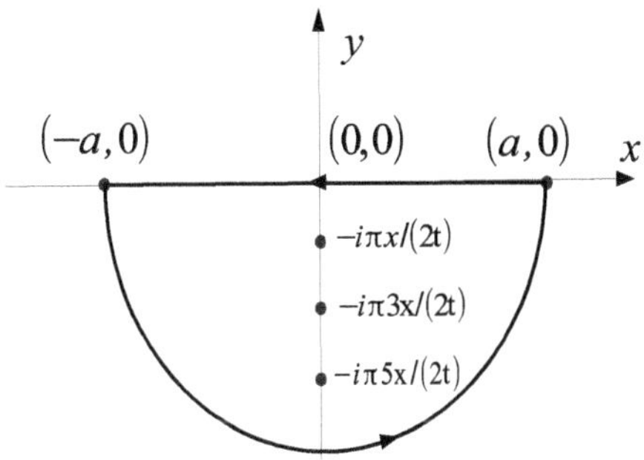

Figure 2.3: *The residues theorem applied on the contour of the lower half-disk of radius a centered at the origin.*

where by Residues Theorem

$$\int_{-a}^{a} f(\lambda)\,d\lambda - \int_{arc} f(\lambda)\,d\lambda = -2\pi i \sum_{|c|\le a} Res(f,c), \tag{2.11.20}$$

where arc $= \{ae^{i\theta}; \pi \le \theta \le 2\pi\}$ is the lower half of a circle of radius a centered at the origin and we consider the poles contained in the lower half of this disk. The negative signs occur due to the clock-wise rotation, see Fig. 2.3.

First we shall show that

$$\lim_{a\to\infty} \int_{arc} f(\lambda)\,d\lambda = 0 \tag{2.11.21}$$

using a standard estimation procedure. To this end, using that $x \ge 0$ and $\theta \in [\pi, 2\pi]$, we have

$$\begin{aligned}
\Big|\int_{arc} f(\lambda)\,d\lambda\Big| &= \Big|\int_{arc} \frac{e^{-i\lambda x}}{\cosh(\lambda t)}\,d\lambda\Big| \le \int_{\pi}^{2\pi} \frac{ae^{ax\sin\theta}}{|\cosh(ae^{i\theta}t)|}\,d\theta \\
&\le \int_{\pi}^{2\pi} \frac{a}{\sqrt{\cosh(2at\cos\theta)+\cos(2at\sin\theta)}}\,d\theta \\
&= \int_{0}^{\pi} \frac{a}{\sqrt{\cosh(2at\cos\theta)+\cos(2at\sin\theta)}}\,d\theta \\
&\le \int_{0}^{\pi} \frac{a}{\sqrt{\cosh(2at\cos\theta)-1}}\,d\theta \\
&= \frac{a\pi}{\sqrt{\cosh(2at\xi)-1}} \to 0, \quad \text{as } a\to\infty,
\end{aligned}$$

where the value $\xi \in [-1,1]\backslash\{0\}$ is obtained from the Intermediate Value Theorem for definite integrals. The fact that $\xi \neq 0$ follows from the fact that the integral is finite.

Using relations (2.11.19), (2.11.20) and (2.11.21) we write the improper integral as a geometric series, which can be evaluated in closed form as

$$\begin{aligned}
\int_{\mathbb{R}} f(\lambda)\,d\lambda &= -2\pi i \sum_{k\geq 0} Res\Big(f, -i\frac{(2k+1)\pi}{2t}\Big) \\
&= -2\pi i\Big(\frac{i}{t}e^{-\frac{\pi}{2t}x} - \frac{i}{t}e^{-\frac{3\pi}{2t}x} + \frac{i}{t}e^{-\frac{5\pi}{2t}x} - \frac{i}{t}e^{-\frac{7\pi}{2t}x} + \cdots\Big) \\
&= \frac{2\pi}{t}e^{-\frac{\pi}{2t}x}\Big(1 - e^{-\frac{\pi}{t}x} + \big(e^{-\frac{\pi}{t}x}\big)^2 - \big(e^{-\frac{\pi}{t}x}\big)^3 + \cdots\Big) \\
&= \frac{2\pi}{t}e^{-\frac{\pi}{2t}x}\frac{1}{1+e^{-\frac{\pi}{t}x}} = \frac{\pi}{t}\frac{1}{\cosh(\frac{\pi x}{2t})},
\end{aligned} \tag{2.11.22}$$

where the convergence of the series is assured by the fact that $e^{-\frac{\pi}{t}x} < 1$, fact implied by the assumption $x > 0$. Thereby, we have obtained the following result.

Proposition 2.11.1 *The probability distribution of the product of two independent one-dimensional Brownian motions, $X_t = B_t W_t$, is*

$$p_{X_t}(x) = \frac{1}{2t}\text{sech}\Big(\frac{\pi x}{2t}\Big), \quad x \in \mathbb{R}.$$

Proof: Substituting (2.11.22) into the Fourier transform formula (2.11.17) yields

$$p_{X_t}(x) = \frac{1}{2\pi}\int_{\mathbb{R}} e^{-i\lambda x}\frac{1}{\cosh(\lambda t)}\,d\lambda = \frac{1}{2t}\frac{1}{\cosh(\frac{\pi x}{2t})} = \frac{1}{2t}\text{sech}\Big(\frac{\pi x}{2t}\Big).$$

■

With a similar proof we can obtain the distribution of the Lévy area introduced in Section 1.11.

Proposition 2.11.2 *The probability distribution of the Lévy area S_t is*

$$p_{S_t}(x) = \frac{1}{2t}\text{sech}\Big(\frac{\pi x}{2t}\Big), \quad x \in \mathbb{R}.$$

Proof: Inverting the expectation formula (1.12.76) by applying a Fourier transform yields

$$p_{S_t}(x) = \frac{1}{2\pi}\int_{\mathbb{R}} e^{-i\lambda x}\frac{1}{\cosh(\lambda t)}\,d\lambda = \frac{1}{2t}\frac{1}{\cosh(\frac{\pi x}{2t})} = \frac{1}{2t}\text{sech}\Big(\frac{\pi x}{2t}\Big).$$

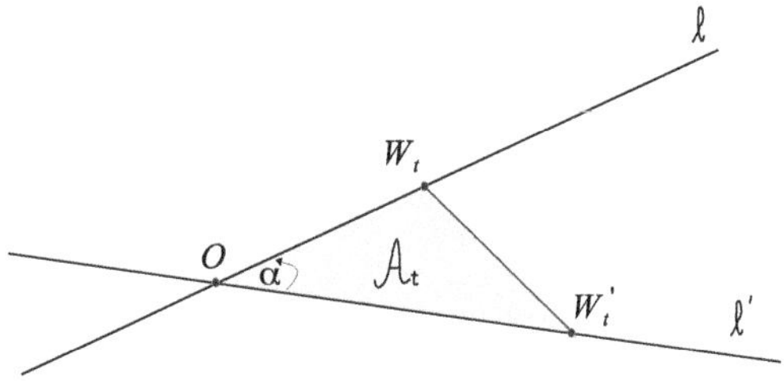

Figure 2.4: *The area $\mathcal{A}_t$ of a triangle with two stochastic vertices described by Brownian motions along the lines ℓ and ℓ'.*

■

We note that S_t and the product $X_t = B_tW_t$ have the same law. This relation becomes even more transparent if we write these processes as Ito integrals as follows

$$S_t = \int_0^t W_s dB_s - B_s dW_s$$
$$X_t = \int_0^t W_s dB_s + B_s dW_s.$$

Corollary 2.11.3 *We have*

$$\int_{\mathbb{R}} y^{2k} \text{sech}(y)\, dy = \frac{\pi^{2k+1}}{2^{4k}} \left(\frac{(2k)!}{k!} \right)^2, \quad k \geq 0.$$

Proof: We compute the even order centered moments of X_t in two ways. First, using the moments of Brownian motion given by Exercise 1.21.9 (a) we have

$$\mathbb{E}[X_t^{2k}] = \mathbb{E}[W_t^{2k}]\mathbb{E}[B_t^{2k}] = \left(\frac{(2k)!}{k!} \frac{t^k}{2^k} \right)^2.$$

Then, by the definition of the $2k$-th moment

$$\mathbb{E}[X_t^{2k}] = \frac{1}{2t} \int_{\mathbb{R}} x^{2k} \text{sech}\Big(\frac{\pi x}{2t}\Big)\, dx = \frac{(2t)^{2k}}{\pi^{2k+1}} \int_{\mathbb{R}} y^{2k} \text{sech}(y)\, dy.$$

Equating the last two relations yields the desired result. ■

Application 2.11.4 We shall provide an application to finding the area of a triangle which has one fixed vertex and the other two stochastic. For this we consider two lines ℓ and ℓ' passing through the point O and making the angle $\alpha \in (0, \pi)$. We consider two independent one-dimensional Brownian motions starting at O, W_t and W_t', along the lines ℓ and ℓ', respectively, as in Fig. 2.4. We shall denote by $\mathcal{A}_t$ the area of the triangle with one of the vertices at O and the other two at the points described by the previous Brownian motions. This area is given by

$$\mathcal{A}_t = \frac{1}{2}|W_t W_t'| \sin\alpha,$$

where $|\cdot|$ represents the absolute value of the product. If $X_t = W_t W_t'$ denotes the product of the Brownian motions, the density of $\mathcal{A}_t$ relates to the density of X_t as

$$\begin{aligned} p_{\mathcal{A}_t}(x) &= \frac{2}{\sin\alpha} p_{|X_t|}\Big(\frac{2x}{\sin\alpha}\Big) = \frac{4}{\sin\alpha} p_{X_t}\Big(\frac{2x}{\sin\alpha}\Big) \\ &= \frac{2}{t\sin\alpha} \operatorname{sech}\Big(\frac{\pi x}{t\sin\alpha}\Big), \qquad x \geq 0, \end{aligned}$$

where in the last equality we used Proposition 2.11.1. The expected triangle area is given by

$$\begin{aligned} \mathbb{E}[\mathcal{A}_t] &= \int_0^\infty x p_{\mathcal{A}_t}(x)\, dx = \frac{2t\sin\alpha}{\pi^2} \int_0^\infty y \operatorname{sech}(y)\, dy \\ &= \frac{4tC\sin\alpha}{\pi^2}, \end{aligned}$$

where C stands for the Catalan's constant $C = \sum_{k=0}^{\infty}(-1)^k \frac{1}{(2k+1)^2}$ and we applied the integral formula (2.10.15).

Similarly, we can compute the p-th moments of the area using Proposition 2.10.1, part *(ii)*

$$\mathbb{E}[\mathcal{A}_t^p] = \int_0^\infty x^p p_{\mathcal{A}_t}(x)\, dx = \frac{4p!t^p \sin^p\alpha}{\pi^{p+1}} \sum_{k\geq 0} (-1)^k \frac{1}{(2k+1)^{p+1}}, \quad p \geq 0.$$

The previous alternating series can be evaluated in closed form in terms of π for p even. For instance, using that

$$\sum_{k\geq 0}(-1)^k \frac{1}{(2k+1)^3} = \frac{\pi^3}{32}$$

we obtain the second moment

$$\mathbb{E}[\mathcal{A}_t^2] = \frac{t^2\sin^2\alpha}{4}.$$

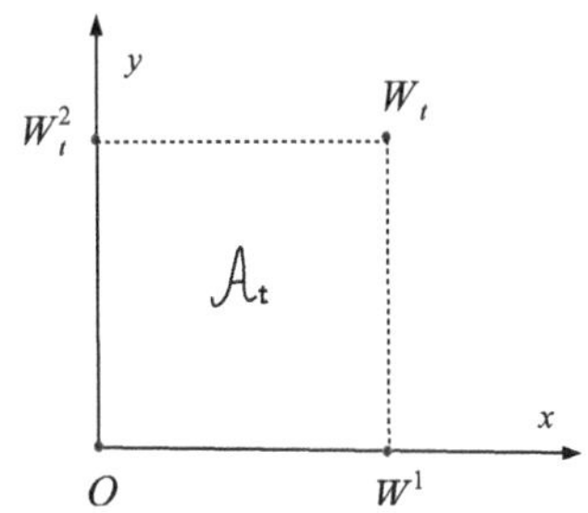

Figure 2.5: *The area $\mathcal{A}_t$ of rectangle with three stochastic vertices described by Brownian motions along the coordinate axes.*

Also, using that

$$\sum_{k\geq 0}(-1)^k \frac{1}{(2k+1)^5} = \frac{5\pi^5}{1536}$$

we obtain the fourth moment

$$\mathbb{E}[\mathcal{A}_t^4] = \frac{5}{16}t^4 \sin^4 \alpha.$$

In the particular case when the lines ℓ and ℓ' are perpendicular, the above formulae simplifies even further as in the following.

Application 2.11.5 We consider now a two-dimensional Brownian motion $W_t = (W_t^1, W_t^2)$ and denote by $\mathcal{A}_t$ the area of the rectangle having vertices at the origin, $(0,0)$, and at the points with coordinates (W_t^1, W_t^2), $(W_t^1, 0)$, $(0, W_t^2)$, see Fig. 2.5. Since this area can be represented as the product of two independent Brownian motions, $\mathcal{A}_t = W_t^1 W_t^2$, its density function is given by Proposition 2.11.1 as

$$p_{\mathcal{A}_t}(x) = \frac{1}{2t}\text{sech}\Big(\frac{\pi x}{2t}\Big), \qquad x \in \mathbb{R}.$$

Using a similar computation with the one done in Application 2.11.4, the area expectation is given by

$$\mathbb{E}[\mathcal{A}_t] = \frac{8tC}{\pi^2}, \qquad t \geq 0,$$

where C is Catalan's constant.

2.12 Inner Product of Brownian Motions

1. Let $W_t = (W_t^1, W_t^2)$ and $B_t = (B_t^1, B_t^2)$ be two independent two-dimensional Brownian motions, and consider their inner product

$$I_t^{(2)} = \langle W_t, B_t\rangle = W_t^1 B_t^1 + W_t^2 B_t^2, \qquad t \geq 0.$$

Using the independence of Brownian motions and the Cameron-Martin argument from formula (2.11.16) we obtain the following expression for the characteristic function of $I_t^{(2)}$

$$\mathbb{E}[e^{i\lambda I_t^{(2)}}] = \mathbb{E}[e^{i\lambda W_t^1 B_t^1}]\,\mathbb{E}[e^{i\lambda W_t^2 B_t^2}] = \frac{1}{\cosh^2(\lambda t)}, \qquad \forall \lambda \in \mathbb{R}. \tag{2.12.23}$$

Proposition 2.12.1 *The probability density of the process* $I_t^{(2)} = \langle W_t, B_t\rangle$ *is given by*

$$p_t(x) = \frac{x}{2t^2}\text{csch}\Big(\frac{\pi x}{2t}\Big), \qquad \forall x \in \mathbb{R}. \tag{2.12.24}$$

Proof: Applying an inverse Fourier transform in relation (2.12.23) and changing the variable yields

$$p_t(x) = \frac{1}{2\pi}\int_{\mathbb{R}} e^{-i\lambda x}\frac{1}{\cosh^2(\lambda t)}\,d\lambda = \frac{1}{2\pi t}\int_{\mathbb{R}} e^{-izx/t}\frac{1}{\cosh^2(z)}\,dz. \tag{2.12.25}$$

This integral can be computed explicitly using the Residues Theorem following a procedure similar to the one used in Section 2.11. We note first that the function $f(z) = e^{-izx/t}\sec^2(z)$ has the poles at $\{\pm i\frac{\pi}{2}, \pm i\frac{3\pi}{2}, \pm i\frac{5\pi}{2}, \dots\}$. Assuming first that $x > 0$, we are interested in the following residues

$$Res\Big(f, -i\frac{\pi}{2}\Big) = ie^{-\frac{\pi x}{2t}}\cdot\frac{x}{t}$$
$$Res\Big(f, -i\frac{3\pi}{2}\Big) = ie^{-\frac{3\pi x}{2t}}\cdot\frac{x}{t}$$
$$\dots\dots\dots\dots\dots\dots\dots\dots\dots$$
$$Res\Big(f, -i\frac{(2k+1)\pi}{2}\Big) = ie^{-\frac{(2k+1)\pi x}{2t}}\cdot\frac{x}{t}, \qquad \forall k \geq 0.$$

Then we shall evaluate the improper integral via the Residues Theorem as a geometric series

$$\begin{aligned}
\int_{\mathbb{R}} f(z)\,dz &= -2\pi i\sum_{k\geq 0} Res\Big(f, -i\frac{(2k+1)\pi}{2}\Big) \\
&= 2\pi e^{-\frac{\pi x}{2t}}\frac{x}{t}\Big(1 + e^{-\frac{\pi x}{t}} + (e^{-\frac{\pi x}{t}})^2 + (e^{-\frac{\pi x}{t}})^3 + \cdots\Big) \\
&= 2\pi e^{-\frac{\pi x}{2t}}\frac{x}{t}\frac{1}{1 - e^{-\frac{\pi x}{t}}} = \frac{\pi x}{t}\text{csch}\frac{\pi x}{2t},
\end{aligned}$$

where we used the convergence condition $e^{-\frac{\pi x}{t}} < 1$, as $x > 0$. Substituting this expression for the integral in relation (2.12.25) yields (2.12.24).

The case $x < 0$ follows from the symmetry condition $p_t(x) = p_t(-x)$, which completes the proof. ■

2. Let $W_t = (W_t^1, W_t^2, W_t^3)$ and $B_t = (B_t^1, B_t^2, B_t^3)$ be two independent 3-dimensional Brownian motions, and consider their inner product

$$I_t^{(3)} = \langle W_t, B_t \rangle = W_t^1 B_t^1 + W_t^2 B_t^2 + W_t^3 B_t^3, \qquad t \geq 0.$$

Using the independence of Brownian motions and a Cameron-Martin's formula argument we obtain the characteristic function of $I_t^{(3)}$

$$\mathbb{E}[e^{i\lambda I_t^{(3)}}] = \mathbb{E}[e^{i\lambda W_t^1 B_t^1}]\,\mathbb{E}[e^{i\lambda W_t^2 B_t^2}]\,\mathbb{E}[e^{i\lambda W_t^3 B_t^3}] = \frac{1}{\cosh^3(\lambda t)}, \qquad \forall \lambda \in \mathbb{R}. \tag{2.12.26}$$

Proposition 2.12.2 *The probability density of the process $I_t^{(3)} = \langle W_t, B_t \rangle$ is given by*

$$p_t(x) = \frac{t^2 + x^2}{4t^3} \operatorname{sech}\Big(\frac{\pi x}{2t}\Big), \qquad \forall x \in \mathbb{R}. \tag{2.12.27}$$

Proof: An inverse Fourier transform applied to (2.12.26) yields

$$\begin{aligned} p_t(x) &= \frac{1}{2\pi} \int_{\mathbb{R}} e^{-i\lambda x} \operatorname{sech}^3(\lambda t)\, d\lambda \\ &= \frac{1}{2\pi t} \int_{\mathbb{R}} e^{-iz\frac{x}{t}} \operatorname{sech}^3(z)\, dz. \end{aligned} \tag{2.12.28}$$

This integral can be evaluated using Residues Theorem as in the previous proofs. Taking $f(z) = e^{-iz\frac{x}{t}} \operatorname{sech}^3(z)$ we have

$$Res\Big(f, -i\frac{(2k+1)\pi}{2}\Big) = (-1)^k \frac{i}{2} e^{-\frac{(2k+1)\pi x}{2t}} \Big(1 + \frac{x^2}{t^2}\Big).$$

We assume $x > 0$ and we evaluate the improper integral as an alternate geometric series

$$\begin{aligned} \int_{\mathbb{R}} f(z)\, dz &= -2\pi i \sum_{k \geq 0} Res\Big(f, -i\frac{(2k+1)\pi}{2}\Big) \\ &= \pi\Big(1 + \frac{x^2}{t^2}\Big) e^{-\frac{\pi x}{2t}} \sum_{k \geq 0} (-1)^k e^{-\frac{k\pi x}{t}} \\ &= \frac{\pi}{2t^2}(t^2 + x^2) \frac{2e^{-\frac{\pi x}{2t}}}{1 + e^{-\frac{\pi x}{t}}} = \frac{\pi}{2t^2}(t^2 + x^2) \operatorname{sech}\Big(\frac{\pi x}{2t}\Big). \end{aligned}$$

Substituting into (2.12.28) yields (2.12.27). The case $x < 0$ follows from the symmetry of the density $p_t(x)$. ■

3. Let W_t and B_t be two independent Brownian motions in $\mathbb{R}^n$, and consider their inner product

$$I_t^{(n)} = \langle W_t, B_t \rangle = \sum_{k=1}^{n} W_t^k B_t^k, \qquad t \geq 0.$$

As before, we obtain the characteristic function of $I_t^{(n)}$ as

$$\mathbb{E}[e^{i\lambda I_t^{(n)}}] = \prod_{k=1}^{n} \mathbb{E}[e^{i\lambda W_t^k B_t^k}] = \frac{1}{\cosh^n(\lambda t)}, \qquad \forall \lambda \in \mathbb{R}. \tag{2.12.29}$$

Inverting yields

$$p_t(x) = \frac{1}{2\pi} \int_{\mathbb{R}} e^{-i\lambda x} \operatorname{sech}^n(\lambda t)\, d\lambda. \tag{2.12.30}$$

This Fourier transform can be computed explicitly using the Residues Theorem as before. The expression of the density $p_t(x)$ depends on the parity of n as follows:

(i) If n is odd, then $p_t(x)$ depends on $\operatorname{sech}\left(\frac{\pi x}{2t}\right)$.

(ii) If n is even, then $p_t(x)$ depends on $\operatorname{csch}\left(\frac{\pi x}{2t}\right)$.

In the case (i) a similar procedure used in the proof of Proposition 2.12.2 leads to the following result:

Proposition 2.12.3 *The density function of $I_t^{(2k+1)}$ is given by*

$$p_t(x) = \frac{1}{2} \frac{1}{(2k)!t^{2k+1}} \prod_{j=1}^{k} \Big((2j-1)t^2 + x^2 \Big) \operatorname{sech}\left(\frac{\pi x}{2t}\right), \quad t > 0,\ x \in \mathbb{R}. \tag{2.12.31}$$

In the case (ii) the factor in front of $\operatorname{csch}\left(\frac{\pi x}{2t}\right)$ is more complex than in formula (2.12.31) and we should not provide it here.

2.13 The Process $Y_t = W_t^2 + B_t^2 + 2c\, W_t B_t$

In this section we consider the process

$$Y_t = W_t^2 + B_t^2 + 2c\, W_t B_t, \tag{2.13.32}$$

where W_t and B_t are two independent one-dimensional Brownian motions and c is a constant with $|c| < 1$. The process Y_t comes up in computing the lengths of sides of a stochastic triangle while using the Cosine Theorem. Since $Y_t \geq (W_t - B_t)^2 \geq 0$, the process takes only non-negative values. To compute the distribution function we consider $s \geq 0$ and evaluate the cumulative distribution function

$$\begin{aligned} F_{Y_t}(s) &= P(Y_t \leq s) = P(W_t^2 + B_t^2 + 2c\,W_t B_t \leq s) \\ &= \iint_{\{x^2+y^2+2cxy \leq s\}} \frac{1}{2\pi t} e^{-\frac{x^2+y^2}{2t}}\, dxdy. \end{aligned}$$

The change of coordinates

$$x = \frac{1}{\sqrt{2}}(u - v), \quad y = \frac{1}{\sqrt{2}}(u + v)$$

rotates the ellipse $\{x^2 + y^2 + 2cxy = s\}$ by an angle value of $\pi/4$ into the standard ellipse $\{(1+c)u^2 + (1-c)v^2 = s\}$, fact that can be verified by a straightforward computation. The integral can be now written as

$$F_{Y_t}(s) = \frac{1}{2\pi t} \iint_{\{(1+c)u^2+(1-c)v^2 \leq s\}} e^{-\frac{u^2+v^2}{2t}}\, dudv.$$

By the change of coordinates

$$u' = \sqrt{1+c}\, u, \qquad v' = \sqrt{1-c}\, v$$

the ellipse $\{(1+c)u^2 + (1-c)v^2 = s\}$ becomes the circle $\{u'^2 + v'^2 = s\}$. Therefore

$$F_{Y_t}(s) = \frac{1}{2\pi t\sqrt{1-c^2}} \iint_{u'^2+v'^2 \leq s} e^{-\frac{1}{2t}\left[\frac{u'^2}{1+c} + \frac{v'^2}{1-c}\right]}\, du'dv'.$$

We continue by employing polar coordinates, $u' = \rho\cos\theta$, $v' = \rho\sin\theta$, with $0 \leq \rho \leq \sqrt{s}$ and $0 \leq \theta < 2\pi$. This implies

$$\begin{aligned} F_{Y_t}(s) &= \frac{1}{2\pi t\sqrt{1-c^2}} \int_0^{\sqrt{s}} \rho \int_0^{2\pi} e^{-\frac{\rho^2}{2t}\left[\frac{\cos^2\theta}{1+c} + \frac{\sin^2\theta}{1-c}\right]}\, d\theta\, d\rho \\ &= \frac{1}{2\pi t\sqrt{1-c^2}} \int_0^{\sqrt{s}} \rho \int_0^{2\pi} e^{-\frac{\rho^2}{2t}\frac{1-c\cos(2\theta)}{1-c^2}}\, d\theta\, d\rho \\ &= \frac{1}{2\pi t\sqrt{1-c^2}} \int_0^{\sqrt{s}} \rho e^{-\frac{\rho^2}{2t(1-c^2)}} \int_0^{2\pi} e^{\frac{\rho^2}{2t}\frac{c}{1-c^2}\cos(2\theta)}\, d\theta\, d\rho. \end{aligned} \tag{2.13.33}$$

Next we let $z = \frac{\rho^2}{2t}\frac{c}{1-c^2}$ and compute the inner integral. Substituting $\xi = 2\theta$ and using $\zeta = \xi - 2\pi$, we obtain

$$\begin{aligned}
\int_0^{2\pi} e^{z\cos(2\theta)}\,d\theta &= \frac{1}{2}\int_0^{4\pi} e^{z\cos\xi}\,d\xi \\
&= \frac{1}{2}\int_0^{2\pi} e^{z\cos\xi}\,d\xi + \frac{1}{2}\int_{2\pi}^{4\pi} e^{z\cos\xi}\,d\xi \\
&= \frac{1}{2}\int_0^{2\pi} e^{z\cos\xi}\,d\xi + \frac{1}{2}\int_0^{2\pi} e^{z\cos\zeta}\,d\zeta \\
&= \int_0^{2\pi} e^{z\cos\xi}\,d\xi = 2\pi I_0(z),
\end{aligned}$$

where I_0 denotes the modified Bessel function of order zero of the first kind, $I_0(z) = \sum_{m\geq 0}\frac{z^{2m}}{2^{2m}}\frac{1}{(m!)^2}$. The last identity used the integral representation of the Bessel function I_0, which can be proved by expanding the exponential in power series and then integrating it term by term, see for instance Calin et al. [29], page 78.
Substituting into (2.13.33) yields

$$F_{Y_t}(s) = \frac{1}{t\sqrt{1-c^2}}\int_0^{\sqrt{s}} \rho e^{-\frac{\rho^2}{2t(1-c^2)}} I_0\Big(\frac{\rho^2}{2t}\frac{c}{1-c^2}\Big)\,d\rho. \tag{2.13.34}$$

Differentiating in s provides the following density function for Y_t

$$f_{Y_t}(s) = \frac{1}{2t\sqrt{1-c^2}}e^{-\frac{s}{2t(1-c^2)}} I_0\Big(\frac{c}{1-c^2}\frac{s}{2t}\Big), \qquad s\geq 0. \tag{2.13.35}$$

Remark 2.13.1 If $c = 0$ then $Y_t = W_t^2 + B_t^2$. Using that $I_0(0) = 1$ then the density (2.13.35) becomes the exponential distribution

$$f_{Y_t}(s) = \frac{1}{2t}e^{-\frac{s}{2t}}, \qquad s\geq 0.$$

2.14 The Length of a Median

We shall provide an application to finding the law of a median length of a triangle with one fixed vertex and the other two stochastic. To this end we consider two lines ℓ and ℓ' passing through the point O and making the angle $\alpha\in(0,\pi)$. Let W_t and B_t be two independent one-dimensional Brownian motions starting at O, along the lines ℓ and ℓ', respectively, as in Fig. 2.6 a. We shall denote by m_t the median length of the triangle which corresponds to the vertex O, namely the distance between the vertex O and the midpoint of the opposite side.

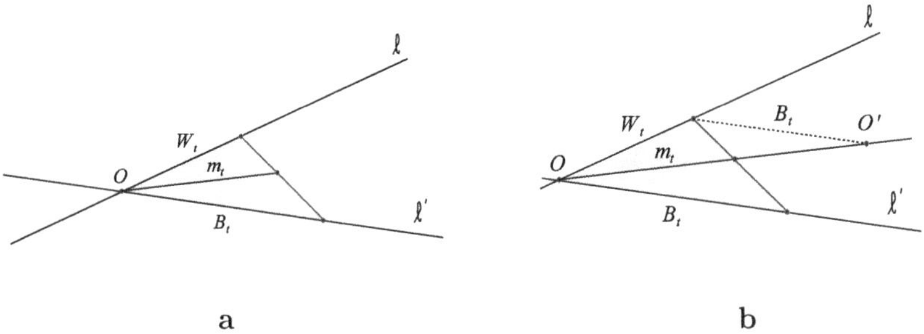

Figure 2.6: *l (a) The median of a triangle with two stochastic vertices described by Brownian motions along the lines ℓ and ℓ'. (b) Completing the triangle to a parallelogram.*

Extending symmetrically the point O on the other side of W_tB_t we obtain a parallelogram, see Fig. 2.6 b. Then applying the Cosine Theorem yields

$$4m_t^2 = W_t^2 + B_t^2 + 2W_tB_t\cos\alpha.$$

Applying formula (2.13.35) with $Y_t = 4m_t^2$ and $c = \cos\alpha$, we obtain

$$f_{Y_t}(s) = \frac{1}{2t\sin\alpha}\, e^{-\frac{s}{2t\sin^2\alpha}}\, I_0\Big(\frac{s\ \cos\alpha}{2t\sin^2\alpha}\Big), \qquad s \geq 0.$$

Using the relation between the distribution functions $F_{m_t}(s) = F_{Y_t}(4s^2)$, we obtain the density of the median length as

$$\begin{aligned} f_{m_t}(s) &= 8s f_{Y_t}(4s^2) \\ &= \frac{4s}{t\sin\alpha}\, e^{-\frac{2s^2}{t\sin^2\alpha}}\, I_0\Big(\frac{2}{t}\frac{\cos\alpha}{\sin^2\alpha}s^2\Big), \qquad s \geq 0. \end{aligned} \tag{2.14.36}$$

The case $\alpha = \pi/2$ We shall deal with the distinguished case when the lines ℓ and ℓ' are perpendicular. In this case relation (2.14.36) simplifies to

$$f_{m_t}(s) = \frac{4s}{t}e^{-\frac{2s^2}{t}}, \qquad s \geq 0.$$

Substituting $\tau = t/4$ this distribution becomes

$$f_\tau(s) = \frac{s}{\tau}e^{-\frac{s^2}{2\tau}}, \qquad s \geq 0,$$

which is Wald's distribution that describes the density function of a two-dimensional Bessel process, $R_t^{(2)}$. We note that this agrees with the geometric fact that in a right triangle the median length is half of the hypotenuse. This is based on the fact that the legs of this right triangle have lengths $|W_t|$ and $|B_t|$ and by the Pythagorean Theorem the hypothenuse has the length $R_t^{(2)} = \sqrt{W_t^2 + B_t^2}$.

2.15 The Length of a Triangle Side

We shall consider again the situation depicted in Fig. 2.6 a, namely we have two independent Brownian motions situated respectively on two lines ℓ and ℓ' passing through the point O and making the angle $\alpha \in (0, \pi)$. We consider that both Brownian motions start at the intersection point O.

Now we are concerned with the law of the distance d_t between the previous Brownian motions. By the Cosine Theorem

$$d_t^2 = W_t^2 + B_t^2 - 2W_t B_t \cos\alpha.$$

Applying formula (2.13.35) with $Y_t = d_t^2$ and $c = -\cos\alpha$, and then using that the Bessel function I_0 is even, we obtain the density of d_t^2 as

$$f_{d_t^2}(s) = \frac{1}{2t\sin\alpha}\, e^{-\frac{s}{2t\sin^2\alpha}}\, I_0\Big(\frac{s}{2t}\frac{\cos\alpha}{\sin^2\alpha}\Big), \qquad s \geq 0.$$

Using the distribution functions relation $F_{d_t}(s) = F_{d_t^2}(s^2)$, the density of the distance d_t becomes

$$\begin{aligned} f_{d_t}(s) &= 2s f_{d_t^2}(s^2) \\ &= \frac{s}{t\sin\alpha}\, e^{-\frac{s^2}{2t\sin^2\alpha}}\, I_0\Big(\frac{s^2}{2t}\frac{\cos\alpha}{\sin^2\alpha}\Big), \qquad s \geq 0. \end{aligned}$$

It is worth noting a few particular cases:
1. If $\alpha = \pi/2$, then the distance between the Brownian motions has the density given by the Wald's distribution

$$f_{d_t}(s) = \frac{s}{t}\, e^{-\frac{s^2}{2t}}, \qquad s \geq 0.$$

2. If $\alpha = \pi/4$, then the distance has the density

$$f_{d_t}(s) = \frac{s\sqrt{2}}{t}\, e^{-\frac{s^2}{t}}\, I_0\Big(\frac{s^2}{t\sqrt{2}}\Big), \qquad s \geq 0.$$

3. If $\alpha = \arccos(\phi - 1)$, where $\phi = \frac{\sqrt{5}+1}{2}$ is the golden ratio, then α satisfies the equation $\cos\alpha = \sin^2\alpha$. In this case the density becomes

$$f_{d_t}(s) = \frac{s}{t\sqrt{3\phi - 1}}\, e^{-\frac{s^2}{2t(3\phi-1)}}\, I_0\Big(\frac{s^2}{2t}\Big), \qquad s \geq 0.$$

2.16 The Brownian Chord

We consider two processes on the unit circle, $X_t = e^{i\alpha_t}$ and $Y_t = e^{i\beta_t}$, with α_t and β_t one-dimensional Brownian motions. The line segment X_tY_t represents a chord with stochastic end points constrained on the unit circle. We shall characterize the law of the Euclidean distance $d_t = d_{Eu}(X_t, Y_t)$. We consider the central angle $\theta_t = \beta_t - \alpha_t$. Then there is a Brownian motion W_t such that $\theta_t = \sqrt{2}W_t$, see point 1 of Section 2.8. Then the Cosine Theorem applied in triangle OX_tY_t provides

$$d_t^2 = 2(1 - \cos\theta_t) = 2\sin^2\theta_t = 2\sin^2(\sqrt{2}W_t).$$

Therefore, there is a one-dimensional Brownian motion, W_t, such that the length of the chord can be represented as $d_t = 2|\sin(\sqrt{2}W_t)|$.

2.17 Brownian Triangles

A triangle $A_tB_tC_t$ in the Euclidean plane is called a *Brownian triangle* if the coordinates of each vertex are independent Brownian motions in the plane. Let $\alpha_t = (\alpha_t^1, \alpha_t^2)$, $\beta_t = (\beta_t^1, \beta_t^2)$ and $\gamma_t = (\gamma_t^1, \gamma_t^2)$ be the plane Brownian motions describing the coordinates of triangle vertices A_t, B_t and C_t, respectively. We shall compute the laws of the area and sum of the squares of the sides of a Brownian triangle.

2.17.1 Area of a Brownian Triangle

In this section we shall compute the law of the area process $\mathcal{A}_t = area(A_tB_tC_t)$. We start by writing

$$\mathcal{A}_t = |\sigma_t|,$$

where σ_t is the *signed area* of the triangle $A_tB_tC_t$, which depends on the orientation of the triangle vertices. This can be described in terms of the vertex coordinates using the following determinant notation

$$\sigma_t = \frac{1}{2}\begin{vmatrix} \alpha_t^1 & \alpha_t^2 & 1 \\ \beta_t^1 & \beta_t^2 & 1 \\ \gamma_t^1 & \gamma_t^2 & 1 \end{vmatrix}.$$

It is convenient to first find the law of σ_t. The discussion of Section 2.8 provides the existence of two two-dimensional Brownian motions, $W_t = (W_t^1, W_t^2)$ and $B_t = (B_t^1, B_t^2)$, such that

$$\alpha_t - \beta_t = \sqrt{2}W_t, \qquad \beta_t - \gamma_t = \sqrt{2}B_t. \tag{2.17.37}$$

Elementary transformations of determinants together with relation (2.17.37) yield

$$\sigma_t = \frac{1}{2}\begin{vmatrix} \alpha_t^1 - \beta_t^1 & \alpha_t^2 - \beta_t^2 \\ \beta_t^1 - \gamma_t^1 & \beta_t^2 - \gamma_t^2 \end{vmatrix} = \begin{vmatrix} W_t^1 & W_t^2 \\ B_t^1 & B_t^2 \end{vmatrix} = W_t^1 B_t^2 - W_t^2 B_t^1.$$

We note from (2.17.37) that the first component of W_t and the second component of B_t are independent. The same independence relation holds for the second component of W_t and the first component of B_t. Then by Ito's lemma the differential of the area is given by

$$d\sigma_t = W_t^1\, dB_t^2 + B_t^2\, dW_t^1 - W_t^2\, dB_t^1 - B_t^1\, dW_t^2, \qquad t \geq 0,$$

with the initial condition $\sigma_0 = 0$, since all Brownian motions start from the origin. Writing the signed area using Ito integrals

$$\sigma_t = \int_0^t W_s^1\, dB_s^1 + \int_0^t B_s^2\, dW_s^1 - \int_0^t W_s^2\, dB_s^1 - \int_0^t B_s^1\, dW_s^2,$$

it follows that σ_t is a continuous martingale with respect to the history $\mathcal{F}_t = \{(W_s, B_s); 0 \leq s \leq t\}$. Since all previous Ito integrals are independent with zero mean and variance

$$\mathbb{E}\Big[\Big(\int_0^t W_s^j\, dB_s^j\Big)^2\Big] = \int_0^t \mathbb{E}[(W_s^j)^2]\, ds = \frac{t^2}{2}, \qquad j = 1, 2$$

then $\mathbb{E}[\sigma_t^2] = 2t^2 < \infty$, and hence σ_t is quadratically integrable. Its quadratic variation is given by

$$\begin{aligned}\langle \sigma \rangle_t &= \int_0^t (d\sigma_t)^2 = \int_0^t \Big((W_t^1)^2 + (W_t^2)^2 + (B_t^1)^2 + (B_t^2)^2\Big)\, ds \\ &= \int_0^t \Big(R_s^{(4)}\Big)^2\, ds,\end{aligned}$$

where $R_t^{(4)}$ is a 4-dimensional Bessel process starting from the origin. Then by Theorem 1.1.11 there is a DDS Brownian motion, η_t, such that

$$\sigma_t = \eta_{\int_0^t (R_s^{(4)})^2\, ds}, \qquad t \geq 0.$$

Computing the characteristic function of σ_t using properties of conditional expectations and Cameron-Martin's formula, we obtain

$$\begin{aligned}
\mathbb{E}[e^{i\lambda\sigma_t}] &= \mathbb{E}[e^{i\lambda\eta\int_0^t (R_s^{(4)})^2\,ds}] \\
&= \mathbb{E}[\,\mathbb{E}[\,e^{i\lambda\eta\int_0^t (R_s^{(4)})^2\,ds}|W[0,t], B[0,t]]\,] \\
&= \mathbb{E}[e^{-\frac{\lambda^2}{2}\int_0^t (R_s^{(4)})^2\,ds}] \\
&= \mathbb{E}[e^{-\frac{\lambda^2}{2}\int_0^t \|W_s\|^2\,ds}]\,\mathbb{E}[e^{-\frac{\lambda^2}{2}\int_0^t \|B_s\|^2\,ds}] \\
&= \frac{1}{\cosh^2(\lambda t)}.
\end{aligned}$$

Comparing to (2.12.23) we note that σ_t has the same characteristic function as the inner product of two Brownian motions, $I_t^{(2)}$, which was studied in Section 2.12. Then following Proposition 2.12.1 the density of the signed area, σ_t, is given by

$$p_t(x) = \frac{x}{2t^2}\operatorname{csch}\Big(\frac{\pi x}{2t}\Big), \quad \forall x \in \mathbb{R}. \tag{2.17.38}$$

Proposition 2.17.1 *The probability density of the area process $\mathcal{A}_t$ is given by*

$$p_{\mathcal{A}_t}(x) = \frac{x}{t^2}\operatorname{csch}\Big(\frac{\pi x}{2t}\Big), \quad \forall x \geq 0. \tag{2.17.39}$$

Proof: Taking the derivative in

$$P(\mathcal{A}_t \leq x) = P(-x \leq \sigma_t \leq x) = \int_{-x}^{x} p_t(u)\,du = F(x) - F(-x),$$

with $F(x)$ the cummulative distribution function of $\sigma(t)$, we obtain

$$p_{\mathcal{A}_t}(x) = p_t(x) + p_t(-x) = 2p_t(x), \qquad x \geq 0$$

where we used that $p_t(x)$ is the even function provided by formula (2.17.38). ∎

The next result shows that the p-th moment of the area of a Brownian triangle is proportional to t^p and depends on the Riemann zeta function $\zeta(p+2)$.

Proposition 2.17.2 *The p-th moment expression of the area of a Brownian triangle is given by*

$$\mathbb{E}[\mathcal{A}_t^p] = \frac{2t^p(p+1)!}{\pi^{p+2}}(2^{p+2}-1)\,\zeta(p+2), \quad t \geq 0.$$

Proof: Using the definition of momentum and changing variables we have

$$\begin{aligned}
\mathbb{E}[\mathcal{A}_t^p] &= \int_0^\infty x^p p_{\mathcal{A}_t}(x)\,dx = \int_0^\infty \frac{x^{p+1}}{t^2}\text{csch}\Big(\frac{\pi x}{2t}\Big)\,dx \\
&= \frac{2^{p+2}t^p}{\pi^{p+2}}\int_0^\infty y^{p+1}\text{csch}(y)\,dy \\
&= \frac{2t^p(p+1)!}{\pi^{p+2}}(2^{p+2}-1)\,\zeta(p+2),
\end{aligned}$$

where for the evaluation of the last integral we used Proposition 2.10.1, part (*i*). ■

In particular, the expectation of the area of a Brownian triangle has the expression

$$\mathbb{E}[\mathcal{A}_t] = \frac{28t}{\pi^3}\,\zeta(3), \quad t \geq 0.$$

Since the second moment is given by

$$\mathbb{E}[\mathcal{A}_t^2] = \frac{180t^2}{\pi^4}\,\zeta(4), \quad t \geq 0,$$

the variance of the triangle area becomes

$$Var[\mathcal{A}_t^2] = \Big(180\zeta(4) - \frac{784}{\pi^2}\zeta(3)^2\Big)\frac{t^2}{\pi^4}.$$

Remark 2.17.3 We have observed that the p-th moment of the area $\mathcal{A}_t$ of a Brownian triangle in the Euclidean plane scales proportionally with t^p. It is worthwhile to study the behavior of the area of Brownian triangles on curved surfaces and determine if it is possible to deduce the curvature of the surface from the moments of the area.

2.17.2 Sum of Squares of Sides

Let $A_tB_tC_t$ be a Brownian triangle in the plane and denote by $\alpha_t = (\alpha_t^1, \alpha_t^2)$, $\beta_t = (\beta_t^1, \beta_t^2)$ and $\gamma_t = (\gamma_t^1, \gamma_t^2)$ three independent Brownian motions describing the coordinates of triangle vertices A_t, B_t and C_t, respectively.

The *sum of the squares of the sides* of triangle $A_tB_tC_t$ is given by the process

$$q_t = a_t^2 + b_t^2 + c_t^2, \tag{2.17.40}$$

where $a_t = d_{Eu}(B_t, C_t)$, $b_t = d_{Eu}(A_t, C_t)$, and $c_t = d_{Eu}(A_t, B_t)$ denote the length of the triangle sides. As pointed out in Section 2.8, there are two

two-dimensional Brownian motions $W_t = (W_t^1, W_t^2)$ and $B_t = (B_t^1, B_t^2)$ such that

$$\begin{aligned} a_t &= |\beta_t - \gamma_t| = \sqrt{2}|W_t| \\ b_t &= |\gamma_t - \alpha_t| = \sqrt{2}|B_t| \\ c_t &= |\alpha_t - \beta_t| = \sqrt{2}|W_t + B_t|. \end{aligned}$$

Then the sum of squares process can be written as

$$\begin{aligned} q_t &= a_t^2 + b_t^2 + c_t^2 = 2\Big(|W_t|^2 + |B_t|^2 + |W_t + B_t|^2\Big) \\ &= 4\Big(|W_t|^2 + |B_t|^2 + \langle W_t, B_t\rangle\Big) = 4(Y_t^1 + Y_t^2), \end{aligned} \tag{2.17.41}$$

where

$$\begin{aligned} Y_t^1 &= (W_t^1)^2 + (B_t^1)^2 + W_t^1 B_t^1 \\ Y_t^2 &= (W_t^2)^2 + (B_t^2)^2 + W_t^2 B_t^2. \end{aligned}$$

The law of these two processes is given by the next result.

Lemma 2.17.4 *(i) The Brownian motions $W_t^j = \frac{1}{\sqrt{2}}(\beta_t^j - \gamma_t^j)$ and $B_t^j = \frac{1}{\sqrt{2}}(\gamma_t^j - \alpha_t^j)$, with $j \in \{1, 2\}$, are negatively correlated with the correlation coefficient $\rho = corr(W_t^j, B_t^j) = -\frac{1}{2}$.*
(ii) The processes Y_t^1 and Y_t^2 are independent and identically exponentially distributed with the law $Y_t^j \sim \frac{2}{3t}e^{-\frac{2}{3t}s}$, $s \geq 0$.

Proof: (*i*) Using the bilinearity of covariance and the Brownian motions independence we have

$$\begin{aligned} Cov(W_t^j, B_t^j) &= Cov\Big(\frac{1}{\sqrt{2}}(\beta_t^j - \gamma_t^j), \frac{1}{\sqrt{2}}(\gamma_t^j - \alpha_t^j)\Big) \\ &= \frac{1}{2}Cov(\beta_t^j - \gamma_t^j, \gamma_t^j - \alpha_t^j) = -\frac{1}{2}Cov(\gamma_t, \gamma_t) = -\frac{t}{2}. \end{aligned}$$

Then the correlation coefficient becomes

$$corr(W_t^j, B_t^j) = \frac{Cov(W_t^j, B_t^j)}{\sqrt{t}\sqrt{t}} = -\frac{1}{2}.$$

(*ii*) The independence of Y_t^1 and Y_t^2 follows from the fact that W_t^1 and B_t^1 are independent of W_t^2 and B_t^2. Since it is obvious that Y_t^1 and Y_t^2 are identically distributed, it suffices to find the law of Y_t^1 only.

Since (W_t^1, B_t^1) has a bivariate normal distribution, with $\rho = corr(W_t^1, B_t^1) = -\frac{1}{2}$, then their joint density is given by

$$f_{W_t^1, B_t^1}(x, y) = \frac{1}{2\pi t\sqrt{1-\rho^2}} e^{-\frac{1}{1-\rho^2}\left(\frac{x^2}{2t} + \frac{y^2}{2t} - \rho\frac{xy}{t}\right)}.$$

We compute next the distribution function of Y_t^1. For $s \geq 0$ we have

$$\begin{aligned}
F_{Y_t^1}(s) &= P\big((W_t^1)^2 + (B_t^1)^2 + W_t^1 B_t^1 \leq s\big) = \\
&= \iint_{\{x^2+y^2+xy\leq s\}} f_{W_t^1, B_t^1}(x, y)\, dxdy \\
&= \frac{1}{2\pi t\sqrt{1-\rho^2}} \iint_{\{x^2+y^2+xy\leq s\}} e^{-\frac{1}{1-\rho^2}\frac{1}{2t}\left(x^2+y^2-2\rho xy\right)}\, dxdy.
\end{aligned}$$

Changing variables

$$x = \frac{1}{\sqrt{2}}(u + v), \quad y = \frac{1}{\sqrt{2}}(u - v)$$

and substituting $\rho = -\frac{1}{2}$ yields

$$F_{Y_t^1}(s) = \frac{1}{\pi\sqrt{3t}} \iint_{\{\frac{3}{2}u^2+\frac{1}{2}v^2\leq s\}} e^{-\frac{2}{3t}(\frac{3}{2}u^2+\frac{1}{2}v^2)}\, dudv.$$

Changing variables again

$$u' = \sqrt{\frac{3}{2}}\, u, \qquad v' = \frac{1}{\sqrt{2}}\, v$$

the integral domain becomes a disk. Then employing polar coordinates we obtain

$$\begin{aligned}
F_{Y_t^1}(s) &= \frac{2}{3\pi t} \iint_{\{u'^2+v'^2\leq s\}} e^{-\frac{2}{3t}(u'^2+v'^2)}\, du'dv' \\
&= \frac{2}{3\pi t} \int_0^{2\pi} \int_0^{\sqrt{s}} e^{-\frac{2}{3t}\rho^2}\, \rho d\rho d\theta \\
&= \frac{4}{3t} \int_0^{\sqrt{s}} \rho e^{-\frac{2}{3t}\rho^2}\, d\rho.
\end{aligned}$$

Differentiating in s we obtain the density of Y_t^1

$$f_{Y_t^1}(s) = F'_{Y_t^1}(s) = \frac{2}{3t} e^{-\frac{2}{3t}s}, \qquad s \geq 0,$$

which is an exponential distribution.

■

Proposition 2.17.5 *The sum of squares of the sides of a Brownian triangle is gamma distributed, with $q_t \sim \Gamma(a, b)$, where $a = 2$ and $b = 6t$.*

Proof: Using relation (2.17.41) and Lemma 2.17.4 we compute the moment generating function of q_t

$$\begin{aligned}\mathbb{E}[e^{\lambda q_t}] &= \mathbb{E}[e^{4\lambda(Y_t^1+Y_t^2)}] = \mathbb{E}[e^{4\lambda Y_t^1}]\,\mathbb{E}[e^{4\lambda Y_t^2}] \\ &= \frac{1}{(1-6\lambda t)^2}.\end{aligned}$$

This matches the moment generating function of a gamma distribution, $\Gamma(a, b)$, with $a = 2$ and $b = 6t$. It follows that q_t is gamma distributed with the law

$$q_t \sim \frac{s}{\Gamma(2)(6t)^2} e^{-\frac{s}{6t}}, \quad s \geq 0.$$

■

Using the properties of gamma distributions we obtain:

Corollary 2.17.6 *The expectation of the sum of squares of the sides of a Brownian triangle is linear in the parameter t*

$$\mathbb{E}[q_t] = 12t.$$

2.18 Brownian Motion on Curves

We start by providing a few examples of Brownian motions on plane curves. These are two-dimensional processes constrained to belong to a curve, which is parametrized by a one-dimensional Brownian motion, W_t.

2.18.1 Brownian Motion on the unit circle

We consider a point on the unit circle whose angular argument is a one-dimensional Brownian motion, W_t. This process is regarded as a Brownian motion on the unit circle, $\{x; x_1^2 + x_2^2 = 1\}$, and can be expressed by

$$X_t = \big(X_1(t), X_2(t)\big) = \big(\cos(\theta_0 + W_t), \sin(\theta_0 + W_t)\big),$$

with $X_0 = (\cos\theta_0, \sin\theta_0)$. An application of Ito's lemma shows that

$$dX_1(t) = -\frac{1}{2}X_1(t)\,dt - X_2(t)\,dW_t \tag{2.18.42}$$

$$dX_2(t) = -\frac{1}{2}X_2(t)\,dt + X_1(t)\,dW_t. \tag{2.18.43}$$

Since the conditions of the existence and uniqueness theorem are satisfied, then X_t is the only solution of the system (2.18.42)-(2.18.43) that starts at $X_0 = (\cos\theta_0, \sin\theta_0)$. We notice that the drift vector and the diffusion matrix of X_t are given by

$$b(x) = -\frac{1}{2}\begin{pmatrix} x_1 \\ x_2 \end{pmatrix}, \qquad \sigma\sigma^T(x) = \begin{pmatrix} x_2^2 & -x_1x_2 \\ -x_1x_2 & x_1^2 \end{pmatrix}, \tag{2.18.44}$$

so the associated generator operator takes the form

$$\begin{aligned}\mathcal{A} &= \frac{1}{2}\sum_{i,j}(\sigma\sigma^T)_{ij}\partial_{x_i}\partial_{x_j} + \sum_k b_k(x)\partial_{x_k} \\ &= \frac{1}{2}[x_2^2\partial_{x_1}^2 + x_1^2\partial_{x_2}^2 - 2x_1x_2\partial_{x_1x_2}^2 - x_1\partial_{x_1} - x_2\partial_{x_2}] \\ &= \frac{1}{2}(x_1\partial_{x_2} - x_2\partial_{x_1})^2 = \frac{1}{2}\partial_\theta^2,\end{aligned}$$

which is the Laplacian on $\mathbb{S}^1$, the parameter θ being the central angle on the unit circle. We note that in the last identity we expressed the operator using the parametrization $x_1 = \cos\theta$, $x_2 = \sin\theta$. The heat kernel of $\mathcal{A}$ will be computed as the transition probability of diffusion X_t. Taking into account periodicity, the distribution function of X_t can be reduced to the distribution of W_t using an infinite series as follows

$$\begin{aligned}F_{X_t}(\theta|\theta_0) &= P(\theta_0 + W_t + 2n\pi < \theta, \forall n \in \mathbf{Z}) \\ &= \sum_{n\in\mathbf{Z}} P(W_t < \theta - \theta_0 + 2n\pi) = \sum_{n\in\mathbf{Z}} F_{W_t}(\theta - \theta_0 + 2n\pi).\end{aligned}$$

Taking the derivative we obtain the transition density of X_t

$$\begin{aligned}p_t(\theta_0, \theta) &= \frac{d}{d\theta}F_{X_t}(\theta|\theta_0) = \sum_{n\in\mathbf{Z}} \frac{1}{\sqrt{2\pi t}}e^{-\frac{(\theta-\theta_0+2n\pi)^2}{2t}} \\ &= \frac{1}{\sqrt{2\pi t}}e^{-\frac{(\theta-\theta_0)^2}{2t}}\left(1 + 2\sum_{n\geq 1} e^{-\frac{2n^2\pi^2}{t}}\cosh\frac{2n\pi(\theta-\theta_0)}{t}\right) \\ &= \frac{1}{\sqrt{2\pi t}}e^{-\frac{(\theta-\theta_0)^2}{2t}}\left(1 + 2\sum_{n\geq 1} e^{-\frac{2n^2\pi^2}{t}}\cos\frac{2n\pi(\theta-\theta_0)i}{t}\right) \\ &= \frac{1}{\sqrt{2\pi t}}e^{-\frac{(\theta-\theta_0)^2}{2t}}\theta_3\Big(\frac{\pi(\theta-\theta_0)i}{t} \mid \frac{2\pi i}{t}\Big),\end{aligned}$$

where the θ_3-function is defined by

$$\theta_3(z|i\tau) = 1 + 2\sum_{n\geq 1} e^{-\pi n^2\tau}\cos(2nz).$$

To conclude, the transition probability of a Brownian motion on $\mathbb{S}^1$ is given by

$$p_t(\theta_0, \theta) = \frac{1}{\sqrt{2\pi t}} e^{-\frac{(\theta-\theta_0)^2}{2t}} \theta_3\Big(\frac{\pi(\theta-\theta_0)i}{t} \mid \frac{2\pi i}{t}\Big). \tag{2.18.45}$$

Remark 2.18.1 Using (2.18.42)-(2.18.43) we can easily check that

$$\int_0^t \frac{1}{2}(X_1 dX_2 - X_2 dX_1) = \frac{1}{2} W_t.$$

The stochastic integral on the left side represents the area swept by the vectorial radius of X_s for $0 \le s \le t$. Also, the following similar relation holds

$$\int_0^t (X_1 dX_1 + X_2 dX_2) = -\frac{1}{2} t.$$

2.18.2 Brownian Motion on an ellipse

This process can be expressed by

$$X_t = \big(X_1(t), X_2(t)\big) = \big(a\cos(\theta_0 + W_t), b\sin(\theta_0 + W_t)\big)$$

and represents a point on the ellipse $\Big\{(x_1, x_2); \frac{x_1^2}{a^2} + \frac{x_2^2}{b^2} = 1\Big\}$. It satisfies the system

$$dX_1(t) = -\frac{1}{2} X_1(t)\, dt - \frac{a}{b} X_2(t)\, dW_t \tag{2.18.46}$$

$$dX_2(t) = -\frac{1}{2} X_2(t)\, dt + \frac{b}{a} X_1(t)\, dW_t. \tag{2.18.47}$$

Since the diffusion matrix is given by

$$\sigma\sigma^T = \begin{pmatrix} (\frac{a}{b})^2 x_2^2 & -x_1 x_2 \\ -x_1 x_2 & (\frac{b}{a})^2 x_1^2 \end{pmatrix},$$

then the associated infinitesimal generator becomes

$$\begin{aligned} \mathcal{A} &= \frac{1}{2}\Big[\Big(\frac{a}{b}\Big)^2 x_2^2 \partial_{x_1}^2 + \Big(\frac{b}{a}\Big)^2 x_1^2 \partial_{x_2}^2 - 2x_1 x_2 \partial_{x_1 x_2}^2 - x_1 \partial_{x_1} - x_2 \partial_{x_2}\Big] \\ &= \frac{1}{2}\Big(\frac{b}{a} x_1 \partial_{x_2} - \frac{a}{b} x_2 \partial_{x_1}\Big)^2 = \frac{1}{2}\partial_\theta^2, \end{aligned}$$

where $x_1 = a\cos\theta$ and $x_2 = b\sin\theta$.

2.18.3 Brownian Motion on a hyperbola

We consider the process

$$X_t = \big(X_1(t), X_2(t)\big) = \big(\cosh(\theta_0 + W_t), \sinh(\theta_0 + W_t)\big),$$

where W_t is a one-dimensional Brownian motion. This is the unique strong solution of the system

$$dX_1(t) = \frac{1}{2}X_1(t)\,dt + X_2(t)\,dW_t \tag{2.18.48}$$

$$dX_2(t) = \frac{1}{2}X_2(t)\,dt + X_1(t)\,dW_t. \tag{2.18.49}$$

with the initial condition $X_0 = (\cosh\theta_0, \sinh\theta_0)$. Using the diffusion matrix

$$\sigma\sigma^T = \begin{pmatrix} x_2^2 & x_1x_2 \\ x_1x_2 & x_1^2 \end{pmatrix},$$

the associated generator operator is

$$\begin{aligned}\mathcal{A} &= \frac{1}{2}[x_2^2\partial_{x_1}^2 + x_1^2\partial_{x_2}^2 + 2x_1x_2\partial_{x_1x_2}^2 + x_1\partial_{x_1} + x_2\partial_{x_2}] \\ &= \frac{1}{2}(x_1\partial_{x_2} + x_2\partial_{x_1})^2 = \frac{1}{2}\partial_t^2,\end{aligned}$$

where we used the hyperbola's parametrization $x_1 = \cosh t$, $x_2 = \sinh t$. Unlike the previous cases of the circle and ellipse that are periodic in the angular parameter θ, the heat kernel of $\mathcal{A}$ in this case is given by the Gaussian

$$p_\tau(t_0, t) = \frac{1}{\sqrt{2\pi\tau}}e^{-\frac{(t-t_0)^2}{2\tau}}, \quad \tau > 0.$$

This means that the Brownian motion on the hyperbola $\{x_1^2 - x_2^2 = 1\}$ is equivalent in law to a one-dimensional Brownian motion parameterized by the parameter t.

Remark 2.18.2 It is worth noting that relations (2.18.48)-(2.18.49) imply

$$\int_0^t X_1dX_1 - X_2dX_2 = \frac{1}{2}t$$

$$\int_0^t X_1dX_2 - X_2dX_1 = W_t.$$

2.18.4 Scaled Brownian Motion

Scaled Brownian motion on a line Let W_t be a one-dimensional Brownian motion and consider the scaled process $Y_t = rW_t$, with $r > 0$ constant. Since Y_t is a continuous martingale with the quadratic variation $\langle Y \rangle_t = \int_0^t r^2\, ds = r^2 t$, by Theorem 1.1.11, there is a DDS Brownian motion B_t such that $Y_t = B_{r^2 t}$. Therefore, a scaled Brownian motion is also a Brownian motion under a time change.

Let us consider how the scaling property applies to Brownian motion between circles.

Scaled Brownian motion between circles We consider the circle of radius r, which is concentric with the unit circle. Any Brownian motion X_t on the unit circle can be projected centrally into a process Y_t onto the circle of radius r by $Y_t = rX_t$. Both processes, X_t and Y_t satisfy the stochastic differential system (2.18.42)-(2.18.43). Since we have

$$vv^T(Y_t) = c\,\sigma\sigma^T(Y_t), \qquad u(Y_t) = c\,b(Y_t)$$

for $c = 1$, it follows from Theorem 1.1.14 that the scaled process Y_t is also a Brownian motion on the circle of radius r. Therefore, by scaling a Brownian motion on the unit circle we obtain again a Brownian motion on a circle.

2.18.5 Brownian Motion on Plane Curves

We consider the plane curve $\gamma(s) = \big(\gamma_1(s), \gamma_2(s)\big)$ with the arc length parameter s, so the curve is unit speed, $|\gamma'(s)| = 1$. The unit tangent vector field along the curve is given by $T(s) = \gamma'(s) = \big(\gamma_1'(s), \gamma_2'(s)\big)$, while the unit normal vector field is defined by $N(s) = \big(-\gamma_2'(s), \gamma_1'(s)\big)$. We note that $|T(s)| = |N(s)| = 1$ and $\langle T(s), N(s)\rangle = 0$. Also, the orthonormal frame $\{T, N\}$ is positively oriented, as the determinant of the components of the vectors is equal to 1.

A *Brownian motion on the curve* γ is the two-dimensional process

$$Y_t = \gamma(W_t) = \big(\gamma_1(W_t), \gamma_2(W_t)\big),$$

where W_t is a one-dimensional Brownian motion. Stated into words, the Brownian motion on the curve γ is obtained by replacing the arc length parameter s by the Brownian motion W_t.

The following processes will play a role in the study of the stochastic kinematics on the curve γ:
The *tangent process*, $T_t = T(W_t) = \big(\gamma_1'(W_t), \gamma_2'(W_t)\big)$;
The *normal process*, $N_t = N(W_t) = \big(-\gamma_2'(W_t), \gamma_1'(W_t)\big)$;

The *infinitesimal increment process*, $dY_t = (dY_1(t), dY_2(t))$, which is given by the Ito's formula

$$dY_1(t) = \frac{1}{2}\gamma_1''(W_t)dt + \gamma_1'(W_t)\,dW_t \tag{2.18.50}$$
$$dY_2(t) = \frac{1}{2}\gamma_2''(W_t)dt + \gamma_2'(W_t)\,dW_t. \tag{2.18.51}$$

The stochastic kinematics of the curve can be studied by investigating the orthogonal projections of dY_t onto the tangent and normal processes as

$$dY_t = \langle T_t, dY_t\rangle T_t + \langle N_t, dY_t\rangle N_t, \tag{2.18.52}$$

where $\langle\,,\,\rangle$ denotes the Euclidean scalar product. The formulas that provide $\langle T_t, dY_t\rangle$ and $\langle N_t, dY_t\rangle$ are generalizations of formulas provided by Remarks (2.18.1) and (2.18.2) for the circle and the hyperbola, respectively.

Proposition 2.18.3 *The tangent and normal components of the increment dY_t are given by*

$$\langle T_t, dY_t\rangle = dW_t$$
$$\langle N_t, dY_t\rangle = \frac{1}{2}\kappa(W_t)dt,$$

where $\kappa(s)$ is the curvature of $\gamma(s)$.

Proof: To prove the first relation, we multiply relations (2.18.50) and (2.18.51) by $\gamma_1'(W_t)$ and $\gamma_2'(W_t)$, respectively, and then add

$$\begin{aligned}\gamma_1'(W_t)dY_1(t) + \gamma_2'(W_t)dY_2(t) &= \frac{1}{2}(\gamma_1'(W_t)\gamma_1''(W_t) + \gamma_2'(W_t)\gamma_2''(W_t))dt\\ &\quad + (\gamma_1'(W_t)^2 + \gamma_2'(W_t)^2)dW_t\\ &= dW_t,\end{aligned}$$

where we used that the curve is unit speed, ${\gamma_1'}^2 + {\gamma_2'}^2 = 1$, as well as its consequence $\gamma_1'\gamma_1'' + \gamma_2'\gamma_2'' = 0$. Using the definition of T_t, we note that the left side of the previous expression is

$$\gamma_1'(W_t)dY_1(t) + \gamma_2'(W_t)dY_2(t) = \langle T_t, dY_t\rangle,$$

which ends the proof.

To show the second relation, we recall the Frenet equations for the unit speed plane curve $\gamma(s)$

$$T'(s) = \kappa(s)N(s)$$
$$N'(s) = -\kappa(s)T(s),$$

then choose one of the equations, let's say the first, and express the curvature as a scalar product

$$\begin{aligned}\kappa(s) &= \langle \kappa(s)N(s), N(s)\rangle = \langle T'(s), N(s)\rangle \\ &= \gamma_1'(s)\gamma_2''(s) - \gamma_1''(s)\gamma_2'(s).\end{aligned}$$

Then replacing the arc length s by the Brownian motion W_t, we obtain

$$\kappa(W_t) = \gamma_1'(W_t)\gamma_2''(W_t) - \gamma_1''(W_t)\gamma_2'(W_t).$$

Now, multiplying (2.18.50) by $\gamma_2'(W_t)$ and (2.18.51) by $\gamma_1'(W_t)$, then subtracting and using the previous curvature relation, we get

$$\gamma_1'(W_t)dY_2(t) - \gamma_2'(t)dY_1(t) = \frac{1}{2}\kappa(W_t)dt.$$

Noting that the left side is equal to $\langle N_t, dY_t\rangle$, we arrive to the desired expression. ∎

From the previous proposition, the cumulative projections are given by the stochastic integrals

$$\int_0^t \langle T_s, dY_s\rangle = W_t \tag{2.18.53}$$

$$\int_0^t \langle N_s, dY_s\rangle = \frac{1}{2}\int_0^t \kappa(W_s)ds. \tag{2.18.54}$$

Therefore, relation (2.18.53) has an intrinsic character, stating that the cumulative contribution of the increment dY_s in the tangent direction is a Brownian motion, while relation (2.18.54) has an extrinsic character, showing that the contribution of dY_s in the normal direction depends on the total stochastic curvature. There are two distinguished cases:

1. If γ is a line, then its curvature is zero, so $\langle N_t, dY_t\rangle = 0$, and hence there is no normal contribution of the increment process in this case.

2. If γ is a circle, then its curvature is constant, $\kappa = c$, so $\langle N_t, dY_t\rangle = c/2$, which corresponds to an equal normal contribution of the increment process.

Corollary 2.18.4 *We have the kinematic decomposition*

$$dY_t = T_t\, dW_t + \frac{1}{2}\kappa(W_t)N_t dt.$$

Proof: Using Proposition 2.18.3 and substituting in relation (2.18.52) we obtain the desired relation. ∎

Remark 2.18.5 We can also arrive at the aforementioned decomposition by writing equations (2.18.50)-(2.18.51) in the following vectorial form

$$dY_t = \gamma'(W_t)dW_t + \frac{1}{2}\gamma''(W_t)dt,$$

and then use the Frenet equation $\gamma'' = T' = \kappa N$ and the fact that $T = \gamma'$. Therefore, the kinematic decomposition is a consequence of Ito's formula and Frenet equations. It is worth noting that the aforementioned decomposition holds only for unit-speed curves.

Now we consider Y_t as an Ito process, which can be written as

$$dY(t) = u(t,\omega)dt + v(t,\omega)dW_t(\omega),$$

whose drift, $u(t,\omega) = \frac{1}{2}\kappa(W_t)N_t$, is normal to the curve, having the magnitude equal to half the curvature. Its dispersion, $v(t,\omega) = T_t$, is given by the unit tangent vector field. Its associated diffusion matrix is

$$vv^t(t,\omega) = T_t T_t^T = \begin{pmatrix} \gamma_1'(W_t(\omega))^2 & \gamma_1'(W_t(\omega))\gamma_2'(W_t(\omega)) \\ \gamma_1'(W_t(\omega))\gamma_2'(W_t(\omega)) & \gamma_2'(W_t(\omega))^2 \end{pmatrix}.$$

The next result states that the diffusion matrix is singular in the normal direction to the curve.

Proposition 2.18.6 *The diffusion matrix, vv^T, has the following properties:*
1. $Trace(vv^T) = 1$;
2. $\det(vv^T) = 0$;
3. Its eigenvalues are $\{0, 1\}$;
4. $vv^T T_t = T_t$ and $vv^T N_t = 0$.

Proof: 1. It is a consequence of the fact that γ is a unit speed curve.
2. It follows from the definition of the determinant.
3. The characteristic equation $\lambda^2 - Trace(vv^T)\lambda + \det(vv^T) = 0$ simplifies to $\lambda(\lambda - 1) = 0$.
4. It follows from a straightforward computation. Part 4 states that the degenerate direction of the diffusion matrix is normal to the curve, while the tangent process, T_t, is an invariant vector. This agrees with the fact that the process Y_t diffuses along the curve and not across the curve. ■

2.19 Extrinsic Theory of Curves

Let $\mathcal{S}$ be a submanifold of $\mathbb{R}^m$. Can we retrieve any geometric properties of $\mathcal{S}$ from the behavior of a diffusion? We should investigate two cases:

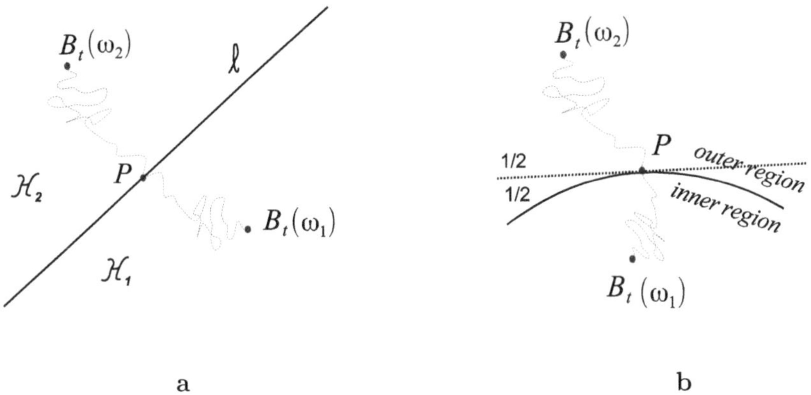

Figure 2.7: (a) *A Brownian motion starting from a line has equal chances to land in any of the half-planes determined by the line.* (b) *A Brownian motion starting from a curve has more chances to diffuse outside rather than inside the curve.*

1. The case when the diffusion starts from a point on the submanifold $\mathcal{S}$ and may diffuse anywhere in $\mathbb{R}^m$. This type of diffusion is called an *extrinsic diffusion* and its study is the subject of the extrinsic theory of diffusions. One question of this theory is to retrieve the curvature of the submanifold from the probability that the diffusion heads outside (or inside) the submanifold.

2. The case when the diffusion starts from a point on a submanifold $\mathcal{S}$ and is constrained to live on $\mathcal{S}$ at all times. This type of diffusion is called an *intrinsic diffusion.* Chapter 6 deals with specific intrinsic diffusions, which can be considered as Brownian motions on the submanifold. Section 2.18 deals also with the case of one-dimensional submanifolds (curves). One type of problem is to use intrinsic diffusions to asses whether the manifold is positively or negatively curved.

To start from a simple example, we consider a line ℓ that divides a plane in two half-planes, $\mathcal{H}_1$ and $\mathcal{H}_2$, see Fig. 2.7 a. By symmetry reasons, a plane Brownian motion, B_t, starting at a point P situated on the line ℓ has the property

$$P(B_t \in \mathcal{H}_1) = P(B_t \in \mathcal{H}_2) = \frac{1}{2}, \qquad \forall t > 0.$$

In the case when the line is replaced by a curve, we distinguish two regions in the plane: the outer region, which is concave and the inner region, which is convex, see Fig. 2.7 b. Since there is more room for the diffusion to move into the outer direction rather than into the inner one, the probability that

B_t is situated in the outer region is larger than $1/2$, at least for small values of time.

The topic of this section is to characterize the probability of diffusion about the point P of a planar Brownian motion in terms of the curvature of the curve at P. We shall obtain an asymptotics of this probability as t approaches zero.

The following result will be useful shortly. It follows from an asymptotic correspondence computed using the Laplace method and by keeping only the first non-zero term in the expansion.

Lemma 2.19.1 *Let $\alpha, \beta \in \mathbb{R}$ with $\beta > -1$. Then the following asymptotic correspondence holds*

$$\int_0^T e^{-\lambda t^\alpha} t^\beta \, dt \sim \frac{\Gamma(\frac{\beta+1}{\alpha})}{\alpha} \lambda^{-\frac{\beta+1}{\alpha}}, \qquad \lambda \to \infty.$$

The following probability density plays a role in our problem solution. We recall that $I_0(x)$ and $I_1(x)$ denote the modified Bessel functions of the first type.

Lemma 2.19.2 *(i) The function*

$$\phi(x) = e^{-x}\big(I_0(x) - I_1(x)\big) \tag{2.19.55}$$

is a probability density on $[0, \infty)$.

(ii) The following asymptotic correspondence holds

$$\phi(x) \sim \frac{1}{2\sqrt{2\pi}} x^{-3/2}, \qquad x \to \infty. \tag{2.19.56}$$

Proof: (*i*) We shall use the following integral representations of the modified Bessel functions[1]

$$I_0(x) = \frac{1}{\pi} \int_0^\pi e^{x\cos\theta} \, d\theta, \qquad I_1(x) = \frac{1}{\pi} \int_0^\pi e^{x\cos\theta} \cos\theta \, d\theta.$$

The non-negativity of $\phi(x)$ follows from the fact that

$$I_1(x) = \frac{1}{\pi} \int_0^\pi e^{x\cos\theta} \cos\theta \, d\theta \leq \frac{1}{\pi} \int_0^\pi e^{x\cos\theta} \, d\theta = I_0(x).$$

[1] In general, we have $I_n(x) = \frac{1}{\pi} \int_0^\pi e^{x\cos\theta} \cos(n\theta) \, d\theta$ for any integer $n \geq 0$, see Abramowitz and Stegun [1], p. 376.

Using Fubini's theorem, we have

$$\begin{aligned}\int_0^\infty \phi(x)\,dx &= \frac{1}{\pi}\int_0^\infty e^{-x}\Big[\int_0^\pi e^{x\cos\theta} - e^{x\cos\theta}\cos\theta\Big]d\theta dx \\ &= \frac{1}{\pi}\int_0^\pi (1-\cos\theta)\int_0^\infty e^{-x(1-\cos\theta)}\,dx\,d\theta \\ &= \frac{1}{\pi}\int_0^\pi d\theta = 1.\end{aligned}$$

Therefore, $\phi(x)$ is a probability density on $[0,\infty)$.

(ii) Using the integral representations of I_0 and I_1 we have

$$\begin{aligned}\phi(x) &= e^{-x}\big(I_0(x) - I_1(x)\big) \\ &= \frac{1}{\pi}\int_0^\pi e^{-x(1-\cos\theta)}(1-\cos\theta)\,d\theta.\end{aligned}$$

To find the limit as $x\to\infty$ we employ an idea that is reminiscent of the Laplace method of approximation. Since the function $\theta\to 1-\cos\theta$ has a global maximum at $\theta=0$, only the behavior of the function about $\theta=0$ matters for the integral as x becomes large. Therefore, using the Taylor series $1-\cos\theta = \frac{\theta^2}{2}+o(\theta^5)$, we have the following correspondence of integrals for x large

$$\frac{1}{\pi}\int_0^\pi e^{-x(1-\cos\theta)}(1-\cos\theta)\,d\theta \sim \frac{1}{\pi}\int_0^\pi e^{-\frac{x}{2}\theta^2}\frac{\theta^2}{2}\,d\theta, \quad x\to\infty.$$

Using Lemma 2.19.1 for $\alpha=\beta=2$, $\lambda=x/2$, and using $\Gamma\big(\frac{3}{2}\big)=\frac{\sqrt{\pi}}{2}$, the last integral can be further approximated as

$$\frac{1}{2\pi}\int_0^\pi e^{-\frac{x}{2}\theta^2}\theta^2\,d\theta \sim \frac{1}{2\pi}\frac{\Gamma\big(\frac{3}{2}\big)}{2}\Big(\frac{x}{2}\Big)^{-3/2} = \frac{1}{2\sqrt{2\pi}}x^{-3/2}, \qquad x\to\infty,$$

which provides formula (2.19.56).

∎

The next result provides the exit probability from a disk of a Brownian motion starting from its boundary using the distribution function of $\phi(x)$.

Lemma 2.19.3 *Let $\Omega=\{x\in\mathbb{R}^2; \|x\|\le r\}$ be a disk of radius r and $x_0\in\partial\Omega=\{x\in\mathbb{R}^2; \|x\|=r\}$ be a point on the circumference of Ω. Consider a planar Brownian motion B_t starting from the point x_0. The exit probability from the disk is given by*

$$P(B_t\notin\Omega) = 1-\frac{1}{2}\int_0^{\frac{r^2}{t}}\phi(v)\,dv,$$

with ϕ given by (2.19.55).

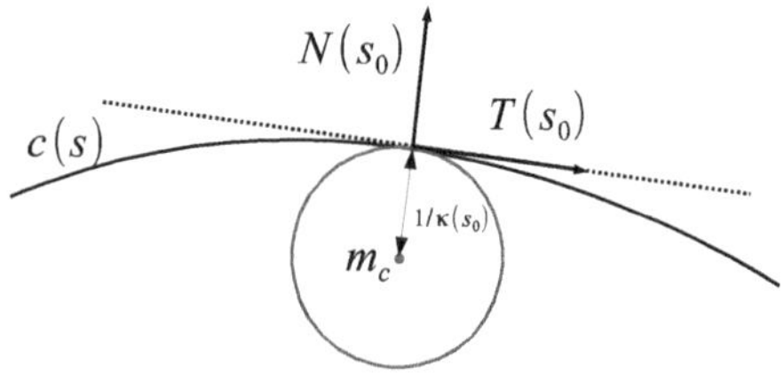

Figure 2.8: *The osculating circle of c at* s_0.

While we shall postpone the proof of the lemma until page 292, we make a few suggestive comments.

Remark 2.19.4 (i) If the disk radius $r \to \infty$, the disk becomes a half-plane and the exit probability will become

$$P(B_t \notin \Omega) = 1 - \frac{1}{2}\int_0^\infty \phi(v)\,dv = \frac{1}{2},$$

which retrieves a familiar result. The same probability is obtained if r is kept fixed and we take $t \to 0$.
(ii) If the disk radius $r \to 0$, i.e. the disk shrinks to a point, the exit probability becomes

$$P(B_t \notin \Omega) = 1 - \frac{1}{2}\int_0^0 \phi(v)\,dv = 1,$$

which checks that the diffusion is almost surely outside the point x_0.

Osculating circle The osculating circle of a curve at a given point on a curve has the same tangent line as well as the same curvature as the curve. The osculating circle is the circle that approximates best the curve at a point. This is similar to the concept of the tangent line, which is the best approximating line to a curve at a point, see Fig. 2.8.

We developed Lemma 2.19.3 regarding diffusions for circles and then we shall transfer it to a curve via the osculating circle approximation. We need to recall first a few differential geometry notions, see Millman and Parker [110], page 39.

Let $c(s) = (c_1(s), c_2(s))$ be a unit speed plane curve, i.e. $\|\dot{c}(s)\| = 1$, which means that the parameter s is the arc length along the curve. The tangent vector field is defined by $T(s) = \dot{c}(s)$ and the curvature is given by function $\kappa(s) = \|\dot{T}(s)\|$, i.e. it is the magnitude of the acceleration along the curve. Then the principal normal vector field is defined by $N(s) = \dot{T}(s)/\kappa(s)$. It is worth noting that $\{T(s), N(s)\}$ form an orthonormal frame along the curve $c(s)$.

Let's consider a fixed point $c(s_0)$ on the curve, such that $\kappa(s_0) \neq 0$. The number $\rho(s_0) = 1/\kappa(s_0)$ is called the *radius of curvature* of c at s_0 and the point $m_c = c(s_0) + \frac{1}{\kappa(s_0)} N(s_0)$ is called the *center of curvature* of c at s_0. The circle of radius $\rho(s_0)$ centered at m_c is called the *osculating circle* of c at s_0.

Local inner and outer regions A plane curve $c(s)$ divides the plane locally, about each point $c(s_0)$, into two regions, the *outer* and the *inner regions.* The outer region is defined by the neighborhood towards which the principal normal $N(s_0)$ points into; this description works as long as the curvature $c(s)$ is different than zero, so $N(s)$ can be defined. The complementary region is the inner region.

We are interested in describing the probability of a Brownian motion, which starts from a point $c(s_0)$ on the curve, to belong to the inner or outer region; we shall call these complementary probabilities the *inner* and the *outer probabilities* near the point $c(s_0)$ and we shall denote them by p_{out} and $p_{in}(= 1 - p_{in})$.

If the curve is a straight line, then both the outer and the inner probabilities are equal, $p_{in} = p_{out} = \frac{1}{2}$. This is due to the symmetry of the Brownian motion about the line. For a general curve these probabilities are not equal. The outer probability will always be larger than the inner probability, $p_{out} > p_{in}$, since there is more room for B_t to diffuse outside rather than inside.

Proposition 2.19.5 *Let $c(s)$ be a plane curve with $\kappa(s_0) \neq 0$. Consider a planar Brownian motion starting at $x_0 = c(s_0)$. The probability that B_t belongs to the outer region satisfies the following asymptotic correspondence*

$$p_{out} \sim \frac{1}{2} + \frac{\kappa(s_0)}{4\sqrt{2\pi}}\sqrt{t}, \qquad t \to 0. \tag{2.19.57}$$

Proof: The main idea of the proof is to approximate the curve $c(s)$ at s_0 by its osculating circle and to apply Lemma 2.19.3, which will provide asymptotics for the inner and outer probabilities for small values of t.

Consider the osculating circle of c at s_0, with radius $r = 1/\kappa(s_0)$. By Lemma 2.19.3 the probability that B_t is outside of the osculating circle is

$$p_{out}^{circle} = 1 - \frac{1}{2}\int_0^{\frac{1}{t\kappa^2(s_0)}} \phi(v)\, dv. \tag{2.19.58}$$

Due to the approximating properties of the osculating circle we have the correspondence

$$p_{out} \sim p_{out}^{circle}, \qquad t \sim 0.$$

It suffices to work a linear approximation near $t = 0$ for the function given by the right side of (2.19.58). Let

$$G(t) = 1 - \frac{1}{2}\int_0^{\frac{1}{t\kappa^2(s_0)}} \phi(v)\, dv.$$

Differentiating and using Lemma 2.19.2, part *(ii)* we have

$$G'(t) = \frac{1}{2t^2\kappa^2(s_0)}\phi\Big(\frac{1}{t\kappa(s_0)^2}\Big) \sim \frac{\kappa(s_0)}{4\sqrt{2\pi}}t^{-1/2}, \qquad t \to 0.$$

Since $G(0) = \frac{1}{2}$, using the approximation formula

$$G(t) \sim G(0) + G'(t)t, \qquad t \sim 0,$$

we obtain the desired relation

$$G(t) \sim \frac{1}{2} + \frac{\kappa(s_0)}{4\sqrt{2\pi}}\sqrt{t}, \qquad t \to 0.$$

■

Remark 2.19.6 Since for any plane curve we have $\kappa(s) \geq 0$, formula (2.19.57) implies that $p_{out} > \frac{1}{2}$, as $t \sim 0$, i.e. the Brownian motion has a tendency of diffusing outwards. This tendency is enhanced by the curvature term and depends on the Brownian motion dispersion, $\sqrt{t}$.

2.20 Dynamic Brownian Motion on a Curve

Let $c(t) = \big(x(t), y(t)\big)$ be a plane curve and consider the two-dimensional process $c_t = \big(x(t) + W_1(t), y(t) + W_2(t)\big)$ obtained by supperposing the planar Brownian motion $\big(W_1(t), W_2(t)\big)$ on the curve $c(t)$. The expectation of c_t is $\mathbb{E}[c_t] = c(t)$, while the covariance matrix is $Cov(c_t) = t\mathbb{I}_2$. The process c_t can be regarded as a Brownian motion along the curve $c(t)$, whose drift is $c(t)$. We shall call this a *dynamical Brownian motion* along the curve $c(t)$.

The questions of concern in this section are: *What is the probability that the process c_t belongs to one side of the curve $c(t)$ at a given time t? How is the curvature of $c(t)$ influencing this probability?*

We shall start with the simplest examples.

The case of a line We shall show that in this case the probability that the process c_t belongs to one of the half-planes determined by the line is 1/2. To this end, we assume the curve is a line given by the equation $c(t) =$

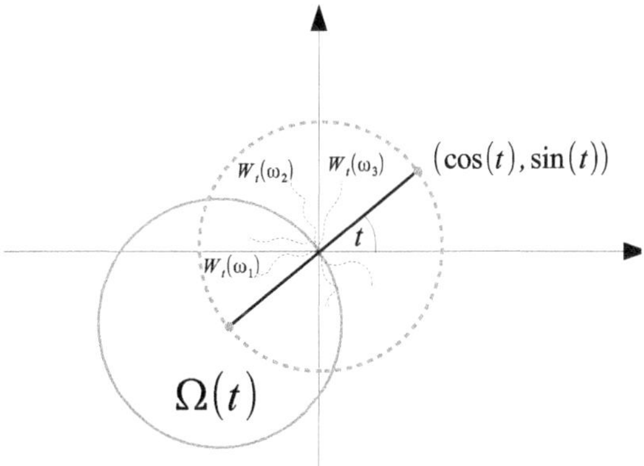

Figure 2.9: *The planar Brownian motion $B(t)$ is radiary invariant.*

$(a_1 t + b_1, a_2 t + b_2)$, with a_i and b_i constants. Using the invariance of a planar Brownian motion to translations, it suffices to assume $b_i = 0$. Then using the invariance of planar Brownian motions to rotations, we may perform a rotation such that the line becomes $c(t) = (mt, 0)$, i.e. its graph is the x-axis. Since the process can be written now as $\tilde{c}_t = \big(mt + B_1(t), B_2(t)\big)$, where $(B_1(t), B_2(t))$ is a planar Brownian motion, by symmetry reasons the probability that $\tilde{c}_t$ is above the x-axis at time t is $P(B_2(t) > 0) = \frac{1}{2}$.

The case of a circle Let $c(t) = (r\cos t, r\sin t)$ be a circle of radius r centered at the origin. The dynamic Brownian motion on this circle is given by

$$c_t = (x_t, y_t) = (r\cos t + W_1(t), r\sin t + W_2(t)), \qquad t \geq 0. \tag{2.20.59}$$

The probability that c_t is outside the circle $c(t)$ at time t is evaluated as

$$\begin{aligned} P(x_t^2 + y_t^2 > r^2) &= P\big((W_1(t) + r\cos t)^2 + W_2(t) + r\sin t)^2 > r^2\big) \\ &= P(W(t) \notin \Omega(t)), \end{aligned}$$

where $\Omega(t) = \{(x + r\cos t)^2 + (y + r\sin t)^2 \leq r^2\}$ is the disk of radius r centered at $(-r\cos t, -r\sin t)$. We note that the circle $\partial\Omega(t)$ passes through the origin for any $t \geq 0$. The planar Brownian motion $W_t = (W_1(t), W_2(t))$ starts at the origin, $W(0) = (0, 0)$, so it belongs to the circle $\partial\Omega(t)$, see Fig. 2.9. Using the radial symmetry of $W(t)$, we obtain that the probability that $W(t)$ belongs to the interior of the disk $\Omega(t)$ is independent of the argument t and depends just on the disk radius, r. By Lemma 2.19.3 we have

$$P(B(t) \in \Omega(t)) = \frac{1}{2}\int_0^{\frac{r^2}{t}} \phi(v)\,dv.$$

Therefore, the dynamical Brownian motion (2.20.59) is outside the circle $c(t)$ with probability

$$P = 1 - \frac{1}{2}\int_0^{\frac{r^2}{t}} \phi(v)dv.$$

It is worth stating that for t small or r large, this probability tends to $1/2$.

2.21 Summary

This chapter delves into the analysis of Brownian motions on various geometric structures, including lines, circles, curves, and planes, and explores their corresponding probability laws. Transformations of Brownian motions are also investigated, such as rotations, reflections, orthogonal and central projections, and inner products.

Interesting findings include the fact that the product of two independent Brownian motions has the same law as the Lévy area, and that the projection of a two-dimensional Brownian motion onto a line also follows a Brownian motion. Furthermore, the correlation of two independent Brownian motions moving along two lines making an angle θ is $\cos\theta$.

The chapter also characterizes all transformations of the plane that leave the Brownian motions of the plane invariant, showing that they are either rotations or flips or their compositions. The probability law and moments of the area of a Brownian triangle, characterized by planar Brownian motions, are also examined.

Finally, the curvature of a plane curve is shown to describe the behavior of a Brownian motion when it moves around the curve for small time t.

2.22 Exercises

Exercise 2.22.1 (*a*) Let U and V be two independent real-valued random variables and consider their quotient $Y = U/V$. Show that Y has the following probability density

$$p_Y(y) = \int_{-\infty}^{\infty} p_U(yz)p_V(z)|z|\,dz,$$

where p_U and p_V are the probability densities of U and V, respectively.

(*b*) Let U and V be two independent standard normal variables. Then the quotient $Y = U/V$ is Cauchy distributed

$$Y \sim \frac{1}{\pi(1+y^2)}, \qquad y \in \mathbb{R}.$$

Exercise 2.22.2 (see [25]) Let $\epsilon \sim \mathcal{N}(0, k^2)$ be a normally distributed random variable with mean zero and variance k^2, where $k > 0$. We define the following two stochastic functions

$$\mathrm{Ssin}(t,k) = e^{k^2/2}\sin(t+\epsilon), \qquad \mathrm{Scos}(t,k) = e^{k^2/2}\cos(t+\epsilon).$$

(*a*) Show that

$$\begin{aligned}\mathrm{Ssin}(t+u,k) &= \mathrm{Scos}(u,k)\sin t + \mathrm{Ssin}(u,k)\cos t \\ &\quad - \mathrm{Scos}(t,k)\sin u + \mathrm{Ssin}(t,k)\cos u \\ \mathrm{Scos}(t+u,k) &= \cos t\,\mathrm{Scos}(u,k) - \sin t\,\mathrm{Ssin}(u,k) \\ &= \cos u\,\mathrm{Scos}(t,k) - \sin u\,\mathrm{Ssin}(t,k).\end{aligned}$$

(*b*) Verify relations

$$\mathbb{E}[\mathrm{Ssin}(t,k)] = \sin(t), \qquad \mathbb{E}[\mathrm{Scos}(t,k)] = \cos(t).$$

(*c*) Prove relation

$$Var(\mathrm{Ssin}(t,k)) = \frac{1}{2}(e^{k^2} - 1).$$

Find a similar relation for $Var(\mathrm{Scos}(t,k))$.

Exercise 2.22.3 Prove Proposition 2.12.3

Exercise 2.22.4 Let G_t the center of mass of a Brownian triangle $A_tB_tC_t$.

(*a*) Show that G_t follows a two-dimensional Brownian motion.

(*b*) Is this also true for the center of the triangle circumcircle?

Exercise 2.22.5 Let B_t be a Brownian motion in $\mathbb{R}^3$ and M_t the orthogonal projection of B_t onto a plane $\mathcal{P}$. Show that M_t is a two-dimensional Brownian motion in $\mathcal{P}$.

Exercise 2.22.6 Let $\mathcal{P}_1$ and $\mathcal{P}_2$ be two non-parallel planes in $\mathbb{R}^3$, which make a dihedral angle equal to θ. The orthogonal projections of a 3-dimensional Brownian motions on $\mathcal{P}_1$ and $\mathcal{P}_2$ are denoted by $A_1(t)$ and $A_2(t)$, respectively. Find the matrix $Cov(A_1(t), A_2(t))$ in terms of θ.

Exercise 2.22.7 Let ℓ_1 and ℓ_2 be two lines in the plane given by the equations $y = a_1 x$ and $y = a_2 x$. We denote by $A_1(t)$ and $A_2(t)$ the projections of a two-dimensional Brownian motion onto the lines ℓ_1 and ℓ_2, respectively. Find the density of the process $X_t = \mathrm{dist}(A_1(t), A_2(t))$.

Exercise 2.22.8 Let $A'(t)$, $B'(t)$ and $C'(t)$ be the projections of a two-dimensional Brownian motion onto the sides of an acute triangle ABC. Show that

$$Corr(A'(t), B'(t))Corr(B'(t), C'(t))Corr(C'(t), A'(t)) \leq \frac{1}{8}.$$

Exercise 2.22.9 Let B_t and W_t be two Brownian motions in $\mathbb{R}^2$ and α_t be the angle made by the vectorial radii OB_t and OW_t, where O is the origin of the plane. Find the density of the process α.

Exercise 2.22.10 Prove Lemma 2.19.1 (Hint: Consider it as a Laplace integral.)

Chapter 3

Hypoelliptic Operators

Hypoelliptic operators obtained by summing squares of vector fields and satisfying the full rank condition exhibit a smooth transition density for the associated diffusion process. This feature makes them particularly useful for the stochastic approach to differential operators. The chapter focuses on two types of hypoelliptic operators, elliptic and sub-elliptic; their main properties will be elaborated in subsequent sections.

3.1 Elliptic Operators

Let $U \subseteq \mathbb{R}^n$ be an open set and consider the second order differential operator

$$L = \sum_{i,j=1}^{n} a_{ij}(x)\partial_{x_i}\partial_{x_j} + \sum_{k=1}^{n} b_k(x)\partial_{x_k} \tag{3.1.1}$$

with $a_{ij}(x) = a_{ji}(x)$ and $a_{ij}(x)$ and $b_k(x)$ continuous functions on the closure of U. The *total symbol* of L is the quadratic polynomial

$$p(x,\xi) = \sum_{i,j=1}^{n} a_{ij}(x)\xi_i\xi_j + \sum_{k=1}^{n} b_k(x)\xi_k. \tag{3.1.2}$$

The *principal symbol* is the highest-degree component of p

$$\sigma_p(x,\xi) = \sum_{i,j=1}^{n} a_{ij}(x)\xi_i\xi_j.$$

The differential operator L is called *elliptic* if there is a constant $k > 0$ such that

$$\sigma_p(x,\xi) \geq k\|\xi\|^2, \qquad \forall \xi \in \mathbb{R}^n,\ x \in \overline{U}. \tag{3.1.3}$$

This condition implies that the coefficients matrix $A(x) = (a_{ij}(x))$ is positive definite for any $x \in \overline{U}$, namely $\sum_{i,j=1}^n a_{ij}(x)\xi_i\xi_j > 0$, for any $\xi \in \mathbb{R}^n\backslash\{0\}$ and $x \in \overline{U}$. By a well-known result of linear algebra (see Theorem 7.2.1 in Horn and Johnson [78]), the matrix $A(x) = (a_{ij}(x))$ is invertible with the inverse $A^{-1}(x) = (a^{ij}(x))$ positive definite. Since both $A(x)$ and $A^{-1}(x)$ are symmetric, it follows that we can consider the Riemannian metric tensor $g(x) = \sum_{i,j} a^{ij}(x)dx_i \otimes dx_j$ and construct the Riemannian space (U, g) with the associated elliptic operator L. Then the study of the operator L can be approached from two distinct perspectives:

(1) *The geometric viewpoint:* The first approach to studying the operator L is through the geometric properties of the Riemannian manifold (U, g). These properties include the behavior of geodesics, Jacobi vector fields, Riemannian distance, classical action, Lagrangian and Hamiltonian formalisms. These concepts are related to the heat kernel of the elliptic operator L through the van Vleck formula, at least for case of small time t. This approach has been explored in previous works such as Calin et al. [36], [26], and [29].

(2) *The stochastic viewpoint:* We can view the Riemannian metric g as a diffusion metric and associate a diffusion process X_t in U with it. This process is an elliptic diffusion, meaning it can diffuse in all directions, and its transition probability gives the heat kernel of the differential operator L. This perspective will be predominantly pursued in this book.

Remark 3.1.1 The ellipticity condition (3.1.3) can be formulated at each point $x \in \overline{D}$ as

$$\langle A(x)\xi, \xi\rangle \geq k\|\xi\|^2, \quad \forall \xi \in \mathbb{R}^n, \tag{3.1.4}$$

where $\langle\, , \rangle$ denotes the Euclidean scalar product on $\mathbb{R}^n$ and $\overline{D}$ the closure of D in $\mathbb{R}^n$. From the Rayleigh equation, we know that for a symmetric $n \times n$ matrix $A(x)$ its minimum eigenvalue is given by

$$\lambda_{\min}(x) = \min_{\xi \in \mathbb{R}^n\backslash\{0\}} \frac{\langle A(x)\xi, \xi\rangle}{\|\xi\|^2}.$$

Therefore, equation (3.1.4) is equivalent to

$$\lambda_{\min}(x) \geq k > 0.$$

This condition is valid whenever all eigenvalues of $A(x)$ are strictly positive, i.e. the matrix $A(x)$ is positive definite.

An example of an elliptic operator, which will play a predominant role in this book, is obtained in the case of a sum of squares of vector fields, as shown in the next section.

3.2 Sum of Squares Laplacian

We consider n linearly independent vector fields on $\mathbb{R}^n$ given by

$$Y_j = \sum_{p=1}^{n} a_{jp}(x)\partial_{x_p}, \qquad j = 1, \dots, n, \tag{3.2.5}$$

with $a_{jp}(x)$ continuous differentiable functions satisfying $\det a_{ij}(x) \neq 0$, for all $x \in \mathbb{R}^n$. We construct the sum of squares Laplacian

$$\Delta_Y = \frac{1}{2}\sum_{j=1}^{n} Y_j^2, \tag{3.2.6}$$

which we shall show it is elliptic. First, an explicit computation provides

$$\Delta_Y = \frac{1}{2}\sum_{j=1}^{n}\sum_{p,k=1}^{n} a_{jp}(x)a_{jk}(x)\partial_{x_p}\partial_{x_k} + \frac{1}{2}\sum_{k=1}^{n}\sum_{p,j=1}^{n} a_{jp}(x)\partial_{x_p}(a_{jk}(x))\partial_{x_k}. \tag{3.2.7}$$

The principal symbol is given by

$$\sigma_P(x,\xi) = \frac{1}{2}\sum_{p,k=1}^{n}\sum_{j=1}^{n} a_{jp}(x)a_{jk}(x)\xi_p\xi_k = \langle A(x)\,\xi, \xi\rangle,$$

where $A(x) = \frac{1}{2}a^T(x)a(x)$ is a symmetric and non-degenerated $n \times n$ matrix. The ellipticity of Δ_Y at a point x is equivalent with the strictly positivity of all eigenvalues of $A(x)$. The details of this statement follow from a proof by contradiction that assumes that $A(x)$ has some non-positive eigenvalues. To this end, let $\lambda \leq 0$ be one of these eigenvalues, which corresponds to a unitary eigenvector η. Then

$$\langle A(x)\eta, \eta\rangle = \langle \lambda\eta, \eta\rangle = \lambda\|\eta\|^2 \leq 0.$$

On the other side

$$\langle A(x)\eta, \eta\rangle = \langle \frac{1}{2}a(x)^T a(x)\eta, \eta\rangle = \frac{1}{2}\langle a(x)\eta, a(x)\eta\rangle = \frac{1}{2}\|a(x)\eta\|^2.$$

From the last two expressions we obtain $\|a(x)\eta\|^2 \leq 0$, which implies $a(x)\eta = 0$. Since $a(x)$ is non-degenerated, it follows that $\eta = 0$, which contradicts the fact that η is a unitary vector. Hence, the sum of squares operator Δ_Y is elliptic.

The associated diffusion We shall deal next with the diffusion associated to the operator Δ_Y. Let's consider an n-dimensional diffusion $X_t = (X_t^1, \dots, X_t^n)$ on $\mathbb{R}^n$ given by

$$dX_t = \sigma(X_t)dW_t + b(X_t)dt,$$

with $\sigma = (\sigma_{ij})_{1\leq i,j,\leq n}$ and $b = (b_1, \dots, b_n)^T$. Comparing the operator (3.2.7) to the associated generator of X_t

$$\begin{aligned}\mathcal{A} &= \frac{1}{2}\sum_{p,k}(\sigma\sigma^T)_{pk}\,\partial_{x_p}\partial_{x_k} + \sum_k b_k(x)\partial_{x_k}\\ &= \frac{1}{2}\sum_{p,k}\sigma_{pj}\sigma_{kj}\,\partial_{x_p}\partial_{x_k} + \sum_k b_k(x)\partial_{x_k}\end{aligned}$$

it makes sense to consider the following expressions for the dispersion and the drift

$$\begin{aligned}\sigma_{pj}(x) &= a_{jp}(x)\\ b_k(x) &= \sum_{p,j=1}^{n} a_{jp}(x)\partial_{x_p}(a_{jk}(x)).\end{aligned}$$

Hence, the diffusion associated to Δ_Y can be expressed in terms of the coefficients (a_{jp}) of the vector fields Y_j as

$$dX_t^k = \sum_j a_{jk}(X_t)dW_t^j + \sum_{p,j=1}^{n} a_{jp}(X_t)\partial_{x_p}(a_{jk})(X_t)dt, \quad k = 1, \dots, n, \tag{3.2.8}$$

with W_t^j independent Brownian motions. This can be considered as the diffusion defined by the vector fields (3.2.6).

The associated geometry The associated Riemannian metric on $\mathbb{R}^n$ induced by diffusion X_t has the covariant coefficients given by

$$g_{ij} = (\sigma\sigma^T)^{ij} = (a^Ta)^{ij},$$

which implies the matrix identity $g = (a^Ta)^{-1} = a^{-1}(a^{-1})^T$, where we used the invertibility of the matrix a.[1] This is also equivalent to $aga^T = \mathbb{I}_n$. We shall use this relation to show that the vector fields $\{Y_j\}$ are orthonormal with respect to the Riemannian metric g. This follows from the following matrix

[1]If the number of vectors Y_j is smaller than the dimension of $\mathbb{R}^n$, then this statement no longer holds true. This is the case we will encounter with a sub-elliptic operator.

computation

$$\begin{aligned} g(Y_j, Y_k) &= g\Big(\sum_p a_{jp}\partial_{x_p}, \sum_r a_{kr}\partial_{x_r}\Big) \\ &= \sum_p \sum_r a_{jp}a_{kr}g(\partial_{x_p}, \partial_{x_r}) = \sum_p \sum_r a_{jp}g_{pr}a_{kr} \\ &= \sum_r (ag)_{jr}a_{kr} = \sum_r (ag)_{jr}(a^T)_{rk} = (aga^T)_{jk} = \delta_{jk}. \end{aligned}$$

Since $\{Y_j\}$ are orthonormal with respect to the diffusion metric g, for a vector field Z on $\mathbb{R}^n$ we can write

$$Z = \gamma_1 Y_1 + \cdots + \gamma_n Y_n,$$

with γ_j smooth functions on $\mathbb{R}^n$. Then the magnitude of Z taken with respect to the metric g is given by the Pythagorean relation

$$\|Z\|_g = \sqrt{\gamma_1^2 + \cdots + \gamma_n^2}.$$

This can be used to find the length of a curve. If $c : [0, 1] \to \mathbb{R}^n$ is a smooth curve, then its tangent vector field can be written as $\dot{c}(s) = \sum_j u_j(s)Y_{j|c(s)}$ and then its length on the Riemannian manifold $(\mathbb{R}^n, g)$ is given by

$$\ell(c) = \int_0^1 \sqrt{u_1(s)^2 + \cdots u_n(s)^2}\, ds.$$

Let $x_0, x \in \mathbb{R}^2$ be two points. Then the Riemannian distance between them is given by

$$d(A, B) = \min\{\ell(c); c(0) = x_0, c(1) = x\}.$$

The curve

$$c = \arg\min\{\ell(c); c(0) = x_0, c(1) = x\}$$

is a geodesic between x_0 and x. This curve is unique if the Euclidean distance $\|x - x_0\|$ is small enough. Using this distance, one can define the Riemannian action between x_0 and x within time t by

$$S(t; x_0, x) = \frac{d(x_0, x)^2}{2t}. \tag{3.2.9}$$

This important concept will reappear multiple times throughout the rest of the book. It is related to both the Lagrangian and the Hamiltonian formalisms. In the former case, it arises as an integral of the energy along the geodesic, while in the latter, it arises as a solution of the Hamilton-Jacobi equation, see Calin and Chang [36].

Hamiltonian formalism We associate a Hamiltonian function, which is given by the principal symbol of Δ_Y, as $H(x,\xi) = \sigma_P(x,\xi)$. The Hamiltonian can be also written using the metric g as

$$\begin{aligned} H(x,\xi) &= \frac{1}{2}\|\xi\|_g^2 = \frac{1}{2}\sum_{i,j} g^{ij}(x)\xi_i\xi_j = \frac{1}{2}\sum_{i,j}(a(x)^T a(x))_{ij}\xi_i\xi_j \\ &= \frac{1}{2}\langle a(x)^T a(x)\xi, \xi\rangle = \frac{1}{2}\|a(x)\xi\|^2, \end{aligned} \tag{3.2.10}$$

where $\|\cdot\|$ stands for the Euclidean norm in $\mathbb{R}^n$ and $\|\cdot\|_g^2$ for the norm with respect to the metric g. We consider the bicharacteristics system, which consists of the following $2n$ differential equations

$$\dot{x}_k = \frac{\partial H(x,\xi)}{\partial \xi_k} \tag{3.2.11}$$

$$\dot{\xi}_k = -\frac{\partial H(x,\xi)}{\partial x_k}, \quad 1 \le k \le n \tag{3.2.12}$$

with $2n$ boundary conditions, $x(0) = x_0$, $x(1) = x$. The aforementioned system contains a great deal of information about the Riemannian geometry associated with the diffusion metric. This relation partially consists of the fact that the x-component of the solution is a geodesic with respect to the diffusion metric.

Proposition 3.2.1 *The system* (3.2.11)-(3.2.12) *can be written in the following matrix form*

$$\dot{x} = a^T a\,\xi \tag{3.2.13}$$

$$\dot{\xi} = -(a\xi)^T \nabla_x a\,\xi, \tag{3.2.14}$$

where $\nabla_x a = (\partial_{x_1} a, \dots, \partial_{x_n} a)$ *and* $a = (a_{ij})_{i,j}$.

Proof: Using the Einstein summation convention over repeated indices, we have

$$\begin{aligned} \frac{\partial H(x,\xi)}{\partial \xi_k} &= \frac{\partial}{\partial \xi_k}\Big(\frac{1}{2}g^{ij}(x)\xi_i\xi_j\Big) \\ &= \frac{1}{2}\big(g^{kj}(x)\xi_j + g^{ik}(x)\xi_i\big) = g^{kj}(x)\xi_j \\ &= \sum_{r,j} a_{rk}(x)a_{rj}(x)\xi_j. \end{aligned}$$

Then equation (3.2.11) becomes

$$\dot{x}_k = \sum_{r,j} a_{rk}(x) a_{rj}(x) \xi_j, \qquad 1 \le k \le n,$$

whose equivalent matrix form is $\dot{x} = a^T a\, \xi$.
For the second identity, we use the product rule to obtain

$$\begin{aligned}\frac{\partial}{\partial x_k} g^{ij}(x) &= \frac{\partial}{\partial x_k}(a^T a)_{ij} = \frac{\partial}{\partial x_k} \sum_r a^T_{ir} a_{rj} \\ &= \sum_r \frac{\partial a^T_{ir}}{\partial x_k} a_{rj} + \sum_r a^T_{ir} \frac{\partial a_{rj}}{\partial x_k} \\ &= \sum_r \frac{\partial a_{ri}}{\partial x_k} a_{rj} + \sum_r a_{ri} \frac{\partial a_{rj}}{\partial x_k}.\end{aligned}$$

Using the symmetry in indices i and j we have

$$\begin{aligned}\frac{\partial}{\partial x_k} H(x,\xi) &= \frac{1}{2} \sum_{i,j} \frac{\partial}{\partial x_k} g^{ij}(x) \xi_i \xi_j \\ &= \sum_r \frac{\partial a_{ri}}{\partial x_k} a_{rj} \xi_i \xi_j = \sum_r a^T_{jr} \frac{\partial a_{ri}}{\partial x_k} \xi_i \xi_j \\ &= \langle a^T \partial_{x_k} a\, \xi, \xi \rangle = \xi^T a^T \partial_{x_k} a\, \xi \\ &= (a\xi)^T \partial_{x_k} a\, \xi.\end{aligned}$$

Hence, the equation (3.2.12) becomes

$$\dot{\xi}_k = -(a\xi)^T \partial_{x_k} a\, \xi, \qquad 1 \le k \le n. \tag{3.2.15}$$

■

Corollary 3.2.2 *The amount $\xi^T \dot{x}$ is preserved along the solution of the bicharacteristics system.*

Proof: Since the Hamiltonian function does not depend explicitly on time, using the system (3.2.11)-(3.2.12) yields

$$\frac{d}{ds} H(x(s), \xi(s)) = \sum_k \frac{\partial H}{\partial x_k} \dot{x}_k(s) + \sum_k \frac{\partial H}{\partial \xi_k} \dot{\xi}_k(s) = 0.$$

Using the expression of the Hamiltonian given by (3.2.10) we obtain that $\|a(x)\xi\|^2$ is preserved along the solution. By equation (3.2.13) we obtain

$$\xi^T \dot{x} = \xi^T a^T a \xi = (a\xi)^T (a\xi) = \|a\xi\|^2,$$

and hence $\xi^T \dot{x}$ is preserved along the solution. ■

Remark 3.2.3 Additional conservation laws can be obtained in the case when the matrix $a_{ij}(x)$ is independent of one, or more variables x_k. In this case equation (3.2.15) implies the corresponding momenta ξ_k are preserved along the solution.

Proposition 3.2.4 *Let $(x(s), \xi(s))$ be a solution of the bicharacteristics system (3.2.13)-(3.2.14). Then*

$$\Big\langle \ddot{x}(s) - a^T(x(s))\frac{d}{ds}a(x(s))\,\xi(s), \xi(s)\Big\rangle = 0.$$

Proof: We take a derivative with respect to s in the expression $\xi^T\dot{x} = \langle \dot{x}, \xi\rangle = c$, constant. Then by the product rule we obtain

$$\begin{aligned}
0 &= \frac{d}{ds}\langle \dot{x}, \xi\rangle = \frac{d}{ds}\sum_{k=1}^n \dot{x}_k\xi_k = \sum_{k=1}^n \ddot{x}_k\xi_k + \sum_{k=1}^n \dot{x}_k\dot{\xi}_k \\
&= \sum_{k=1}^n \ddot{x}_k\xi_k - \sum_{k=1}^n \dot{x}_k(a\xi)^T\partial_{x_k}a\,\xi = \sum_{k=1}^n \ddot{x}_k\xi_k - (a\xi)^T\Big(\sum_{k=1}^n \dot{x}_k\partial_{x_k}a\Big)\,\xi \\
&= \sum_{k=1}^n \ddot{x}_k\xi_k - (a\xi)^T\frac{d}{ds}a(x)\,\xi = \sum_{k=1}^n \ddot{x}_k\xi_k - \xi^Ta^T\frac{d}{ds}a(x)\,\xi \\
&= \langle \ddot{x}, \xi\rangle - \langle a^T\frac{d}{ds}a(x)\,\xi, \xi\rangle = \langle \ddot{x} - a^T\frac{d}{ds}a(x)\,\xi, \xi\rangle.
\end{aligned}$$

■

Remark 3.2.5 It is worth noting how the aforementioned formulas simplify in the case of an orthogonal coefficients matrix, $a(x) \in \mathcal{O}(n)$, $\forall x \in \mathbb{R}^n$. Since $a^Ta = aa^T = \mathbb{I}_n$, then the metric is Euclidean, $g_{ij} = \delta_{ij}$, and the Hamiltonian is given by $H(x, \xi) = \frac{1}{2}\|\xi\|^2$, fact that implies that the geodesic $x(s)$ is a straight line with constant velocity ξ.

Hamilton-Jacobi equation The Riemannian action $S(t; x_0, x)$ given by (3.2.9) satisfies the nonlinear equation

$$\partial_t S(t; x_0, x) + \frac{1}{2}\|a(x)\partial_x S(t; x_0, x)\|^2 = 0$$

subject to the initial condition $S(t; x_0, x_0) = 0$. This equation can be solved for some particular cases of vector fields $\{Y_j\}$ and can be used for finding the Riemannian distance between x_0 and x via formula (3.2.9).

Example 3.2.6 We shall write explicitly the equation of the diffusion (3.2.8) and the expression of the elliptic operator (3.2.7) in the case of an orthogonal coefficients matrix, $a(x) \in \mathcal{O}(2)$, $\forall x \in \mathbb{R}^2$. To this end, let $a = \begin{pmatrix} u & v \\ t & w \end{pmatrix}$, where $u = u(x)$, $v = v(x)$, $t = t(x)$, and $w = w(x)$ are smooth functions of $x \in \mathbb{R}^2$. Then $a^T a = \mathbb{I}_2$ implies

$$u^2 + v^2 = 1, \quad t^2 + w^2 = 1, \quad ut + vw = 0.$$

The first two relations state that (u, v) and (t, w) are unit vectors, while the third relation states that these vectors are orthogonal, for any fixed value x. Hence, $\mathcal{B} = \{(u, v), (t, w)\}$ forms an orthonormal basis of $\mathbb{R}^2$. This can be obtained from a rotation of a standard orthonormal basis as follows. Let $e_1 = (1, 0)$ and $e_2 = (0, 1)$. We have two cases:

1. If the basis $\mathcal{B}$ is positively oriented, then it can be obtained by a rotation of the standard basis $\{e_1, e_2\}$ by the angle θ, where θ denotes the angle between e_1 and (u, v), i.e.

$$a = \begin{pmatrix} \cos\theta & -\sin\theta \\ \sin\theta & \cos\theta \end{pmatrix}.$$

2. If the basis $\mathcal{B}$ is negatively oriented, then it can be obtained by a rotation of the standard basis $\{e_2, e_1\}$ by the angle θ, where θ denotes the angle between e_1 and (t, w), i.e.

$$a = \begin{pmatrix} -\sin\theta & \cos\theta \\ \cos\theta & \sin\theta \end{pmatrix}.$$

In the case 1 the vector fields (3.2.6) take the form

$$\begin{aligned} Y_1 &= \cos\theta(x)\partial_{x_1} - \sin\theta(x)\partial_{x_2} \\ Y_2 &= \sin\theta(x)\partial_{x_1} + \cos\theta(x)\partial_{x_2}, \end{aligned}$$

with $\theta(x)$ smooth function of $x \in \mathbb{R}^2$. By a straightforward computation involving basic trigonometric formulas and chain rule we obtain

$$\begin{aligned} \Delta_Y &= \frac{1}{2}(Y_1^2 + Y_2^2) = \frac{1}{2}(\partial_{x_1}^2 + \partial_{x_2}^2) + \frac{1}{2}\partial_{x_2}\theta(x)\,\partial_{x_1} - \frac{1}{2}\partial_{x_1}\theta(x)\,\partial_{x_2} \\ &= \Delta_2 + U, \end{aligned}$$

where Δ_2 is the Laplacian on $\mathbb{R}^2$ and $U = \frac{1}{2}\partial_{x_2}\theta(x)\,\partial_{x_1} - \frac{1}{2}\partial_{x_1}\theta(x)\,\partial_{x_2}$ is a divergence-free vector field on $\mathbb{R}^2$, namely $\operatorname{div} U = \frac{1}{2}\big(\partial_{x_1}\partial_{x_2}\theta(x) - \partial_{x_2}\partial_{x_1}\theta(x)\big) = 0$. The associated diffusion (3.2.8) becomes

$$\begin{aligned} dX_t^1 &= \cos\theta(X_t)dW_t^1 - \sin\theta(X_t)dW_t^2 + \frac{1}{2}\partial_{x_2}\theta(X_t)dt \\ dX_t^2 &= \sin\theta(X_t)dW_t^1 + \cos\theta(X_t)dW_t^2 - \frac{1}{2}\partial_{x_1}\theta(X_t)dt. \end{aligned}$$

By Proposition 2.3.1 there are two independent Brownian motions B_t^1 and B_t^2 such that

$$dX_t^1 = dB_t^1 + \frac{1}{2}\partial_{x_2}\theta(X_t)dt$$
$$dX_t^2 = dB_t^2 - \frac{1}{2}\partial_{x_1}\theta(X_t)dt.$$

We note that the diffusion $X_t = (X_t^1, X_t^2)$ has the divergence-free drift vector

$$b(x) = \frac{1}{2}\Big(\partial_{x_2}\theta(x), -\partial_{x_1}\theta(x)\Big)^T. \tag{3.2.16}$$

Remark 3.2.7 It is worth noting that any divergence-free, C^2-smooth vector field $b(x) = \big(b_1(x), b_2(x)\big)$ on $\mathbb{R}^2$ can be written in the form (3.2.16). This is a consequence of the Poincaré's lemma applied to the one-form $\omega = b_1 dx_2 - b_2 dx_1$. Therefore, the associated diffusion writes as $dX_t = dB_t + b(X_t)dt$, with $\operatorname{div} b(x) = 0$.

Remark 3.2.8 The drift-less case, namely the case $b(x) = 0$, is obtained for functions $\theta(x)$ satisfying $\partial_{x_1}\theta(x) = 0$, $\partial_{x_2}\theta(x) = 0$, which implies $\theta(x) = \theta$, constant. In this case $\Delta_Y = \Delta_2$ and X_t becomes a two-dimensional Brownian motion on $\mathbb{R}^2$.

The case 2, which corresponds to a negatively oriented basis, can be treated in a similar way. The vector fields (3.2.6) take the form

$$Y_1 = -\sin\theta(x)\partial_{x_1} + \cos\theta(x)\partial_{x_2}$$
$$Y_2 = \cos\theta(x)\partial_{x_1} + \sin\theta(x)\partial_{x_2},$$

with $\theta(x)$ smooth function of $x \in \mathbb{R}^2$. By a straightforward computation we obtain

$$\Delta_Y = \frac{1}{2}(\partial_{x_1}^2 + \partial_{x_2}^2) - \frac{1}{2}\partial_{x_2}\theta(x)\,\partial_{x_1} + \frac{1}{2}\partial_{x_1}\theta(x)\,\partial_{x_2}.$$

And the diffusion is given by

$$dX_t^1 = dB_t^1 - \frac{1}{2}\partial_{x_2}\theta(X_t)dt$$
$$dX_t^2 = dB_t^2 + \frac{1}{2}\partial_{x_1}\theta(X_t)dt,$$

with B_t^1 and B_t^2 independent Brownian motions. It is worth noting that in this case the drift is equal to the opposite of the drift of the diffusion discussed in the case 1.

3.3 Hypoelliptic Operators

We are concerned with the second order differential operators of the type

$$P(x,\partial) = \sum_{i,j=1}^{n} a_{ij}(x)\partial_{x_i}\partial_{x_j} + \sum_{k=1}^{n} b_k(x)\partial_{x_k}, \tag{3.3.17}$$

with $a_{ij} \in \mathcal{C}^\infty(U)$ and $b_k \in \mathcal{C}^\infty(U)$ for some open set U in $\mathbb{R}^n$. The operator $P(x,\partial)$ is called *hypoelliptic* on U if for any open subset $V \subseteq U$ we have

$$P(x,\partial)u \in \mathcal{C}^\infty(V) \Rightarrow u \in \mathcal{C}^\infty(V),$$

for any generalized function u. This means that u must be a $\mathcal{C}^\infty$-function in every open set where $P(x,\partial)u$ is a $\mathcal{C}^\infty$-function. An equivalent formulation can be expressed in terms of the singular support as

$$\text{sing supp}\, P(x,\partial)u = \text{sing supp}\, u,$$

where sing supp u stands for the complement of the largest open set on which u is $\mathcal{C}^\infty$-smooth.

In particular, any solution u of the equation $P(x,\partial)u = 0$ must be $\mathcal{C}^\infty$-smooth. Consequently, if one can construct a solution $u \notin \mathcal{C}^\infty(V)$, for some open set V, then the operator $P(x,\partial)$ is not hypoelliptic. This will be used as the main method for showing that an operator is not hypoelliptic.

The term of hypoelliptic operator was introduced by L. Schwartz [126] in his treatise on distributions, to describe certain types of differential operators that possess a significant regularization property.

We shall consider in the following a few classical examples.

Example 3.3.1 The Euclidean Laplacian, $\Delta_n = \frac{1}{2}\sum_{k=1}^{n} \partial_{x_k}^2$, is hypoelliptic on $\mathbb{R}^n$. In fact, any elliptic operator with constant coefficients is hypoelliptic.

Example 3.3.2 A result of L. Schwartz states that an operator with constant coefficients is hypoelliptic on $\mathbb{R}^n$ if and only if it has an elementary solution $E \in \mathcal{C}^\infty(\mathbb{R}^n \backslash \{0\})$. By this result it follows that $P = \partial_t - \Delta_n$ is hypoelliptic since it has an elementary solution[2] $E(x,t) = \frac{H(t)}{(2\pi t)^{n/2}} e^{-\frac{\|x\|^2}{2t}} \in \mathcal{C}(\mathbb{R}^{n+1} \backslash \{0\})$, where $H(t)$ denotes the Heaviside step function. Also, the Laplacian $\Delta_2 = \frac{1}{2}(\partial_{x_1}^2 + \partial_{x_2}^2)$ is hypoelliptic because its elementary solution $E(x) = \frac{1}{\pi} \ln \|x\|$

[2] This is also a heat kernel for the operator Δ_n.

has sing supp $E = \{0\}$. However, the wave operator $\partial_t^2 - \partial_x^2$ is not hypoelliptic, because it can be shown that it has an elementary solution E_+ with supp $E_+ \subset \{t \geq 0\}$, which has the singular support sing supp $E_+ = \{(x,t) \in \mathbb{R}^2; x = t\}$.

The fact that the wave operator is not hypoelliptic can be also shown directly. To this end, we consider

$$u(t,x) = \int_{\mathbb{R}} \cos\left((t+x)\sqrt{1+s^2}\right) \frac{1}{(1+s^2)^2}\, ds$$

and note that $u \in \mathcal{C}^2(\mathbb{R}^2) \backslash \mathcal{C}^\infty(\mathbb{R}^2)$ and $(\partial_t^2 - \partial_x^2)u(t,x) = 0$.

Example 3.3.3 The Kolmogorov operator $K(x,y,t,\partial) = \partial_x^2 + x\partial_y - \partial_t$ is hypoelliptic on $\mathbb{R}^3$, see [90].

Example 3.3.4 The Fedii operator, which is given by

$$F(x,y,\partial) = \frac{1}{2}(\partial_x^2 + \varphi(x)^2 \partial_y^2)$$

with $\varphi \in \mathcal{C}^\infty(\mathbb{R})$, $\varphi(0) = 0$ and $\varphi(x) > 0$ for $x \neq 0$, is hypoelliptic on $\mathbb{R}^2$, see Fedii [55]. In particular, the Grushin operator $G = \frac{1}{2}(\partial_x^2 + \frac{x^4}{4}\partial_y^2)$ is hypoelliptic on $\mathbb{R}^2$.

Example 3.3.5 Let $\sigma \in \mathcal{C}^\infty(\mathbb{R})$ be an even function, nondecreasing on $[0,\infty)$, which vanishes only at $x = 0$. Kusuoka and Stroock [11] proved that the operator

$$L_\sigma = \partial_{x_1}^2 + \sigma^2(x_1)\partial_{x_2}^2 + \partial_{x_3}^3$$

is hypoelliptic on $\mathbb{R}^3$ if and only if σ satisfies the limit condition

$$\lim_{s \to 0+} s \ln \sigma(s) = 0.$$

In particular, taking $\sigma(s) = s^2/2$, we obtain that

$$L_\sigma = \partial_{x_1}^2 + \frac{1}{4}x_1^4 \partial_{x_2}^2 + \partial_{x_3}^3$$

is hypoelliptic on $\mathbb{R}^3$. Also, considering $\sigma(s) = e^{-|s|^p}$ with $p \in (-1,0)$, we obtain a hypoelliptic operator, which does not satisfy the Hörmander condition, see the upcoming definition 3.4.1.

Example 3.3.6 One can construct more hypoelliptic operators from other hypoelliptic operators using composition. It follows from the definition of hypoellipticity that if $P(x,\partial)$ and $Q(x,\partial)$ are two differential operators with coefficients $\mathcal{C}^\infty$- functions, which are hypoelliptic on the open set U, then the differential operators $P \circ Q$ and $Q \circ P$ are hypoelliptic. Conversely, if $P \circ Q$ and $Q \circ P$ are hypoelliptic, then $P(x, \partial)$ and $Q(x, \partial)$ are hypoelliptic. As an application, the operators $\partial_t - \partial_x$ and $\partial_t + \partial_x$ are not hypoelliptic because their composition $(\partial_t - \partial_x)(\partial_t + \partial_x) = \partial_t^2 - \partial_x^2$ is the wave operator, which is not hypoelliptic. However, the Cauchy-Riemann operators, $\partial_{\bar{z}} = \frac{1}{2}(\partial_{x_1} + i\partial_{x_2})$ and $\partial_z = \frac{1}{2}(\partial_{x_1} - i\partial_{x_2})$ are hypoelliptic because $4\partial_{\bar{z}}\partial_{\bar{z}} = \partial_{x_1}^2 + \partial_{x_2}^2$ is hypoelliptic.

The next result shows that hypoellipticity is inherited while passing from higher dimensions to lower dimensions.

Proposition 3.3.1 *If $L(x, t, \partial) = P(x, \partial) - \partial_t$ is hypoelliptic on $\mathbb{R}^{n+1}$, then $P(x, \partial)$ is hypoelliptic on $\mathbb{R}^n$.*

Proof: By contradiction, we assume that $P(x, \partial)$ is not hypoelliptic on $\mathbb{R}^n$, so we have $P(x, \partial)u(x) \in \mathcal{C}^\infty(\mathbb{R}^n)$ for some $u \notin \mathcal{C}^\infty(\mathbb{R}^n)$. Let $v(x, t) = u(x)$. Then $L(x, t, \partial)v = P(x, \partial)v \in \mathcal{C}^\infty(\mathbb{R}^n \times \mathbb{R})$ but $v \notin \mathcal{C}^\infty(\mathbb{R}^{n+1})$. ■

Remark 3.3.2 It is worth noting that the reciprocal of the previous proposition is in general false, namely if $P(x, \partial)$ is hypoelliptic, then it is not necessarily true that $L(x, t, \partial)$ is hypoelliptic. To show this, we consider a nontrivial solution $u(x)$ for $P(x, \partial)u = 0$. Then construct $v(x, t) = H(t)u(x)$, where $H(t)$ is the Heaviside step function. Then $Lv = Pv = 0$, but $v \notin \mathcal{C}^\infty(\mathbb{R}^{n+1})$. We shall get back to this property in Proposition 3.6.2, where we show that in fact it holds true for certain particular operators.

Remark 3.3.3 Hypoellipticity is a property of differential operators that is invariant under changes of coordinates. Hence, the hypoellipticity can be defined on differentiable manifolds, by asking for hypoellipticity to hold in a local chart. This invariance property is a consequence of both the chain rule and the definition of hypoellipticity as follows. Let U and V be two open sets in $\mathbb{R}^n$ and $\phi : U \to V$ be a $\mathcal{C}^\infty$-diffeomeorphism, and denote $y = \phi(x)$. Consider a function $v : V \to \mathbb{R}$ and let $u : U \to \mathbb{R}$, with $u(x) = v(\phi(x)) = v(y)$. Then the operator (3.3.17) acts on u as follows

$$
\begin{aligned}
P(x, \partial_x)u &= \langle A(x)\partial_x u, \partial_x u\rangle + \langle B(x), \partial_x u\rangle \\
&= \langle \tilde{A}(y)\partial_y v, \partial_y v\rangle \circ \phi + \langle \tilde{B}(y), \partial_y v\rangle \circ \phi \\
&= \big(P(y, \partial_y)v\big) \circ \phi,
\end{aligned}
$$

with $\tilde{A}(y) = \Big(\dfrac{\partial\phi^{-1}}{\partial y}\Big)^{-1} A(\phi^{-1}(y))\Big(\dfrac{\partial\phi^{-1}}{\partial y}\Big)^{-1}$ and $\tilde{B}(y) = \Big(\dfrac{\partial\phi^{-1}}{\partial y}\Big)^{-1} B(\phi^{-1}(y))$. Let $P(y, \partial_y)$ be hypoelliptic on V. Then the fact $P(x, \partial_x)u \in \mathcal{C}^\infty(U)$ implies $P(y, \partial_y)v \in \mathcal{C}^\infty(V)$, and hence $v \in \mathcal{C}^\infty(V)$. This is equivalent to $u \in \mathcal{C}^\infty(U)$, via the diffeomorphism ϕ.

As an example, we consider the wave operator $P(x, \partial_x) = \partial_{x_1}^2 - \partial_{x_2}^2$, which can be transformed into $P(y, \partial_y) = 4\partial_{y_1}\partial_{y_2}$ via the diffeomorphism $x_1 = y_1 + y_2$, $x_2 = y_1 + y_2$. Since $v(y) = H(y_1) + H(y_2)$ is a non-smooth solution of $P(y, \partial_y)v = 0$, then $P(y, \partial_y)$ is not hypoelliptic. Therefore, $P(x, \partial_x)$ is not hypoelliptic either.

In this book we are concerned with hypoelliptic operators provided as sum of squares of vector fields. Hypoellipticity results for second order differential operators that cannot be expressed as a sum of squares can be found in Oleinic and Radkevic [115].

3.4 Hörmander's Theorem

Hörmander's starting point in the study of the hypoellipticity of operators given by sums of squares of vector fields was the Kolmorov's example of a hypoelliptic operator introduced in 1934

$$K = \partial_x^2 - x\partial_y - \partial_t.$$

This operator can be written as $K = X_1^2 - X_0$, where $X_0 = x\partial_y + \partial_t$ and $X_1 = \partial_x$. We note that $[X_1, X_0] = \partial_y$, and hence the space span $\{X_0, X_1, [X_1, X_0]\}$ has dimension 2 everywhere. Hörmander's result extends this example, relating the maximal dimension of the aforementioned space to the hypoellipticity of the operator.

Let $X_0, \ldots, X_r$ be a collection of vector fields with $\mathcal{C}^\infty$ coefficients, defined on an open set U of $\mathbb{R}^n$, and $c : U \to \mathbb{R}$ a $\mathcal{C}^\infty$-function. Then consider the second order differential operator

$$L = \sum_{i=1}^{r} X_i^2 + X_0 + c. \tag{3.4.18}$$

Let $\text{Lie}(X_0, \ldots, X_r)$ be the Lie algebra generated by the vector fields $X_0, \ldots, X_r$. This is the vector space generated by the collection of vector fields $X_0, \ldots, X_r$ together with their iterative Lie brackets, where the Lie bracket of two vector fields is defined by $[X, Y] = XY - YX$. This also satisfies
(i) Anticommutativity: $[X, Y] = -[Y, X]$;
(ii) Jacobi identity: $[X, [Y, Z]] + [Y, [Z, X]] + [Z, [X, Y]] = 0$, for all X, Y, Z smooth vector fields on U.

Definition 3.4.1 *We say the vector fields $X_0, \ldots, X_r$ satisfy the bracket-generating condition if the associated Lie algebra has maximal rank everywhere, namely, the vector space $\text{Lie}(X_0, \ldots, X_r)$ has dimension n at every $x \in U$.*

The aforementioned bracket-generating condition is equivalent to the fact that among the vector fields

$$X_{j_1}, [X_{j_1}, X_{j_2}], [X_{j_1}, [X_{j_2}, X_{j_3}]], \cdots, [X_{j_1}, [X_{j_2}, [X_{j_3}, \dots]]], \cdots \quad, 0 \le j_k \le r$$

there are always n, which are linearly independent at any given point $x \in U$. The number of bracket iterations plus 1 is called the *step* of the operator L at x.

Definition 3.4.2 *The differential operator* (3.4.18) *is regular at a point x if the algebra* $\mathrm{Lie}(X_0, \dots, X_r)$ *has a constant rank on a neighborhood of x. The operator is called regular on an open set U if any point $x \in U$ is regular.*

The next result provides the bracket-generating condition as a necessarily condition of hypoellipticity.

Proposition 3.4.3 *Let X_j be vector fields on $U \subseteq \mathbb{R}^n$ with $\mathcal{C}^\infty$ coefficients. If the operator*

$$L = \sum_{i=1}^{r} X_i^2 + X_0 \tag{3.4.19}$$

is hypoelliptic and regular on U, then $\{X_0, \dots, X_r\}$ satisfy the bracket-generating condition on U.

Proof: By contradiction, we assume that the bracket-generating condition does not hold at a point $x \in U$. Then the dimension of the vector space $\mathrm{Lie}(X_0, \dots, X_r)$ is equal to k on a neighborhood of the point x, with $k < n$. Since for any two vector fields $X, Y \in \mathrm{Lie}(X_0, \dots, X_r)$ we have $[X, Y] \in \mathrm{Lie}(X_0, \dots, X_r)$, then the subbundle $p \to \mathrm{Lie}(X_0, \dots, X_r)_p$ is involutive on U. Then by Frobenius' theorem, $\mathrm{Lie}(X_0, \dots, X_r)$ is locally integrable, namely there is a submanifold $\mathcal{S}$ of dimension k passing through x such that the vector fields X_j and their iterated brackets are tangent to $\mathcal{S}$. The local explicit representation of the k-submanifold $\mathcal{S}$ about x can be written as

$$x_{k+1} = \varphi_{k+1}(x_1, \dots, x_k)$$
$$\dots\dots\dots\dots$$
$$x_n = \varphi_n(x_1, \dots, x_k)$$

with φ_j real-valued smooth functions on a neighborhood of x. In this coordinate system the vector fields can be written as

$$X_j = \sum_{i=1}^{n} \alpha_{ji}(x_1, \dots, x_n)\partial_{x_i} = \sum_{p=1}^{k} \beta_{jp}(x_1, \dots, x_k)\partial_{x_p},$$

and hence, locally about x, the operator (3.4.19) can be written in terms of the first k coordinates as

$$\tilde{L} = \sum_{i,j=1}^{k} \gamma_{ij}(x_1, \dots, x_k)\partial_{x_i}\partial_{x_j} + \sum_{i=1}^{k} \gamma_i(x_1, \dots, x_k)\partial_{x_i}.$$

Since L is hypoelliptic about x, then $\tilde{L}$ is too, see Remark 3.3.3. To arrive to a contradiction, it suffices to construct a nontrivial function $u = u(x_{k+1}, \dots, x_n)$ such that $\tilde{L}u = 0$ and $u \notin \mathcal{C}^\infty$ on a neighborhood of x. One of these functions is for instance, $u(x) = H(x_n)$, where H denotes the Heaviside step function. Therefore, the bracket-generating condition holds at the arbitrary chosen point $x \in U$. ■

It is worth noting that the regularity condition imposed on the operator is essential. For a counterexample, see Remark 3.5.2 on page 143. However, the regularity condition can be dropped if we assume the vector fields X_j have analytic coefficients, see Derridj [48]. For a particular case, see the forthcomming Proposition 3.5.1.

A converse implication of the previous result holds true and represents the main result of the paper [77]; this can be stated as:

Theorem 3.4.4 (Hörmander) *If the vector fields $\{X_0, \dots, X_r\}$ satisfy the bracket-generating condition on the open set U, then the operator L given by* (3.4.18) *is hypoelliptic on U.*

Hörmander's original proof was formulated in terms of second-order differential operators and was quite complex and relied heavily on norms estimations, Fourier transforms, as well as a repeated use of the Campbell-Hausdorff formula. However, Kohn [89] simplified the proof considerably by incorporating the theory of pseudo-differential operators.

Alternatively, Hörmander's theorem can also be proven through a probabilistic approach using the stochastic calculus of variations developed by Paul Malliavin [104]. This method is now referred to as the *Malliavin calculus* and can be found in Nualart [113] for a comprehensive explanation. The general idea of the proof is to demonstrate that a specific Ito's map is weakly differentiable and then show that, under Hörmander's conditions, the derivative is non-degenerate.

A systematic exposition of a relevant part of the theory of Hörmander operators, together with the necessary background and prerequisites can be found in Bramanti and Brandolini [19].

3.5 Non Bracket-generating Vector Fields

There are differential operators, which are hypoelliptic but they are not bracket generating. By Proposition 3.4.3, this can occur only in a neighborhood of non-regular points.

Proposition 3.5.1 *Let $\phi \in \mathcal{A}(\mathbb{R})$ be a real analytic nontrivial function on $\mathbb{R}$. Then the operator*

$$L = \partial_x^2 + \phi^2(x)\partial_y^2$$

is hypoelliptic on $\mathbb{R}^2$.

Proof: It suffices to show that the vector fields $X = \partial_x$, $Y = \phi(x)\partial_y$ satisfy the bracket-generating condition. By contradiction, we assume there is a point $(x_0, y_0) \in \mathbb{R}^2$ such that $\dim \mathrm{Lie}\{X, Y\}_{(x_0,y_0)} < 2$, namely

$$\dim \mathrm{span}_{(x_0,y_0)}\{\partial_x, \phi(x_0)\partial_y, \dots, \phi^{(j)}(x_0)\partial_y, \dots\} = 1,$$

fact that implies $\phi(x_0) = \phi'(x_0) = \cdots = \phi^{(j)}(x_0) = \cdots = 0$. By analyticity,

$$\phi(x) = \sum_{j\geq 0} \frac{\phi^{(j)}(x_0)}{j!} x^j = 0, \quad \forall x \in \mathbb{R}$$

which is a contradiction. ■

Remark 3.5.2 In a similar way, we can show that if ϕ is analytic and nonzero then the operator

$$L = \partial_x^2 + \phi^2(x)\partial_y^2 + \partial_z^2 \tag{3.5.20}$$

is hypoelliptic on $\mathbb{R}^3$. If we choose

$$\phi(x) = \begin{cases} e^{-\frac{1}{\sqrt{|x|}}}, & \text{if } x \neq 0 \\ 0, & \text{otherwise} \end{cases}, \tag{3.5.21}$$

then $\phi \in \mathcal{C}^\infty(\mathbb{R})\backslash\mathcal{A}(\mathbb{R})$. Since

$$\phi(0) = \phi'(0) = \cdots = \phi^{(j)}(0) = \cdots = 0,$$

the vector fields $\{\partial_x, \phi(x)\partial_y, \partial_z\}$ do not satisfy the bracket-generating condition along the plane $\{x = 0\}$. However, for this choice of the function ϕ the operator (3.5.20) is still hypoelliptic, see Kusuoka and Stroock [11]. Hence, the hypoellipticity does not imply the bracket-generating condition, unless the rank of the generated Lie algebra is locally constant. In our case the rank drops to 1 along the plane $\{x = 0\}$ and hence the operator is not regular along this plane. This example shows that the regularity condition is essential for Proposition 3.4.3.

Remark 3.5.3 Proposition 3.5.1 holds true even for non-analytic functions ϕ. Choosing ϕ given by (3.5.21), we obtain that the bracket-generating condition does not hold for $L = \partial_x^2 + \phi^2(x)\partial_y^2$, but the operator is hypoelliptic because $\phi(0) = 0$ and $\phi(x) > 0$ for $x \neq 0$, see Example 3.3.4.

Remark 3.5.4 The vector fields on $\mathbb{R}^3$

$$X_1 = \partial_x + ae^{(x^2+y^2)^{-a}} x(x^2+y^2)^{-2(a+1)}\partial_z$$
$$X_2 = \partial_y + ae^{(x^2+y^2)^{-a}} y(x^2+y^2)^{-2(a+1)}\partial_z,$$

with $a > 0$, are not bracket-generating at the origin. However, it can be proved that the sub-elliptic operator $\Delta_X = \frac{1}{2}(X_1^2 + X_2^2)$ is hypoelliptic on $\mathbb{R}^3$ if a is sufficiently small.

3.6 Some Applications of Hörmander's Theorem

In this section we shall present a few applications of Theorem 3.4.4 regarding the smoothness of heat kernels.

Heat operators The following result is related to the smoothness of the heat kernel.

Proposition 3.6.1 *Let*

$$L = \frac{1}{2}\sum_{j=1}^{r} X_j^2 + X_0$$

with $\{X_0, \dots, X_r\}$ *satisfying the bracket-generating condition on an open domain* U *in* $\mathbb{R}^n$*. Then the heat kernel,* $p_t(x)$*, of* L *is a* $\mathcal{C}^\infty$*-smooth function for any* $t > 0$.

Proof: The heat kernel satisfies $(L - \partial_t)p_t(x) = 0$, for all $t > 0$. To show that $p_t(x)$ is C^∞-smooth, it suffices to prove that $L - \partial_t$ is hypoelliptic. As in the proof of Proposition 3.6.2, we can show that

$$\text{span}\,\{\partial_t, X_0, \dots, X_r\}_{(x,t)} = T_x\mathbb{R}^n \otimes T_t\mathbb{R} = T_{(x,t)}\mathbb{R}^n_x \times \mathbb{R}_t, \quad \forall (x,t) \in \mathbb{R}^n_x \times \mathbb{R}_+.$$

By Theorem 3.4.4 the operator $L - \partial_t$ becomes hypoelliptic on $\mathbb{R}^n_x \times \mathbb{R}_+$. ■

It is transparent now why in the case of operators represented as sums of squares of vector fields, which satisfy the maximal rank condition, it makes sense to look for elementary solutions and heat kernels of a smooth function type.

He have seen that the reciprocal of Proposition 3.3.1 does not hold true in general, see Remark 3.3.2. However, if the operator is of sums of squares type, then the reciprocal holds true on domains where the Lie algebra has a constant rank.

Proposition 3.6.2 *Consider the differential operator*

$$P = \sum_{j=1}^{r} X_j^2 + X_0$$

and assume that P is hypoelliptic and regular on an open set $U \subseteq \mathbb{R}^n$. Then the associate heat operator, $L = P - \partial_t$, is also hypoelliptic on $U \times \mathbb{R}_t$.

Proof: If P is hypoelliptic then by Proposition 3.4.3 the vector fields $\{X_0, \dots, X_r\}$ satisfy the bracket-generating condition, namely span $\{X_0, \dots, X_r\}_x = T_x\mathbb{R}^n$ for any $x \in U$. Since the vector field ∂_t commutes with all the other vector fields, $[\partial_t, X_j] = 0$, then

$$\text{span}\,\{\partial_t, X_0, \dots, X_r\}_{(x,t)} = T_x\mathbb{R}^n \otimes T_t\mathbb{R} = T_{(x,t)}\mathbb{R}^n_x \times \mathbb{R}_t, \quad \forall (x,t) \in U \times \mathbb{R}_t.$$

By Theorem 3.4.4 the operator $L = P - \partial_t$ is hypoelliptic on $U \times \mathbb{R}_t$. ∎

Ito diffusions The next application is related to Ito diffusions. Let

$$dZ_t^j = \sum_{p=1}^{k} \sigma_{jp}(Z_t)dW_p(t), \qquad 1 \le j \le n$$

be a drift-less Ito diffusion on $\mathbb{R}^n$, with $k < n$ and $W_p(t)$, $1 \le p \le k$, independent Brownian motions. Then the associated generator operator can be written as

$$L = \frac{1}{2}\sum_{i,j}(\sigma\sigma^T)_{ij}\partial_{x_i}\partial_{x_j} = \frac{1}{2}\sum_{j=1}^{k} X_j^2 - \frac{1}{2}\sum_{j} \nabla^0_{X_j} X_j,$$

where $X_j = \sum_{i=1}^{n} \sigma_{ij}(x)\partial_{x_i}$ are vector fields on $\mathbb{R}^n$ and ∇^0 denotes the Levi-Civita connection on the Euclidean space $\mathbb{R}^n$, see page 411 for more computational details. Then $X_0 = -\frac{1}{2}\sum_j \nabla^0_{X_j} X_j$ is also a vector field on $\mathbb{R}^n$ and hence the generator can be written under the familiar form $L = \frac{1}{2}\sum_{j=1}^{k} X_j^2 + X_0$.

Proposition 3.6.3 *If the vector fields $\{X_0, X_1, \dots, X_k\}$ satisfy the bracket-generating condition, then the transition probability of the diffusion Z_t is $\mathcal{C}^\infty$-smooth.*

Proof: The transition density of Z_t is the heat kernel of the operator L. Then using Proposition 3.6 yields the desired result. ∎

3.7 Elements of Sub-Riemannian Geometry

Holonomic and nonholonomic systems In the early 1900s' the notion of *holonomy* was introduced to describe certain features of mechanical systems (Vranceanu [135]). A mechanical system is said to be *nonholonomic* with respect to a given constrained motion (involving positions and velocities) if the system can move between any two states of the configuration space without violating the constraint. Otherwise, the system is said to be *holonomic* with respect to the constraint. Mathematically speaking, the non-holonomic systems are the systems which are given by nonintegrable constraints, while the holonomic systems are systems for which all constraints are integrable.

Many mechanical systems can be described by a set of coordinates of a manifold subject to differential constraints. A few examples are: a bike, a two-wheel cart robot, the knife edge, skating, rolling a disk or a sphere on a plane, etc.

There are two equivalent ways to introduce constraints on a differentiable manifold; one variant involves constraints in the tangent bundle, while the the other uses constraints in the cotangent bundle. The constraints using tangent vectors involve the concept of *differentiable distribution of vectors*, which is a subbundle of the tangent bundle. The constraints using cotangent vectors use the concept of *Pfaff systems*, which are systems of differential one-forms. The reader interested in the latter variant can consult Calin and Chang [26]. In this book we shall mainly use the former constraint approach.

Differentiable distributions of vectors Let M denote a differentiable manifold and denote by T_pM the tangent space of M at $p \in M$. A *regular distribution* $\mathcal{D}$ of rank k on M assigns to each point $p \in M$ a k-dimensional subspace $\mathcal{D}_p$ of T_pM. Thus, the distribution $\mathcal{D}$ is a vector subbundle of the tangent bundle TM.

The distribution $\mathcal{D}$ is called *differentiable* if every point p has an open neighborhood U in M such that there are k linearly independent $\mathcal{C}^\infty$-vector fields on U that generate $\mathcal{D}$, namely

$$\mathcal{D}_q = \text{span}_q\{X_1, \cdots, X_k\}, \qquad \forall q \in U.$$

The vector fields that are contained in the distribution $\mathcal{D}$ are called *horizontal vector fields*. A curve $c : [a, b] \to U$ is called *horizontal* if its velocity is tangent to the distribution, i.e., $\dot{c}(s) \in \mathcal{D}_{c(s)}$, for any $s \in [a, b]$. Thus, any motion of a constrained system can be modeled as a horizontal curve on the coordinate space.

The distribution $\mathcal{D}$ is called *involutive* if it is closed with respect to the Lie bracket, namely, for any two horizontal vector fields, $X, Y \in \mathcal{D}$, their Lie bracket $[X, Y] = XY - YX$ belongs to $\mathcal{D}$. This means that $[X_i, X_j]$ is a linear combination of vector fields $\{X_1, \dots, X_k\}$. We note that the distributions of maximal rank are always involutive. This follows from the fact that in this case $\mathcal{D}_p = T_p\mathbb{R}^n$, for all $p \in U$. Distributions of rank 1 are also involutive, since $[X_1, X_1] = 0$.

An *integral manifold* of the distribution $\mathcal{D}$ is a connected k-dimensional submanifold N of M such that $T_pN = \mathcal{D}_p$, for all $p \in N$. A distribution $\mathcal{D}$ is called *integrable* if through any point $x \in M$ there is an integral manifold. The integrability means that for every point $p \in M$ there exists a local chart $(U, \{x_1, \dots, x_k\})$ on N about p such that, for every $q \in U$, the space $\mathcal{D}_q$ is spanned by the coordinate vectors $\partial_{x_1}{}_{|q}, \dots, \partial_{x_k}{}_{|q}$.

The following classical result of Frobenius [59] states the relation between the aforementioned concepts.

Theorem 3.7.1 *A distribution is involutive if and only if it is integrable.*

Sub-Riemannian metric The dimension of the vector space $\mathcal{D}_p$ is given by the rank of the matrix of coefficients of the vectors $X_j(p)$ as

$$\dim \mathcal{D}_p = \text{rank}A(p) \le k \le n,$$

where $A(p) = (a_{ij}(p))_{i,j}$ and $X_i(p) = \sum_{i=1}^{n} a_{ij}(p)\partial_{x_j}$. The principal symbol of the sum of squares operator $L = \frac{1}{2}\sum_{j=1}^{k} X_j^2$ is then given by the following quadratic form in ξ

$$\sigma_p(x, \xi) = \frac{1}{2}\xi^T A(x)A^T(x)\xi = \frac{1}{2}\sum_{i,j=1}^{n} g^{ij}(x)\xi_i\xi_j, \quad \forall x \in U, \xi \in \mathbb{R}^n.$$

The contravariant matrix $g^{ij}(x)$ is symmetric and has rank

$$\text{rank } g^{ij}(x) = \text{rank}A(x)A^T(x) = \text{rank}A(x) = k \le n.$$

The case rank $g^{ij}(x) = n$ corresponds to a Riemannian metric on U. The case $k < n$ corresponds to a *sub-Riemannian metric.* In the latter case, the matrix $g^{ij}(x)$ is degenerated and hence there is no inverse covariant matrix $g_{ij}(x)$. Consequently, the principal symbol of the operator L is a semi-definite quadratic form. Thus, there are $n - k$ *missing directions* which cannot be controlled from the metrical point of view. In this case, there is hope that these directions can be retrieved and controlled by an iterative application of the Lie bracket, as in the following.

We also note that all distributions $\mathcal{D}$ encountered so far had constant rank, i.e. were regular. However, there are cases when the distribution rank drops along some sets, case in which the distribution is called *irregular.*

Bracket-generating distributions The distribution $\mathcal{D}$ is called *bracket-generating* at x if

$$\dim \mathrm{Lie}_x(X_1, \cdots, X_r) = n,$$

namely the vectors $X_{j|x}$ together with their iterated Lie brackets generate the tangent space $T_x\mathbb{R}^n$. The distribution $\mathcal{D}$ is called bracket-generating on U if it is bracket-generating at each x in U. The bracket-generating concept is applied to both regular and irregular distributions. If m is the number of iterated Lie brackets needed to span the tangent space at a point x, then the distribution is called of step $m+1$ at x. This is because the tangent space space can be spanned in $m+1$ steps.

A *sub-Riemannian manifold* $(M, \mathcal{D})$ is a differentiable manifold M endowed with a horizontal distribution $\mathcal{D}$ on M. Sub-Riemannian manifolds are suitable mathematical models for mechanical systems with constraints, where the manifold M represents the coordinates space, while the distribution $\mathcal{D}$ represent the velocity constraints.

The following connectedness result was proved independently by both Rashevskii [119] and Chow [46] in late 1930s.

Theorem 3.7.2 (Chow-Rashevskii) *Any two points of a connected sub-Riemannian manifold, endowed with a bracket-generating distribution, are connected by a horizontal piece-wise smooth path in the manifold.*

The physical significance of the aforementioned result is that a mechanical system with constraints can be steered from any given position into any final position, while satisfying a set of nonholonomic constraints. A familiar application of this type of system is the process of parking a car.

We note that bracket-generating distributions of rank $k < \dim M$ are not involutive, and hence not integrable. Therefore, nonholonomic systems are not integrable.

Remark 3.7.3 If the bracket-generating condition is not satisfied, then the connectivity property does not necessarily hold. To this end, we consider the distribution generated by the vector fields $X = xy\partial_x + \partial_z$ and $Y = \partial_y$ in $\mathbb{R}^3$. The bracket-generating condition fails along the plane $\{x = 0\}$, since $[X, Y] = -x\partial_x$, $[X, [X, Y]] = 0$, $[Y, [X, Y]] = 0$ and all other Lie brackets vanish. We shall show that there are no horizontal curves connecting points situated on the opposite half-spaces $\{x > 0\}$ and $\{x < 0\}$. To this end, we shall first find the horizontality constraint. Let $c(s) = (x(s), y(s), z(s))$, $s \in [0, T]$ be a curve in $\mathbb{R}^3$. Since its velocity can be written as

$$\dot{c}(s) = (\dot{x}(s), \dot{y}(s), \dot{z}(s)) = \dot{z}X_{c(s)} + \dot{y}(s)Y_{c(s)} + (\dot{x} - xy\dot{z})\partial_x,$$

the curve $c(s)$ is horizontal if and only if its components satisfy the constraint $\dot{x} - xy\dot{z} = 0$. This can be integrated using the integrating factor $\mu = e^{-\int y dz}$. The exact equation obtained is $\frac{d}{ds}(x\mu) = 0$, with solution $x(s) = x_0 e^{\int_0^s y(t)\dot{z}(t)dt}$. If $x(T) = x_T$ then $x_T = x_0 e^{\int_0^T y(t)\dot{z}(t)dt}$. Since x_0 and x_T have the same sign, the horizontal curve $c(s)$ cannot connect points situated on opposite sides of the plane $\{x = 0\}$.

In general, if two points, $p, q \in M$ cannot be joined by a horizontal curve on a connected sub-Riemannian manifold $(M, \mathcal{D})$, then:
(i) there is a non-empty region $\mathcal{R} \subset M$, where $\mathcal{D}$ is not bracket-generating;
(ii) the region $\mathcal{R}$ separates the points p and q (in the sense that any continuous curve in M joining p and q must intersect $\mathcal{R}$).

To show this we assume that $\mathcal{R}$ does not separate the points p and q, so there is a continuous curve between p and q, which does not intersect $\mathcal{R}$. Therefore, $N = M \backslash \mathcal{R}$ is connected and $\mathcal{D}$ is bracket-generating on N. Then Chow-Rashevskii's Theorem yields that p and q can be joined by a piece-wise continuous horizontal curve, which leads to a contradiction.

Remark 3.7.4 The converse of Chow-Rashevskii's Theorem, that is, if the manifold $(M, \mathcal{D})$ is horizontally connected, then the bracket-generating condition holds, is in general, not true. Two simple counter-examples were provided by Sussmann [130] as follows. Consider $M = \mathbb{R}^2$ and the vector fields $X_1 = \partial_x$, $X_2 = \psi(x)\partial_y$, with $\psi \in \mathcal{C}^\infty(\mathbb{R})$, satisfying $\psi(x) = 0$ for $x \leq 0$ and $\psi(x) > 0$ for $x > 0$. The bracket-generating condition does not hold on the set $\{x < 0\}$. To show the connectivity between two given points, $p, q \in \mathbb{R}^2$, we first transport them into the points $p', q' \in \{x > 0\}$, following the integral curves of the vector field X_1, such that $x(p') = x(q')$. If $y(p') < y(q')$, then we transport the point p' towards q' along the integral curve of X_2.

Moreover, a slight modification provides an example where the bracket-generating condition is nowhere satisfied, while the connectivity by horizontal curves still holds. For this we set $M = \mathbb{R}^3$, and consider the vector fields $X_1 = \partial_x$, $Y = \rho(x)\partial_y$, $X_3 = \sigma(x)\partial_z$, with $\rho, \sigma \in \mathcal{C}^\infty(\mathbb{R})$ satisfying

$$\rho(x) = 0 \text{ for } x \leq 0, \qquad \sigma(x) = 0 \text{ for } x \geq -1$$

$$\rho(x) > 0 \text{ for } x > 0, \qquad \sigma(x) > 0 \text{ for } x < -1.$$

3.8 Examples of Distributions

We shall introduce a few distributions that will be useful in later chapters.

Example 3.8.1 (Grushin distribution) Let $M = \mathbb{R}^2$ and consider the distribution generated by the vector fields $X = \partial_x$ and $Y = x\partial_y$. The distribution is regular of rank 2 on $\mathbb{R}^2\backslash\{x = 0\}$ and irregular along the line $\{x = 0\}$, of rank 1. Since $[X, Y] = \partial_y$, the distribution is bracket-generating everywhere of step 2.

Example 3.8.2 (Heisenberg distribution) Let $M = \mathbb{R}^3$ and consider the vector fields $X = \partial_x + 2y\partial_z$ and $Y = \partial_y - 2x\partial_z$. The distribution generated by $\{X, Y\}$ is regular of rank 2 everywhere on $\mathbb{R}^3$. Since $[X, Y] = -4\partial_z$, the distribution is bracket-generating everywhere on $\mathbb{R}^3$ of step 2. Since the distribution is not integrable, there is no surface in $\mathbb{R}^3$ such that X and Y are tangent to.

Example 3.8.3 (Martinet distribution) Consider on $M = \mathbb{R}^3$ the vector fields $X = \partial_x + \frac{1}{2}y^2\partial_z$ and $Y = \partial_y$. The distribution generated by $\{X, Y\}$ is regular of rank 3 along $\mathbb{R}^3\backslash\{y = 0\}$ and irregular of rank 2 along the plane $\{y = 0\}$. Since $[X, Y] = -y\partial_z$ and $[[X, Y], Y] = \partial_z$, the distribution is step 2 along $\{y = 0\}$ and step 3 in rest.

Example 3.8.4 (Engel distribution) In this case $M = \mathbb{R}^4$. Let $X = \partial_x$ and $Y = \partial_y + x\partial_z + \frac{1}{2}x^2\partial_u$ be two vector fields on $\mathbb{R}^4$. The distribution generated by $\{X, Y\}$ is regular and bracket-generating of step 3 everywhere.

Example 3.8.5 (Distribution on $\mathbb{S}^3$) Let $M = \mathbb{S}^3 = \{x \in \mathbb{R}^4; \|x\| = 1\}$ and consider the vector fields

$$\begin{aligned} X &= x_2\partial_{x_1} - x_1\partial_{x_2} - x_4\partial_{x_3} + x_3\partial_{x_4} \\ Y &= x_4\partial_{x_1} - x_3\partial_{x_2} + x_2\partial_{x_3} - x_1\partial_{x_4} \\ Z &= x_3\partial_{x_1} + x_4\partial_{x_2} - x_1\partial_{x_3} - x_2\partial_{x_4}, \end{aligned}$$

which are tangent to $\mathbb{S}^3$ and form a basis of the tangent space $T_x\mathbb{S}^3$ at each point $x \in \mathbb{S}^3$. The distribution generated by $\{X, Y\}$ is regular and bracket-generating of step 2, since $[X, Y] = 2Z$.

It is worth noting that the distribution generated by all three vector fields, $\{X, Y, Z\}$, in $\mathbb{R}^4$ is regular and not bracket-generating on $\mathbb{R}^4$, because $[X, Y] = 2Z$, $[Y, Z] = 2X$, and $[Z, X] = 2Y$. However, the distribution is involutive and hence integrable. The integral manifold is the unit sphere $\mathbb{S}^3$.

3.9 Examples of Nonholonomic Systems

We shall present in the following a few examples of nonholonomic systems and their relation with sub-Riemannian geometry and diffusions.

The rolling disk This simple model was studied in Calin et al. [26], [40] and Zhou and Chirikjian [144]. Since the model satisfies the bracket-generating condition, it follows that the disk can be moved continuously between any two given positions.

A disk of radius R rolls without slipping on a horizontal plane and is constrained to be vertical at all times. The position of the disk is parameterized by coordinates $(x, y, \phi, \psi) \in \mathbb{R}^2 \times \mathbb{S}^1 \times \mathbb{S}^1$, where (x, y) serves as parameters for the center of the disk, ϕ is the angle made by some fixed radius on the disk with the vertical, and ψ is the angle made by the plane of the disk with the x-axis.

The motion of the disk corresponds to a curve on the space $M = \mathbb{R}^2 \times \mathbb{S}^1 \times \mathbb{S}^1$ subject to the constraints given by the following one-forms

$$\omega_1 = dx - R\cos\psi\, d\phi, \quad \omega_2 = dy - R\sin\psi\, d\phi.$$

These constraints define a rank 2 distribution $\mathcal{D}$ generated by the linearly independent vector fields

$$X_1 = \partial_\psi, \qquad X_2 = R(\cos\psi\, \partial_x + \sin\psi\, \partial_y) + \partial_\phi.$$

If we consider two more vector fields

$$X_3 = R(-\sin\psi\, \partial_x + \cos\psi\, \partial_y), \quad X_4 = -R(\cos\psi\, \partial_x + \sin\psi\, \partial_y),$$

the following commutation relations hold

$$[X_1, X_2] = -[X_1, X_4] = X_3, \quad [X_1, X_3] = X_4, \quad [X_2, X_3] = [X_2, X_4] = 0.$$

Since $\{X_1, X_, X_3, X_4\}$ forms a basis of the tangent space of $M = \mathbb{R}^2 \times \mathbb{S}^1 \times \mathbb{S}^1$ at each point, then by Chow-Rashevskii's Theorem we obtain the following connectivity result, which states that a coin can be rolled continuously between any two given states:

Proposition 3.9.1 *Given two states $(x_0, y_0, \phi_0, \psi_0)$, $(x_1, y_1, \phi_1, \psi_1)$ in $M = \mathbb{R}^2 \times \mathbb{S}^1 \times \mathbb{S}^1$, there is at least one piece-wise smooth trajectory of the disk that starts with the contact point (x_0, y_0) and initial angles ϕ_0, ψ_0 and ends at the contact point (x_1, y_1) having final angles ϕ_1 and ψ_1.*

It can be shown that the curve $\gamma(s) = \big(x(s), y(s), \phi(s), \psi(s)\big)$ is horizontal if and only if its velocity satisfies $\dot\gamma(s) = \dot\psi(s)X_1 + \dot\phi(s)X_2$. Hence, the dynamics of the rolling coin can be described by the following Lagrangian with constraints

$$\begin{aligned}\mathcal{L} &= \frac{1}{2}\|\dot\gamma(s)\|_X^2 + \lambda_1\omega_1(\dot\gamma(s)) + \lambda_2\omega_2(\dot\gamma(s)) \\ &= \frac{1}{2}\big(\dot\psi(s)^2 + \dot\phi(s)^2\big) + \lambda_1(\dot x - R\dot\phi\cos\psi) + \lambda_2(\dot y - R\dot\phi\sin\psi),\end{aligned}$$

where λ_j are Lagrange multipliers and $\|\cdot\|_X$ is the sub-Riemannian norm associated with $\{X_1, X_2\}$. The energy-minimizing horizontal curves optimize the action $\int \mathcal{L}\,ds$ and are solutions of the Euler-Lagrange equations. The explicit solutions involve expressions involving theta functions, elliptic functions and inverse hyperbolic functions.

Since the bracket-generating condition holds, then Hörmander's Theorem, Theorem 3.4.4, implies that the differential operator $L = \frac{1}{2}(X_1^2+X_2^2)$ is hypoelliptic and its heat kernel is $\mathcal{C}^\infty$-smooth. The associated sub-elliptic diffusion will be approached in Chapter 7.

The two-wheel cart We consider a cart with two equal wheels of radius R that can roll at different speeds on a plane, so the orientation of the cart might change at any time. Being connected by an axle of constant length L, the wheels steer together. The state of the cart can be parameterized by the following five coordinates: the coordinates $(x, y) \in \mathbb{R}^2$ of the axle center projection on the (x, y)-plane; the angle $\theta \in \mathbb{S}^1$ made by the plane of the wheels with the x-axis; the angles $\phi_1, \phi_2 \in \mathbb{S}^1$ made by some fixed radii of the wheels with the vertical direction. These coordinates are related by the following three nonholonomic constraints due to rolling without slipping and rotation angle of the axle

$$\omega_1 = dx - \frac{1}{2}R\cos\theta(d\phi_1 + d\phi_2)$$

$$\omega_2 = dy - \frac{1}{2}R\sin\theta(d\phi_1 + d\phi_2)$$

$$\omega_3 = d\theta - \frac{R}{L}\cos\theta(d\phi_1 - d\phi_2).$$

The motion of the cart can be described by a curve $(x(s), y(s), \theta(s), \phi_1(s), \phi_2(s))$ on the space $M = \mathbb{R}^2\times\mathbb{S}^1\times\mathbb{S}^1\times\mathbb{S}^1$ subject to the constraints $\omega_j = 0, 1 \le j \le 3$. This is equivalent by stating that the curve is tangent to the rank 2 distribution $\mathcal{D} = \bigcap_{j=1}^3 \ker\omega_j$. A basis of the distribution $\mathcal{D}$ is given by the vector fields

$$\begin{aligned}X_1 &= \frac{R}{2}\cos\theta\,\partial_x + \frac{R}{2}\sin\theta\,\partial_y + \frac{R}{L}\partial_\theta + \partial_{\phi_1} \\ X_2 &= \frac{2R}{L}\partial_\theta + \partial_{\phi_1} - \partial_{\phi_2}.\end{aligned}$$

Considering two more vector fields

$$X_3 = \sin\theta\,\partial_x - \cos\theta\,\partial_y, \quad X_4 = \cos\theta\,\partial_x + \sin\theta\,\partial_y,$$

we note the following commutation relations

$$[X_1, X_2] = \frac{R^2}{L}X_3, \qquad [X_1, X_3] = \frac{R}{L}X_4, \qquad [X_2, X_3] = \frac{2R}{L}X_4$$
$$[X_1, X_4] = -\frac{R}{L}X_3, \qquad [X_2, X_4] = -\frac{2R}{L}X_3.$$

Therefore the rank of $\mathrm{Lie}(X_1, X_2)$ is 4, while $\dim M = 5$. Hence, the bracket-generating condition does not hold anywhere on M. Even if Chow-Rashevskii's Theorem cannot be applied, the global connectivity by smooth horizontal curves still holds, see Alshamary et al. [4]. In fact, it has been shown in [4] that the cart system can be moved continuously between any two given states such that the total energy is minimized. The dynamics of the system is described by a Lagrangian with constraints, which is the total kinetic energy of the system plus the nonholonomic constraints as follows

$$\begin{aligned}\mathcal{L} &= \frac{1}{2}m(\dot{x}^2 + \dot{y}^2) + \frac{1}{2}I\dot{\theta}^2 + \frac{1}{2}J(\dot{\phi}_1^2 + \dot{\phi}_2^2) \\ &\quad + \mu_1\big(\dot{x} - \frac{1}{2}R\cos\theta(\dot{\phi}_1 + \dot{\phi}_2)\big) + \mu_2\big(\dot{y} - \frac{1}{2}R\sin\theta(\dot{\phi}_1 + \dot{\phi}_2)\big) \\ &\quad + \mu_3\big(\dot{\theta} - \frac{R}{L}(\dot{\phi}_1 - \dot{\phi}_2)\big),\end{aligned}$$

where m denotes the mass of the cart, I is the moment of inertia of the cart about the symmetry axis perpendicular to the rolling plane and passing through the center of the axle, and J is the moment of inertia about an axis passing through the centers of the wheels. The Lagrange multipliers are denoted by μ_i and the Euler-Lagrange system of equation can be explicitly solved and global connectivity by horizontal curves can be proved.

The knife edge This example deals with a nonholonomic system called knife edge or skate on a horizontal plane. To describe the problem, we let (x, y) be the coordinates of the contact point of the knife edge with the plane. Let θ denote the orientation angle between the knife edge and the xy-axes. The skating motion can be described by a curve $\big(x(s), y(s), \theta(s)\big) \in \mathbb{R}^2 \times \mathbb{S}^1$ subject to the nonholonomic constraint $\dot{x}\sin\theta = \dot{y}\cos\theta$. This is equivalent to stating that the curve is tangent to the rank 2 distribution $\mathcal{D}$ generated by the vector fields

$$X_1 = \cos\theta\,\partial_x + \sin\theta\,\partial_y, \qquad X_2 = \partial_\theta.$$

Since $[X_1, X_2] = \sin\theta\,\partial_x - \cos\theta\,\partial_y$, it follows that the distribution $\mathcal{D}$ is bracket-generating. Hence, Chow-Rashevskii's Theorem can be applied and the global connectivity by horizontal curves holds. This means that given the initial and final positions $(x_0, y_0, \theta_0), (x_1, y_1, \theta_1) \in \mathbb{R}^2 \times \mathbb{S}^1$, the knife edge can move continuously between these two positions. Since any smooth curve $\gamma(s) = (x(s), y(s), \theta(s)) \in \mathbb{R}^2 \times \mathbb{S}^1$ can be written as

$$\dot{\gamma} = (\dot{x}\cos\theta + \dot{y}\sin\theta)X + \dot{\theta}Y + (\dot{x}\sin\theta - \dot{y}\cos\theta)[X, Y],$$

it follows that the curve $\gamma(s)$ is tangent to the distribution $\mathcal{D}$ if and only if $\dot{x}\sin\theta = \dot{y}\cos\theta$. More remarks on the geometry and mechanics of this model can be found in Bellaïche [12] and Bloch [14], while a characterization of the geodesics using the Hamiltonian formalism can be found in Calin and Chang [26], page 256.

The sum of squares operator

$$\Delta_X = \frac{1}{2}(X_1^2 + X_2^2)$$

is hypoelliptic by Theorem 3.4.4 and hence its heat kernel is $\mathcal{C}^\infty$. An integral closed form expression of this kernel, involving Mathieu funtions, can be found in Agrachev et al. [3] A study of the diffusion associated to Δ_X will be done in Chapter 7.

The following necessary condition of hypoellipticity states the horizontal connectedness property.

Proposition 3.9.2 *Let X_j be vector fields on the connected open set $U \subseteq \mathbb{R}^n$ with $\mathcal{C}^\infty$ coefficients. If $L = \sum_{i=1}^r X_i^2 + X_0$ is hypoelliptic and regular on U, then any two points $p, q \in U$ can be joined by a piece-wise curve tangent to the horizontal subbundle $x \to \mathrm{span}\{X_0, \dots, X_r\}_x \subset T_x\mathbb{R}^n$*

Proof: By Proposition 3.4.3 we obtain that $\{X_0, \dots, X_r\}$ satisfy the bracket-generating condition on U. Applying Chow-Rashevskii's theorem, Theorem 3.7.2, we obtain the desired conclusion. ■

It is worth noting that the previous result still holds if we assume that $L = \sum_{i=1}^r X_i^2 + X_0$ is hypoelliptic and the coefficients of the vector fields X_j are analytic functions.

3.10 Summary

This chapter explores the topic of hypoelliptic operators and sub-Riemannian geometry. The chapter begins by introducing elliptic operators that are represented as sums of squares of vector fields, and discussing the corresponding Riemannian structure on the coordinate space. It then moves on to define hypoelliptic operators, which are differentiable operators with a smooth regularity condition. Hörmander's theorem is utilized to establish the connection between the bracket-generating condition and hypoelliptic operators.

The latter half of this chapter delves into the definition of differential distributions as subbundles of the tangent bundle, exploring their characteristics such as integrability, involutivity, rank, dimension, and step. The chapter examines a particular class of distributions, namely the bracket-generating distributions, which align with the Hörmander condition. These distributions are recognized for their horizontal connectivity through curves that are tangent to the distribution, as demonstrated by Chow-Rashewskii's theorem. The chapter provides various examples of distributions and their applications to model nonholonomic systems like the rolling disk, 2-wheel cart, knife-edge, and more.

3.11 Exercises

Exercise 3.11.1 Let $E(x) = \frac{1}{\pi}\ln\|x\|$, $x \in \mathbb{R}^2$ and denote by $\delta(x)$ the Dirac function. Show that $\Delta_2 E(x) = \delta(x)$, namely

$$\iint_{\mathbb{R}^2} \phi(x)\Delta_2 E(x)\, dx_1 dx_2 = \phi(0,0), \qquad \forall \phi \in C_0^\infty(\mathbb{R}^2),$$

where $\Delta_2 = \frac{1}{2}(\partial_{x_1}^2 + \partial_{x_2}^2)$.

Exercise 3.11.2 Let $u(t,x) = \displaystyle\int_{\mathbb{R}} \cos\left((t+x)\sqrt{1+s^2}\right)\frac{1}{(1+s^2)^2}\, ds.$

(*a*) Show that $u \in \mathcal{C}^2(\mathbb{R}^2)\backslash\mathcal{C}^\infty(\mathbb{R}^2)$.

(*b*) Verify that $(\partial_t^2 - \partial_x^2)u(t,x) = 0$.

(*c*) Deduce that the operator $\partial_t^2 - \partial_x^2$ is not hypoelliptic.

Exercise 3.11.3 (*a*) For any real constants a, b, c show that the operator $P = a\dfrac{d^2}{dx^2} + b\dfrac{d}{dx} + c$ is hypoelliptic on $\mathbb{R}$.

(*b*) The dynamics of an *RLC*-circuit (resistor-inductor-capacitor circuit) is given by Kirchhoff's equation

$$L\frac{d^2q}{dx^2} + R\frac{dq}{dx} + \frac{1}{C}q = f,$$

where f is the exterior voltage source and q represents the charge on the capacitor, while x denotes time. Provide a physical interpretation of the hypoellipticity proved in part (*a*) in the context of *RLC*-circuits.

Exercise 3.11.4 Let P and Q be two differential operators on $\mathbb{R}^n$.

(a) Show that if P and Q are hypoelliptic, then the compositions $P \circ Q$ and $Q \circ P$ are both hypoelliptic.

(b) Prove that if $P \circ Q$ and $Q \circ P$ are hypoelliptic then P and Q are hypoelliptic.

Exercise 3.11.5 Let $X = \sum_{j=1}^n a_j(x)\partial_{x_j}$ be a non-zero C^∞-vector field on $\mathbb{R}^n$.

(a) Show that there is a function $u \in C^1(\mathbb{R}^n) \backslash C^\infty(\mathbb{R}^n)$ such that $\operatorname{grad} u$ is normal to the vector field X.

(b) Show that X is not hypoelliptic on $\mathbb{R}^n$.

(c) Is the operator $Q = X^2$ hypoelliptic on $\mathbb{R}^n$?

(d) Give an example of two operators that are not hypoelliptic whose sum is hypoelliptic.

Exercise 3.11.6 Let $K = [-1, 1] \times [0, 1] \subset \mathbb{R}^2$ and consider the complete metric space $(\mathcal{C}^2(K), d)$, endowed with the distance

$$d(u, v) = \max_{x \in K} \Big\{ (|u(x) - v(x)| + |u'(x) - v'(x)| + |u''(x) - v''(x)|) e^{-k(x_1 + x_2)} \Big\}, \ k > 0.$$

(a) Let $a \in C^\infty(K)$ and choose $k > 1 + \max_{x \in K} |a(x)|$. Show that the integral operator $T : \mathcal{C}^2(K) \to \mathcal{C}^2(K)$

$$(Tu)(x) = |x_1|^3 - \int_0^{x_1} \int_0^{x_2} a(y_1, y_2) u(y_1, y_2)\, dy_1 dy_2,$$

is a contraction of the metric space $(\mathcal{C}^2(K), d)$.

(b) Using part (a) and the fixed point theorem, show that the boundary value problem

$$\begin{aligned} \partial_{x_1}\partial_{x_2} u + a(x) u &= 0 \\ u(0, x_2) &= 0 \\ u(x_1, 0) &= |x_1|^2 \end{aligned}$$

has a unique solution $u \in \mathcal{C}^2(K) \backslash \mathcal{C}^\infty(K)$.

(c) Is the operator $P = \partial_{x_1}\partial_{x_2} + a(x)$ hypoelliptic?

(d) Show that the operator $Q = \partial_{y_1}^2 - \partial_{y_2}^2 + b(y)$ is not hypoelliptic for any function $b \in \mathcal{C}^\infty(K)$.

Exercise 3.11.7 Let $P = \partial_{x_1}^2 - 2\partial_{x_1}\partial_{x_2} + \partial_{x_2}^2$ be a differential operator on $\mathbb{R}^2$.

(a) Show that the principal symbol of P is non-negative definite.

(b) Prove that P is not hypoelliptic.

Exercise 3.11.8 Let $P = \sum_{i,j=1}^{n} a_{ij}\partial_{x_i}\partial_{x_j} + \sum_{k=1}^{n} b_k \partial_{x_k}$ be a differential operator on $\mathbb{R}^2$ with constant coefficients.

(a) Show that the principal symbol of P satisfies

$$\sigma_p(x,\xi) = \sum_{i,j=1}^{n} a_{ij}\xi_i\xi_j \geq 0 \text{ (or } \xi_j \leq 0), \ \forall \xi_i \in \mathbb{R}^2$$

(b) Does the converse hold true?

Exercise 3.11.9 Consider the differential operator $L = \frac{1}{2}(\cos x)^2 \frac{d^2}{dx^2} - (\sin x)\frac{d}{dx}$.

(a) Using Hörmander's theorem show that L is hypoelliptic.

(b) Show that the one-dimensional diffusion

$$\begin{aligned} dX_t &= \cos X_t \, dW_t - \sin X_t \, dt \\ X_0 &= 0 \end{aligned}$$

has a unique strong solution and its density is C^∞-smooth.

(c) Using Gronwall's lemma prove that $\mathbb{E}[X_t] = 0$.

Exercise 3.11.10 (Hairer, [72]) Let U and V be two vector fields on $\mathbb{R}^n$ and consider a curve $x_n(t)$ such that

$$\dot{x}_n(t) = \dot{u}_n(t)U(x(t)) + \dot{v}_n(t)V(x(t)),$$

where $u_n(t) = \frac{1}{n}\cos(n^2 t)$ and $v_n(t) = \frac{1}{n}\sin(n^2 t)$. Assume the limit $x(t) = \lim_{n\to\infty} x_n(t)$ exists. Show that $\dot{x}(t) = \frac{1}{2}[U,V](x(t))$. It can be inferred from this that by combining movements in the directions of U and V, it is possible to achieve an approximation of the motion in the direction of the bracket $[U,V]$, with a high degree of accuracy.

Exercise 3.11.11 Let $X_1, \ldots, X_k, Y$ be smooth vector fiends on $\mathbb{R}^n$ and denote $L = \frac{1}{2}\sum_{i=1}^{k} X_i^2 + Y$. We introduce an extra variable, x_{n+1}, and a non-vanishing function $\rho(x_{n+1}) = 2 + \sin x_{n+1}$. Then define new vector fields $\hat{X}_i = \sqrt{\rho(x_{n+1})}\, X_i$, $1 \leq i \leq k$, $\hat{X}_{n+1} = \frac{\partial}{\partial x_{n+1}}$, and $\hat{Y} = \rho(x_{n+1})Y$, and let $\hat{L} = \frac{1}{2}\sum_{i=1}^{k+1} \hat{X}_i^2 + \hat{Y}$.

(a) Show that if $\hat{L}$ is hypoelliptic, then L is hypoelliptic.

(b) Show that the vector fields $\{X_1, \cdots, X_k, Y\}$ satisfy the bracket generating condition in $\mathbb{R}^n$ if and only if $\{\hat{X}_1, \cdots, \hat{X}_k, \hat{X}_{k+1}, \hat{Y}\}$ satisfy the bracket generating condition in $\mathbb{R}^{n+1}$.

(c) If the vector fields $\{X_1, \cdots, X_k, Y\}$ satisfy the bracket generating condition in $\mathbb{R}^n$, then $\hat{L}$ is hypoelliptic if and only if L is hypoelliptic.

Exercise 3.11.12 Consider the distribution $\mathcal{D}$ in $M = \mathbb{R}^4_x \times \mathbb{R}^3_y$ generated by the vector fields

$$\begin{aligned} X_1 &= \partial_{x_1} + x_2\partial_{y_1} - x_4\partial_{y_2} - x_3\partial_{y_3} \\ X_2 &= \partial_{x_2} - x_1\partial_{y_1} - x_3\partial_{y_2} + x_4\partial_{y_3} \\ X_3 &= \partial_{x_3} + x_4\partial_{y_1} + x_2\partial_{y_2} + x_1\partial_{y_3} \\ X_4 &= \partial_{x_4} - x_3\partial_{y_1} + x_1\partial_{y_2} - x_2\partial_{y_3}. \end{aligned}$$

(a) Show that $\mathcal{D}$ is of step 2 everywhere on M.

(b) Find three one-forms ω_j on M such that $\mathcal{D} = \bigcap_{j=1}^3 ker\ \omega_j$.

(c) Is the operator $L = \frac{1}{2}\sum_{j=1}^4 X_j^2$ hypoelliptic?

Exercise 3.11.13 Consider the vector fields $X_1 = \partial_{x_1} + e^{x_2}\partial_y$ and $X_2 = \partial_{x_2}$ on $\mathbb{R}^3 = \mathbb{R}^2_x \times \mathbb{R}_y$. Show that the heat kernel of the operator $L = \frac{1}{2}(X_1^2 + X_2^2)$ is a C^∞-function.

Exercise 3.11.14 Let $X_1 = \partial_{x_1} + 2x_2|x|^{2k}\partial_y$ and $X_2 = \partial_{x_2} - 2x_1|x|^{2k}\partial_y$ be vector fields on $\mathbb{R}^3$, where $|x|^2 = x_1^2 + x_2^2$. Show that the heat kernel of the operator $L = \frac{1}{2}(X_1^2 + X_2^2)$ is a C^∞-function.

Chapter 4

Heat Kernels with Applications

The focus of this chapter is on calculating heat kernels for various operators, consisting of sums of squares of vector fields, through the use of stochastic methods. Stochastic techniques are applicable in discovering the heat kernel for elliptic or sub-elliptic operators such as the Kolmogorov operator, Grushin operator, Heisenberg operator, and Casimir operator, among others. Since the heat kernel represents the transition probability of the corresponding diffusion, the smoothness of the transition densities associated with the aforementioned operators is a result of Hörmander's theorem.

4.1 Kolmogorov Operators

Studying the heat kernel of Kolmogorov operators has a financial incentive, particularly in pricing Asian options. The classical Black-Scholes equation emerges when a European option is based on a stock. However, when a stock moving average replaces the plain value of the stock, more intricate variants of the Black-Scholes equation arise. For example, if the underlying variable of an Asian option is a geometric moving average of a stock, a Kolmogorov-type operator is obtained. The first step in acquiring closed-form solutions for the associated Asian option is finding an explicit form for its heat kernel. This concept is explored further in Section 4.2. For more details, the reader is referred to Calin et al. [37].

In addition, Kolmogorov's description of the heat kernel as a transition density for the associated diffusion of this operator makes it historically relevant. Hörmander later extended the study of these types of operators to sum of squares hypoelliptic operators.

The joint distribution of W_t and $Z_t = \int_0^t W_s\,ds$ The following result will be useful in the computation leading to the heat kernel.

Proposition 4.1.1 *Let W_t be a Brownian motion and $Z_t = \int_0^t W_s\,ds$ be the associated integrated Brownian motion process. Then the joint density of (W_t, Z_t) is given by*

$$P(W_t \in dw, Z_t \in dz) = \frac{\sqrt{3}}{\pi t^2} e^{-\frac{2}{t^3}[w^2 t + 3z^2 - 3wzt]}\, dz dw, \quad t > 0.$$

Proof: We show first that (W_t, Z_t) has a bivariate normal distribution. For this it suffices to prove that any linear combination $\alpha_1 W_t + \alpha_2 Z_t$ with α_1, $\alpha_2 \in \mathbb{R}$ has a one-dimensional normal distribution. Using integration by parts for Wiener integrals we have

$$\begin{aligned}
\alpha_1 W_t + \alpha_2 Z_t &= \alpha_1 W_t + \alpha_2 \Big(tW_t - \int_0^t s\,dW_s\Big) = (\alpha_1 + \alpha_2 t)W_t - \alpha_2 \int_0^t s\,dW_s \\
&= \int_0^t (\alpha_1 + \alpha_2 t - \alpha_2 s)\,dW_s,
\end{aligned}$$

which is a normally distributed Wiener integral with zero mean.

Since (W_t, Z_t) is bivariate normally distributed, its density is determined by $\rho = corr(W_t, Z_t)$, $\sigma_1^2 = Var(W_t) = t$, $\sigma_2^2 = Var(Z_t) = t^3/3$ and is given by

$$f_{(W_t,Z_t)}(w,z) = \frac{1}{2\pi\sigma_1\sigma_2\sqrt{1-\rho^2}}\, e^{-\frac{1}{1-\rho^2}\left[\frac{w^2}{2\sigma_1^2} + \frac{z^2}{2\sigma_2^2} - \rho\frac{wz}{\sigma_1\sigma_2}\right]}, \tag{4.1.1}$$

since $\mathbb{E}[W_t] = \mathbb{E}[Z_t] = 0$. To find the correlation ρ, we compute first the covariance between W_t and Z_t. Using the expression $Z_t = tW_t - \int_0^t s\,dW_s$ and the properties of Wiener integrals we have

$$\begin{aligned}
Cov(W_t, Z_t) &= t^2 - Cov\Big(W_t, \int_0^t s\,dW_s\Big) = t^2 - Cov\Big(\int_0^t dW_s, \int_0^t s\,dW_s\Big) \\
&= t^2 - \mathbb{E}\Big(\int_0^t dW_s \int_0^t s\,dW_s\Big) = t^2 - \int_0^t s\,ds = \frac{t^2}{2}.
\end{aligned}$$

The correlation coefficient

$$\rho = corr(W_t, Z_t) = \frac{Cov(W_t, Z_t)}{\sigma_1\sigma_2} = \frac{\sqrt{3}}{2}$$

is independent of t. Substituting in (4.1.1) we obtain the following density

$$f_{(W_t,Z_t)}(w,z) = \frac{\sqrt{3}}{\pi t^2} e^{-\frac{2}{t^3}[w^2 t + 3z^2 - 3wzt]}.$$

■

The one-dimensional Kolmogorov operator The one-dimensional Kolmogorov operator is given by

$$L_1 = \frac{1}{2}\partial_{x_1}^2 + ax_1\partial_{x_2}, \tag{4.1.2}$$

where $a \in \mathbb{R}$ is a non-vanishing parameter. Its heat kernel is given by the transition probability of the associated diffusion $X_t = (X_1(t), X_2(t))$, which satisfies the stochastic differential equation

$$\begin{aligned} dX_1(t) &= dW_t \\ dX_2(t) &= aX_1(t)dt \end{aligned}$$

with the initial conditions $X_1(0) = x_1^0$, $X_2(0) = x_2^0$. Integrating we obtain the following expression for the *Kolmogorov diffusion* starting at (x_1^0, x_2^0)

$$\begin{aligned} X_1(t) &= x_1^0 + W_t \\ X_2(t) &= x_2^0 + ax_1^0 t + a\int_0^t W_s ds = x_2^0 + ax_1^0 t + aZ_t. \end{aligned}$$

In order to find the joint density of $(X_1(t), X_2(t))$ we notice first that any linear combination $\alpha_1 X_1(t) + \alpha_2 X_2(t)$ is normally distributed. This follows a similar procedure as the one used in the proof of Proposition 4.1.1. Then $(X_1(t), X_2(t))$ is bivariate normally distributed. The joint density depends on the following parameters

$$\begin{aligned} \sigma_1^2 &= Var(X_1(t)) = t, \quad \sigma_2^2 = Var(X_2(t)) = a^2\frac{t^3}{3} \\ \mu_1 &= \mathbb{E}[X_1(t)] = x_1^0, \quad \mu_2 = \mathbb{E}[X_2(t)] = x_2^0 + ax_1^0 t \end{aligned}$$

and $\rho = corr(X_1(t), X_2(t))$, which can be computed as previously

$$\rho = corr(W_t, aZ_t) = sign(a)\frac{\sqrt{3}}{2}.$$

Then the joint distribution of $(X_1(t), X_2(t))$ becomes

$$\begin{aligned} p_t(x_1^0, w; x_2^0, z) &= \frac{1}{2\pi\sigma_1\sigma_2\sqrt{1-\rho^2}} e^{-\frac{1}{1-\rho^2}\left[\frac{(w-\mu_1)^2}{2\sigma_1^2} + \frac{(z-\mu_2)^2}{2\sigma_2^2} - \rho\frac{(w-\mu_1)(z-\mu_2)}{\sigma_1\sigma_2}\right]} \\ &= \frac{\sqrt{3}}{|a|\pi t^2} e^{-4\left[\frac{(w-x_1^0)^2}{2t} + \frac{3(z-x_2^0-ax_1^0t)^2}{2a^2t^3} - sign(a)\frac{\sqrt{3}}{2}\frac{(w-x_1^0)(z-x_2^0-ax_1^0t)}{|a|t^2/\sqrt{3}}\right]} \\ &= \frac{\sqrt{3}}{|a|\pi t^2} e^{-\frac{2}{t}\left[(w-x_1^0)^2 + \frac{3(z-x_2^0-ax_1^0t)^2}{a^2t^2} - \frac{3(w-x_1^0)(z-x_2^0-ax_1^0t)}{at}\right]} \\ &= \frac{\sqrt{3}}{|a|\pi t^2} e^{-\frac{2}{t}\left[(w-x_1^0)^2 + \frac{3}{a^2t^2}(z-x_2^0-ax_1^0t)(z-x_2^0-awt)\right]}. \end{aligned}$$

Replacing w by x_1 and z by x_2 we obtain the expression for the heat kernel of operator (4.1.2) given by

$$p_t(x_1^0, x_1; x_2^0, x_2) = \frac{\sqrt{3}}{|a|\pi t^2} e^{-\frac{2}{t}\left[(x_1-x_1^0)^2+\frac{3}{a^2t^2}(x_2-x_2^0-ax_1^0t)(x_2-x_2^0-ax_1t)\right]}, \quad t>0. \tag{4.1.3}$$

In order to obtain an expression compatible with the one given in [37], we shall perform further computations. Writing

$$x_2 - x_2^0 - ax_1^0 t = x_2 - x_2^0 - \frac{a}{2}(x_1^0 + x_1)t + \frac{a}{2}(x_1 - x_1^0)t$$
$$x_2 - x_2^0 - ax_1 t = x_2 - x_2^0 - \frac{a}{2}(x_1^0 + x_1)t - \frac{a}{2}(x_1 - x_1^0)t,$$

the difference of squares formula yields

$$(x_2 - x_2^0 - ax_1^0 t)(x_2 - x_2^0 - ax_1 t) = \left(x_2 - x_2^0 - \frac{a}{2}(x_1^0 + x_1)t\right)^2 - \frac{a^2t^2}{4}(x_1 - x_1^0)^2.$$

Multiplying by $\frac{3}{a^2t^2}$ and adding $(x_1 - x_1^0)^2$ we obtain

$$(x_1 - x_1^0)^2 + \frac{3}{a^2t^2}(x_2 - x_2^0 - ax_1^0 t)(x_2 - x_2^0 - ax_1 t) = \frac{1}{4}(x_1 - x_1^0)^2$$
$$+ \frac{3}{a^2t^2}\left(x_2 - x_2^0 - \frac{a}{2}(x_1^0 + x_1)t\right)^2.$$

Multiplying by $-\frac{2}{t}$, we obtain the following equivalent form of formula (4.1.3)

$$p_t(x_1^0, x_1; x_2^0, x_2) = \frac{\sqrt{3}}{|a|\pi t^2} e^{-\frac{(x_1-x_1^0)^2}{2t} - \frac{6}{a^2t^3}[x_2 - x_2^0 - \frac{a}{2}(x_1^0+x_1)t]^2}, \quad t>0. \tag{4.1.4}$$

This expression involves squares in the exponent and it is more desired than the expression given by (4.1.3) because relates to the geometric concept of distance. It is worth noting that the heat kernel of the Kolmogorov operator L_1 is of a function type and this function is C^∞-smooth.

We note that if $\mathcal{X}_1 = \partial_{x_1}$ and $\mathcal{X}_2 = x_1\partial_{x_2}$ are vector fields on $\mathbb{R}^2$, then the Kolmogorov operator writes as $L_1 = \frac{1}{2}\mathcal{X}_1^2 + a\mathcal{X}_2$. Since $[\mathcal{X}_1, \mathcal{X}_2] = \partial_{x_2}$, then $\text{span}\{\mathcal{X}_1, \mathcal{X}_2, [\mathcal{X}_1, \mathcal{X}_2]\} = \text{span}\{\partial_{x_1}, \partial_{x_2}\} = \mathbb{R}^2$, so the maximal rank condition is satisfied. By Hörmander's theorem the operator L_1 is hypoelliptic.

The n-dimensional Kolmogorov operator The n-dimensional Kolmogorov operator is given by

$$L_n = \frac{1}{2}\sum_{j=1}^n \partial_{x_j}^2 + \Big(\sum_{j=1}^n a_j x_j\Big)\partial_y. \tag{4.1.5}$$

The dimension n refers to the number of squared vector fields. If $\mathcal{X}_j = \partial_{x_j}$, $j = 1, \dots, n$ and $\mathcal{Y} = \Big(\sum_{j=1}^n a_j x_j\Big)\partial_y$ are $n+1$ vector fields on $\mathbb{R}^{n+1}$, then

$$L_n = \frac{1}{2}(\mathcal{X}_1^2 + \cdots + \mathcal{X}_n^2) + \mathcal{Y}.$$

Since we have the following commutation relations

$$[\mathcal{X}_i, \mathcal{X}_j] = 0, \qquad [\mathcal{X}_j, \mathcal{Y}] = a_j \partial_y$$

it follows that $\text{span}\{\mathcal{X}_j, \mathcal{Y}, [\mathcal{X}_j, \mathcal{Y}]\} = \mathbb{R}^{n+1}$ as long as $\sum_{j=1}^n a_j^2 \neq 0$ (i.e., the vector field $\mathcal{Y}$ does not vanish). From now on we shall assume this nonvanishing condition satisfied. Then applying Hörmander's theorem the operator L_n becomes hypoelliptic and we shall find its heat kernel.

Since $L_n = \Delta_n + Y$ is the sum between the Laplacian on $\mathbb{R}^n$ and a vector field Y, we expect the associated Kolmogorov diffusion on $\mathbb{R}^{n+1}$ to be a Brownian motion on the first n components and a drifty diffusion along the $(n+1)$th component. To construct the associated diffusion we note first that the diffusion matrix and the drift vector are respectively given by

$$\sigma\sigma^T = \begin{pmatrix} \mathbb{I}_n & 0 \\ 0 & 0 \end{pmatrix}, \qquad b = \begin{pmatrix} 0 \\ \sum_{j=1}^n a_j x_j \end{pmatrix}.$$

The Kolmogorov diffusion on $\mathbb{R}^{n+1}$ is given by the stochastic process $X_t = (X_1(t), \dots, X_n(t), X_{n+1}(t))^T$, which satisfies the following system of stochastic differential equations

$$\begin{aligned} dX_1(t) &= dW_1(t) \\ &\cdots\cdots\cdots \\ dX_n(t) &= dW_n(t) \\ dX_{n+1}(t) &= \Big(\sum_{j=1}^n a_j X_j(t)\Big)dt, \end{aligned}$$

where $W_i(t)$ are independent one-dimensional Brownian motions. The initial condition is given by $X(0) = (x_1^0, \dots, x_n^0, x_{n+1}^0)^T \in \mathbb{R}^{n+1}$. Standard uniqueness results, see Øksendal [114] page 66, provide the following unique strong solution

$$\begin{aligned} X_1(t) &= x_1^0 + W_1(t) \\ &\cdots\cdots\cdots \\ X_n(t) &= x_n^0 + W_n(t) \\ X_{n+1}(t) &= x_{n+1}^0 + \Big(\sum_{j=1}^n a_j x_j^0\Big)t + \sum_{j=1}^n a_j \int_0^t W_j(s)\,ds. \end{aligned}$$

It can be shown that the aforementioned Kolmogorov diffusion X_t has a multivariate Gaussian distribution. In order to write an explicit expression for its density we need some information about the covariance matrix and its inverse.

Proposition 4.1.2 *Let $A_{ij} = Cov(X_i(t), X_j(t))$ be the covariance matrix of the process X_t. Then*

$$A = \begin{pmatrix} t & 0 & 0 & \dots & \frac{a_1 t^2}{2} \\ 0 & t & 0 & \dots & \frac{a_2 t^2}{2} \\ 0 & 0 & t & \dots & \vdots \\ 0 & 0 & \dots & t & \frac{a_n t^2}{2} \\ \frac{a_1 t^2}{2} & \frac{a_2 t^2}{2} & \dots & \frac{a_n t^2}{2} & \frac{t^3}{3}\|a\|^2 \end{pmatrix}, \tag{4.1.6}$$

where $\|a\|^2 = \sum_{j=1}^n a_j^2$.

Proof: Since the computation of the first n rows are similar, we shall perform the computation only for the first row. Using the symmetry of the matrix A we can reduce the computation just to the entries above the diagonal. Then we compute the entry $A_{n+1,n+1}$. We have

$$\begin{aligned} A_{11} &= Var(X_1(t)) = Var(W_1(t)) = t \\ A_{1k} &= Cov(X_1(t), X_k(t)) = Cov(W_1(t), W_k(t)) = 0, \quad 2 \le k \le n \\ A_{1,n+1} &= Cov(X_1(t), X_{n+1}(t)) = Cov\Big(W_1(t), \sum_{j=1}^n a_j \int_0^t W_j(s)\, ds\Big) \\ &= a_1 Cov\Big(W_1(t), \int_0^t W_1(s)\, ds\Big) = a_1 \frac{t^2}{2}, \end{aligned}$$

where we used the independence of $W_1(t)$ and $W_k(t)$, $k \ge 2$, and a computation similar to the one done in Proposition 4.1.1. For the computation of the last entry we have

$$\begin{aligned} A_{n+1,n+1} &= Cov(X_{n+1}(t), X_{n+1}(t)) \\ &= Cov\Big(\sum_{j=1}^n a_j \int_0^t W_j(s)\, ds, \sum_{i=1}^n a_i \int_0^t W_i(s)\, ds\Big) \\ &= \sum_{i,j} a_i a_j Cov\Big(\int_0^t W_j(s)\, ds, \int_0^t W_i(s)\, ds\Big) \\ &= \sum_{j=1}^n a_j^2 Var\Big(\int_0^t W_j(s)\, ds\Big) = \frac{t^3}{3}\|a\|^2, \end{aligned}$$

where we used that $\int_0^t W_j(s)\,ds \sim \mathcal{N}(0, \frac{t^3}{3})$, i.e. the integral of a Brownian motion is normally distributed.

■

The expression for the multivariate Gaussian distribution that describes the joint distribution of $(X_1(t), \dots, X_{n+1}(t))$ is given by

$$p_t(x_1, \dots, x_{n+1}) = \frac{1}{\sqrt{(2\pi)^{n+1} \det A}}\, e^{-\frac{1}{2}(\mathbf{x}-\mu)^T A^{-1}(\mathbf{x}-\mu)}, \tag{4.1.7}$$

where $\mathbf{x} = (x, x_{n+1})^T = (x_1, \dots, x_{n+1})^T$ and

$$\mu = (\mathbb{E}[X_1(t)], \dots, \mathbb{E}[X_{n+1}(t)])^T = (x_1^0, \dots, x_n^0, x_{n+1}^0 + (\Sigma_j a_j x_j^0)t)^T.$$

Therefore

$$\begin{aligned} \mathbf{x} - \mu &= \left(x_1 - x_1^0, \dots, x_n - x_n^0, x_{n+1} - x_{n+1}^0 - (\Sigma_j a_j x_j^0)t\right)^T \\ &= (x - x^0, y - y^0 - \langle a, x^0 \rangle t), \end{aligned}$$

where we adopted the notation $y = x_{n+1}$, used $\langle\ ,\ \rangle$ to denote Euclidean inner product and employed the notation ^{T} for transposition.

Proposition 4.1.3 *Let A be the matrix given by (4.1.6). Then A is invertible and*

$$\det A = \frac{\|a\|^2}{12} t^{n+3}, \qquad \forall t > 0. \tag{4.1.8}$$

Proof: Factorizing the factor t^2 out from the last row and last column of $\det A$ we obtain $\det A = \frac{t^4}{4}\Delta(a_1, \dots, a_n)$, where

$$\Delta(a_1, \dots, a_n) = \begin{vmatrix} t & 0 & 0 & \dots & \frac{a_1}{2} \\ 0 & t & 0 & \dots & \frac{a_2}{2} \\ 0 & 0 & t & \dots & \vdots \\ 0 & 0 & \dots & t & \frac{a_n}{2} \\ \frac{a_1}{2} & \frac{a_2}{2} & \dots & \frac{a_n}{2} & \frac{4}{3}\frac{\|a\|^2}{t} \end{vmatrix}.$$

Expanding over the first row and then over the first column we obtain the following recurrence relation

$$\Delta(a_1, \dots, a_n) = t\Delta(a_2, \dots, a_n) + a_1^2 t^{n-1}.$$

Iterating the recursive relation provides

$$\begin{aligned}\Delta(a_1,\dots,a_n) &= t^2\Delta(a_3,\dots,a_n) - a_2^2t^{n-1} - a_1^2t^{n-1}\\ &= t^3\Delta(a_4,\dots,a_n) - a_3^2t^{n-1} - a_2^2t^{n-1} - a_1^2t^{n-1}\\ &\dots\dots\dots\dots\dots\dots\dots\dots\dots\dots\\ &= t^k\Delta(a_{k+1},\dots,a_n) - a_k^2t^{n-1} - a_{k-1}^2t^{n-1} - \dots - a_2^2t^{n-1} - a_1^2t^{n-1}\\ &= t^{n-1}\Delta(a_n) - \sum_{j=1}^{n-1} a_j^2t^{n-1}\\ &= t^{n-1}\begin{vmatrix} t & a_n \\ a_n & \frac{4}{3}\frac{\|a\|^2}{t}\end{vmatrix} - \sum_{j=1}^{n-1} a_j^2t^{n-1} = \frac{1}{3}\|a\|^2t^{n-1}.\end{aligned}$$

Therefore,

$$\det A = \frac{t^4}{4}\Delta(a_1,\dots,a_n) = \frac{\|a\|^2}{12}t^{n+3}.$$

The matrix A is invertible since not all coefficients a_j are equal to zero. ∎

We make the remark that the matrix A is equivalent to the following upper triangular form

$$U = \begin{pmatrix} t & 0 & 0 & \dots & \frac{a_1t^2}{2}\\ 0 & t & 0 & \dots & \frac{a_2t^2}{2}\\ 0 & 0 & t & \dots & \vdots\\ 0 & 0 & \dots & t & \frac{a_nt^2}{2}\\ 0 & 0 & \dots & 0 & \frac{1}{12}\|a\|^2t^3\end{pmatrix}.$$

This can be achieved by multiplying the jth row of matrix A by $a_jt/2$ and then subtract the result from the $(n+1)$th row. The determinant can be retrieved as the product of diagonal elements, which are actually the eigenvalues of the matrix A.

In order to write the normal multivariate density (4.1.7) in explicit form, we need to compute the exponent $(\mathbf{x}-\mu)^TA^{-1}(\mathbf{x}-\mu)$. Instead of computing the inverse A^{-1} explicitly, we shall proceed as follows. We denote $u = \mathbf{x}-\mu$ and $w = A^{-1}u$, so we need to compute the inner product $\langle u, A^{-1}u\rangle = \langle u, w\rangle = \sum_{j=1}^{n+1} u_jw_j$. We start by writing the system $Aw = u$ in the matrix form

$$\begin{pmatrix} t & 0 & 0 & \dots & \frac{a_1t^2}{2}\\ 0 & t & 0 & \dots & \frac{a_2t^2}{2}\\ 0 & 0 & t & \dots & \vdots\\ 0 & 0 & \dots & t & \frac{a_nt^2}{2}\\ \frac{a_1t^2}{2} & \frac{a_2t^2}{2} & \dots & \frac{a_nt^2}{2} & \frac{t^3}{3}\|a\|^2\end{pmatrix}\begin{pmatrix} w_1\\ w_2\\ \vdots\\ w_n\\ w_{n+1}\end{pmatrix} = \begin{pmatrix} u_1\\ u_2\\ \vdots\\ u_n\\ u_{n+1}\end{pmatrix}.$$

Multiplying the jth row by $a_j t/2$, $1 \le j \le n$, and subtracting it from the $(n+1)$th row yields the system

$$\begin{aligned} tw_1 + \frac{a_1}{2}t^2 w_{n+1} &= u_1 \\ tw_2 + \frac{a_2}{2}t^2 w_{n+1} &= u_2 \\ \dots\dots\dots\dots\dots\dots\dots \\ tw_n + \frac{a_n}{2}t^2 w_{n+1} &= u_n \\ \frac{1}{12}\|a\|^2 t^3 w_{n+1} &= u_{n+1} - \langle \tilde{u}, a\rangle \frac{t}{2}, \end{aligned}$$

where $\tilde{u} = (u_1, \dots, u_n)^T$ and $\tilde{w} = (w_1, \dots, w_n)^T$. Since we are interested in just computing the inner product $\langle u, w\rangle = \langle \tilde{u}, \tilde{w}\rangle + u_{n+1}w_{n+1}$, we shall multiply the jth equation by u_j and add the first n equations to obtain

$$\langle \tilde{u}, \tilde{w}\rangle t + \frac{t^2}{2} w_{n+1}\langle a, \tilde{u}\rangle = \|\tilde{u}\|^2. \tag{4.1.9}$$

Using the last equation of the previous system we have

$$\begin{aligned} \frac{t^2}{2} w_{n+1} &= \frac{1}{12}\|a\|^2 t^3 w_{n+1} \cdot \frac{6}{t\|a\|^2} \\ &= \Big(\frac{u_{n+1}}{t} - \langle a, \tilde{u}\rangle \frac{1}{2}\Big) \cdot \frac{6}{\|a\|^2}. \end{aligned}$$

Substituting this relation into equation (4.1.9) and solving for $\langle \tilde{u}, \tilde{w}\rangle$ yields

$$\langle \tilde{u}, \tilde{w}\rangle = \frac{\|\tilde{u}\|^2}{t} - \frac{6\langle a, \tilde{u}\rangle}{\|a\|^2}\Big[\frac{u_{n+1}}{t^2} - \frac{1}{2t}\langle a, \tilde{u}\rangle\Big]. \tag{4.1.10}$$

We shall compute next the product $u_{n+1}w_{n+1}$ by multiplying the last equation of the previous system by u_{n+1}

$$\frac{1}{12}\|a\|^2 t^3 u_{n+1} w_{n+1} = u_{n+1}^2 - \langle \tilde{u}, a\rangle \frac{t}{2} u_{n+1},$$

which implies

$$u_{n+1}w_{n+1} = \frac{12}{\|a\|^2 t^3} u_{n+1}^2 - \frac{6}{\|a\|^2 t^2}\langle a, \tilde{u}\rangle u_{n+1}.$$

Using together with (4.1.10) we obtain

$$\begin{aligned} \langle u, w\rangle &= \langle \tilde{u}, \tilde{w}\rangle + u_{n+1}w_{n+1} \\ &= \frac{\|\tilde{u}\|^2}{t} + \frac{12}{t\|a\|^2}\Big[\frac{u_{n+1}^2}{t^2} - \frac{\langle a, \tilde{u}\rangle u_{n+1}}{t} + \frac{\langle a, \tilde{u}\rangle^2}{4}\Big] \\ &= \frac{\|\tilde{u}\|^2}{t} + \frac{12}{t\|a\|^2}\Big(\frac{u_{n+1}}{t} - \frac{\langle a, \tilde{u}\rangle}{2}\Big)^2. \end{aligned}$$

Since $\langle u, A^{-1}u\rangle = \langle u, w\rangle$, we get to

$$\langle u, A^{-1}u\rangle = \frac{\|\tilde{u}\|^2}{t} + \frac{12}{t^3\|a\|^2}\Big(u_{n+1} - \frac{1}{2}\langle a, \tilde{u}\rangle t\Big)^2. \tag{4.1.11}$$

Using $u = (\tilde{u}, u_{n+1}) = (x - x^0, y - y^0 - \langle a, x^0\rangle t)$, we obtain

$$\frac{1}{2}u^T A^{-1} u = \frac{\|x - x^0\|^2}{2t} + \frac{6}{t^3\|a\|^2}\Big((y - y^0) - \frac{t}{2}\langle a, x + x^0\rangle\Big)^2. \tag{4.1.12}$$

The heat kernel of the Kolmogorov operator is provided by the next result.

Proposition 4.1.4 *The heat kernel of the operator*

$$L_n = \frac{1}{2}\sum_{j=1}^{n} \partial^2_{x_j} + \Big(\sum_{j=1}^{n} a_j x_j\Big)\partial_y$$

is given by

$$K(x^0, y^0, x, y; t) = \frac{\sqrt{6}}{2^{\frac{n}{2}}\pi^{\frac{n+1}{2}}\|a\| t^{\frac{n+3}{2}}} e^{-\frac{\|x-x^0\|^2}{2t} - \frac{6}{t^3\|a\|^2}[(y-y^0) - \frac{t}{2}\langle a, x+x^0\rangle]^2}, \quad \forall t > 0. \tag{4.1.13}$$

Proof: The heat kernel of L_n is given by the transition probability density of the associated diffusion process X_t. Substituting relations (4.1.8) and (4.1.12) into the multivariate density formula (4.1.7), after some algebraic manipulations, we obtain formula (4.1.13). ■

4.2 Options on a Geometric Moving Average

As specified in the begining of the previous section, the study of the heat kernel for Kolmogorov operators has applications to pricing Asian options. In the following we shall delve into this topic following Calin et. al [37].

The Geometric Moving Average We start by dividing the time interval $[a, b]$ into n equal subintervals of equal length $\Delta t = t_{k+1} - t_k = \frac{b-a}{n}$. Then we consider the discrete geometric average of the stock values S_{t_k} at time instances t_k, which is given by

$$G(t_1, \ldots, t_n) = \Big(\prod_{k=1}^{n} S_{t_k}\Big)^{1/n} = e^{\ln\left(\prod_{k=1}^{n} S_{t_k}\right)^{1/n}} = e^{\frac{1}{b-a}\sum_{k=1}^{n} \ln S_{t_k} \frac{b-a}{n}}.$$

Taking the limit $n \to \infty$ and using the definition of the integral we obtain

$$G_{a,b} = \lim_{n\to\infty}\Big(\prod_{k=1}^{n} S_{t_k}\Big)^{1/n} = e^{\frac{1}{b-a}\lim_{n\to\infty}\sum_{k=1}^{n} \ln S_{t_k}\frac{b-a}{n}} = e^{\frac{1}{b-a}\int_a^b \ln S_u\, du}.$$

$G_{a,b}$ represents the *continuous geometric moving average* of a stock over the time interval $[a, b]$.

The associated Black-Scholes equation In the following we shall obtain a parabolic partial differential equation, which describes the price of an Asian option on a geometric moving average. Let $n \geq 1$ be a positive integer. We assume the value $v = v(\tau, s_1, s_2, s_3)$ of the Asian option depends on the following variables: time variable τ, stock values $s_1 = S_\tau$, $s_2 = S_{\tau-n}$ at instances τ and $\tau - n$, respectively, and geometric moving average on the interval $[\tau - n, \tau]$

$$s_3 = G_{\tau-n,\tau} = e^{\frac{1}{n}\int_{\tau-n}^{\tau} \ln s_1(u)\, du}.$$

Denoting the stock return by μ and the volatility by σ, we assume the stock satisfies a geometric Brownian motion. Then

$$ds_1 = \mu s_1 d\tau + \sigma s_1 dW_1, \quad ds_2 = \mu s_2 d\tau + \sigma s_2 dW_2, \quad ds_3 = \frac{s_3}{n} \ln\left(\frac{s_1}{s_2}\right) d\tau,$$

where W_1 and W_2 are two Brownian motions with increments correlated by a factor ρ and ds_3 was computed by applying Ito's formula. μ and σ are considered positive constants.
Since $(ds_1)^2 = \sigma^2 s_1^2 d\tau$, $(ds_2)^2 = \sigma^2 s_2^2 d\tau$, $ds_1 ds_2 = \sigma^2 s_1 s_2 \rho d\tau$ and $(ds_3)^2 = 0$, applying Ito's formula we obtain

$$\begin{aligned} dv &= \partial_\tau v\, d\tau + \partial_{s_1} v\, ds_1 + \partial_{s_2} v\, ds_2 + \partial_{s_3} v\, ds_3 \\ &\quad + \frac{1}{2}\partial_{s_1}^2 v\, (ds_1)^2 + \frac{1}{2}\partial_{s_2}^2 v\, (ds_2)^2 + \frac{1}{2}\partial_{s_1}\partial_{s_2} v\, ds_1 ds_2 \\ &= [\partial_\tau v + \mu s_1 \partial_{s_1} v + \mu s_2 \partial_{s_2} v + \frac{1}{2}\sigma^2 s_1^2 \partial_{s_1}^2 v + \frac{1}{2}\sigma^2 s_2^2 \partial_{s_2}^2 v + \sigma^2 s_1 s_2 \rho \partial_{s_1}\partial_{s_2} v \\ &\quad + \frac{s_3}{n} \ln\left(\frac{s_1}{s_2}\right) \partial_{s_3} v] d\tau + \sigma s_1\, \partial_{s_1} v\, dW_1 + \sigma s_2\, \partial_{s_2} v\, dW_2. \end{aligned}$$

In order to reduce the stochastic increments dW_1 and dW_2 we form the portfolio $P = v - a_1 s_1 - a_2 s_2$ and choose the coefficients a_1 and a_2 such that P is risk-free. This means that the return on the portfolio P invested in a bank at interest rate r will grow to $rPd\tau$ in a time $d\tau$ and this should equal the return on the portfolio, i.e., $dP = rPd\tau$. Since $dP = dv - a_1 ds_1 - a_2 ds_2$, the riskless condition implies $a_1 = \partial_{s_1} v$ and $a_2 = \partial_{s_2} v$. Then the condition $dP = rPd\tau$ implies

$$\partial_\tau v + \frac{1}{2}\sigma^2 s_1^2 \partial_{s_1}^2 v + \frac{1}{2}\sigma^2 s_2^2 \partial_{s_2}^2 v + \sigma^2 s_1 s_2 \rho \partial_{s_1}\partial_{s_2} v + \frac{s_3}{n} \ln\left(\frac{s_1}{s_2}\right) \partial_{s_3} v = rP,$$

which is equivalent to

$$\begin{aligned} \partial_\tau v + \frac{1}{2}\sigma^2 s_1^2 \partial_{s_1}^2 v + \frac{1}{2}\sigma^2 s_2^2 \partial_{s_2}^2 v + r s_1 \partial_{s_1} v + r s_2 \partial_{s_2} v + \sigma^2 s_1 s_2 \rho \partial_{s_1}\partial_{s_2} v \\ + \frac{s_3}{n} \ln\left(\frac{s_1}{s_2}\right) \partial_{s_3} v = rv. \qquad (4.2.14) \end{aligned}$$

With the exception of the logarithmic term, the above expression looks like a Black-Scholes equation in two variables, s_1 and s_2.

Simplifying the equation In the following we shall make a change of variables which will make the coefficients of the derivatives in the left side simpler. Let

$$s_1 = e^{x_1}, \quad s_2 = e^{x_2}, \quad s_3 = e^{\frac{y}{\omega}}, \quad \tau = T - \frac{2t}{\sigma^2},$$

with the term ω to be chosen later and denote $w(t, x_1, x_2, y) = v(\tau, s_1, s_2, s_3)$. Substituting in equation (4.2.14) and using relations

$$\partial_\tau = -\frac{1}{2}\sigma^2\partial_t, \ \partial_{x_j} = s_j\partial_{s_j}, \ \partial^2_{x_j} = s_j\partial_{s_j} + s_j^2\partial^2_{s_j}, \ \partial_y = \frac{1}{\omega}s_3\partial_{s_3}, \ j = 1, 2$$

and notation $w(t, x_1, x_2, y) = v(\tau, s_1, s_2, s_3)$ we obtain the following parabolic equation

$$\partial_t w = \partial^2_{x_1}w + \partial^2_{x_2}w + (k-1)\partial_{x_1}w + (k-1)\partial_{x_2}w + \frac{k\omega}{nr}(x_1 - x_2)\partial_y w - kw + \rho\sigma^2\partial_{x_1}\partial_{x_2}w, \tag{4.2.15}$$

where $k = 2r/\sigma^2$. Choosing $\omega = nr/k$ makes the coefficient of $(x_1 - x_2)$ equal to 1. We remove next the convection terms by considering

$$w(t, x_1, x_2, y) = e^{\alpha x_1 + \beta x_2 + \gamma t}u(t, x_1, x_2, y)$$

with α, β and γ to be specified later. To further simplify the equation we also consider the correlation coefficient $\rho = 0$. We obtain the equation

$$\begin{aligned}\gamma u + \partial_t u = {} & \partial^2_{x_1}u + \partial^2_{x_2}u \\ & + (2\alpha + k - 1)\partial_{x_1}u + (2\beta + k - 1)\partial_{x_2}u \\ & + [\alpha^2 + \beta^2 + \alpha(k-1) + \beta(k-1) - k]u \\ & + (x_1 - x_2)\partial_y u.\end{aligned} \tag{4.2.16}$$

The coefficients of $\partial_{x_1}u$ and $\partial_{x_2}u$ vanish for $\alpha = \beta = \frac{1-k}{2}$. Equating the coefficients of u in the previous equation yields $\gamma = -\frac{(1-k)^2}{2} - k$. With these choices of α, β and γ equation (4.2.16) takes the following simple form

$$\partial_t u = \partial^2_{x_1}u + \partial^2_{x_2}u + (x_1 - x_2)\partial_y u. \tag{4.2.17}$$

If furthermore, we replace x_1 by $x_1/\sqrt{2}$ and x_2 by $x_2/\sqrt{2}$, we arrive at the more familiar form

$$\partial_t u = \frac{1}{2}\partial^2_{x_1}u + \frac{1}{2}\partial^2_{x_2}u + \sqrt{2}(x_1 - x_2)\partial_y u,$$

which represents the heat equation for the two-dimensional Kolmogorov-type operator

$$L_2 = \frac{1}{2}\partial_{x_1}^2 + \frac{1}{2}\partial_{x_2}^2 + \sqrt{2}(x_1 - x_2)\partial_y.$$

Particularizing formula (4.1.13) for this case we obtain the following heat kernel for L_2

$$K(x^0, y^0, x, y; t) = \frac{\sqrt{6}}{4\pi^{3/2}t^{5/2}} e^{-\frac{\|x-x^0\|^2}{2t} - \frac{3}{2t^3}\left[y - y^0 - \frac{t}{\sqrt{2}}(x_1 + x_1^0 - x_2 - x_2^0)\right]^2}, \quad t > 0.$$

In particular, if the derivative price depends only on the stock price s_1 and the geometric moving average s_3, i.e., $v = v(\tau, s_1, s_3)$, we obtain a heat equation for the one-dimensional Kolmogorov operator

$$L_1 = \frac{1}{2}\partial_{x_1}^2 + \sqrt{2}x_1\partial_y.$$

Similarly, if the derivative price depends on s_2 and s_3, i.e., $v = v(\tau, s_2, s_3)$, we obtain a heat equation for the operator

$$\tilde{L}_1 = \frac{1}{2}\partial_{x_2}^2 - \sqrt{2}x_2\partial_y,$$

whose heat kernel was studied for the first time by Kolmogorov [90] as a transition probability of a diffusion process.

4.3 The Grushin Operator

The Grushin operator on $\mathbb{R}^2$ is given by $\Delta_G = \frac{1}{2}(\partial_x^2 + x^2\partial_y^2)$. The operator is degenerate along the line $\{x = 0\}$ and elliptic on $\mathbb{R}^2\backslash\{x = 0\}$. Since we can write $\Delta_G = \frac{1}{2}(X^2 + Y^2)$, with $X = \partial_x$, $Y = x\partial_y$ and we have $[X, Y] = \partial_y$, the vector fields $\{X, Y\}$ satisfy the bracket generating condition on $\mathbb{R}^2$, so by Hörmander's theorem, Theorem 3.4.4, the operator Δ_G is hypoelliptic.

The diffusion associated with the Grushin operator Δ_G, called the *two-dimensional Grushin diffusion*, is denoted by $G_t = (X(t), Y(t))$ and satisfies

$$\begin{aligned} dX(t) &= dW_1(t) \\ dY(t) &= X(t)dW_2(t). \end{aligned}$$

We denote by $p_t(x_0, y_0; x, y)$ the transition probability density of G_t, which starts at (x_0, y_0) and reaches (x, y) in time t. Since there is no group law on $\mathbb{R}^2$ under which Δ_G is left invariant, the diffusion should be considered as starting from any point $(x_0, y_0) \in \mathbb{R}^2$, so that an integration provides

$$\begin{aligned} X(t) &= x_0 + W_1(t) \\ Y(t) &= y_0 + x_0 W_2(t) + \int_0^t W_1(t)dW_2(t), \quad \forall t \geq 0. \end{aligned}$$

The process $U_t = \int_0^t W_1(t)dW_2(t)$, which is represented as an Ito integral, is a martingale. Since

$$\begin{aligned}\mathbb{E}[U_t^2] &= \mathbb{E}\Big[\Big(\int_0^t W_1(s)dW_2(s)\Big)^2\Big] = \mathbb{E}\Big[\int_0^t W_1(s)^2 ds\Big] \\ &= \int_0^t \mathbb{E}[W_1(s)^2]ds = \int_0^t s\,ds = \frac{t^2}{2} < \infty,\end{aligned}$$

then U_t is a square integrable martingale for any $0 \le t < \infty$. Since

$$(dU_t)^2 = W_1(t)^2(dW_2(t))^2 = W_1(t)^2\,dt,$$

its associated quadratic variation process is

$$Z_t = \langle U, U\rangle_t = \int_0^t W_1(s)^2\,ds.$$

By Theorem 1.1.11, there is a DDS Brownian motion B_t, such that the martingale U_t can be written as a time-transformed Brownian motion, $U_t = B_{Z_t}$. Therefore, the associated diffusion can be written as

$$\begin{aligned}X(t) &= x_0 + W_1(t) \\ Y(t) &= y_0 + x_0 W_2(t) + B_{Z_t}, \quad \forall t \ge 0.\end{aligned}$$

Since the vector fields X and Y are translation invariant in the y-direction, it suffices to compute the heat kernel just for the case $y_0 = 0$. Then the general heat kernel is retrieved by

$$p_t(x_0, y_0; x, y) = p_t(x_0, 0; x, y - y_0).$$

For the sake of simplicity in the next computation we assume $x_0 = 0$ and $y_0 = 0$, i.e., the diffusion starts from the origin. Then using the definition of the expectation and Fubini's theorem, we have on one side

$$\mathbb{E}[e^{i\lambda_1 X_t}e^{i\lambda_2 Y_t}] = \int_{\mathbb{R}} e^{i\lambda_1 x}\Big(\int_{\mathbb{R}} e^{i\lambda_2 y} p_t(0,0;x,y)\,dy\Big)\,dx. \tag{4.3.18}$$

On the other side, using the tower property by conditioning over the history $W_1[0,t]$ of $W_1(s)$ with $0 \le s \le t$, we obtain

$$\begin{aligned}\mathbb{E}[e^{i\lambda_1 X_t}e^{i\lambda_2 Y_t}] &= \mathbb{E}[e^{i\lambda_1 W_1(t)}e^{i\lambda_2 B_{Z_t}}] \\ &= \mathbb{E}[\,\mathbb{E}[e^{i\lambda_1 W_1(t)}e^{i\lambda_2 B_{Z_t}} \,|\, W_1[0,t]]\,] \\ &= \mathbb{E}[\,e^{i\lambda_1 W_1(t)}\mathbb{E}[e^{i\lambda_2 B_{Z_t}} \,|\, W_1[0,t]]\,] \\ &= \mathbb{E}[\,e^{i\lambda_1 W_1(t)}e^{-\frac{1}{2}\lambda_2^2 Z_t}].\end{aligned}$$

For computing the expectation $\mathbb{E}[\, e^{i\lambda_1 W_1(t)} e^{-\frac{1}{2}\lambda_2^2 Z_t}]$ we shall use the joint distribution of $W_1(t)$ and $Z_t = \int_0^t W_1(s)^2\, ds$. We denote for the sake of simplicity $W_t = W_1(t)$. The joint density of (W_t, Z_t) is given by the law

$$P(Z_t \in dy, W_t \in dz) = (2\pi)^{-1/2} \mathrm{ee}_y\Big(\frac{1}{2}, t, \frac{z^2}{2}, 0\Big)\, dydz, \tag{4.3.19}$$

where the density function $\mathrm{ee}_y(\cdot)$ is characterized by its Laplace transform

$$\int_0^\infty e^{-\gamma y}\, \mathrm{ee}_y\Big(\frac{1}{2}, t, \frac{z^2}{2}, 0\Big)\, dy = \left(\frac{\sqrt{2\gamma}}{\sinh(t\sqrt{2\gamma})}\right)^{1/2} e^{-\frac{\sqrt{2\gamma}}{2} z^2 \coth(t\sqrt{2\gamma})}, \qquad \forall \gamma > 0,$$

see Borodin and Salminen [15], pages 175 and 653. It is worth noting that the previous Laplace transform can be inverted obtaining an expression using an infinite power series with cylindric functions coefficients. Fortunately, this cumbersome expression won't be needed here, since it is the Laplace transform expression that will be used in the next computation. Using the definition of expectation, Fubini's theorem and the substitution $\gamma = \lambda_2^2/2$, we have

$$\begin{aligned}
\mathbb{E}[\, e^{i\lambda_1 W_1(t)} e^{-\frac{1}{2}\lambda_2^2 Z_t}] &= \int_{\mathbb{R}} \int_0^\infty e^{i\lambda_1 z} e^{-\frac{1}{2}\lambda_2^2 y} (2\pi)^{-1/2} \mathrm{ee}_y\Big(\frac{1}{2}, t, \frac{z^2}{2}, 0\Big)\, dydz \\
&= \int_{\mathbb{R}} e^{i\lambda_1 z} (2\pi)^{-1/2} \Big(\int_0^\infty e^{-\frac{1}{2}\lambda_2^2 y} \mathrm{ee}_y\Big(\frac{1}{2}, t, \frac{z^2}{2}, 0\Big)\, dy\Big)\, dz \\
&= \int_{\mathbb{R}} e^{i\lambda_1 z} (2\pi)^{-1/2} \sqrt{\frac{\lambda_2}{\sinh(\lambda_2 t)}}\, e^{-\frac{\lambda_2}{2} z^2 \coth(\lambda_2 t)}\, dz \\
&= \int_{\mathbb{R}} e^{i\lambda_1 x} \Phi_t(\lambda_2, x)\, dx, \qquad (4.3.20)
\end{aligned}$$

where

$$\Phi_t(\lambda_2, x) = \sqrt{\frac{\lambda_2}{2\pi \sinh(\lambda_2 t)}}\, e^{-\frac{\lambda_2}{2} x^2 \coth(\lambda_2 t)}$$

and we substituted x in lieu of z. Comparing (4.3.18) and (4.3.20) yields

$$\int_{\mathbb{R}} e^{i\lambda_2 y} p_t(0, 0; x, y)\, dy = \Phi_t(\lambda_2, x).$$

Using the inverse Fourier transform formula we obtain the heat kernel

$$\begin{aligned}
p_t(0, 0; x, y) &= \frac{1}{2\pi} \int_{\mathbb{R}} e^{-i\lambda_2 y}\, \Phi_t(\lambda_2, x)\, d\lambda_2 \\
&= \frac{1}{(2\pi)^{3/2}} \int_{\mathbb{R}} \sqrt{\frac{\lambda_2}{\sinh(\lambda_2 t)}}\, e^{-i\lambda_2 y - \frac{\lambda_2}{2} x^2 \coth(\lambda_2 t)}\, d\lambda_2.
\end{aligned}$$

Using the behavior with respect to the y-translation, we obtain

$$p_t(0, y_0; x, y) = \frac{1}{(2\pi)^{3/2}} \int_{\mathbb{R}} \sqrt{\frac{\xi}{\sinh(\xi t)}} e^{-i\xi(y-y_0) - \frac{\xi}{2}x^2 \coth(\xi t)} \, d\xi. \tag{4.3.21}$$

This represents the integral representation of the transition density for the diffusions starting on the y-axis. This is the familiar integral representation of the heat kernel for the Grushin operator which appears for instance in Calin et al. [29].

Formula (4.3.21) contains the imaginary number i in the exponent, which is reminiscent from the Fourier transform definition. Can we find an integral formula which avoids this peculiarity?

To this end, we shall denote

$$a_t(z, y) = (2\pi)^{-1/2} \mathrm{ee}_y\Big(\frac{1}{2}, t, \frac{z^2}{2}, 0\Big),$$

so formula (4.3.19) becomes

$$P(Z_t \in dy, W_t \in dz) = a_t(z, y)\, dydz. \tag{4.3.22}$$

Assuming $x_0 = y_0 = 0$, then for any two bounded Borel functions f and g we have

$$\mathbb{E}[f(X_t)g(Y_t)] = \iint_{\mathbb{R}^2} f(x)g(y)p_t(x, y)\, dxdy, \tag{4.3.23}$$

where $p_t(x, y) = p_t(0, 0; x, y)$. We shall represent the transition density of the Grushin diffusion, $p_t(x, y)$, in terms of the joint density $a_t(z, y)$. For this we compute the expectation $\mathbb{E}[f(X_t)g(Y_t)]$ in a different way and then compare the result with formula (4.3.23).

The tower property for conditional expectations provides

$$\begin{aligned} \mathbb{E}[f(X_t)g(Y_t)] &= \mathbb{E}[f(W_t)g(B_{Z_t})] \\ &= \mathbb{E}[\,\mathbb{E}[f(W_t)g(B_{Z_t}) \,|\, W[0, t]\,]\,] \\ &= \mathbb{E}[\,f(W_t)\mathbb{E}[g(B_{Z_t}) \,|\, W[0, t]\,]\,]. \end{aligned} \tag{4.3.24}$$

Since $Z_t = \int_0^t W_s^2\, ds$ is determined by the history of the Brownian motion $W[0, t]$, we can write

$$\mathbb{E}[g(B_{Z_t}) \,|\, W[0, t]\,] = \int_{\mathbb{R}} g(y) \frac{1}{\sqrt{2\pi Z_t}} e^{-\frac{y^2}{2Z_t}}\, dy.$$

Substituting back in (4.3.24) and then swapping the integral and the exponential operators yields

$$\begin{aligned}\mathbb{E}[f(X_t)g(Y_t)] &= \int_{\mathbb{R}} g(y)\mathbb{E}\big[\,f(W_t)\frac{1}{\sqrt{2\pi Z_t}}\,e^{-\frac{y^2}{2Z_t}}\,\big]\,dy \\ &= \int_{\mathbb{R}} g(y)\int_0^\infty \int_{\mathbb{R}} f(x)\frac{1}{\sqrt{2\pi u}}\,e^{-\frac{y^2}{2u}}\,a_t(x,u)dxdudy \\ &= \iint_{\mathbb{R}^2} f(x)g(y)\Big(\int_0^\infty \frac{1}{\sqrt{2\pi u}}\,e^{-\frac{y^2}{2u}}\,a_t(x,u)\,du\Big)\,dxdy. \qquad (4.3.25)\end{aligned}$$

From (4.3.23) and (4.3.25) we infer

$$p_t(x,y) = \int_0^\infty \frac{1}{\sqrt{2\pi u}}\,e^{-\frac{y^2}{2u}}\,a_t(x,u)\,du. \qquad (4.3.26)$$

This integral representation of the heat kernel of the Grushin operator can be written in terms of the operator T defined by (1.3.15) as

$$p_t(x,y) = T(a_t(x,\cdot))(y). \qquad (4.3.27)$$

Using the properties of the operator T, formula (4.3.27) states that the marginal density $y \to p_t(x,y)$, which describes the diffusion component Y_t, is the density of a Brownian motion stopped at a random time with law $a_t(x,\cdot)$. This random time is actually the process $Z_t = \int_0^t W_s^2\,ds$. This statement is consistent with the fact that $Y_t = B_{Z_t}$, with B_s one-dimensional Brownian motion.

In the following we shall rewrite formula (4.3.26) in a more convenient form using a Laplace transform. Substituting $v = 1/u$ yields

$$p_t(x,y) = \frac{1}{\sqrt{2\pi}}\int_0^\infty e^{-\frac{y^2}{2}v}\,\frac{1}{v^{3/2}}a_t(x,1/v)\,dv. \qquad (4.3.28)$$

To continue the computation we shall use Lemma 1.6.3, page 23, which provides the following formula for the inverse Laplace transform

$$\mathcal{L}_\gamma^{-1}\Big(F(\sqrt{\gamma})\Big)(v) = \frac{1}{2\sqrt{\pi}}\frac{1}{v^{3/2}}\int_0^\infty re^{-\frac{r^2}{4v}}f(r)\,dr.$$

Using

$$F(\sqrt{\gamma}) = \int_0^\infty e^{-\sqrt{\gamma}s}f(s)\,ds,$$

the previous formula writes as

$$\frac{1}{2\sqrt{\pi}}\int_0^\infty e^{-\gamma v}\frac{1}{v^{3/2}}\left(\int_0^\infty re^{-\frac{r^2}{4v}}f(r)\,dr\right)dv = \int_0^\infty e^{-\sqrt{\gamma}s}f(s)\,ds.$$

Then we use the substitution $\gamma = y^2/2$ to obtain

$$\frac{1}{2\sqrt{\pi}}\int_0^\infty e^{-\frac{y^2}{2}v}\frac{1}{v^{3/2}}\left(\int_0^\infty re^{-\frac{r^2}{4v}}f(r)\,dr\right)dv = \int_0^\infty e^{-ys/\sqrt{2}}f(s)\,ds. \quad (4.3.29)$$

By comparison with formula (4.3.28) we make the following claim regarding the integral representation of a_t:

Ansatz: There is a continuous function $f(r) = f_{t,x}(r)$ such that

$$a_t(x, 1/v) = \int_0^\infty re^{-\frac{r^2}{4v}}f(r)\,dr. \quad (4.3.30)$$

Therefore, formulas (4.3.29) and (4.3.28) imply

$$\begin{aligned} p_t(x,y) &= \sqrt{2}\int_0^\infty e^{-ys/\sqrt{2}}f(s)\,ds \quad (4.3.31)\\ &= \int_0^\infty e^{-yu}g(u)\,du, \end{aligned}$$

where $g(u) = 2f(\sqrt{2}u)$ and we used substitution $u = s/\sqrt{2}$. This is a representation of the probability density $p_t(x,y)$ as a Laplace transform of the function g, which is subject to be determined next.

We shall show now the existence of the function f satisfying (4.3.30). Writing the expression in the variable $u = 1/v$ and then using substitutions $\rho = r^2$ and $s = \rho/4$, we have

$$\begin{aligned} a_t(x,u) &= \int_0^\infty re^{-\frac{r^2}{4}u}f(r)\,dr = \frac{1}{2}\int_0^\infty e^{-\frac{\rho}{4}u}f(\sqrt{\rho})\,d\rho\\ &= 2\int_0^\infty e^{-su}f(2\sqrt{s})\,ds = \mathcal{L}(\psi(s))(u), \end{aligned}$$

where $\mathcal{L}$ stands for the Laplace transform applied to function $\psi(s) = 2f(2\sqrt{s})$. Therefore, $\psi(s)$ is the inverse partial Laplace transform

$$\psi(s) = \mathcal{L}_u^{-1}(a_t(x,u))(s),$$

and then $f(r) = \psi(r^2/2)/2$, for $r \geq 0$, which is the function needed in (4.3.31). We note that the expression of f is reduced to an inverse Laplace transform of a_t, which was initially defined in terms of f. This creates a cycle, which does not provide a closed form expression for f. Therefore, this procedure shows just the existence of the aforementioned function f.

The next section presents the financial application of the Grushin operator.

4.4 Financial Interpretation of Grushin Diffusion

This section deals with an unexpected relation between the two-dimensional Grushin diffusion, G_t, introduced in the previous section and a particular version of the Heston's model of a stock price, see Heston [75]. To this end, we shall denote by Y_t the price of a stock at time t, which follows a geometric Brownian motion, having a stochastic volatility, $\sqrt{V_t}$, that follows a CIR process as follows

$$dY_t = \mu Y_t dt + \sqrt{V_t} Y_t \, dB_t \tag{4.4.32}$$

$$dV_t = (a - bV_t)dt + \sigma\sqrt{V_t}\, dZ_t, \tag{4.4.33}$$

where B_t and Z_t are two Brownian motions and a and b are positive constants. The constants μ and σ represent respectively the drift rate of the stock and the volatility of volatility. In the case of a highly frequency traded stock, the drift can be assumed to be zero, $\mu = 0$. If we further assume $a = 1$, $b = 0$ and $\sigma = 2$, we obtain the following "toy" Heston model

$$dY_t = \sqrt{V_t} Y_t \, dB_t \tag{4.4.34}$$

$$dV_t = dt + 2\sqrt{V_t}\, dZ_t. \tag{4.4.35}$$

The state of the stock can be characterized at time t by the pair (Y_t, V_t). We shall show next that this is related to a Grushin diffusion. We will use the following result shortly.

Lemma 4.4.1 *Let B_t be a one-dimensional Brownian motion and consider the process $X_t = \int_0^t sgn(B_s)\, dB_s$, where $sgn(x) = \begin{cases} 1, & if\ x > 0 \\ -1, & if\ x \le 0. \end{cases}$ Then X_t is a Brownian motion.*

Proof: Since $dX_t = \text{sgn}(B_t)\, dB_t$, then $(dX_t)^2 = dt$ and hence the quadratic variation is $\langle X \rangle_t = \int_0^t ds = t$. As an Ito process, X_t is a continuous martingale with respect to the filtration $\mathcal{F}_t = B[0, t]$. Since

$$\mathbb{E}[X_t^2] \le \mathbb{E}\Big[\Big(\int_0^t dB_s\Big)^2\Big] = t < \infty,$$

then X_t is square integrable. By Lévy's theorem, see Theorem 1.1.6, X_t becomes a Brownian motion. ∎

In the following we consider the Grushin operator on $\mathbb{R}^2$

$$\Delta_G = \frac{1}{2}\partial_x^2 + \frac{k^2}{2} x^2 \partial_y^2,$$

which depends on the parameter $k > 0$. The associated Grushin diffusion, $G_t = (X_t, Y_t)$, satisfies the system

$$\begin{aligned} dX_t &= dW_1(t) \\ dY_t &= kX_t\, dW_2(t), \end{aligned}$$

with $W_1(t)$ and $W_2(t)$ independent Brownian motions. We let $V_t = X_t^2$ and rewrite the stochastic differential of Y_t as

$$\begin{aligned} dY_t &= kX_t\, dW_2(t) = k|X_t|\,\mathrm{sgn}(X_t)\, dW_2(t) \\ &= k\sqrt{V_t}\,\mathrm{sgn}(X_t)\, dW_2(t). \end{aligned}$$

By Lemma 4.4.1 there is a Brownian motion, B_t, such that $dB_t = \mathrm{sgn}(X_t)\, dW_2(t)$. Then the previous equation becomes

$$dV_t = 2\sqrt{V_t}\, dB_t.$$

On the other side, using Ito's formula we have

$$\begin{aligned} dV_t &= 2X_t\, dX_t + (dX_t)^2 \\ &= 2|X_t|\mathrm{sgn}(X_t)\, dW_1(t) + dt. \end{aligned}$$

Applying again Lemma 4.4.1, there is a Brownian motion, Z_t, such that $dZ_t = \mathrm{sgn}(X_t)\, dW_1(t)$. The previous equation becomes

$$dV_t = 2\sqrt{V_t}\, dZ_t + dt.$$

Since $W_1(t)$ and $W_2(t)$ are independent Brownian motions, so they are Z_t and B_t. To conclude, the process (Y_t, V_t) satisfies the system

$$dY_t = k\sqrt{V_t}\, dB_t \tag{4.4.36}$$

$$dV_t = dt + 2\sqrt{V_t}\, dZ_t, \tag{4.4.37}$$

which is exactly the system (4.4.34)-(4.4.35). Therefore, the Grushin diffusion $G_t = (X_t, Y_t)$ characterizes the state of a stock with volatility X_t and price Y_t.

To conclude, the study of the heat kernel for the Grushin operator is significant in developing closed-form pricing formulas for options on stocks that satisfy the toy Heston model because it relates to the transition density of the associated Grushin diffusion.

4.5 The Generalized Grushin Operator

In the upcoming sections, we will explore the extension of the Grushin operator to multiple variables in the sum of squares and multiple missing directions.

4.5.1 The Case of One Missing Direction

The differential operator on $\mathbb{R}^{n+1} = \mathbb{R}^n_x \times \mathbb{R}_y$ given by

$$\Delta_G^{(n,1)} = \frac{1}{2}\sum_{k=1}^{n} \partial^2_{x_k} + 2\|x\|^2\partial^2_y, \tag{4.5.38}$$

where $\|x\|^2 = x_1^2 + \cdots + x_n^2$, is called the $(n+1)$-*dimensional Grushin operator with one missing direction.* The operator is degenerate along the y-axis and elliptic in rest. We shall see that integrating along the missing direction a certain expression will supply the heat kernel.

If (4.5.38) is considered as the infinitesimal generator of a diffusion, then the associated diffusion and dispersion matrices are given by

$$\sigma\sigma^T = \begin{pmatrix} \mathbb{I}_n & 0 \\ O & 4\|x\|^2 \end{pmatrix}, \quad \sigma = \begin{pmatrix} \mathbb{I}_n & 0 \\ O & 2\|x\| \end{pmatrix}$$

and hence the associated diffusion on $\mathbb{R}^{n+1}$ satisfies the system

$$\begin{aligned} dX_1(t) &= dW_1(t) \\ &\cdots\cdots\cdots\cdots \\ dX_n(t) &= dW_n(t) \\ dY(t) &= 2\|X(t)\|dB(t), \end{aligned}$$

with $\|X(t)\| = (\sum_{k=1}^n X_k(t)^2)^{1/2}$ and $W_k(t)$, $B(t)$ independent one-dimensional Brownian motions. The diffusion starting at the origin is given by

$$\begin{aligned} X_1(t) &= W_1(t) \\ &\cdots\cdots\cdots\cdots \\ X_n(t) &= W_n(t) \\ Y(t) &= 2\int_0^t \|X(s)\|dB(s). \end{aligned}$$

Since the process $R_t = \|X(t)\|$ is the n-dimensional Bessel process on $\mathbb{R}^n$ starting from the origin, then the last component of the diffusion can be written as the martingale

$$Y(t) = 2\int_0^t R_s\, dB(s)$$

satisfying

$$\mathbb{E}[Y(t)^2] = 2\int_0^t \mathbb{E}[R_s^2]\, ds = 2nt^2 < \infty.$$

Since the quadratic variation is $\langle Y\rangle_t = 4\int_0^t R_s^2 ds$, By Theorem 1.1.11 there is a DDS Brownian motion, β_t, independent of $W_k(t)$ and $B(t)$, such that $Y(t)$ can be represented as β_t under a stochastic clock, $Y(t) = \beta_{Z_t}$, with $Z_t = 4\int_0^t R_s^2 ds$.

If the transition density from the origin to (x, y) within time t is denoted by $p_t(x, y)$, then by the definition of the expectation we have

$$\mathbb{E}[e^{i\lambda_1 X_1(t)} \cdots e^{i\lambda_n X_n(t)} e^{i\lambda_{n+1} Y(t)}] = \int_{\mathbb{R}^n} e^{i\lambda_1 x_1} \cdots e^{i\lambda_n x_n} \Big(\int_{\mathbb{R}} e^{i\lambda_{n+1} y} p_t(x, y) dy \Big) dx, \tag{4.5.39}$$

with $dx = dx_1 \dots dx_n$ and $x = (x_1, \dots, x_n)$. We denote by $W[0, t]$ the history of $W_1(s), \dots, W_n(s)$, $0 \le s \le t$, and note that the diffusion is determined at time t by $W[0, t]$. Then the tower property of conditional expectations provides

$$\begin{aligned}
\mathbb{E}[e^{i\lambda_1 X_1(t)} \cdots e^{i\lambda_n X_n(t)} e^{i\lambda_{n+1} Y(t)}] &= \mathbb{E}[e^{i\lambda_1 W_1(t)} \cdots e^{i\lambda_n W_n(t)} e^{i\lambda_{n+1}\beta_{Z_t}}] \\
&= \mathbb{E}[\,\mathbb{E}[e^{i\lambda_1 W_1(t)} \cdots e^{i\lambda_n W_n(t)} e^{i\lambda_{n+1}\beta_{Z_t}} | W[0, t]]\,] \\
&= \mathbb{E}[e^{i\lambda_1 W_1(t)} \cdots e^{i\lambda_n W_n(t)} \mathbb{E}[e^{i\lambda_{n+1}\beta_{Z_t}} | W[0, t]]\,] \\
&= \mathbb{E}[e^{i\lambda_1 W_1(t)} \cdots e^{i\lambda_n W_n(t)} e^{-\frac{1}{2}\lambda_{n+1}^2 Z_t}].
\end{aligned} \tag{4.5.40}$$

Using the definition formula for Z_t and substituting $\xi = 2\lambda_{n+1}$, we have

$$e^{-\frac{1}{2}\lambda_{n+1}^2 Z_t} = e^{-\frac{1}{2}\xi^2 \int_0^t R_s^2 ds} = e^{-\frac{1}{2}\xi^2 \int_0^t W_1(s)^2 ds} \cdots e^{-\frac{1}{2}\xi^2 \int_0^t W_n(s)^2 ds}.$$

Then using the independence of Brownian motions, we infer that relation (4.5.40) can be written as a product of n identical expectations

$$\mathbb{E}[e^{i\lambda_1 W_1(t)} e^{-\frac{1}{2}\xi^2 \int_0^t W_1(s)^2 ds}] \cdots \mathbb{E}[e^{i\lambda_1 W_n(t)} e^{-\frac{1}{2}\xi^2 \int_0^t W_n(s)^2 ds}].$$

Each of these expectations have been already computed and are given by formula (4.3.20) as follows

$$\mathbb{E}[e^{i\lambda_k W_k(t)} e^{-\frac{1}{2}\xi^2 \int_0^t W_k(s)^2 ds}] = \int_{\mathbb{R}} e^{-i\lambda_k x_k} \Phi_t(\xi, x_k)\, dx_k, \qquad 1 \le k \le n.$$

where

$$\Phi_t(\xi, x) = \sqrt{\frac{\xi}{2\pi \sinh(\xi t)}} e^{-\frac{\xi}{2}x^2 \coth(\xi t)}.$$

Using Fubini's theorem we put the aforementioned integrals together and hence (4.5.40) becomes

$$\int_{\mathbb{R}^n} e^{i\lambda_1 x_1} \cdots e^{i\lambda_n x_n} \Phi_t(\xi, x_1) \cdots \Phi_t(\xi, x_n)\, dx_1 \cdots dx_n.$$

Comparing with (4.5.39) yields

$$\int_{\mathbb{R}} e^{i\lambda_{n+1}y} p_t(x,y)dy = \Phi_t(\xi, x_1)\cdots\Phi_t(\xi, x_n)$$
$$= \Phi_t(2\lambda_{n+1}, x_1)\cdots\Phi_t(2\lambda_{n+1}, x_n).$$

This integral can be inverted using the inverse Fourier transform to obtain

$$p_t(x,y) = \frac{1}{2\pi}\int_{\mathbb{R}} e^{-i\lambda_{n+1}y}\Phi_t(2\lambda_{n+1}, x_1)\cdots\Phi_t(2\lambda_{n+1}, x_n)\, d\lambda_{n+1}$$
$$= \frac{1}{2\pi}\int_{\mathbb{R}} e^{-i\lambda_{n+1}y}\left(\frac{2\lambda_{n+1}}{2\pi\sinh(2\lambda_{n+1}t)}\right)^{n/2} e^{-\lambda_{n+1}(x_1^2+\cdots+x_n^2)\coth(2\lambda_{n+1}t)} d\lambda_{n+1}.$$

Substituting $s = \lambda_{n+1}$ we obtain

$$p_t(x,y) = \frac{1}{(2\pi)^{1+n/2}}\int_{\mathbb{R}}\left(\frac{2s}{\sinh(2st)}\right)^{n/2} e^{-isy-s\|x\|^2\coth(2st)}\, ds. \tag{4.5.41}$$

To conclude, the heat kernel of the generalized Grushin operator (4.5.38) is given by the integral formula (4.5.41). We note that a different approach for finding the heat kernel for the operator (4.5.38) can be found in Calin et al. [29], page 85.

4.5.2 The Case of Two Missing Directions

We consider the case of a 3-dimensional Grushin operator with two missing directions

$$\Delta_G^{(1,2)} = \frac{1}{2}\partial_x^2 + \frac{1}{2}x^2(\partial_{y_1}^2 + y_{y_2}^2). \tag{4.5.42}$$

Its hypoellipticity follows from the Hörmander's theorem, Theorem 3.4.4. To this end we write the operator as a sum of squares

$$\Delta_G^{(1,2)} = \frac{1}{2}X^2 + \frac{1}{2}(Y_1^2 + Y_2^2)$$

and noting that one of the vectors and the following two Poisson brackets

$$X = \partial_x, \quad [X, Y_1] = \partial_{y_1}, \quad [X, Y_2] = \partial_{y_2}$$

generate the space $\mathbb{R}^3$ at each point, $\operatorname{span}\{\partial_x, \partial_{y_1}, \partial_{y_2}\} = \mathbb{R}^3$. The heat kernel for this operator has been treated from the Hamiltonian formalism point of view in Calin et al. [34]. In this section we shall find the heat kernel by investigating the transition density of the associate diffusion $\big(X(t), Y_1(t), Y_2(t)\big)$, which is defined by the system of stochastic differential equations

$$\begin{aligned} dX(t) &= dW(t) \\ dY_1(t) &= X(t)dB_1(t) \\ dY_2(t) &= X(t)dB_2(t), \end{aligned}$$

with $W(t)$, $B_1(t)$ and $B_2(t)$ independent one-dimensional Brownian motions. Assuming the diffusion starting at the origin, $\big(X(0), Y_1(0), Y_2(0)\big) = (0,0,0)$, integrating yields

$$\begin{aligned} X(t) &= W(t) \\ Y_1(t) &= \int_0^t W(s)dB_1(s) \\ Y_2(t) &= \int_0^t W(s)dB_2(s). \end{aligned}$$

We note that the components $Y_1(t)$ and $Y_2(t)$ are asymmetric area processes as described in Section 1.13, with the characteristic function given by (1.13.79)

$$\phi_{Y_k(t)}(\lambda) = \frac{1}{\sqrt{\cosh(\lambda t)}}.$$

In the following we show that $Y_1(t)$ and $Y_2(t)$ are uncorrelated processes. To this end we recall first the product rule

$$\int_0^t W(s)\,dB_k(s) = W(t)B_k(t) - \int_0^t B_k(s)\,dW(s)$$

and use the properties of stochastic integrals and Brownian motions to obtain

$$\begin{aligned} Corr\big(Y_1(t), Y_2(t)\big) &= \mathbb{E}[Y_1(t)Y_2(t)] - \mathbb{E}[Y_1(t)]\mathbb{E}[Y_2(t)] = \mathbb{E}[Y_1(t)Y_2(t)] \\ &= \mathbb{E}\Big[\int_0^t W(s)dB_1(s) \int_0^t W(s)dB_2(s)\Big] \\ &= \mathbb{E}\Big[\Big(W(t)B_1(t) - \int_0^t B_1(s)dW(s)\Big)\Big(W(t)B_2(t) - \int_0^t B_2(s)dW(s)\Big)\Big] \\ &= t\mathbb{E}[B_1(t)]\mathbb{E}[B_2(t)] - \mathbb{E}[W(t)]\mathbb{E}[B_1(t)]\mathbb{E}\Big[\int_0^t B_2(s)dW(s)\Big] \\ &\quad - \mathbb{E}[W(t)]\mathbb{E}[B_2(t)]\mathbb{E}\Big[\int_0^t B_1(s)dW(s)\Big] \\ &\quad + \mathbb{E}\Big[\int_0^t B_1(s)dW(s) \int_0^t B_2(s)dW(s)\Big] \\ &= \mathbb{E}\Big[\int_0^t B_1(s)B_2(s)\,ds\Big] = 0. \end{aligned}$$

By the argument of Section 1.13 there are two Brownian motions, β_t and γ_t, independent of $W(t)$, such that

$$Y_1(t) = \beta_{A_t}, \qquad Y_2(t) = \gamma_{A_t}, \tag{4.5.43}$$

where $A_t = \int_0^t W(s)^2\,ds$. Since β_t is independent of $W(t)$, it should depend on $B_1(t)$. Similarly, since γ_t is independent of $W(t)$, it should depend on $B_2(t)$. Since $B_1(t)$ and $B_2(t)$ are independent, if follows that β_t and γ_t are independent Brownian motions. This independence will be used later when computing the joint expectation (4.5.46).

In order to compute the transition probability, we shall compare, as usual, two representations of the characteristic function. The first one is

$$\mathbb{E}\Big[e^{i\lambda_0 X(t)}e^{i\lambda_1 Y_1(t)}e^{i\lambda_2 Y_2(t)}\Big] = \int_{\mathbb{R}} e^{i\lambda_0 x}\Big(\iint_{\mathbb{R}^2} e^{i\lambda_1 y_1}e^{i\lambda_2 y_2}p_t(x, y_1, y_2)\,dy_1 dy_2\Big)\,dx. \tag{4.5.44}$$

For the second representation, we have

$$\begin{aligned}
\mathbb{E}\Big[e^{i\lambda_0 X(t)}e^{i\lambda_1 Y_1(t)}e^{i\lambda_2 Y_2(t)}\Big] &= \mathbb{E}\Big[e^{i\lambda_0 W(t)}e^{i\lambda_1 \beta_{A_t}}e^{i\lambda_2 \gamma_{A_t}}\Big] \\
&= \mathbb{E}\Big[\,\mathbb{E}\Big[e^{i\lambda_0 W(t)}e^{i\lambda_1 \beta_{A_t}}e^{i\lambda_2 \gamma_{A_t}}|W[0,t]\Big]\Big] \\
&= \mathbb{E}\Big[e^{i\lambda_0 W(t)}\mathbb{E}\Big[e^{i\lambda_1 \beta_{A_t}}e^{i\lambda_2 \gamma_{A_t}}|W[0,t]\Big]\Big].
\end{aligned} \tag{4.5.45}$$

We compute next the conditional expectation $\mathbb{E}\Big[e^{i\lambda_1 \beta_{A_t}}e^{i\lambda_2 \gamma_{A_t}}|W[0,t]\Big]$. Since $A_t = \int_0^t W(s)^2 ds$ is determined by the history $W[0,t]$ and β_t and γ_t are independent, we obtain

$$\mathbb{E}\Big[e^{i\lambda_1 \beta_{A_t}}e^{i\lambda_2 \gamma_{A_t}}|W[0,t]\Big] = e^{-\frac{1}{2}(\lambda_1^2+\lambda_2^2)A_t}. \tag{4.5.46}$$

Therefore, (4.5.45) becomes

$$\begin{aligned}
\mathbb{E}\Big[e^{i\lambda_0 W(t)}\mathbb{E}\Big[e^{i\lambda_1 \beta_{A_t}}e^{i\lambda_2 \gamma_{A_t}}|W[0,t]\Big]\Big] &= \mathbb{E}\Big[e^{i\lambda_0 W(t)}e^{-\frac{1}{2}(\lambda_1^2+\lambda_2^2)A_t}\Big] \\
&= \int_{\mathbb{R}} e^{i\lambda_0 x}\Phi_t(\sqrt{\lambda_1^2+\lambda_2^2},\, x)dx,
\end{aligned}$$

where we employed the computation done in relation (4.3.20) and used the function Φ_t that was introduced at page 173. Comparing to (4.5.44) yields

$$\iint_{\mathbb{R}^2} e^{i\lambda_1 y_1}e^{i\lambda_2 y_2}p_t(x, y_1, y_2)\,dy_1 dy_2 = \Phi_t\Big(\sqrt{\lambda_1^2+\lambda_2^2},\, x\Big).$$

Inverting the double integral we obtain the following density formula

$$p_t(x, y_1, y_2) = \frac{1}{(2\pi)^2} \iint_{\mathbb{R}^2} e^{-i\lambda_1 y_1} e^{-i\lambda_2 y_2} \Phi_t\Big(\sqrt{\lambda_1^2 + \lambda_2^2}, x\Big)\, d\lambda_1 d\lambda_2 \qquad (4.5.47)$$

where we used the notations $\lambda = (\lambda_1, \lambda_2)$, $|\lambda| = \sqrt{\lambda_1^2 + \lambda_2^2}$ and $y = (y_1, y_2)$.

Using the formula of Φ_t we obtain that the heat kernel of the 3-dimensional Grushin operator with two missing directions, $\Delta_G^{(1,2)}$, is given by

$$p_t(x, y_1, y_2) = \frac{1}{(2\pi)^{5/2}} \iint_{\mathbb{R}^2} \sqrt{\frac{|\lambda|}{\sinh(|\lambda| t)}}\; e^{-i\langle \lambda, y\rangle - \frac{1}{2}|\lambda| x^2 \coth(|\lambda| t)}\, d\lambda_1 d\lambda_2,$$

where we employed notations $\lambda = (\lambda_1, \lambda_2)$, $|\lambda| = \sqrt{\lambda_1^2 + \lambda_2^2}$ and $y = (y_1, y_2)$.

4.5.3 The Case of m Missing Directions

We consider now the case of the $(m+1)$-dimensional Grushin operator with m missing directions

$$\Delta_G^{(1,m)} = \frac{1}{2}\partial_x^2 + \frac{1}{2}x^2(\partial_{y_1}^2 + \cdots + \partial_{y_m}^2). \qquad (4.5.48)$$

Its analysis is similar with the one applied to the operator $\Delta_G^{(1,2)}$. This means it is hypoelliptic and the associated diffusion on $\mathbb{R}^{m+1}$ is given by

$$\begin{aligned} dX(t) &= dW(t) \\ dY_1(t) &= X(t)dB_1(t) \\ &\cdots\cdots\cdots\cdots\cdots \\ dY_m(t) &= X(t)dB_m(t), \end{aligned}$$

with $W(t)$, $B_k(t)$ independent one-dimensional Brownian motions. The diffusion starting at the origin is given by

$$\begin{aligned} X(t) &= W(t) \\ Y_1(t) &= \int_0^t W(s)dB_1(s) = \beta^{(1)}_{A_t} \\ &\cdots\cdots\cdots\cdots\cdots\cdots\cdots\cdots \\ Y_m(t) &= \int_0^t W(s)dB_m(s) = \beta^{(m)}_{A_t}, \end{aligned}$$

with $\beta_t^{(k)}$ mutual independent DDS Brownian motions, which are also independent of $W(t)$. A similar computation with the one performed at page 183 provides the heat kernel of $\Delta_G^{(1,m)}$ under the form

$$p_t(x,y) = \frac{1}{(2\pi)^{2m+1/2}} \int \cdots \int_{\mathbb{R}^m} \sqrt{\frac{|\lambda|}{\sinh(|\lambda| t)}}\, e^{-i\langle \lambda, y\rangle - \frac{1}{2}|\lambda| x^2 \coth(|\lambda| t)}\, d\lambda_1 \ldots \lambda_m, \tag{4.5.49}$$

where we used notations $\lambda = (\lambda_1, \ldots, \lambda_m)$, $|\lambda| = \sqrt{\lambda_1^2 + \cdots + \lambda_m^2}$ and $y = (y_1, \ldots, y_m)$.

4.6 The Exponential-Grushin Operator

The Grushin operator on $\mathbb{R}^2$, $\Delta_G = \frac{1}{2}(\partial_x^2 + x^2\partial_y^2)$, is not left invariant under a group law on $\mathbb{R}^2$, fact that makes the computation of the heat kernel more elaborate, depending on the initial point. However, if the coefficient x^2 is replaced by e^{2x}, then we obtain a differential operator, $\Delta_{EG} = \frac{1}{2}(\partial_x^2 + e^{2x}\partial_y^2)$, which is left invariant with respect to a group law on $\mathbb{R}^2$. This section deals with the heat kernel of this operator.

Consider the group $G = (\mathbb{R}^2, \circ)$, with the law

$$(x_1, y_1) \circ (x_2, y_2) = (x_1 + x_2, y_1 + y_2 e^{x_1}), \tag{4.6.50}$$

which is a Lie group of dimension 2, with the neutral element $(0,0)$ and inverse element $(x,y)^{-1} = (-x, -ye^{-x})$. The vector fields

$$X = \partial_x, \qquad Y = e^x \partial_y$$

are left invariant on G and span its Lie algebra. Therefore, the *exponential Grushin operator*

$$\Delta_{EG} = \frac{1}{2}(X^2 + Y^2) = \frac{1}{2}(\partial_x^2 + e^{2x}\partial_y^2)$$

is elliptic on $\mathbb{R}^2$ and left invariant with respect to the law (4.6.50). Therefore, it suffices to compute the heat kernel of Δ_{EG} just from the origin, i.e., to find $p_t(x,y) = p_t(0,0;x,y)$. This problem has been approached geometrically using the Hamiltonian formalism in Calin et al. [39]. In the following we shall solve the problem using diffusions following a variant of the proof presented in Urban [131].

The diffusion process, (X_t, Y_t), associated with the operator Δ_{EG} satisfies the system

$$\begin{aligned} dX_t &= dW_t^1 \\ dY_t &= e^{X_t} dW_t^2, \end{aligned}$$

where W_t^1, W_t^2 are two independent, one-dimensional Brownian motions. Due to the group law invariance it suffices to study the diffusion that starts at the origin, which is given by

$$X_t = W_t^1$$
$$Y_t = \int_0^t e^{W_s^1} dW_s^2.$$

Remark 4.6.1 The component $Y_t = \int_0^t e^{W_s^1} dW_s^2$ is a martingale with the quadratic variation

$$\langle Y \rangle_t = \int_0^t e^{2W_s}\, ds,$$

where we denoted $W_s = W_s^1$. By Lévy's theorem, see Theorem 1.1.6, there is a one-dimensional Brownian motion β_t such that

$$Y_t = \beta_{A_t}, \qquad A_t = \int_0^t e^{2W_s}\, ds,$$

namely, Y_t is a Brownian motion under a time clock given by A_t. By Bougerol's identity, see Proposition 1.14.4, there is a Brownian motion B_t such that $Y_t = \beta_{A_t} = \sinh(B_t)$. Therefore, the law of Y_t is

$$P(Y_t \in dy) = \rho_t(y)\, dy,$$

where

$$\rho_t(y) = \frac{1}{\sqrt{2\pi t}} \frac{1}{\sqrt{y^2+1}} e^{-\frac{(\sinh^{-1} y)^2}{2t}}, \qquad \forall y \in \mathbb{R},\ t \geq 0.$$

We continue with the computation of the transition density of (X_t, Y_t). By the definition of expectation and Fubini's theorem

$$\begin{aligned} \mathbb{E}[e^{i\lambda_1 X_t} e^{i\lambda_2 Y_t}] &= \iint_{\mathbb{R}^2} e^{i\lambda_1 x} e^{i\lambda_2 y} p_t(x,y)\, dx dy \\ &= \int_{\mathbb{R}} e^{i\lambda_1 x} \Big(\int_{\mathbb{R}} e^{i\lambda_2 y} p_t(x,y)\, dy \Big) dx. \end{aligned} \tag{4.6.51}$$

Let $W[0,t] = \{W_s; 0 \leq s \leq t\}$ be the history of the Brownian motion W_s up to time t. Then we use a conditional expectation over $W[0,t]$ and the tower property to compute the aforementioned expectation as follows

$$\begin{aligned} \mathbb{E}[e^{i\lambda_1 X_t} e^{i\lambda_2 Y_t}] &= \mathbb{E}[e^{i\lambda_1 W_t} e^{i\lambda_2 \beta_{A_t}}] = \mathbb{E}[\,\mathbb{E}[e^{i\lambda_1 W_t} e^{i\lambda_2 \beta_{A_t}} \mid W[0,t]]\,] \\ &= \mathbb{E}[e^{i\lambda_1 W_t}\, \mathbb{E}[e^{i\lambda_2 \beta_{A_t}} \mid W[0,t]]\,] \\ &= \mathbb{E}[e^{i\lambda_1 W_t}\, e^{-\frac{1}{2}\lambda_2^2 A_t}\,]. \end{aligned} \tag{4.6.52}$$

We recall that $A_t = \int_0^t e^{2W_s}\,ds$ and W_t have the joint distribution

$$P(A_t \in du, W_t \in dx) = \frac{1}{u} e^{-\frac{1+e^{2x}}{2u}} \theta_{e^x/u}(t)\, dudx, \tag{4.6.53}$$

where $t \to \theta_r(t)$ is the Hartman-Watson density, which is characterized by its Laplace transform via formula

$$\int_0^\infty e^{-\frac{1}{2}\gamma^2 t}\theta_r(t)\, dt = I_\gamma(r),$$

where I_γ is the usual modified Bessel function. The previous Laplace transform can be inverted, see Yor [140], obtaining the following integral representation

$$\theta_r(t) = \frac{r}{\sqrt{2\pi^3 t}} e^{\frac{\pi^2}{2t}} \int_0^\infty e^{-\frac{y^2}{2t}} e^{-r\cosh y} \sinh y \sin\frac{\pi y}{t}\, dy. \tag{4.6.54}$$

These formulas can be also found in Borodin and Salmilen [15], page 82.

Then the expectation (4.6.52) can be computed using relation (4.6.53) as follows

$$\begin{aligned}
\mathbb{E}[e^{i\lambda_1 W_t} e^{-\frac{1}{2}\lambda_2^2 A_t}] &= \int_{\mathbb{R}}\int_0^\infty e^{i\lambda_1 x} e^{-\frac{1}{2}\lambda_2^2 u}\frac{1}{u} e^{-\frac{1+e^{2x}}{2u}} \theta_{e^x/u}(t)\, dudx \\
&= \int_{\mathbb{R}} e^{i\lambda_1 x}\Big(\int_0^\infty \frac{1}{u}\theta_{e^x/u}(t) e^{-\frac{1}{2}\lambda_2^2 u} e^{-\frac{1+e^{2x}}{2u}}\, du\Big)\, dx \\
&= \int_{\mathbb{R}} e^{i\lambda_1 x}\Psi_t(x,\lambda_2)\, dx,
\end{aligned} \tag{4.6.55}$$

with

$$\Psi_t(x,\lambda_2) = \int_0^\infty \frac{1}{u}\theta_{e^x/u}(t) e^{-\frac{1}{2}\lambda_2^2 u} e^{-\frac{1+e^{2x}}{2u}}\, du.$$

We substitute (4.6.55) into (4.6.52) and then compare to (4.6.51). Using the injectivity of the Fourier transform yields

$$\int_{\mathbb{R}} e^{i\lambda_2 y} p_t(x,y)\, dy = \Psi_t(x,\lambda_2).$$

Inverting, we obtain the following integral formula for the heat kernel

$$\begin{aligned}
p_t(x,y) &= \frac{1}{2\pi}\int_{\mathbb{R}} e^{-i\lambda_2 y}\,\Psi_t(x,\lambda_2) d\lambda_2 \\
&= \frac{1}{2\pi}\int_{\mathbb{R}} e^{-i\lambda_2 y}\Big(\int_0^\infty \frac{1}{u}\theta_{e^x/u}(t) e^{-\frac{1}{2}\lambda_2^2 u} e^{-\frac{1+e^{2x}}{2u}}\, du\Big) d\lambda_2 \\
&= \frac{1}{2\pi}\int_0^\infty \frac{1}{u}\theta_{e^x/u}(t) e^{-\frac{1+e^{2x}}{2u}}\Big(\int_{\mathbb{R}} e^{-\frac{1}{2}\lambda_2^2 u - i\lambda_2 y}\, d\lambda_2\Big)\, du.
\end{aligned} \tag{4.6.56}$$

Using formula

$$\int_{\mathbb{R}} e^{-a\lambda^2+i\xi\lambda}\,d\lambda = \sqrt{\frac{\pi}{a}}\,e^{-\frac{\xi^2}{4a}}, \qquad a\in\mathbb{R}^*$$

the inner integral becomes

$$\int_{\mathbb{R}} e^{-\frac{1}{2}\lambda_2^2 u - i\lambda_2 y}\,d\lambda_2 = \sqrt{\frac{2\pi}{u}}e^{-\frac{y^2}{2u}}.$$

Substituting back into relation (4.6.56) provides the following formula for the heat kernel starting from the origin

$$p_t(x,y) = \frac{1}{\sqrt{2\pi}}\int_0^\infty u^{-3/2}\,\theta_{e^x/u}(t)\,e^{-\frac{1}{2u}(1+e^{2x}+y^2)}\,du. \tag{4.6.57}$$

The general formula for the heat kernel is given by the next result.

Proposition 4.6.2 *The heat kernel of the operator* $\Delta_{EG} = \frac{1}{2}(\partial_x^2 + e^{2x}\partial_y^2)$ *is given by*

$$p_t(x,y;x',y') = \frac{1}{\sqrt{2\pi}}\int_0^\infty u^{-3/2}\,\theta_{e^{x'-x}/u}(t)\,e^{-\frac{1}{2u}(1+e^{2(x'-x)}+e^{-2x}(y'-y)^2)}\,du,\ t>0. \tag{4.6.58}$$

Proof: Since $(x,y)^{-1}\circ(x',y') = (x'-x, e^{-x}(y'-y))$, using the group law invariance of the heat kernel together with formula (4.6.57) we obtain

$$\begin{aligned} p_t(x,y;x',y') &= p_t\big(0,0;x'-x,e^{-x}(y'-y)\big) = p_t\big(x'-x,e^{-x}(y'-y)\big)\\ &= \frac{1}{\sqrt{2\pi}}\int_0^\infty u^{-3/2}\,\theta_{e^{x'-x}/u}(t)\,e^{-\frac{1}{2u}(1+e^{2(x'-x)}+e^{-2x}(y'-y)^2)}\,du. \end{aligned}$$

■

Corollary 4.6.3 *The Laplace transform of the heat kernel is given by*

$$\int_0^\infty e^{-\frac{1}{2}\gamma^2 t}p_t(x,y)\,dt = \frac{1}{\sqrt{2\pi}}\int_0^\infty u^{-3/2}\,I_\gamma(e^x/u)\,e^{-\frac{1}{2u}(1+e^{2x}+y^2)}\,du.$$

where I_γ *denotes the modified Bessel function of the first kind.*

Proof: Applying the Laplace transform to formula (4.6.57) and using Fubini's theorem yields

$$\begin{aligned} \int_0^\infty e^{-\frac{1}{2}\gamma^2 t}p_t(x,y)\,dt &= \frac{1}{\sqrt{2\pi}}\int_0^\infty u^{-3/2}\int_0^\infty e^{-\frac{1}{2}\gamma^2 t}\theta_{e^x/u}(t)\,dt\,e^{-\frac{1}{2u}(1+e^{2x}+y^2)}\,du\\ &= \frac{1}{\sqrt{2\pi}}\int_0^\infty u^{-3/2}\,I_\gamma(e^x/u)\,e^{-\frac{1}{2u}(1+e^{2x}+y^2)}\,du, \end{aligned}$$

where we used the Laplace transform formula for the Hartman-Watson density

$$\int_0^\infty e^{-\frac{1}{2}\gamma^2 t}\theta_r(t)\,dt = I_\gamma(t).$$

■

4.7 The Exponential-Kolmogorov Operator

Given the resemblance with the Kolmogorov operator (4.1.2), the differential operator

$$\Delta_{EK} = \frac{1}{2}\partial_x^2 + e^{2x}\partial_y \tag{4.7.59}$$

will be called the one-dimensional *Exponential-Kolmogorov operator*. Since the vector fields ∂_x and $e^{2x}\partial_y$ obviously satisfy the bracket-generating condition, by Hörmander's theorem the operator Δ_{EK} is hypoelliptic, and hence its heat kernel is a smooth function. We shall find in the following this heat kernel.

To this end, we note that the operator Δ_{EK} is the generator of the diffusion (X_t, Y_t)

$$\begin{aligned} dX_t &= dW_t \\ dY_t &= e^{2X_t}dt, \end{aligned}$$

with W_t standard Brownian motion. Assuming the diffusion starts at the point $(X_0, Y_0) = (x_0, y_0) \in \mathbb{R}^2$, we have

$$X_t = x_0 + W_t \tag{4.7.60}$$

$$Y_t = y_0 + e^{2x_0}\int_0^t e^{2W_s}\,ds. \tag{4.7.61}$$

We shall denote the heat kernel of the operator Δ_{EK} by $p_t(x_0, y_0; x, y)$. Also, we denote by $a_t(u, x)$ the joint distribution of $A_t = \int_0^t e^{2W_s}\,ds$ and W_t, which is given in terms of the Hartman-Watson density by formula (4.6.53). Then we have

$$\begin{aligned} P(X_t \le x, Y_t \le y) &= P(W_t \le x - x_0, A_t \le (y - y_0)e^{-2x_0}) \\ &= \int_0^{\frac{y-y_0}{e^{2x_0}}}\int_{-\infty}^{x-x_0} a_t(u,\xi)\,d\xi du. \end{aligned}$$

Taking the mixed derivative $\partial_x\partial_y$ we obtain the following relation for the transition density

$$\begin{aligned} p_t(x_0, y_0; x, y) &= e^{-2x_0}a_t((y-y_0)e^{-2x_0}, x - x_0) \\ &= e^{-2x_0}\frac{1}{u}e^{-\frac{1+e^{2\xi}}{2u}}\theta_{e^{\xi}/u}(t)\Big|_{\xi = x - x_0,\, u = (y-y_0)e^{-2x_0}} \\ &= \frac{1}{y - y_0}e^{-\frac{e^{2x_0}+e^{2x}}{2(y-y_0)}}\theta_{e^{x+x_0}/(y-y_0)}(t), \end{aligned} \tag{4.7.62}$$

where $0 \le y_0 < y$ and

$$\theta_r(t) = \frac{r}{\sqrt{2\pi^3 t}}e^{\frac{\pi^2}{2t}}\int_0^\infty e^{-\frac{y^2}{2t}}e^{-r\cosh y}\sinh y \sin\frac{\pi y}{t}\,dy.$$

Therefore, the heat kernel of the operator (4.7.59) is given by (4.7.62).

The multidimensional case is more complex. The n-dimensional Exponential-Kolmogorov operator is defined by

$$\Delta_{EK}^n = \frac{1}{2}\sum_{j=1}^n \partial_{x_j}^2 + \Big(\sum_{j=1}^n a_j e^{2x_j}\Big)\partial_y, \tag{4.7.63}$$

with $a_j \in \mathbb{R}$. The associated diffusion, $(X_1(t), \dots, X_n(t), Y(t))$, satisfies

$$\begin{aligned} dX_1(t) &= dW_1(t) \\ &\dots\dots\dots\dots \\ dX_n(t) &= dW_n(t) \\ dY(t) &= \sum_{j=1}^n a_j e^{2X_j(t)} dt. \end{aligned}$$

Integrating, yields

$$\begin{aligned} X_1(t) &= x_1^0 + W_1(t) \\ &\dots\dots\dots\dots \\ X_n(t) &= x_n^0 + W_n(t) \\ Y(t) &= y_0 + \sum_{j=1}^n a_j e^{2x_j^0} \int_0^t e^{2W_j(s)}\, ds, \end{aligned}$$

with $W_1(t)$, $\cdots$, $W_n(t)$ independent one-dimensional Brownian motions. The characteristic function of the aforementioned diffusion can be computed as follows

$$\begin{aligned} \varphi(t; \lambda, \eta) &= \mathbb{E}[e^{i\lambda_1 X_1(t)} \cdots e^{i\lambda_n X_n(t)} e^{i\eta Y(t)}] \\ &= e^{i(\sum_{j=1}^n \lambda_j x_j^0 + \eta y_0)} \mathbb{E}[e^{i\sum_{j=1}^n \lambda_j W_j(t)} e^{i\eta \sum_{j=1}^n a_j e^{2x_j^0} \int_0^t e^{2W_j(s)} ds}] \\ &= e^{i(\sum_{j=1}^n \lambda_j x_j^0 + \eta y_0)} \prod_{j=1}^n \mathbb{E}[e^{i\lambda_j W_j(t) + i\eta a_j e^{2x_j^0} \int_0^t e^{2W_j(s)} ds}] \\ &= e^{i(\sum_{j=1}^n \lambda_j x_j^0 + \eta y_0)} \prod_{j=1}^n G(\lambda_j, \eta, a_j, x_j^0; t), \end{aligned} \tag{4.7.64}$$

where

$$G(\lambda, \eta, a, x^0; t) = \mathbb{E}[e^{i\lambda W(t) + i\eta a e^{2x^0} \int_0^t e^{2W(s)} ds}] = \mathbb{E}[e^{i\lambda W(t) + i\eta a e^{2x^0} A_t}],$$

with $\lambda, \eta, a, x \in \mathbb{R}$ and $W(t)$ one-dimensional Brownian motion. Using formula (4.6.53)

$$P(A_t \in du, W(t) \in dx) = a_t(u,x)dudx, \qquad u > 0, x \in \mathbb{R}$$

then

$$G(\lambda, \eta, a, x^0; t) = \int_{\mathbb{R}} \int_0^{\infty} e^{i\lambda x} e^{i\eta a u e^{2x^0}} a_t(u,x)\, dudx.$$

Substituting into (4.7.64) and using Fubini's theorem yields

$$\varphi(t; \lambda, \eta) = e^{i(\sum_{j=1}^n \lambda_j x_j^0 + \eta y_0)} \int_{\mathbb{R}^n} e^{i\sum_{j=1}^n \lambda_j x_j} \int_{\mathbb{R}_+^n} e^{i\eta \sum_{j=1}^n a_j u_j e^{2x_j^0}} \Pi_{j=1}^n a_t(u_j, x_j)\, dudx, \tag{4.7.65}$$

with $du = du_1 \cdots du_n$ and $dx = dx_1 \cdots dx_n$.

On the other hand, we can express the characteristic function in terms of the transition density $p_t(x_0, y_0; x, y)$ as

$$\begin{aligned}\varphi(t; \lambda, \eta) &= \mathbb{E}[e^{i\lambda_1 X_1(t)} \cdots e^{i\lambda_n X_n(t)} e^{i\eta Y(t)}] \\ &= \int_{\mathbb{R}^n} \int_0^{\infty} e^{i\sum_{j=1}^n \lambda_j x_j} e^{i\eta y} p_t(x_0, y_0; x, y)\, dxdy \\ &= \int_{\mathbb{R}^n} e^{i\sum_{j=1}^n \lambda_j x_j} \int_0^{\infty} e^{i\eta y} p_t(x_0, y_0; x, y)\, dy\, dx. \end{aligned} \tag{4.7.66}$$

Using the Fourier transform injectivity, equations (4.7.65) and (4.7.66) imply

$$\int_0^{\infty} e^{i\eta y} p_t(x_0, y_0; x, y)\, dy = e^{i(\sum_{j=1}^n \lambda_j x_j^0 + \eta y_0)} \int_{\mathbb{R}_+^n} e^{i\eta \sum_{j=1}^n u_j a_j e^{2x_j^0}} \Pi_{j=1}^n a_t(u_j, x_j)\, du.$$

The left side can be considered as a partial Laplace transform evaluated at $-i\eta$.

4.8 Options on Arithmetic Moving Average

One of the reasons for the study of the Exponential-Kolmogorov operator is its financial application. More precisely, if the underlying variable of an Asian option is the arithmetic moving average of a stock, then its associated Black-Scholes equation is the heat equation of an Exponential-Kolmogorov operator. This has serious implications to the derivation of explicit analytic expressions for the value of Asian options on the arithmetic stock average.

The reader interested in financial mathematics can find similar applications to the European and the Asian options in Yor [142], [143], Matsumoto and Yor [106], Carr and Schröder [42], and Donati-Martin et al. [50].

The Arithmetic Moving Average We consider the time interval $[a, b]$ divided into n subintervals of equal length $\Delta t = t_{k+1} - t_k = \frac{b-a}{n}$. We consider the *discrete arithmetic average* of the stock values S_{t_k} at time instances t_k to be given by

$$A(t_1, \ldots, t_n) = \frac{1}{n} \sum_{k=1}^{n} S_{t_k}.$$

Taking the limit $n \to \infty$ and using the definition of the integral as a Riemann sum we have

$$A_{a,b} = \lim_{n\to\infty} \Big(\frac{1}{n} \sum_{k=1}^{n} S_{t_k}\Big) = \lim_{n\to\infty} \Big(\frac{1}{b-a} \sum_{k=1}^{n} S_{t_k} \frac{b-a}{n}\Big) = \frac{1}{b-a} \int_a^b S_u \, du.$$

$A_{a,b}$ represents the *continuous arithmetic moving average* of a stock over the time interval $[a, b]$.

The associated Black-Scholes equation The equation describing the trading equilibrium of the value of a financial contract on an arithmetic average is given by a parabolic partial differential equation. Let $n > 0$ be a positive integer. We assume the value of an Asian option, $v = v(\tau, s_1, s_2, s_3)$, depends on the following variables: the time variable τ, the stock values $s_1 = S_\tau$, $s_2 = S_{\tau-n}$ at instances τ and $\tau - n$, respectively, and the arithmetic moving average on the interval $[\tau - n, \tau]$

$$s_3 = A_{\tau-n,\tau} = \frac{1}{n} \int_{\tau-n}^{\tau} S_u \, du.$$

We assume the stock follows a geometric Brownian motion, has the return μ and the volatility σ, both positive constants. Then

$$ds_1 = \mu s_1 d\tau + \sigma s_1 dW_1, \quad ds_2 = \mu s_2 d\tau + \sigma s_2 dW_2, \quad ds_3 = \frac{1}{n}(s_1 - s_2) d\tau,$$

where W_1 and W_2 are two Brownian motions with increments correlated by a factor $\rho \in [0, 1]$.
Using $(ds_1)^2 = \sigma^2 s_1^2 d\tau$, $(ds_2)^2 = \sigma^2 s_2^2 d\tau$, $ds_1 ds_2 = \sigma^2 s_1 s_2 \rho d\tau$ and $(ds_3)^2 = 0$, applying Ito's formula we get

$$\begin{aligned}
dv &= \partial_\tau v \, d\tau + \partial_{s_1} v \, ds_1 + \partial_{s_2} v \, ds_2 + \partial_{s_3} v \, ds_3 \\
&\quad + \frac{1}{2} \partial_{s_1}^2 v \, (ds_1)^2 + \frac{1}{2} \partial_{s_2}^2 v \, (ds_2)^2 + \frac{1}{2} \partial_{s_1} \partial_{s_2} v \, ds_1 ds_2 \\
&= [\partial_\tau v + \mu s_1 \partial_{s_1} v + \mu s_2 \partial_{s_2} v + \frac{1}{2} \sigma^2 s_1^2 \partial_{s_1}^2 v + \frac{1}{2} \sigma^2 s_2^2 \partial_{s_2}^2 v + \sigma^2 s_1 s_2 \rho \partial_{s_1} \partial_{s_2} v \\
&\quad + \frac{1}{n}(s_1 - s_2) \partial_{s_3} v] d\tau + \sigma s_1 \, \partial_{s_1} v \, dW_1 + \sigma s_2 \, \partial_{s_2} v \, dW_2.
\end{aligned}$$

We choose the portfolio $P = v - a_1 s_1 - a_2 s_2$, consisting in one long option v, a_1 units short of s_1 and a_2 units short of s_2. If let $a_1 = \partial_{s_1} v$ and $a_2 = \partial_{s_2} v$ the portfolio becomes riskless, i.e., the terms in dP involving dW_1 and dW_2 vanish. Since the return on the portfolio P invested in a bank at interest rate r will grow to $rPd\tau$ in a time $d\tau$, then the absence of arbitrage opportunities yields $dP = rPd\tau$. This condition is equivalent to the following parabolic equation

$$\partial_\tau v + \frac{1}{2}\sigma^2 s_1^2 \partial_{s_1}^2 v + \frac{1}{2}\sigma^2 s_2^2 \partial_{s_2}^2 v + rs_1 \partial_{s_1} v + rs_2 \partial_{s_2} v + \sigma^2 s_1 s_2 \rho \partial_{s_1} \partial_{s_2} v + \frac{1}{n}(s_1 - s_2)\partial_{s_3} v = rv. \quad (4.8.67)$$

This represents the Black-Scholes equation for Asian options on the arithmetic average of the stock.

Simplifying the equation We employ the substitutions

$$s_1 = e^{x_1}, \quad s_2 = e^{x_2}, \quad s_3 = \frac{1}{n} y, \quad \tau = T - \frac{2t}{\sigma^2},$$

and denote $w(t, x_1, x_2, y) = v(\tau, s_1, s_2, s_3)$. Using

$$\partial_\tau = -\frac{1}{2}\sigma^2 \partial_t, \quad \partial_{x_j} = s_j \partial_{s_j}, \quad \partial_{x_j}^2 = s_j \partial_{s_j} + s_j^2 \partial_{s_j}^2, \quad \partial_y = \frac{1}{n}\partial_{s_3}, \ j = 1, 2$$

equation (4.8.67) becomes

$$\partial_t w = \partial_{x_1}^2 w + \partial_{x_2}^2 w + (k-1)\partial_{x_1} w + (k-1)\partial_{x_2} w + (e^{x_1} - e^{x_2})\partial_y w - kw + \rho\sigma^2 \partial_{x_1}\partial_{x_2} w, \quad (4.8.68)$$

with $k = 2r/\sigma^2$. We remove next the convection terms by considering

$$w(t, x_1, x_2, y) = e^{\alpha x_1 + \beta x_2 + \gamma t} u(t, x_1, x_2, y)$$

with α, β and γ to be specified later. For the sake of simplicity we also assume $\rho = 0$. Substituting in (4.8.68) and simplifying by the exponential factor, we obtain the equation

$$\begin{aligned} \gamma u + \partial_t u = {} & \partial_{x_1}^2 u + \partial_{x_2}^2 u \\ & + (2\alpha + k - 1)\partial_{x_1} u + (2\beta + k - 1)\partial_{x_2} u \\ & + [\alpha^2 + \beta^2 + \alpha(k-1) + \beta(k-1) - k]u \\ & + (e^{x_1} - e^{x_2})\partial_y u. \end{aligned} \quad (4.8.69)$$

The coefficients of $\partial_{x_1} u$ and $\partial_{x_2} u$ vanish for $\alpha = \beta = \frac{1-k}{2}$. Equating the coefficients of u yields $\gamma = -\frac{(1-k)^2}{2} - k$. With these choices of α, β and γ equation (4.8.69) takes the following simple form

$$\partial_t u = \partial_{x_1}^2 u + \partial_{x_2}^2 u + (e^{x_1} - e^{x_2})\partial_y u. \quad (4.8.70)$$

Finally, making the substitutions $x_1 \to 2x_1$, $x_2 \to 2x_2$, $t \to 2t$, and $y \to 2y$, we arrive at the more familiar form

$$\partial_t u = \frac{1}{2}(\partial_{x_1}^2 u + \partial_{x_2}^2 u) + (e^{2x_1} - e^{2x_2})\partial_y u,$$

which is the heat equation for a two dimensional Exponential-Kolmogorov operator. It is worth noting that if the derivative price depends only on the stock price s_1 and the geometric moving average s_3, i.e., $v = v(\tau, s_1, s_3)$, we obtain a heat equation for the one-dimensional Exponential-Kolmogorov operator

$$\Delta_{EK} = \frac{1}{2}\partial_x^2 + e^{2x}\partial_y, \tag{4.8.71}$$

which was treated in Section 4.7.

4.9 The Heisenberg Group

The 3-dimensional Heisenberg group, $\mathbb{H}^1 = (\mathbb{R}^3, \circ)$, consists of the set $\mathbb{R}^3$ endowed with the noncommutative group law

$$(x_1, x_2, x_3) \circ (x_1', x_2', x_3') = \big(x_1 + x_1', x_2 + x_2', x_3 + x_3' + 2(x_2 x_1' - x_1 x_2')\big).$$

$\mathbb{H}^1$ can be considered as a Lie group, having the Lie algebra generated by the following left invariant vector fields

$$X = \partial_{x_1} + 2x_2\partial_{x_3}, \quad Y = \partial_{x_2} - 2x_1\partial_{x_3}, \quad Z = \partial_{x_3}. \tag{4.9.72}$$

This means that if $L_g x = g \circ x$ is the left translation on $\mathbb{H}^1$, then $(L_g)_* X_{g'} = X_{g \circ g'}$, $(L_g)_* Y_{g'} = Y_{g \circ g'}$ and $(L_g)_* Z_{g'} = Z_{g \circ g'}$, where $(L_g)_*$ denotes the differential map defined by $(L_g)_*(v)(f) = v(f(L_g))$, $\forall v \in \mathbb{R}^3$ and $f : \mathbb{R}^3 \to \mathbb{R}$ differentiable function.

The Lie brackets of the aforementioned vector fields satisfy the noncommutative relations

$$[X, Y] = -4Z, \qquad [X, Z] = [Y, Z] = 0.$$

We can verify that $\{X, Y, Z\}$ are linearly independent at each point in $\mathbb{R}^3$ and that $Z \notin span\{X, Y\}$. By Frobenius' theorem, see Theorem 3.7.1, the distribution generated by the vector fields $\{X, Y\}$ is not integrable, so there is no two-dimensional surface tangent (locally) to both X and Y. Since $\{X, Y, [X, Y]\}$ generate $\mathbb{R}^3$ at each point, then by Chow-Rashevskii theorem, Theorem 3.7.2, any two points in $\mathbb{H}^1$ can be joint by a piece-wise curve tangent to the distribution generated by $\{X, Y\}$. It can be shown that this curve can be of polynomial type (of degrees 2, 3 and 5 in x_1, x_2 and x_3, respectively), see Calin et al. [38], page 12.

The elliptic differential operator

$$\Delta_{Cas} = \frac{1}{2}(X^2 + Y^2 + Z^2) \tag{4.9.73}$$

is called the *Casimir operator*. The sub-elliptic operator

$$\Delta_H = \frac{1}{2}(X^2 + Y^2) \tag{4.9.74}$$

is called the *Heisenberg Laplacian operator*.

4.10 The Heat Kernel for Heisenberg Laplacian

By Hörmander's theorem, Theorem 3.4.4, the Heisenberg Laplacian operator Δ_H is hypoelliptic, and hence its associated diffusion has a smooth density. We shall find a closed form expression for the density, which in turn will provide the heat kernel for Δ_H. To this end, we first write the operator in local coordinates as

$$\begin{aligned}
\Delta_H &= \frac{1}{2}(X^2 + Y^2) \\
&= \frac{1}{2}\Big(\partial_{x_1}^2 + \partial_{x_2}^2 + 4(x_1^2 + x_2^2)\partial_{x_3}^2 + 4x_2\partial_{x_1}\partial_{x_3} - 4x_1\partial_{x_2}\partial_{x_3}\Big) \\
&= \frac{1}{2}\sum_{i,j=1}^{3} a_{ij}\partial_{x_i}\partial_{x_j},
\end{aligned}$$

where

$$(a_{ij}) = \begin{pmatrix} 1 & 0 & 2x_2 \\ 0 & 1 & -2x_1 \\ 2x_2 & -2x_1 & 4(x_1^2 + x_2^2) \end{pmatrix}$$

is a singular matrix, which can also be written as $a = \sigma\sigma^T$, with

$$(\sigma_{ij}) = \begin{pmatrix} 1 & 0 & 0 \\ 0 & 1 & 0 \\ 2x_2 & -2x_1 & 0 \end{pmatrix}.$$

Then the associated diffusion, $H_t = (H_t^1, H_t^2, H_t^3)$, satisfies the stochastic differential equation

$$\begin{aligned}
dH_t^1 &= dW_t^1 \\
dH_t^2 &= dW_t^2 \\
dH_t^3 &= 2H_t^2 dW_t^1 - 2H_t^1 dW_t^2, \quad t \geq 0,
\end{aligned}$$

where W_t^1 and W_t^2 are two one-dimensional independent Brownian motions. Due to the group law invariance, it suffices to make the simplifying assumption $H_0 = (0, 0, 0)$, i.e., the diffusion starts at the origin. In this case, the aforementioned system can be integrated as

$$\begin{aligned} H_t^1 &= W_t^1 \\ H_t^2 &= W_t^2 \\ H_t^3 &= 2S_t, \quad \forall t \geq 0 \end{aligned}$$

where

$$S_t = \int_0^t W_s^2 \, dW_s^1 - \int_0^t W_s^1 \, dW_s^2$$

is the Lévy area (1.11.71). The heat kernel of Δ_H will be computed as the density of the diffusion H_t that starts from the origin. Let $p_t(x_1, x_2, x_3)$ be given by

$$P(H_t^1 \in dx_1, H_t^2 \in dx_2, H_t^3 \in dx_3) = p_t(x_1, x_2, x_3) \, dx_1 dx_2 dx_3.$$

Next we shall find $p_t(x_1, x_2, x_3)$ by comparing two equivalent expressions of the expectation

$$\mathbb{E}[e^{i\lambda_1 H_t^1} e^{i\lambda_2 H_t^2} e^{i\lambda_3 H_t^3}], \qquad \lambda_1, \lambda_2, \lambda_3 \in \mathbb{R}.$$

Using that H_t^3 depends on H_t^1 and H_t^2, we shall use the tower property and then Lévy's area formula given by Theorem 1.11.3

$$\mathbb{E}[e^{2i\lambda_3 S_t} \mid W_t^1, W_t^2] = \frac{2\lambda_3 t}{\sinh(2\lambda_3 t)} e^{-\frac{(W_t^1)^2 + (W_t^2)^2}{2t}(2\lambda_3 t \coth(2\lambda_3 t) - 1)} \tag{4.10.75}$$

to express the aforementioned expectation as

$$\begin{aligned} \mathbb{E}[e^{i\lambda_1 H_t^1} e^{i\lambda_2 H_t^2} e^{i\lambda_3 H_t^3}] &= \mathbb{E}[e^{i\lambda_1 H_t^1} e^{i\lambda_2 H_t^2} e^{2i\lambda_3 S_t}] \\ &= \mathbb{E}[\,\mathbb{E}[e^{i\lambda_1 H_t^1} e^{i\lambda_2 H_t^2} e^{2i\lambda_3 S_t} \mid W_t^1, W_t^2]\,] \\ &= \mathbb{E}[\, e^{i\lambda_1 H_t^1} e^{i\lambda_2 H_t^2} \mathbb{E}[e^{2i\lambda_3 S_t} \mid W_t^1, W_t^2]\,] \\ &= \frac{2\lambda_3 t}{\sinh(2\lambda_3 t)} \mathbb{E}[e^{i\lambda_1 W_t^1} e^{-\frac{(W_t^1)^2}{2t}(2\lambda_3 t \coth(2\lambda_3 t) - 1)}] \\ &\quad \mathbb{E}[e^{i\lambda_1 W_t^2} e^{-\frac{(W_t^2)^2}{2t}(2\lambda_3 t \coth(2\lambda_3 t) - 1)}], \end{aligned}$$

where we used that W_t^1 and W_t^2 are independent processes. Since

$$\mathbb{E}[e^{i\lambda_1 W_t^1} e^{-\frac{(W_t^1)^2}{2t}(2\lambda_3 t \coth(2\lambda_3 t) - 1)}] = \frac{1}{\sqrt{2\pi t}} \int_{\mathbb{R}} e^{i\lambda_1 x_1} e^{-\frac{x_1^2}{2t} 2\lambda_3 t \coth(2\lambda_3 t)} \, dx_1, \tag{4.10.76}$$

then the previous expression becomes a double Fourier transform with respect to (x_1, x_2)

$$\begin{aligned}
&\mathbb{E}[e^{i\lambda_1 H_t^1} e^{i\lambda_2 H_t^2} e^{i\lambda_3 H_t^3}] \\
&= \frac{2\lambda_3 t}{\sinh(2\lambda_3 t)} \frac{1}{2\pi t} \iint_{\mathbb{R}^2} e^{i\lambda_1 x_1} e^{i\lambda_2 x_2} e^{-\frac{1}{2t}(x_1^2+x_2^2)\, 2\lambda_3 t \coth(2\lambda_3 t)} dx_1 dx_2 \\
&= \frac{2\lambda_3 t}{\sinh(2\lambda_3 t)} \frac{1}{2\pi t} \mathcal{F}_{x_1,x_2}\Big(e^{-\frac{1}{2t}(x_1^2+x_2^2)\, 2\lambda_3 t \coth(2\lambda_3 t)}\Big)(\lambda_1, \lambda_2). \qquad (4.10.77)
\end{aligned}$$

On the the other side, we have using Fubini's theorem

$$\begin{aligned}
\mathbb{E}[e^{i\lambda_1 H_t^1} e^{i\lambda_2 H_t^2} e^{i\lambda_3 H_t^3}] &= \iiint_{\mathbb{R}^3} e^{i\lambda_1 x_1} e^{i\lambda_2 x_2} e^{i\lambda_3 x_3} p_t(x_1, x_2, x_3)\, dx_1 dx_2 dx_3 \\
&= \iint_{\mathbb{R}^2} e^{i\lambda_1 x_1} e^{i\lambda_2 x_2} \Big(\int_{\mathbb{R}} e^{i\lambda_3 x_3} p_t(x_1, x_2, x_3)\, dx_3\Big)\, dx_1 dx_2 \\
&= \mathcal{F}_{x_1,x_2}\Big(\int_{\mathbb{R}} e^{i\lambda_3 x_3} p_t(x_1, x_2, x_3)\, dx_3\Big)(\lambda_1, \lambda_2). \qquad (4.10.78)
\end{aligned}$$

Comparing (4.10.77) and (4.10.78) yields

$$\int_{\mathbb{R}} e^{i\lambda_3 x_3} p_t(x_1, x_2, x_3)\, dx_3 = \frac{2\lambda_3 t}{\sinh(2\lambda_3 t)} \frac{1}{2\pi t} e^{-\frac{1}{2t}(x_1^2+x_2^2)\, 2\lambda_3 t \coth(2\lambda_3 t)}.$$

We can retrieve the density $p_t(x_1, x_2, x_3)$ by applying an inverse Fourier transform

$$p_t(x_1, x_2, x_3) = \frac{1}{2\pi} \int_{\mathbb{R}} \frac{2\lambda_3 t}{2\pi t \sinh(2\lambda_3 t)} e^{-i\lambda_3 x_3 - \lambda_3 (x_1^2+x_2^2) \coth(2\lambda_3 t)}\, d\lambda_3. \quad (4.10.79)$$

We arrived to the formula

$$p_t(x_1, x_2, x_3) = \frac{1}{4\pi^2} \int_{\mathbb{R}} V(t, \xi) e^{-i\xi x_3 - \xi(x_1^2+x_2^2) \coth(2\xi t)}\, d\xi, \qquad (4.10.80)$$

where

$$V(t, \xi) = \frac{2\xi}{\sinh(2\xi t)} \qquad (4.10.81)$$

is called the *volume element.* The stochastic interpretation of the volume element as a conditional expectation is obtained by substituting $(W_t^1, W_t^2) = (0, 0)$ in Lévi's area formula (4.10.75)

$$\mathbb{E}[e^{2i\xi S_t} \mid W_t^1 = 0, W_t^2 = 0] = \frac{2\xi}{\sinh(2\xi t)}.$$

Thus, the heat kernel $p_t(x_1, x_2, x_3)$ is obtained by integrating the volume element $V(t, \xi)$ with respect to an exponential measure. The exponent is the complex action that depends on the length of the subRiemannian geodesics on the Heisenberg group. The reader interested in the geometry underlying this topic may consult the book Calin et al. [38].

In general, the improper integral in formula (4.10.80) cannot be computed explicitly. This occurs naturally for most sub-elliptic operators, case in which the heat kernel is represented as an integral formula. However, if $(x_1, x_2) = (0, 0)$, this integral can be computed explicitly. Taking the Fourier transform, we have

$$\begin{aligned} p_t(0,0,x_3) &= \frac{1}{4\pi^2} \int_{\mathbb{R}} e^{-i\xi x_3} \frac{2\xi}{\sinh(2\xi t)}\, d\xi \\ &= \frac{1}{16t^2} \operatorname{sech}^2\Big(\frac{\pi x_3}{4t}\Big). \end{aligned}$$

This represents the amount of heat transferred from the origin to a point $(0, 0, x_3)$ on the vertical direction within time t.

Remark 4.10.1 The vertical axis $\{(0, 0, x_3); x_3 \in \mathbb{R}\}$ represents the *missing direction* on the Heisenberg group $\mathbb{H}^1$, i.e., the direction that is missing from the horizontal distribution generated by $\{X, Y\}$. It can be shown that it can be also described as the conjugate locus of the origin along the subRiemannian geodesics emerging from $(0, 0, 0)$. It can be shown that the point $(0, 0, x_3)$ can be reached from the origin by a rotational invariant family of subRiemannian geodesics, see [38]. This rotational invariance might be the reason for being able to compute the previous explicit formula.

Formula (4.10.80) provides the expression of the heat kernel for Δ_H from the origin to the point $x = (x_1, x_2, x_3)$. Since the vector fields X and Y are left invariant with respect to the group law on $\mathbb{H}^1$, then so will be the operator Δ_H and its heat kernel. Denoting the heat kernel from y to x within time t by $p_t(y; x)$, the group law invariance yields

$$p_t(y; x) = p_t(0; y^{-1} \circ x) = p_t(y^{-1} \circ x).$$

Since $y^{-1} = (-y_1, -y_2, -y_3)$ and

$$y^{-1} \circ x = (x_1 - y_1, x_2 - y_2, x_3 - y_3 + 2(y_1 x_2 - y_2 x_1)),$$

applying formula (4.10.80) we obtain

$$p_t(y; x) = \frac{1}{4\pi^2} \int_{\mathbb{R}} V(t, \xi) e^{-i\xi(x_3 - y_3 + 2(y_1 x_2 - y_2 x_1)) - \xi[(x_1 - y_1)^2 + (x_2 - y_2)^2] \coth(2\xi t)}\, d\xi. \tag{4.10.82}$$

Remark 4.10.2 If substitute $\xi = -\tau/t$ in formula (4.10.80) we obtain

$$p_t(x_1, x_2, x_3) = \frac{1}{4\pi^2 t^2} \int_{\mathbb{R}} \frac{2\tau}{\sinh(2\tau)} e^{-f(x;\tau)/t}\, d\tau,$$

where $f(x;\tau) = -ix_3\tau + \tau(x_1^2 + x_2^2)\cosh(2\tau)$ is called the *modified action function.* This form of the heat kernel appears in the work of Beals, Gaveau and Greiner [10], where it is obtained by means of complex Hamiltonian mechanics.

4.11 The Heat Kernel for Casimir Operator

In this section we shall find the heat kernel for the Casimir operator introduced by formula (4.9.73)

$$\Delta_{Cas} = \frac{1}{2}(X^2 + Y^2 + Z^2).$$

The approach is very similar with the one used in the case of the Heisenberg Laplacian. By Hörmander's theorem, this heat kernel is a smooth function.

To find the heat kernel, we start by writing the operator as

$$\begin{aligned}
\Delta_H &= \frac{1}{2}(X^2 + Y^2 + Z^2) \\
&= \frac{1}{2}\Big(\partial_{x_1}^2 + \partial_{x_2}^2 + 4(x_1^2 + x_2^2)\partial_{x_3}^2 + 4x_2\partial_{x_1}\partial_{x_3} - 4x_1\partial_{x_2}\partial_{x_3} + \partial_{x_3}^2\Big) \\
&= \frac{1}{2}\sum_{i,j=1}^{3} b_{ij}\partial_{x_i}\partial_{x_j},
\end{aligned}$$

where

$$(b_{ij}) = \begin{pmatrix} 1 & 0 & 2x_2 \\ 0 & 1 & -2x_1 \\ 2x_2 & -2x_1 & 4(x_1^2 + x_2^2) + 1 \end{pmatrix}$$

is a non-singular matrix with $\det(b_{ij}) = 1$. We can write $b = \sigma\sigma^T$, with

$$(\sigma_{ij}) = \begin{pmatrix} 1 & 0 & 0 \\ 0 & 1 & 0 \\ 2x_2 & -2x_1 & 1 \end{pmatrix}.$$

Then the associated diffusion, $C_t = (C_t^1, C_t^2, C_t^3)$, satisfies the stochastic differential equation

$$\begin{aligned}
dC_t^1 &= dW_t^1 \\
dC_t^2 &= dW_t^2 \\
dC_t^3 &= 2C_t^2 dW_t^1 - 2C_t^1 dW_t^2 + dW_t^3, \quad t \geq 0
\end{aligned}$$

where W_t^1, W_t^2 and W_t^3 are one-dimensional independent Brownian motions. Due to the group law, it suffices to make the simplifying assumption $C_0 = (0,0,0)$. In this case we have

$$\begin{aligned} C_t^1 &= W_t^1 \\ C_t^2 &= W_t^2 \\ C_t^3 &= 2S_t + W_t^3, \quad \forall t \geq 0 \end{aligned}$$

where $S_t = \displaystyle\int_0^t W_s^2\, dW_s^1 - \int_0^t W_s^1\, dW_s^2$ denotes the Lévy's area. We have the following computation

$$\begin{aligned} \mathbb{E}[e^{i\lambda_1 C_t^1} e^{i\lambda_2 C_t^2} e^{i\lambda_3 C_t^3}] &= \mathbb{E}[e^{i\lambda_1 C_t^1} e^{i\lambda_2 C_t^2} e^{2i\lambda_3 S_t} e^{i\lambda_3 W_t^3}] \\ &= \mathbb{E}[\,\mathbb{E}[e^{i\lambda_1 C_t^1} e^{i\lambda_2 C_t^2} e^{2i\lambda_3 S_t} e^{i\lambda_3 W_t^3} \mid W_t^1, W_t^2, W_t^3]\,] \\ &= \mathbb{E}[\, e^{i\lambda_1 C_t^1} e^{i\lambda_2 C_t^2} e^{i\lambda_3 W_t^3} \mathbb{E}[e^{2i\lambda_3 S_t} \mid W_t^1, W_t^2, W_t^3]\,] \\ &= \mathbb{E}[\, e^{i\lambda_1 C_t^1} e^{i\lambda_2 C_t^2} e^{i\lambda_3 W_t^3} \mathbb{E}[e^{2i\lambda_3 S_t} \mid W_t^1, W_t^2]\,] \\ &= \frac{2\lambda_3 t}{\sinh(2\lambda_3 t)} \mathbb{E}[e^{i\lambda_1 W_t^1} e^{-\frac{(W_t^1)^2}{2t}(2\lambda_3 t\coth(2\lambda_3 t)-1)}] \\ &\quad \mathbb{E}[e^{i\lambda_1 W_t^2} e^{-\frac{(W_t^2)^2}{2t}(2\lambda_3 t\coth(2\lambda_3 t)-1)}]\, \mathbb{E}[e^{i\lambda_3 W_t^3}], \end{aligned}$$

where we first used the tower property to condition over $\{W_t^1, W_t^2, W_t^3\}$, then we used that C_t^1, C_t^2, C_t^3 are determined by $\{W_t^1, W_t^2, W_t^3\}$, so we could take them out of the inner expectation. Then we dropped the W_t^3 condition since it is independent of S_t. Then we used the Lévy's area formula (4.10.75) and the independence of the Brownian motions. Since $\mathbb{E}[e^{i\lambda_3 W_t^3}] = e^{-\frac{\lambda_3^2}{2}t}$, using (4.10.76) we obtain

$$\begin{aligned} &\mathbb{E}[e^{i\lambda_1 C_t^1} e^{i\lambda_2 C_t^2} e^{i\lambda_3 C_t^3}] \\ &= \frac{2\lambda_3 t}{\sinh(2\lambda_3 t)} \frac{1}{2\pi t} e^{-\frac{\lambda_3^2}{2}t} \iint_{\mathbb{R}^2} e^{i\lambda_1 x_1} e^{i\lambda_2 x_2} e^{-\frac{1}{2t}(x_1^2+x_2^2)\, 2\lambda_3 t\coth(2\lambda_3 t)}\, dx_1 dx_2 \\ &= \frac{2\lambda_3 t}{\sinh(2\lambda_3 t)} \frac{1}{2\pi t} e^{-\frac{\lambda_3^2}{2}t} \mathcal{F}_{x_1,x_2}\Big(e^{-\frac{1}{2t}(x_1^2+x_2^2)\, 2\lambda_3 t\coth(2\lambda_3 t)}\Big)(\lambda_1, \lambda_2). \end{aligned} \tag{4.11.83}$$

Comparing with a formula similar to (4.10.78)

$$\mathbb{E}[e^{i\lambda_1 C_t^1} e^{i\lambda_2 C_t^2} e^{i\lambda_3 C_t^3}] = \mathcal{F}_{x_1,x_2}\Big(\int_{\mathbb{R}} e^{i\lambda_3 x_3} p_t(x_1,x_2,x_3)\, dx_3\Big)(\lambda_1, \lambda_2)$$

yields

$$\int_{\mathbb{R}} e^{i\lambda_3 x_3} p_t(x_1,x_2,x_3)\, dx_3 = \frac{2\lambda_3 t}{\sinh(2\lambda_3 t)} \frac{1}{2\pi t} e^{-\frac{1}{2}\lambda_3^2 t - \frac{1}{2t}(x_1^2+x_2^2)\, 2\lambda_3 t\coth(2\lambda_3 t)}.$$

We can retrieve the density $p_t(x_1, x_2, x_3)$ by applying an inverse Fourier transform

$$p_t(x_1, x_2, x_3) = \frac{1}{2\pi} \int_{\mathbb{R}} \frac{2\lambda_3 t}{2\pi t \sinh(2\lambda_3 t)} e^{-\frac{1}{2}\lambda_3^2 t - i\lambda_3 x_3 - \lambda_3 (x_1^2 + x_2^2) \coth(2\lambda_3 t)} \, d\lambda_3.$$

We arrived to the formula

$$p_t(x_1, x_2, x_3) = \frac{1}{4\pi^2} \int_{\mathbb{R}} V(t, \xi) e^{-\frac{1}{2}\xi^2 t - i\xi x_3 - \xi(x_1^2 + x_2^2) \coth(2\xi t)} \, d\xi, \tag{4.11.84}$$

with $V(t, \xi)$ given by (4.10.81). The general formula representing the heat kernel between any two points is using the group law invariance

$$p_t(y_1, y_2, y_3; x_1, x_2, x_3) = p_t(z_1, z_2, z_3),$$

where $z = y^{-1} \circ x$.

4.12 The Nonsymmetric Heisenberg Group

The vector fields on $\mathbb{R}^3$

$$X = \partial_{x_1}, \quad Y = \partial_{x_2} + x_1 \partial_{x_3}, \qquad Z = \partial_{x_3}$$

are left invariant with respect to the noncommutative Lie group law on $\mathbb{R}^3$

$$(x_1, x_2, x_3) * (x_1', x_2', x_3') = (x_1 + x_1', x_2 + x_2', x_3 + x_3' + x_1 x_2').$$

The pair $(\mathbb{R}^3, *)$ is regarded as the *noncommutative 3-dimensional Heisenberg group*. The neutral element of this group is $e = (0, 0, 0)$ and the law of the inverse is given by $(x_1, x_2, x_3)^{-1} = (-x_1, -x_2, x_1 x_2 - x_3)$.
The commutator relations

$$[X, Y] = Z, \quad [X, Z] = 0, \quad [Y, Z] = 0$$

show that $\{X, Y, [X, Y]\}$ span the tangent space of $\mathbb{R}^3$ at each point and hence the maximal rank condition is satisfied. Then the associated sub-elliptic operator

$$\Delta_{NH} = \frac{1}{2}(X^2 + Y^2) = \frac{1}{2}(\partial_{x_1}^2 + \partial_{x_2}^2 + 2x_1 \partial_{x_2} \partial_{x_3} + x_1^2 \partial_{x_3}^2)$$

is hypoelliptic and has a smooth heat kernel. Since the associated diffusion and dispersion matrices are respectively given by

$$a = \sigma\sigma^T = \begin{pmatrix} 1 & 0 & 0 \\ 0 & 1 & x_1 \\ 0 & x_1 & x_1^2 \end{pmatrix}, \qquad \sigma = \begin{pmatrix} 1 & 0 & 0 \\ 0 & 1 & 0 \\ 0 & x_1 & 0 \end{pmatrix},$$

then the associated diffusion $X_t = (X_1(t), X_2(t), X_3(t))$ satisfies the system

$$\begin{aligned} dX_1(t) &= dW_1(t) \\ dX_2(t) &= dW_2(t) \\ dX_3(t) &= X_1(t)\, dW_2(t), \end{aligned}$$

where $W_1(t)$ and $W_2(t)$ represent two independent standard Brownian motions. Due to the Lie group invariance, it suffices to assume the diffusion starting at the origin, $X_1(0) = X_2(0) = X_3(0) = 0$. Therefore,

$$X_1(t) = W_1(t) \tag{4.12.85}$$

$$X_2(t) = W_2(t) \tag{4.12.86}$$

$$X_3(t) = \int_0^t W_1(s)\, dW_2(s), \tag{4.12.87}$$

namely the first two components of the diffusion are independent standard Brownian motions, while the third component is the asymmetric area process introduced in Section 1.13. Therefore, there is a DDS Brownian motion B_t, such that $X_3(t) = B_{A_t}$, where $A_t = \int_0^t W_1(s)^2\, ds$.

The heat kernel of Δ_{NH} Finding the heat kernel for the subelliptic operator Δ_{NH} is similar with the case of the Heisenberg Laplacian, with the difference that in the present case we use the formula provided by Proposition 1.13.1

$$\mathbb{E}[e^{i\lambda \int_0^t W_1(s) dW_2(s)} | W_1, W_2] = \frac{\lambda t}{2\sinh(\lambda t/2)} e^{\frac{i\lambda}{2} W_1 W_2 - \frac{\|W\|^2}{2t}\left(\frac{\lambda t}{2}\coth(\frac{\lambda t}{2}) - 1\right)}.$$

Due to the Lie group invariance, it suffices to find the heat kernel starting from the origin. We denote $F(\lambda_1, \lambda_2, \lambda) = \mathbb{E}[e^{i\lambda_1 X_1(t)} e^{i\lambda_2 X_2(t)} e^{i\lambda X_3(t)}]$. Using the tower property we have

$$\begin{aligned} F(\lambda_1, \lambda_2, \lambda) &= \mathbb{E}[\, \mathbb{E}[e^{i\lambda_1 W_1(t)} e^{i\lambda_2 W_2(t)} e^{i\lambda X_3(t)} | W_1, W_2]\,] \\ &= \mathbb{E}[\, e^{i\lambda_1 W_1(t)} e^{i\lambda_2 W_2(t)} \mathbb{E}[e^{i\lambda X_3(t)} | W_1, W_2]\,] \\ &= \mathbb{E}\Big[\, e^{i\lambda_1 W_1(t)} e^{i\lambda_2 W_2(t)} \frac{\lambda t}{2\sinh(\lambda t/2)} e^{\frac{i\lambda}{2} W_1 W_2 - \frac{\|W\|^2}{2t}\left(\frac{\lambda t}{2}\coth(\frac{\lambda t}{2}) - 1\right)} \Big]. \end{aligned}$$

Using that $W_1(t), W_2(t) \sim N(0, t)$, we obtain

$$F(\lambda_1, \lambda_2, \lambda_3) = \frac{1}{2\pi t} \frac{\lambda t}{2\sinh(\lambda t/2)} \iint_{\mathbb{R}^2} e^{i\lambda_1 x_1} e^{i\lambda_2 x_2} e^{\frac{i\lambda}{2} x_1 x_2 - \frac{\|\tilde{x}\|^2}{2t} \frac{\lambda t}{2} \coth(\frac{\lambda t}{2})} dx_1 dx_2, \tag{4.12.88}$$

where $\|\tilde{x}\|^2 = x_1^2 + x_2^2$. If $p_t(x_1, x_2, x_3) = p_t(0, 0, 0; x_1, x_2, x_3)$ represents the transition density of diffusion $\big(X_1(t), X_2(t), X_3(t)\big)$ from the origin, then we have

$$\begin{aligned}F(\lambda_1,\lambda_2,\lambda) &= \iiint_{\mathbb{R}^3} e^{i\lambda_1 x_1} e^{i\lambda_2 x_2} e^{i\lambda x_3} p_t(x_1,x_2,x_3)\,dx_1 dx_2 dx_3\\ &= \iint_{\mathbb{R}^2} e^{i\lambda_1 x_1} e^{i\lambda_2 x_2}\Big(\int_{\mathbb{R}} e^{i\lambda x_3} p_t(x_1,x_2,x_3)\,dx_3\Big) dx_1 dx_2.\end{aligned}$$

Comparing to (4.12.88) we get

$$\int_{\mathbb{R}} e^{i\lambda x_3} p_t(x_1,x_2,x_3)\,dx_3 = \frac{\lambda}{4\pi\sinh(\lambda t/2)} e^{\frac{i\lambda}{2}x_1x_2 - \frac{\|\bar{x}\|^2}{2}\cdot\frac{\lambda}{2}\coth(\frac{\lambda t}{2})}.$$

Inverting yields

$$p_t(x_1,x_2,x_3) = \frac{1}{2\pi}\int_{\mathbb{R}} e^{-i\lambda x_3} \frac{\lambda}{4\pi\sinh(\lambda t/2)} e^{\frac{i\lambda}{2}x_1x_2 - \frac{\|\bar{x}\|^2}{2}\cdot\frac{\lambda}{2}\coth(\frac{\lambda t}{2})}\,d\lambda.$$

Therefore, the heat kernel from the origin of Δ_{NH} is

$$\begin{aligned}p_t(x_1,x_2,x_3) &= \frac{1}{8\pi^2}\int_{\mathbb{R}} e^{-i\lambda x_3} \frac{\lambda}{\sinh(\lambda t/2)} e^{\frac{i\lambda}{2}x_1x_2 - \frac{\|\bar{x}\|^2}{2}\cdot\frac{\lambda}{2}\coth(\frac{\lambda t}{2})}\,d\lambda\\ &= \frac{1}{2\pi^2}\int_{\mathbb{R}} \frac{2\xi}{\sinh(\xi t)} e^{-i\xi(2x_3 - x_1x_2) - \frac{\xi}{2}(x_1^2+x_2^2)^2\coth(\xi t)}\,d\xi, \qquad (4.12.89)\end{aligned}$$

where we used substitution $\xi = \lambda/2$. Now, the heat kernel can be computed from any point using the group law invariance. Using

$$x^{-1} * y = (x_1,x_2,x_3)^{-1} * (y_1,y_2,y_3) = (y_1 - x_1, y_2 - x_2, y_3 - x_3 + x_1(x_2 - y_2))$$

we obtain the heat kernel for Δ_{NH} by direct computation from (4.12.89)

$$\begin{aligned}&p_t(x;y) = p_t(0, x^{-1} * y) = p_t(x^{-1} * y)\\ &= \frac{1}{2\pi^2}\int_{\mathbb{R}} \frac{2\xi}{\sinh(\xi t)} e^{-i\xi[2(y_3 - x_3) - (y_2 - x_2)(x_1 + y_1)} e^{-\frac{\xi}{2}[(y_1 - x_1)^2 + (y_2 - x_2)^2]\coth(\xi t)}\,d\xi.\end{aligned}$$

4.13 The Saddle Laplacian

In this section we consider the following vector fields on $\mathbb{R}^3 = \mathbb{R}_x^2 \times \mathbb{R}_z$

$$X_1 = \partial_{x_1} + x_2\partial_z, \qquad X_2 = \partial_{x_2} + x_1\partial_z, \qquad (4.13.90)$$

which look similar to the Heisenberg vector fields (4.9.72) up to a sign difference. However, here there is a major difference, that consists of the fact that the vector fields commute

$$[X_1, X_2] = X_1X_2 - X_2X_1 = 0.$$

This means that the distribution generated by the vector fields $\{X_1, X_2\}$ is integrable, so there is a two-dimensional surface that is tangent to the aforementioned vector fields. This surface is actually given by the saddle surface

$$\phi : \mathbb{R}^2 \to \mathbb{R}^3, \qquad \phi(x_1, x_2) = (x_1, x_2, x_1x_2), \tag{4.13.91}$$

as it is implied by the relations $X_1 = \partial_{x_1}\phi$ and $X_2 = \partial_{x_2}\phi$. We are interested in the sum of squares operator

$$\begin{aligned} L &= \frac{1}{2}(X_1^2 + X_2^2) \\ &= \frac{1}{2}(\partial_{x_1}^2 + \partial_{x_2}^2 + |x|^2\partial_z^2) + (x_2\partial_{x_1}\partial_z + x_1\partial_{x_2}\partial_z), \end{aligned} \tag{4.13.92}$$

with $|x|^2 = x_1^2 + x_2^2$. The operator given by (4.13.92) will be called the *saddle Laplacian*. It is worth noting that the operator which appears in the first term corresponds to the two-dimensional Grushin operator with one missing direction, which is given by relation (4.5.38).

In the following we shall find the heat kernel for the operator L by finding the transition probability of a diffusion on the saddle surface (4.13.91).

Diffusion on a saddle We consider the 3-dimensional process

$$X_t = \phi\big(W_1(t), W_2(t)\big) = \big(W_1(t),\, W_2(t),\, W_1(t)W_2(t)\big),$$

with $W_1(t)$ and $W_2(t)$ two independent Brownian motions on $\mathbb{R}$. Since X_t satisfies the system

$$\begin{aligned} dX_1(t) &= dW_1(t) \\ dX_2(t) &= dW_2(t) \\ dX_3(t) &= W_1(t)dW_2(t) + W_2(t)dW_1(t), \qquad X_0 = (0,0,0), \end{aligned}$$

we can write

$$dX_t = \sigma(X_t)dW_t,$$

with dispersion

$$\sigma(x) = \begin{pmatrix} 1 & 0 \\ 0 & 1 \\ x_2 & x_1 \end{pmatrix}$$

and $dW_t = \big(dW_1(t), dW_2(t)\big)^T$. The process X_t is an Ito diffusion living on the saddle surface (4.13.91) and starting at the origin. The diffusion metric is given by

$$\sigma\sigma^T(x) = \begin{pmatrix} 1 & 0 & x_2 \\ 0 & 1 & x_1 \\ x_2 & x_1 & |x|^2 \end{pmatrix}.$$

It can be shown that the normal unit vector to the saddle (4.13.91) is given by

$$N = \frac{X_1 \times X_2}{|X_1 \times X_2|} = \frac{(-x_2, -x_1, 1)}{\sqrt{1+|x|^2}},$$

where " $\times$ " stands for the vectorial product in $\mathbb{R}^3$. Since we can verify that $\sigma\sigma^T N = 0$, it follows that N is the singular direction of the diffusion matrix in which the diffusion does not occur. This is natural, since the diffusion occurs only along the saddle surface. We note now, using the expression of $\sigma\sigma^T$, that the associated generator operator with diffusion X_t, which is given by $\frac{1}{2}\langle \sigma\sigma^T \partial_{\mathbf{x}}, \partial_{\mathbf{x}}\rangle$, where $\partial_{\mathbf{x}} = (\partial_{x_1}, \partial_{x_2}, \partial_z)^T$, is exactly the operator L given by formula (4.13.92). We shall find the transition probability $p_t(x_1, x_2, x_3)$ of diffusion X_t that starts at the origin, $X_0 = (0, 0, 0)$.

The transition probability We recall from Section 2.11 that the product of two independent Brownian motions, $W_1(t)$ and $W_2(t)$, can be written as a Brownian motion β_t under a stochastic time as

$$W_1(t)W_2(t) = \beta_{\int_0^t R_s^2\, ds}, \quad t \geq 0,$$

where $R_s^2 = W_1(s)^2 + W_2(s)^2$. We shall compute next the expectation on the left side of the expression

$$\mathbb{E}[e^{i\lambda_1 X_1(t)} e^{i\lambda_2 X_2(t)} e^{i\lambda_3 X_3(t)}] = \int_{\mathbb{R}^3} e^{i(\lambda_1 x_1 + \lambda_2 x_2 + \lambda_3 x_3)} p_t(x_1, x_2, x_3)\, dx_1 dx_2 dx_3. \tag{4.13.93}$$

Since $\int_0^t R_s^2\, ds$ depends on the history $W[0,t] = \{(W_1(s), W_2(s)); 0 \leq s \leq t\}$, we evaluate the expectation using tower property and the independence of Brownian motions as

$$\begin{aligned}
&\mathbb{E}[e^{i\lambda_1 X_1(t)} e^{i\lambda_2 X_2(t)} e^{i\lambda_3 X_3(t)}] && (4.13.94)\\
&\quad = \mathbb{E}[e^{i\lambda_1 W_1(t)} e^{i\lambda_2 W_2(t)} e^{i\lambda_3 W_1(t) W_2(t)}]\\
&\quad = \mathbb{E}[e^{i\lambda_1 W_1(t)} e^{i\lambda_2 W_2(t)} e^{i\lambda_3 \beta_{\int_0^t R_s^2\, ds}}]\\
&\quad = \mathbb{E}[\,\mathbb{E}[e^{i\lambda_1 W_1(t)} e^{i\lambda_2 W_2(t)} e^{i\lambda_3 \beta_{\int_0^t R_s^2\, ds}} | W[0,t]]\,]\\
&\quad = \mathbb{E}[e^{i\lambda_1 W_1(t)} e^{i\lambda_2 W_2(t)} \mathbb{E}[e^{i\lambda_3 \beta_{\int_0^t R_s^2\, ds}} | W[0,t]]\,]\\
&\quad = \mathbb{E}[e^{i\lambda_1 W_1(t)} e^{i\lambda_2 W_2(t)} e^{-\frac{1}{2}\lambda_3^2 \int_0^t R_s^2\, ds}]\\
&\quad = \mathbb{E}[e^{i\lambda_1 W_1(t)} e^{-\frac{1}{2}\lambda_3^2 \int_0^t W_1(s)^2\, ds}]\, \mathbb{E}[e^{i\lambda_2 W_2(t)} e^{-\frac{1}{2}\lambda_3^2 \int_0^t W_2(s)^2\, ds}]. && (4.13.95)
\end{aligned}$$

To compute the last expectations we need the joint probability density of the pair $\left(B_t, \int_0^t B_s^2 ds\right)$, which can be found in [15], page 175

$$P_0\Big(B_t \in dz, \int_0^t B_s^2 ds \in dy\Big) = \frac{1}{\sqrt{2\pi}}\, \mathrm{ee}_y\big(1/2, t, z^2/2, 0\big) dy dz,$$

where [1]

$$\mathrm{ee}_y\big(1/2, t, z^2/2, 0\big) = \mathcal{L}^{-1}\bigg(\Big(\frac{\sqrt{2\gamma}}{\sinh(t\sqrt{2\gamma})}\Big)^{1/2} \exp\Big(-\frac{1}{2}z^2\sqrt{2\gamma}\coth(t\sqrt{2\gamma})\Big)\bigg),$$

where $\mathcal{L}^{-1}$ stands for the inverse Laplace transform. The previous expression can also be written equivalently in the integral form as

$$\int_0^\infty e^{-\gamma y}\mathrm{ee}_y\big(1/2, t, z^2/2, 0\big)\, dy = \Big(\frac{\sqrt{2\gamma}}{\sinh(t\sqrt{2\gamma})}\Big)^{1/2} \exp\Big(-\frac{1}{2}z^2\sqrt{2\gamma}\coth(t\sqrt{2\gamma})\Big).$$

For our next computation it is useful to substitute $\gamma = -\lambda^2/2$, which leads to

$$\int_0^\infty e^{-\frac{1}{2}\lambda^2 y}\mathrm{ee}_y\big(1/2, t, z^2/2, 0\big)\, dy = \Big(\frac{|\lambda|}{\sinh(t|\lambda|)}\Big)^{1/2} \exp\Big(-\frac{1}{2}z^2|\lambda|\coth(t|\lambda|)\Big). \tag{4.13.96}$$

Then for B_t Brownian motion and any $\lambda, \eta \in \mathbb{R}$ we have

$$\begin{aligned}
\mathbb{E}[e^{i\eta B_t}e^{-\frac{1}{2}\lambda^2\int_0^t B_s^2\, ds}] &= \frac{1}{\sqrt{2\pi}}\int_{-\infty}^\infty\int_0^\infty e^{i\eta z}e^{-\frac{1}{2}\lambda^2 y}\mathrm{ee}_y\big(1/2, t, z^2/2, 0\big) dy dz \\
&= \frac{1}{\sqrt{2\pi}}\int_{-\infty}^\infty e^{i\eta z}\int_0^\infty e^{-\frac{1}{2}\lambda^2 y}\mathrm{ee}_y\big(1/2, t, z^2/2, 0\big) dy dz \\
&= \frac{1}{\sqrt{2\pi}}\int_{-\infty}^\infty e^{i\eta z}\Big(\frac{|\lambda|}{\sinh(t|\lambda|)}\Big)^{1/2}\exp\Big(-\frac{1}{2}z^2|\lambda|\coth(t|\lambda|)\Big) dz \\
&= \frac{1}{\sqrt{2\pi}}\Big(\frac{\lambda}{\sinh(t\lambda)}\Big)^{1/2}\int_{-\infty}^\infty e^{i\eta z}\exp\Big(-\frac{1}{2}z^2\lambda\coth(t\lambda)\Big) dz \\
&= \frac{1}{\sqrt{2\pi}}\Big(\frac{\lambda}{\sinh(t\lambda)}\Big)^{1/2}\int_{-\infty}^\infty e^{i\eta z-\frac{1}{2}z^2\lambda\coth(t\lambda)}\, dz
\end{aligned}$$

[1]The general definition formula is

$$\mathrm{ee}_y\big(v, t, z, x\big) = \mathcal{L}_\gamma^{-1}\bigg(\Big(\frac{\sqrt{2\gamma}}{\sinh(t\sqrt{2\gamma})}\Big)^{v}\exp\Big(-z\sqrt{2\gamma}\coth(t\sqrt{2\gamma}) - \frac{x\sqrt{2\gamma}}{\sinh(t\sqrt{2\gamma})}\Big)\bigg),$$

see [15], p 653.

$$= \frac{1}{\sqrt{2\pi}}\left(\frac{\lambda}{\sinh(t\lambda)}\right)^{1/2}\frac{\sqrt{2\pi}}{(\lambda\coth(\lambda t))^{1/2}}e^{-\frac{1}{2}\eta^2\tanh(\lambda t)/\lambda}$$

$$= \frac{1}{\sqrt{\cosh(\lambda t)}}e^{-\frac{1}{2}\eta^2\tanh(\lambda t)/\lambda}, \tag{4.13.97}$$

where we have used formula (4.13.96) for the inner integral and applied the improper integral formula

$$\int_{-\infty}^{\infty} e^{i\eta z - az^2}\,dz = \sqrt{\frac{\pi}{a}}e^{-\frac{\eta^2}{4a}}, \qquad \eta\in\mathbb{R},\ a>0 \tag{4.13.98}$$

for the integral with respect to z.

Now, we shall go back to the computation of the expectations in (4.13.95). Substituting (4.13.97) in (4.13.95) twice we obtain

$$\begin{aligned}\mathbb{E}[e^{i\lambda_1 X_1(t)}e^{i\lambda_2 X_2(t)}e^{i\lambda_3 X_3(t)}] &= \frac{1}{\cosh(\lambda_3 t)}e^{-\frac{1}{2}\lambda_1^2\tanh(\lambda_3 t)/\lambda_3}e^{-\frac{1}{2}\lambda_2^2\tanh(\lambda_3 t)/\lambda_3}\\ &= \operatorname{sech}(\lambda_3 t)e^{-\frac{1}{2\lambda_3}(\lambda_1^2+\lambda_2^2)\tanh(\lambda_3 t)}.\end{aligned}$$

Comparing to relation (4.13.93) and applying the inverse Fourier transform we can retrieve the probability density as

$$p_t(x_1,x_2,x_3) = \frac{1}{(2\pi)^3}\int_{\mathbb{R}^3} e^{-i\langle\mathbf{\Lambda},\mathbf{x}\rangle}\operatorname{sech}(\lambda_3 t)e^{-\frac{1}{2\lambda_3}(\lambda_1^2+\lambda_2^2)\tanh(\lambda_3 t)}d\lambda_1 d\lambda_2 d\lambda_3, \tag{4.13.99}$$

where $\langle\mathbf{\Lambda},\mathbf{x}\rangle = \lambda_1 x_1+\lambda_2 x_2+\lambda_3 x_3$. This provides the integral representation of the heat kernel of the saddle Laplacian L given by (4.13.92).

4.14 Operators with Potential

This section will provide explicit formulas for the heat kernels of operators with potential of the form $L = \mathcal{A}_x - U(x)$, where $\mathcal{A}_x$ denotes the infinitesimal generator of a diffusion on $\mathbb{R}^n$ and $U:\mathbb{R}^n\to(0,+\infty)$ is a smooth potential function. These operators appear in quantum mechanics as propagators of the Schrödinger's equation. Their heat kernels will be computed using the Laplace transforms of certain functionals of Brownian bridges.

4.14.1 The Cumulative Expectation

We introduce first a few notations. If X_t is a process starting at $X_0 = x$ and A is an event, we define

$$\mathbb{E}_x[X_t(\omega);A] = \mathbb{E}_x[X_t(\omega)1_A(\omega)] = \int_A X_t(\omega)dP(\omega),$$

where 1_A stands for the characteristic function of the set A. If Y is another real-valued random variable, we consider

$$\mathbb{E}_x[X_t(\omega); Y(\omega) \in dy] = \frac{\partial}{\partial y}\mathbb{E}_x[X_t(\omega); Y(\omega) < y]\, dy.$$

The expression $\mathbb{E}_x[X_t(\omega); Y(\omega) < y]$ is the *cumulative expectation* of X_t with respect to Y and its derivative in y is its associate density.

Lemma 4.14.1 *If $\delta(x - a)$ represents the Dirac delta function centered at a, then we have*

$$\mathbb{E}_x[X_t(\omega); Y(\omega) \in dy] = \mathbb{E}_x[X_t(\omega)\delta(Y(\omega) - y)]\, dy.$$

Proof: Since the functions $\varphi_\epsilon(\omega) = \frac{1}{\epsilon}1_{(y,y+\epsilon)}(\omega)$ are non-negative and have the integral $\int_{\mathbb{R}} \varphi_\epsilon(\omega)d\omega = 1$, then we have the point-wise convergence $\varphi_\epsilon(\omega) \to \delta(\omega - y)$ as $\epsilon \to 0$. Then

$$\begin{aligned}
\mathbb{E}_x[X_t(\omega); Y(\omega) \in dy] &= \frac{\partial}{\partial y}\mathbb{E}_x[X_t(\omega); Y(\omega) < y]\, dy \\
&= \lim_{\Delta y \to 0} \frac{1}{\Delta y}\mathbb{E}_x[X_t(\omega); y < Y(\omega) < y + \Delta y]\, dy \\
&= \lim_{\Delta y \to 0} \frac{1}{\Delta y}\int_\Omega X_t(\omega)1_{y<Y<y+\Delta y}(\omega)dP(\omega)\, dy \\
&= \int_\Omega X_t(\omega)\delta(Y(\omega) - y)\, dP(\omega)\, dy \\
&= \mathbb{E}_x[X_t(\omega)\delta(Y(\omega) - y)]\, dy,
\end{aligned}$$

where we used the dominated convergence theorem, which can be applied under the condition that X_t is integrable. ∎

Theorem 4.14.2 *Let $U(x)$ be a non-negative, continuous function. We assume the differential operator*

$$L = \frac{1}{2}\partial_x^2 - U(x)$$

has a heat kernel, which is denoted by $p_t(x, y)$. Then we have

$$\mathbb{E}_x[e^{-\int_0^t U(W_s)\, ds}; W_t \in dy] = p_t(x, y)dy,$$

where W_s is a one-dimensional Brownian motion starting at $W_0 = x$.

Proof: By Feynman-Kac's formula the heat kernel of the operator L is given by

$$p_t(x, y) = \mathbb{E}_x[e^{-\int_0^t U(W_s)\, ds}\delta(W_t - y)].$$

Using Lemma 4.14.1 we obtain

$$p_t(x, y)dy = \mathbb{E}_x[e^{-\int_0^t U(W_s)\, ds}; W_t \in dy],$$

which is the desired relation. ∎

By analogy with the classical calculus of variations, the expression

$$S(x, y; t)(\omega) = \int_0^t U\big(W_s(\omega)\big)\, ds \tag{4.14.100}$$

can be considered as a *stochastic action* along the Brownian path W_s under the influence of the potential U, between $W_0 = x$ and $W_t = y$, within time t. Then the expectation

$$\mathbb{E}_x[e^{-S(x,y;t)(\omega)}; W_t \in dy] = p_t(x, y)dy \tag{4.14.101}$$

represents the average of all exponential contributions of the stochastic action $S(x, y; t)(\omega)$ for the cases when W_t belongs to an infinitesimal neighborhood of y.

The expression $\mathbb{E}_x[e^{-S(x,y;t)(\omega)}; W_t \in dy]$ can be computed explicitly for a number of cases, fact that will provide heat kernels associated with different operators with potentials. A particular type of stochastic action will be used later in the study of LC-circuits at page 507.

4.14.2 The Linear Potential

In this case we take $U(x) = \gamma x$, with $\gamma \in \mathbb{R}$. Formula (1.8.7) from Borodin and Salminen [15], page 174, provides

$$\mathbb{E}_x[e^{-\gamma \int_0^t W_s ds}; W_t \in dy] = \frac{1}{\sqrt{2\pi t}} e^{-\frac{(y-x)^2}{2t} - \frac{\gamma}{2}(x+y)t + \frac{\gamma^2}{24}t^3}\, dy, \tag{4.14.102}$$

where $W_0 = x$. By Theorem 4.14.2 we obtain:

Proposition 4.14.3 *The heat kernel of the operator $L = \frac{1}{2}\frac{\partial^2}{\partial x^2} - \gamma x$, with $\gamma \in \mathbb{R}$, is given by*

$$p_t(x, y) = \frac{1}{\sqrt{2\pi t}} e^{-\frac{(y-x)^2}{2t} - \frac{\gamma}{2}(x+y)t + \frac{\gamma^2}{24}t^3}.$$

For a deduction of the above result using the Hamiltonian formalism, the reader is referred to Calin and Chang [36], page 215.

4.14.3 The Quadratic Potential

We consider the potential $U(x) = \frac{\gamma^2}{2}x^2$, with $\gamma \in \mathbb{R}$. The differential operator $L = \frac{1}{2}\partial_x^2 - \frac{\gamma^2}{2}x^2$ is called the *Hermite operator*. Formula (1.9.7) from Borodin and Salminen [15], page 174, provides

$$\mathbb{E}_x[e^{-\frac{\gamma^2}{2}\int_0^t W_s^2 ds}; W_t \in dy] = \frac{\sqrt{\gamma}}{\sqrt{2\pi \sinh(t\gamma)}} e^{-[\frac{\gamma}{2}(x^2+y^2)\coth(t\gamma) - \gamma\frac{xy}{\sinh(t\gamma)}]} dy,$$

where $W_0 = x$. By Theorem 4.14.2 we obtain:

Proposition 4.14.4 *The heat kernel of the Hermite operator*

$$L = \frac{1}{2}\frac{\partial^2}{\partial x^2} - \frac{\gamma^2}{2}x^2, \qquad x \in \mathbb{R}$$

with $\gamma > 0$, is given by

$$p_t(x,y) = \frac{\sqrt{\gamma}}{\sqrt{2\pi \sinh(t\gamma)}} e^{-[\frac{\gamma}{2}(x^2+y^2)\coth(t\gamma) - \gamma\frac{xy}{\sinh(t\gamma)}]}, \quad t > 0.$$

The aforementioned heat kernel can be also computed by a method involving the Hamiltonian formalism, see [36], page 182. For a direct method, see Exercise 4.17.8.

4.14.4 The Inverse Quadratic Potential

We take $U(x) = \dfrac{\gamma^2}{2x^2}$, $x > 0$. In this case we consider the contributions along the Brownian motions which do not vanish. Since we consider $y > 0$, then we look for simulations of all Brownian motions that satisfy $\inf\limits_{0 \le s \le t} W_s > 0$.

Formula (1.20.7) from Borodin and Salminen [15], page 186, provides

$$\mathbb{E}_x[e^{-\frac{\gamma^2}{2}\int_0^t \frac{ds}{W_s^2}}; W_t \in dy, \inf_{s\le t} W_s > 0] = \frac{\sqrt{xy}}{t} e^{-\frac{x^2+y^2}{2t}} I_{\sqrt{\gamma^2+1/4}}\Big(\frac{xy}{t}\Big) dy,$$

where I_r denotes the modified Bessel function of order r of the first kind. By Theorem 4.14.2 we obtain:

Proposition 4.14.5 *The heat kernel of the operator*

$$L = \frac{1}{2}\frac{\partial^2}{\partial x^2} - \frac{\gamma^2}{2}\frac{1}{x^2}, \quad x > 0$$

with $\gamma > 0$, is given by

$$p_t(x,y) = \frac{\sqrt{xy}}{t} e^{-\frac{x^2+y^2}{2t}} I_{\sqrt{\gamma^2+1/4}}\Big(\frac{xy}{t}\Big), \quad t > 0.$$

This recovers the result of Theorem 3.18.1 from Calin et al. [29], page 68.

Remark 4.14.6 We note that for the particular value $\gamma = \sqrt{2}$ the formula of the heat kernel can be written in terms of elementary functions, due to the following representation of the modified Bessel function

$$I_{3/2}(x) = \frac{\sqrt{2}}{\sqrt{\pi}} x^{-3/2}(x \cosh x - \sinh x).$$

We obtain that the heat kernel of the operator

$$L = \frac{1}{2}\frac{\partial^2}{\partial x^2} - \frac{1}{x^2}, \qquad x > 0$$

is given by

$$\begin{aligned}
p_t(x,y) &= \frac{\sqrt{xy}}{t} e^{-\frac{x^2+y^2}{2t}} I_{3/2}\Big(\frac{xy}{t}\Big) \\
&= \sqrt{\frac{2t}{\pi}}\frac{1}{xy} e^{-\frac{x^2+y^2}{2t}} \Big[\frac{xy}{t}\cosh\Big(\frac{xy}{t}\Big) - \sinh\Big(\frac{xy}{t}\Big)\Big] \\
&= \frac{2}{\sqrt{2\pi t}} e^{-\frac{x^2+y^2}{2t}} \Big[\cosh\Big(\frac{xy}{t}\Big) - \frac{t}{xy}\sinh\Big(\frac{xy}{t}\Big)\Big] \\
&= \frac{1}{\sqrt{2\pi t}}\Big[e^{-\frac{(x-y)^2}{2t}} + e^{-\frac{(x+y)^2}{2t}}\Big]\Big[1 - \frac{t}{xy}\tanh\Big(\frac{xy}{t}\Big)\Big], \qquad t > 0.
\end{aligned}$$

Remark 4.14.7 By taking $\gamma = 0$ in the previous formula we obtain the heat kernel of the operator $L_+ = \frac{1}{2}\frac{d^2}{dx^2}$ on the half-line $(0, +\infty)$. Substituting the explicit formula $I_{1/2}(x) = \sqrt{\frac{2}{\pi x}}\sinh x$, we have

$$\begin{aligned}
p_t(x,y) &= \frac{\sqrt{xy}}{t} e^{-\frac{x^2+y^2}{2}} I_{1/2}\Big(\frac{xy}{t}\Big) = \sqrt{\frac{2}{\pi t}} e^{-\frac{x^2+y^2}{2t}} \sinh\Big(\frac{xy}{t}\Big) \\
&= \frac{1}{\sqrt{2\pi t}}\Big(e^{-\frac{(x-y)^2}{2t}} - e^{-\frac{(x+y)^2}{2t}}\Big), \qquad x, y > 0. \qquad (4.14.103)
\end{aligned}$$

It is worth noting that relation (4.14.103) represents also the transition density for a Brownian motion on $[0, +\infty)$ absorbed at 0.

4.14.5 The Combo Potential

In this case the potential is a combination between the quadratic and the inverse quadratic potential, $U(x) = \frac{p^2}{2}x^2 + \frac{q^2}{2}\frac{1}{x^2}$, with $x > 0$ and $p, q \in \mathbb{R}$. Only the simulations for which $W_s > 0$ does matter. Formula (1.21.17) from Borodin and Salminen [15], page 187, provides

$$\mathbb{E}_x\Big[e^{-\int_0^t\left(\frac{p^2}{2W_s^2}+\frac{q^2W_s^2}{2}\right)ds}; W_t\in dy, \inf_{s\le t} W_s>0\Big] = \frac{q\sqrt{xy}}{\sinh(tq)}e^{-\frac{q}{2}(x^2+y^2)\coth(tq)}I_{\sqrt{p^2+1/4}}\Big(\frac{xyq}{\sinh(tq)}\Big)dy.$$

We obtain:

Proposition 4.14.8 *The heat kernel of the operator*

$$L=\frac{1}{2}\frac{\partial^2}{\partial x^2}-\frac{p^2}{2}\frac{1}{x^2}-\frac{q^2}{2}x^2, \quad x>0$$

with $\gamma>0$, *is given by*

$$p_t(x,y)=\frac{q\sqrt{xy}}{\sinh(tq)}e^{-\frac{q}{2}(x^2+y^2)\coth(tq)}I_{\sqrt{p^2+1/4}}\Big(\frac{xyq}{\sinh(tq)}\Big), \quad t>0. \tag{4.14.104}$$

It is worth noting that in quantum mechanics the Schrödinger equation on the half line $\{x>0\}$ with potential $U(x)=\frac{p^2}{2}x^2+\frac{q^2}{2}\frac{1}{x^2}$ is also solvable by path integrals, obtaining a similar formula to (4.14.104), see Schulman [125], page 344. The relation between this potential and the one-dimensional 3-body problem via path integrals is treated in Khandekar and Lawande [86], [87].

4.14.6 The Exponential Potential

We consider the potential $U(x)=\gamma e^{2\beta x}$, depending on two parameters, $\beta\neq 0$ and $\gamma>0$. The problem of finding the heat kernel for the operator

$$L=\frac{1}{2}\frac{\partial^2}{\partial x^2}-\gamma e^{2\beta x}$$

has been approached geometrically in Calin and Chang [36], page 229. This is an example where the difficulties of the Hamiltonian method were overcome by the stochastic method. For this we use formula (1.10.7) from Borodin and Salminen [15], page 175 and the definition of the special function ki from page 654

$$\mathbb{E}_x[e^{-\gamma\int_0^t e^{2\beta W_s}ds}; W_t\in dy]=|\beta|\,\mathrm{ki}_{\beta^2t/2}\Big(\frac{\sqrt{2\gamma}}{|\beta|}e^{\beta x}, \frac{\sqrt{2\gamma}}{|\beta|}e^{\beta y}\Big)dy,$$

where

$$\mathrm{ki}_y(\mu,\eta)=\mathcal{L}_\gamma^{-1}\Big(K_{\sqrt{\gamma}}(\mu)\,I_{\sqrt{\gamma}}(\eta)\Big), \qquad \text{if } \mu\ge\eta$$

is the inverse Laplace transform of the product of the modified Bessel functions of the first and second kind. If $\eta \geq \mu$, then the arguments swap

$$\mathrm{ki}_y(\mu, \eta) = \mathrm{ki}_y(\eta, \mu).$$

The aforementioned inverse Laplace transform can be represented in the following integral form

$$\mathrm{ki}_y(\mu, \eta) = \frac{\mu\eta e^{\pi^2/4y}}{\pi\sqrt{\pi y}} \int_0^\infty \frac{K_1(\rho)}{\rho} e^{-u^2/4y} \sinh(u) \sin\big(\frac{\pi u}{2y}\big)\, du$$

where $\rho = \mu^2 + \eta^2 + 2\mu\eta \cosh u$, K_1 is a modified Bessel function of the second kind and we considered $\mu \geq \eta$.

Proposition 4.14.9 *The heat kernel of the operator*

$$L = \frac{1}{2}\frac{\partial^2}{\partial x^2} - \gamma e^{2\beta x}, \qquad x \in \mathbb{R}$$

is given by

$$p_t(x, y) = |\beta|\, \mathrm{ki}_{\beta^2 t/2}\Big(\frac{\sqrt{2\gamma}}{|\beta|} e^{\beta x}, \frac{\sqrt{2\gamma}}{|\beta|} e^{\beta y}\Big). \tag{4.14.105}$$

Remark 4.14.10 However, the method employed by this section cannot provide heat kernels for all types of potentials. For instance, if the potential is quartic, $U(x) = x^4$, then there is no closed form expression for the expectation

$$\mathbb{E}_x[e^{-\int_0^t W_s^4 ds}; W_t \in dy].$$

More reference on this expectation can be found in the study of Martinet distribution at page 405.

4.14.7 Other Operators

The following result is a generalization of Theorem 4.14.2 and is a consequence of Feynman-Kac's formula.

Theorem 4.14.11 *Let X_t be a diffusion with the infinitesimal generator $\mathcal{A}_x$, and $U(x)$ be a non-negative, continuous function. We assume the differential operator*

$$L = \mathcal{A}_x - U(x)$$

has a heat kernel, which is denoted by $p_t(x, y)$. Then we have

$$\mathbb{E}_x[e^{-\int_0^t U(X_s)\, ds}; X_t \in dy] = p_t(x, y) dy,$$

where diffusion X_s starts at $X_0 = x$.

Bessel operator with quadratic potential

Let $X_t = R_t^{(n)}$ be the n-dimensional Bessel process, where[2] $n \geq 3$. Its infinitesimal generator is

$$\mathcal{A}_x = \frac{1}{2}\partial_x^2 + \frac{n-1}{2x}\partial_x, \qquad x > 0.$$

We consider the quadratic potential $U(x) = \frac{\gamma^2}{2}x^2$. Formula (1.9.7) from Borodin and Salminen [15], page 384, provides

$$\mathbb{E}_x[e^{-\frac{\gamma^2}{2}\int_0^t (R_s^{(n)})^2 ds}; R_t^{(n)} \in dy] = \frac{\gamma y^{\nu+1} x^{-\nu}}{\sinh(t\gamma)} e^{-\frac{\gamma}{2}(x^2+y^2)\coth(t\gamma)} I_\nu\Big(\frac{xy\,\gamma}{\sinh(t\gamma)}\Big) dy,$$

where $\nu = \frac{n}{2} - 1 \geq 0$ is the index of the Bessel process.

Proposition 4.14.12 *The heat kernel for the Bessel operator with quadratic potential*

$$L = \frac{1}{2}\partial_x^2 + \frac{n-1}{2x}\partial_x - \frac{\gamma^2}{2}x^2, \quad x > 0$$

is given by

$$p_t(x,y) = \frac{\gamma y^{\nu+1} x^{-\nu}}{\sinh(t\gamma)} e^{-\frac{\gamma}{2}(x^2+y^2)\coth(t\gamma)} I_\nu\Big(\frac{xy\,\gamma}{\sinh(t\gamma)}\Big), \quad t > 0, \qquad (4.14.106)$$

with $\nu = \frac{n}{2} - 1$.

For dimensions $n \in \{3, 5\}$ the heat kernel formula can be written in terms of elementary functions as follows.

The case $n = 3$ In this case $\nu = 1/2$. Using $I_{1/2}(x) = \sqrt{\frac{2}{\pi x}}\sinh x$, equation (4.14.106) becomes

$$p_t(x,y) = \frac{2}{\sqrt{2\pi t}}\frac{y}{x}\sqrt{\frac{\gamma t}{\sinh(\gamma t)}} e^{-\frac{\gamma}{2}(x^2+y^2)\coth(t\gamma)} \sinh\Big(\frac{xy}{t}\frac{\gamma t}{\sinh(\gamma t)}\Big), \qquad t > 0.$$

We note that taking $\gamma \to 0$ in the previous relation we obtain

$$p_t(x,y) = \frac{2}{\sqrt{2\pi t}}\frac{y}{x} e^{-\frac{1}{2}(x^2+y^2)} \sinh\Big(\frac{xy}{t}\Big), \qquad t > 0,$$

[2]We choose $n \geq 3$ because in this case there is a zero probability that $R_s^{(n)}$ reaches the origin for $t > 0$.

which is the transition density of the 3-dimensional Bessel process $R_t^{(3)}$.

The case $n = 5$ In this case $\nu = 3/2$. Using the formula for the modified Bessel function $I_{3/2}(x) = \frac{\sqrt{2}}{\sqrt{\pi} x^{3/2}}(x \cosh x - \sinh x)$, equation (4.14.106) becomes

$$p_t(x, y) = \sqrt{\frac{2}{\pi\gamma}} \frac{y}{x^3} \sqrt{\sinh(\gamma t)} e^{-\frac{\gamma}{2}(x^2+y^2)\coth(t\gamma)} \times \left(\frac{xy\gamma}{\sinh(\gamma t)} \cosh\left(\frac{xy\gamma}{\sinh(\gamma t)}\right) - \sinh\left(\frac{xy\gamma}{\sinh(\gamma t)}\right)\right), \quad t > 0.$$

Bessel operator with inverse quadratic potential

We consider the inverse quadratic potential $U(x) = \frac{\gamma^2}{2}\frac{1}{x^2}$. Formula (1.20.7) from Borodin and Salminen [15], page 392, provides

$$\mathbb{E}_x\left[e^{-\frac{\gamma^2}{2}\int_0^t \frac{1}{(R_s^{(n)})^2} ds}; R_t^{(n)} \in dy\right] = \frac{y^{\nu+1}}{x^\nu t} e^{-\frac{x^2+y^2}{2t}} I_{\sqrt{\gamma^2+\nu^2}}\left(\frac{xy}{t}\right) dy.$$

where $\nu = \frac{n}{2} - 1 \geq 0$ is the index of the Bessel process $R_t^{(n)}$. Theorem 4.14.11 yields:

Proposition 4.14.13 *The heat kernel for the Bessel operator with inverse quadratic potential*

$$L = \frac{1}{2}\partial_x^2 + \frac{n-1}{2x}\partial_x - \frac{\gamma^2}{2}\frac{1}{x^2}, \quad x > 0$$

is given by

$$p_t(x, y) = \frac{y^{\nu+1}}{x^\nu t} e^{-\frac{x^2+y^2}{2t}} I_{\sqrt{\gamma^2+\nu^2}}\left(\frac{xy}{t}\right), \quad t > 0, \tag{4.14.107}$$

with $\nu = \frac{n}{2} - 1$.

All the previous heat kernels were computed using closed form expressions of the expectation $\mathbb{E}_x[e^{-\int_0^t U(W_s)ds}; W_t \in dy]$ available in the book [15] for several cases of potentials U. We shall show how those formulas can be obtained using the Brownian bridge process between x and y. To this end, we shall start with some basic facts regarding Brownian bridges.

4.14.8 Brownian Bridge

Heuristically speaking, a Brownian bridge is a Brownian motion started at x and conditioned to be at y at time t. More precisely, we have:

Definition 4.14.14 *Let $x, y \in \mathbb{R}$ be fixed. A continuous Gaussian process $(X_s)_{0 \leq s \leq t}$ such that $X_0 = x$ and characterized by*

$$\mathbb{E}[X_s] = x + (y - x)\frac{s}{t}$$
$$Cov(X_u, X_s) = \min(u, s) - \frac{us}{t}$$

is called a Brownian bridge from x to y within time t.

Proposition 4.14.15 *A Brownian bridge $(X_s)_{0 \leq s \leq t}$ from x to y within time t satisfies $X_t = y$ almost surely.*

Proof: The definition implies $\mathbb{E}[X_t] = y$ and $VarX_t = Cov(X_t, X_t) = 0$. Since

$$VarX_t = \mathbb{E}[(X_t - \mathbb{E}[X_t])^2] = \mathbb{E}[(X_t - y)^2],$$

then $\mathbb{E}[(X_t - y)^2] = 0$, which implies $X_t = y$ almost surely. ■

Because $X_0 = x$ and $X_t = y$, the process X_s is also called *pinned Brownian motion.*

The following result provides a possible representation of a Brownian bridge.

Proposition 4.14.16 *Let B_s be a standard one-dimensional Brownian motion starting at $B_0 = 0$. Then the process*

$$X_s = x + B_s - \frac{s}{t}B_t + (y - x)\frac{s}{t}, \quad 0 \leq s \leq t \tag{4.14.108}$$

is a Brownian bridge from x to y within time t.

Proof: We shall verify the conditions of the aforementioned definition. The conditions $X_0 = x$ and $\mathbb{E}[X_s] = x + (y - x)\frac{s}{t}$ are easy to verify.
Since $X_s - \mathbb{E}[X_s] = B_s - \frac{s}{t}B_t$, we have

$$\begin{aligned}
Cov(X_u, X_s) &= \mathbb{E}\Big[(X_u - \mathbb{E}[X_u])(X_s - \mathbb{E}[X_s])\Big] \\
&= \mathbb{E}\Big[\Big(B_u - \frac{u}{t}B_t\Big)\Big(B_s - \frac{s}{t}B_t\Big)\Big] \\
&= \mathbb{E}[B_uB_s - \frac{u}{t}B_sB_t - \frac{s}{t}B_uB_t + \frac{us}{t^2}B_t^2] \\
&= \min(s, u) - \frac{us}{t},
\end{aligned}$$

where we used $\mathbb{E}[B_sB_u] = \min(s, u)$ and $\mathbb{E}[B_t^2] = t$. ■

Remark 4.14.17 There are also other representations of the Brownian bridge, which will not be used here, such as the one given by Ikeda and Watanabe [80]

$$X_s = x + (y - s)\frac{s}{t} + (t - s)\int_0^s \frac{dB_u}{t - u}.$$

4.15 Deduction of Formulas

In order to compute the expectation $\mathbb{E}[e^{-\int_0^t U(W_s)ds}; W_t \in dy]$ we shall use Lemma 4.14.1 to write it as an expectation in terms of the delta function and then use Lemma 1.11.2, page 44, to split it into a product of two expectations as follows

$$\begin{aligned}\mathbb{E}[e^{-\int_0^t U(W_s)ds}; W_t \in dy] &= \mathbb{E}[e^{-\int_0^t U(W_s)ds}\delta(W_t - y)]\,dy \\ &= \mathbb{E}[\delta(W_t - y)]\mathbb{E}[e^{-\int_0^t U(W_s)ds}|W_t = y]\,dy.\end{aligned}$$

Using that

$$\begin{aligned}\mathbb{E}[\delta(W_t - y)] = p_{W_t}(y) &= \frac{1}{\sqrt{2\pi t}}e^{-\frac{(x-y)^2}{2t}} \\ \mathbb{E}[e^{-\int_0^t U(W_s)ds}|W_t = y] &= \mathbb{E}[e^{-\int_0^t U(X_s)ds}],\end{aligned}$$

with X_s Brownian bridge between x and y within time t, it follows that

$$\mathbb{E}[e^{-\int_0^t U(W_s)ds}; W_t \in dy] = \frac{1}{\sqrt{2\pi t}}e^{-\frac{(x-y)^2}{2t}}\mathbb{E}[e^{-\int_0^t U(X_s)ds}]\,dy. \tag{4.15.109}$$

Therefore, it suffices to compute the expectation $\mathbb{E}[e^{-\int_0^t U(X_s)ds}]$, which involves an integral functional of the Brownian bridge X_s. For the sake of simplicity, we shall cover in the following only the case when the potential is linear, $U(x) = \gamma x$.

To this end, we compute the integral of X_s using formula of the Brownian bridge (4.14.108)

$$\begin{aligned}\int_0^t X_s ds &= \int_0^t \Big(x + B_s - \frac{s}{t}B_t + (y-x)\frac{s}{t}\Big)ds \\ &= \frac{(x+y)t}{2} - \frac{t}{2}B_t + \int_0^t B_s ds \\ &= \frac{(x+y)t}{2} - \frac{t}{2}B_t + Z_t,\end{aligned}$$

where $Z_t = \int_0^t B_s ds \sim N\big(0, \frac{t^3}{3}\big)$ is normally distributed. The Laplace transform can now be computed as

$$\mathbb{E}[e^{-\gamma\int_0^t X_s ds}] = \mathbb{E}[e^{-\gamma(\frac{(x+y)t}{2} - \frac{t}{2}B_t + Z_t)}] = e^{-\gamma\frac{(x+y)t}{2}}\mathbb{E}[e^{-\gamma(Z_t - \frac{t}{2}B_t)}]. \tag{4.15.110}$$

By Proposition 4.1.1, page 160, the joint density of (B_t, Z_t) is given by

$$f_{(B_t,Z_t)}(w,z) = \frac{\sqrt{3}}{\pi t^2}e^{-\frac{2}{t^3}[w^2t + 3z^2 - 3wzt]}, \quad t > 0.$$

Then the previous expectation can be computed explicitly as

$$\begin{aligned}\mathbb{E}[e^{-\gamma(Z_t-\frac{t}{2}B_t)}] &= \int_{\mathbb{R}^2} e^{-\gamma(z-\frac{t}{2}w)} f_{(B_t,Z_t)}(w,z)\,dwdz\\ &= \frac{\sqrt{3}}{\pi t^2}\int_{\mathbb{R}^2} e^{-\frac{2}{t^3}[w^2t+3z^2-3wzt]-\gamma(z-\frac{t}{2}w)}\,dwdz\\ &= e^{\frac{\gamma^2}{24}t^3},\end{aligned}$$

where we used the multivariate Gaussian integral formula[3]

$$\int_{\mathbb{R}^n} e^{-\frac{1}{2}\langle Ax,x\rangle+b^Tx}\,dx = \frac{(2\pi)^{n/2}}{\sqrt{\det A}}e^{\frac{1}{2}b^TA^{-1}b}.$$

Substituting in (4.15.110) yields the following formula for the Laplace transform of the integral of the Brownian bridge X_s

$$\mathbb{E}[e^{-\gamma\int_0^t X_s ds}] = e^{-\gamma\frac{(x+y)t}{2}+\frac{\gamma^2}{24}t^3}. \tag{4.15.111}$$

Substituting now into (4.15.109) yields

$$\begin{aligned}\mathbb{E}[e^{-\gamma\int_0^t W_s ds}; W_t\in dy] &= \frac{1}{\sqrt{2\pi t}}e^{-\frac{(x-y)^2}{2t}}\mathbb{E}[e^{-\gamma\int_0^t X_s ds}]\,dy\\ &= \frac{1}{\sqrt{2\pi t}}e^{-\frac{(x-y)^2}{2t}}e^{-\gamma\frac{(x+y)t}{2}+\frac{\gamma^2}{24}t^3}.\end{aligned} \tag{4.15.112}$$

We thus arrived at the formula (4.14.102).

We can also compute the heat kernel at the origin of the operator with potential

$$L = \frac{1}{2}\partial_x^2 - U(x)$$

using an alternative approach involving the Fourier transform. To do this, we introduce an additional variable, y, and augment L to create the two-dimensional operator

$$\tilde{L} = \frac{1}{2}\partial_x^2 - U(x)\partial_y.$$

The resulting operator is the infinitesimal generator of the diffusion process given by

[3] See, for instance, *https://en.wikipedia.org/wiki/Gaussian_integral*

$$dX_t = dW_t \quad \Rightarrow \quad X_t = W_t$$
$$dY_t = -U(X_t)dt \quad \Rightarrow \quad Y_t = -\int_0^t U(W_s)\, ds.$$

If we have the joint density $(W_t, \int_0^t U(W_s)ds)$ available, we can determine the transition density $p_t(x,y)$ of (X_t, Y_t) and, consequently, the heat kernel of $\tilde{L}$ at the origin.

The heat kernel of L, denoted by $\rho_t(x)$, can be obtained from $p_t(x,y)$ as follows. By applying the partial Fourier transform with respect to y to the equation

$$\partial_t p_t(x,y) = \frac{1}{2}\partial_x^2 p_t(x,y) - U(x)\partial_y p_t(x,y)$$

and denoting $\rho_t(x,\omega) = \int_{\mathbb{R}} e^{-iy\omega} p_t(x,y)\, dy$, we obtain

$$\partial_t \rho_t(x,\omega) = \frac{1}{2}\partial_x^2 \rho_t(x,\omega) - i\omega U(x)\partial_y \rho_t(x,\omega).$$

Setting $\omega = -i$, we have $\rho_t(x) = \rho_t(x,-i)$, which satisfies

$$\partial_t \rho_t(x) = \frac{1}{2}\partial_x^2 \rho_t(x) - U(x)\rho_t(x).$$

Since $p_t(x,y)$ is the heat kernel of $\tilde{L}$, we have

$$\begin{aligned}
\lim_{t\to 0} \rho_t(x) &= \lim_{t\to 0} \rho_t(x,-i) = \lim_{t\to 0} \int_{\mathbb{R}} e^{-y} p_t(x,y)\, dy \\
&= \int_{\mathbb{R}} e^{-y} \lim_{t\to 0} p_t(x,y)\, dy = \int_{\mathbb{R}} e^{-y}\delta(x)\delta(y)\, dy \\
&= \delta(x) \int_{\mathbb{R}} e^{-y}\delta(y)\, dy = \delta(x).
\end{aligned}$$

Therefore, $\rho_t(x) = \int_{\mathbb{R}} e^{-y} p_t(x,y), dy$ becomes the heat kernel of the operator L.

It is important to highlight that for the linear, quadratic, and exponential cases, namely $U(x) \in \{\gamma x, \gamma x^2, e^{2\beta x}\}$, the aforementioned joint density $(W_t, \int_0^t U(W_s)ds)$ is available.

4.16 Summary

This chapter is a significant part of the book that presents precise quantitative formulas for heat kernels. It covers the computation of heat kernels for various types of hypoelliptic operators, including but not limited to the Kolmogorov operator, Exponential Kolmogorov operator, Grushin operator, Exponential Grushin operator, Heisenberg operator, saddle Laplacian, and Casimir operator. Additionally, the chapter explores heat kernels for operators that have a potential, which can be linear, quadratic, inverse quadratic or exponential. The approach taken involves finding the transition density of the related diffusion process. Some heat kernels (such as those for the Kolmogorov operator, Grushin operator, and Exponential Kolmogorov operator) have applied financial interpretations. Furthermore, the study of heat kernels for operators with potential was based on properties of the Brownian bridge process. The outcomes of the study reproduce previously established results and known formulas in the literature.

4.17 Exercises

Exercise 4.17.1 We shall find the heat kernel $K(t, x_1, x_2)$ for the Kolmogorov operator $L_1 = \frac{1}{2}\partial_{x_1}^2 + ax_1\partial_{x_2}$ using the method of partial Fourier transform. Let $u(t, x, \xi) = \mathcal{F}_y K(x, \xi) = \int_{\mathbb{R}} e^{-iy\xi} K(t, x, y)\, dy$.

(a) Show that u satisfies a heat equation with linear potential.

(b) Using the result of Proposition 4.14.3 and applying an inverse partial Fourier transform, retrieve the expression for $K(t, x_1, x_2)$.

Exercise 4.17.2 Fill in the computational details for the computation of the heat kernel of $\Delta_G^{(1,m)}$ given by formula (4.5.49).

Exercise 4.17.3 (n-dim Heisenberg Laplacian) Let

$$X_j = \frac{\partial}{\partial x_j} + 2y_j\frac{\partial}{\partial z}, \qquad Y_j = \frac{\partial}{\partial y_j} - 2x_j\frac{\partial}{\partial z}, \quad j = 1, \ldots, n.$$

be $2n$ vector fields on $\mathbb{R}^{2n+1}$.

(a) Show that $\{X_i, Y_j\}$ satisfy the bracket-generating condition;

(b) Let

$$\Delta_H = \frac{1}{2}\sum_{j=1}^{n} X_j^2 + \frac{1}{2}\sum_{j=1}^{n} Y_j^2$$

be the n-dimensional Heisenberg Laplacian. Show that Δ_H is hypoelliptic;

(c) Show that $\left\{X_i, Y_j, \dfrac{\partial}{\partial_z}\right\}$ are left invariant with respect to the following Lie group law on $\mathbb{R}^{2n+1}$

$$(\mathbf{x}, z) \circ (\mathbf{x}', z') = \Big(\mathbf{x} + \mathbf{x}', z + z' + 2\sum_{j=1}^{n}(x_{2j}x'_{2j-1} - x'_{2j}x_{2j-1})\Big),$$

where $\mathbf{x} = (x_1, \ldots, x_n)^T$.

(d) Find the diffusion on $\mathbb{R}^{2n+1}$ associated to Δ_H starting from the origin;

(e) Use Gaveau's formula given in Remark 1.11.4 to obtain the heat kernel for Δ_H, first from the origin, and then from any other point.

In the next exercise you will use the idea of Section 4.13 to compute the heat kernel for the n-dimensional saddle Laplacian.

Exercise 4.17.4 Consider the following $2n$ vector fields on $\mathbb{R}^{2n+1}$

$$X_j = \frac{\partial}{\partial x_j} + y_j \frac{\partial}{\partial z}, \qquad Y_j = \frac{\partial}{\partial y_j} + x_j \frac{\partial}{\partial z}, \quad j = 1, \ldots, n.$$

(a) Show that $[X_i, X_j] = [Y_i, Y_j] = [X_i, Y_j] = 0$;

(b) Verify that $\{X_i, Y_j\}$ are tangent vector fields to the surface $\phi : \mathbb{R}^{2n} \to \mathbb{R}^{2n+1}$

$$\phi(x_1, \ldots, x_n, y_1, \ldots, y_n) = (x_1, \ldots, x_n, y_1, \ldots, y_n, \sum_{i+1}^{m} x_i y_i).$$

(c) Find the unit normal to the surface ϕ;

(d) Verify that the diffusion associated with the operator

$$L = \frac{1}{2} \sum_{j=1}^{n} X_j^2 + \frac{1}{2} \sum_{j=1}^{n} Y_j^2$$

is given by

$$X_t = \Big(W_1(t), \ldots, W_n(t), B_1(t), \ldots, B_n(t), \sum_{j=1}^{n} W_j(t) B_j(t) \Big),$$

where $W_j(t)$ and $B_j(t)$ are independent one-dimensional Brownian motions.

(e) Show that the process $I_t^{(n)} = \sum_{j=1}^{n} W_j(t) B_j(t)$ is a continuous martingale and that there is a Brownian motion β_t such that $I_t^{(n)} = \beta_{R_t^{2n}}$, where R_t^{2n} is a $2n$-dimensional Bessel process;

(f) Compute the heat kernel of the operator L from the origin.

Exercise 4.17.5 (Exponential Brownian motion) Consider the operator $L = \dfrac{1}{2}\sigma^2 x^2 \dfrac{d^2}{dx^2} + \mu x \dfrac{d}{dx}$, with $x > 0$, $\sigma \neq 0$ and $\mu \in \mathbb{R}$.

(a) Show that the associate diffusion satisfies

$$dX_t = \sigma X_t dW_t + \mu X_t dt, \qquad X_0 = x_0.$$

(b) Solve the equation from (b) to obtain the unique strong solution

$$X_t = x_0 e^{\sigma W_t + (\mu - \frac{1}{2}\sigma^2)t}, \qquad t > 0.$$

(c) Find the heat kernel of L by computing the transition density of diffusion X_t.

Exercise 4.17.6 (Linear noise) Let $L = \frac{1}{2}x^2 \frac{d^2}{dx^2}$, with $x > 0$. Show that the associated diffusion is given by $X_t = X_0 e^{W_t - \frac{t}{2}}$ and then compute the heat kernel of L.

Exercise 4.17.7 (Linear noise with drift) Let $L = \dfrac{1}{2}\alpha^2 x^2 \dfrac{d^2}{dx^2} + r\dfrac{d}{dx}$, with $x > 0$, $\alpha \neq 0$ and $r \in \mathbb{R}$.

(a) verify that the associated diffusion satisfies

$$dX_t = \alpha X_t dW_t + r dt, \quad X_0 = x_0.$$

(b) Consider $\rho_t = e^{\frac{1}{2}\alpha^2 t - \alpha W_t}$. Use Ito's formula to show that

$$d\rho_t = \rho_t(\alpha^2 dt - \alpha dW_t).$$

(c) Multiply the differential stochastic equation from part (a) by the integrating factor ρ_t to obtain the exact equation

$$d(\rho_t X_t) = r\rho_t dt.$$

(d) Solve by integration the equation from part (c) to get

$$X_t = e^{\alpha W_t - \frac{1}{2}\alpha^2 t}\Big(x_0 + r\int_0^t e^{\frac{1}{2}\alpha^2 s - \alpha W_s}\, ds\Big).$$

(e) Find the heat kernel of the operator L. For this, use the following formula for the joint distribution of $V_t = e^{\sigma^2 \nu s + \sigma W_s}$ and $\int_0^t V_s ds$, $\sigma > 0$

$$P_x\Big(\int_0^t V_s ds \in dy, V_t \in dz\Big) = \frac{z^{\nu-1}}{4yx^\nu} e^{-\frac{1}{2}\nu^2\sigma^2 t - 2(x+z)/\sigma^2 y} \mathbf{i}_{\sigma^2 t/8}\Big(\frac{4\sqrt{xz}}{\sigma^2 y}\Big) dydz,$$

with $\mathbf{i}_y z = 2\theta_z(2y)$, where $\theta_r(t)$ denotes the Hartman-Watson density defined by the Yor's integral representation (4.6.54).

Exercise 4.17.8 Let $L = \dfrac{1}{2}\dfrac{d^2}{dx^2} - a^2x^2$, for $a > 0$. Find smooth functions $a(t)$, $b(t)$ and $c(t)$ such that

$$p_t(x_0, x) = e^{a(t)\frac{x^2}{2} + b(t)xx_0 + a(t)x_0^2 + c(t)}$$

satisfies

$$\begin{aligned} \partial_t p_t(x_0, x) =& L_x p_t(x_0, x) \\ \lim_{t \to 0} p_t(x_0, x) =& \delta(x - x_0). \end{aligned}$$

Exercise 4.17.9 Provide a proof for Theorem 4.14.11.

Chapter 5

Fundamental Solutions

In this chapter, we will explicitly construct fundamental solutions for several second-order differential operators, both elliptic and sub-elliptic. The study of the recurrence and transitivity of the associated Brownian motion is closely related to these fundamental solutions. Our focus is on differential operators of the form

$$\mathcal{A} = \frac{1}{2}\sum_{i,j} g^{ij}(x)\partial_{x_i}\partial_{x_j} + \sum_k b_k(x)\partial_k$$

where $g^{ij} = (\sigma\sigma^T)_{ij}$. These operators act as generators for n-dimensional Ito diffusions described by

$$dX_t = b(X_t)dt + \sigma(X_t)dW_t.$$

Since the heat kernel of $\mathcal{A}$ serves as the transition density of the diffusion X_t, we will leverage this relationship to develop our fundamental solutions.

5.1 Definition and Construction

Let δ denote the Dirac delta function. A solution G of the equation

$$\mathcal{A}G(x_0, x) = \delta(x - x_0) \tag{5.1.1}$$

is called a *fundamental solution* of the operator $\mathcal{A}$. This means two things:

(*i*) $\mathcal{A}_x G(x_0, x) = 0$ for $x \neq x_0$, where the index x in $\mathcal{A}_x$ denotes that the operator acts in the variable x; and
(*ii*) $\int_{\mathbb{R}^n} \mathcal{A}_x G(x_0, x)\phi(x)\, dx = \phi(x_0)$, for any compact supported smooth function $\phi \in \mathcal{C}_0^\infty(\mathbb{R}^n)$.

For all cases of operators considered in this chapter, which are either elliptic or sub-elliptic, the fundamental solution is a smooth function outside of x_0, namely $G(x_0, x) \in \mathcal{C}^\infty(\mathbb{R}^n \backslash \{x_0\})$.

The fundamental solution construction A possible explicit construction of the fundamental solution can be achieved starting form the transition probability $p_t(x_0, x)$ of the Ito diffusion X_t. As a heat kernel, $p_t(x_0, x)$ satisfies

$$\begin{aligned}\partial_t p_t(x_0, x) &= \mathcal{A}_x p_t(x_0, x)\\ \lim_{t \searrow 0} p_t(x_0, x) &= \delta(x - x_0).\end{aligned}$$

Then integrating out the time variable in the heat kernel, we claim that

$$G(x_0, x) = -\int_0^\infty p_t(x_0, x)\, dt \tag{5.1.2}$$

is a fundamental solution for $\mathcal{A}$, provided the integral converges. The verification of (5.1.1) follows from an application of the aforementioned formulas and an application of the Fundamental Theorem of Calculus

$$\begin{aligned}\mathcal{A}_x G(x_0, x) &= -\int_0^\infty \mathcal{A}_x p_t(x_0, x)\, dt = -\int_0^\infty \partial_t p_t(x_0, x)\, dt\\ &= \lim_{t \searrow 0} p_t(x_0, x) - \lim_{t \to \infty} p_t(x_0, x)\\ &= \delta(x - x_0).\end{aligned}$$

We shall apply formula (5.1.2) to find explicit formulas for fundamental solutions for a few familiar operators. However, it is worth noting that the ability to exactly integrate the heat kernel with respect to t and obtain an elementary function is the exception rather than the rule.

Remark 5.1.1 The integral in (5.1.2) is finite only in the case of some Riemannian manifolds (M, g), which are called *non-parabolic*. Conversely, if the integral in (5.1.2) is infinite for some $x, y \in M$, namely

$$\int_0^\infty p_t(x, y)\, dt = \infty,$$

then the manifold is called *parabolic*. The property of being parabolic is equivalent to the fact that the Brownian motion on M is recurrent. Since Brownian motion is recurrent on $\mathbb{R}^n$ for $n = 1, 2$ but transient for $n \geq 3$, it follows that $\mathbb{R}$ and $\mathbb{R}^2$ are parabolic spaces, and as a result, the fundamental solution of the Laplacian on these spaces cannot be obtained by integrating the heat kernel with respect to the t variable.

5.2 The Euclidean Laplacian

The transition density of the n-dimensional Brownian motion $X_t = (W_t^1, \dots, W_t^n)$, is given by the Gaussian density

$$p_t(x_0, x) = \frac{1}{(2\pi t)^{n/2}} e^{-\frac{\|x-x_0\|^2}{2t}}, \qquad t > 0.$$

We note that the function $t \to p_t(x_0, x)$ is integrable on $(0, \infty)$ for $n \geq 3$. This fact is equivalent to the recurrence of the Brownian motion on these Euclidean spaces. Assuming $n \geq 3$, the fundamental solution of the Euclidean Laplacian

$$\Delta_n = \frac{1}{2} \sum_{j=1}^{n} \partial_{x_j}^2$$

is given by the formula (5.1.2). Using the substitution $u = \frac{\|x-x_0\|^2}{2t}$ and the definition of the Gamma function, we have

$$\begin{aligned} G(x_0, x) &= -\int_0^\infty \frac{1}{(2\pi t)^{n/2}} e^{-\frac{\|x-x_0\|^2}{2t}} \, dt \\ &= -\frac{1}{2\pi^{n/2}} \frac{1}{\|x - x_0\|^{n-2}} \int_0^\infty u^{\frac{n}{2}-2} e^{-u} \, du \\ &= -\frac{\Gamma(\frac{n}{2} - 1)}{2\pi^{n/2}} \frac{1}{\|x - x_0\|^{n-2}} \\ &= -\frac{2}{(n-2)\omega_n} \frac{1}{\|x - x_0\|^{n-2}}, \end{aligned} \tag{5.2.3}$$

where $\omega_n = \frac{2\pi^{n/2}}{\Gamma(\frac{n}{2})}$ is the area of the unit hypersphere in $\mathbb{R}^n$ and we used the property of the Gamma function $\Gamma(\frac{n}{2}) = (\frac{n}{2} - 1)\Gamma(\frac{n}{2} - 1)$. In the particular case $n = 3$ we obtain

$$G(x_0, x) = -\frac{1}{2\pi} \frac{1}{\|x - x_0\|}. \tag{5.2.4}$$

It is worth noting that the fundamental solution is represented in terms of the Euclidean distance and the area of the unit sphere. We shall see that a similar behavior will be observed in the case of other operators.

5.3 A Fundamental Solution for $\Delta_{\mathcal{H}^3}$

We start with a result that provides explicit expressions for two improper integrals, which will be used in the computation of the fundamental solution.

Lemma 5.3.1 *For any $x > 0$ we have*

(i) $\displaystyle\int_0^\infty \frac{1}{\sqrt{t}} e^{-\frac{1}{2}(\frac{x^2}{t}+t)}\,dt = \sqrt{2\pi}e^{-x};$

(ii) $\displaystyle\int_0^\infty \frac{1}{t^{3/2}} e^{-\frac{1}{2}(\frac{x^2}{t}+t)}\,dt = \sqrt{2\pi}\,\frac{e^{-x}}{x}.$

Proof: (*i*) We consider the integrals as a function of variable x and let

$$\varphi(x) = \int_0^\infty \frac{1}{\sqrt{t}} e^{-\frac{1}{2}(\frac{x^2}{t}+t)}\,dt, \qquad \phi(x) = \int_0^\infty \frac{1}{t^{3/2}} e^{-\frac{1}{2}(\frac{x^2}{t}+t)}\,dt.$$

We shall show that $\varphi(x)$ satisfies the following initial value problem

$$\varphi'(x) = -\varphi(x) \tag{5.3.5}$$

$$\varphi(0) = \sqrt{2\pi}. \tag{5.3.6}$$

Since its solution is $\varphi(x) = \sqrt{2\pi}e^{-x}$, applying the uniqueness theorem of solutions of initial value problems we obtain the desired result.

To check the initial condition (5.3.6), using the Gamma function we obtain

$$\begin{aligned}\varphi(0) &= \int_0^\infty \frac{1}{\sqrt{t}} e^{-\frac{1}{2}t}\,dt = \sqrt{2}\int_0^\infty u^{-\frac{1}{2}}e^{-u}du \\ &= \sqrt{2}\,\Gamma\Big(\frac{1}{2}\Big) = \sqrt{2\pi}.\end{aligned}$$

Using the substitution $\tau = \frac{x^2}{t}$ we can express $\varphi(x)$ in terms of $\phi(x)$

$$\varphi(x) = x\int_0^\infty \frac{1}{\tau^{3/2}} e^{-\frac{1}{2}(\tau+\frac{x^2}{\tau})}\,d\tau = x\phi(x).$$

Differentiating with respect to x in the definition expression of $\varphi(x)$ we obtain

$$\begin{aligned}\varphi'(x) &= \frac{d}{dx}\int_0^\infty \frac{1}{\sqrt{t}} e^{-\frac{1}{2}(\frac{x^2}{t}+t)}\,dt \\ &= x\int_0^\infty \frac{1}{\tau^{3/2}} e^{-\frac{1}{2}(\tau+\frac{x^2}{\tau})}\,d\tau = -x\phi(x).\end{aligned} \tag{5.3.7}$$

Equating the last two expressions yields (5.3.5).

(*ii*) Combining relations (5.3.7) and (5.3.5) we get

$$\phi(x) = -\frac{1}{x}\varphi'(x) = \frac{1}{x}\varphi(x) = \sqrt{2\pi}\frac{1}{x}e^{-x}.$$

∎

Remark 5.3.2 A generalization of the previous lemma can be found in Watson [136], page 183

$$\int_0^\infty \frac{1}{t^{\nu+1}} e^{-(\frac{x^2}{4t}+t)}\, dt = \frac{2^{\nu+1}}{x^\nu} K_\nu(x), \tag{5.3.8}$$

where $K_\nu(x)$ is the modified Bessel function of the second kind, which is the solution of

$$x^2 K''(x) + xK'(x) - (x^2 + \nu^2)K(x) = 0$$

with singularity at $x = 0$.

We shall continue with the computation of the fundamental solution for the Laplace-Beltrami operator on the upper-half space $\mathcal{H}^3 = \{(x_1, x_2, y); y > 0\}$, which is

$$\Delta_{\mathcal{H}^3} = \frac{1}{2} y^2 \left(\partial_{x_1}^2 + \partial_{x_2}^2 + \partial_y^2\right) - \frac{1}{2} y \partial_y. \tag{5.3.9}$$

We start from the closed form expression of the transition density for the associated diffusion

$$p_t(z_0, z) = \frac{1}{(2\pi t)^{3/2}} \frac{r}{\sinh r} e^{-\frac{r^2}{2t} - \frac{t}{2}}, \qquad t > 0, \tag{5.3.10}$$

where

$$r = d(z_0, z) = \cosh^{-1}\left(1 + \frac{\|z - z_0\|^2}{2y}\right)$$

is the hyperbolic distance between $z_0 = (0, 0, 1)$ and $z = (x_1, x_2, y)$. The fundamental solution centered at z_0 is obtained by integrating out the variable t in (5.3.10)

$$\begin{aligned} G(z_0, z) &= -\int_0^\infty p_t(z_0, z)\, dt = -\frac{r}{\sinh r} \frac{1}{(2\pi)^{3/2}} \int_0^\infty \frac{1}{t^{3/2}} e^{-\frac{1}{2}(\frac{r^2}{t}+t)}\, dt \\ &= -\frac{r}{\sinh r} \frac{1}{(2\pi)^{3/2}} \sqrt{2\pi}\, \frac{e^{-r}}{r} = -\frac{1}{2\pi} \frac{1}{e^r \sinh r} \\ &= \frac{1}{\pi(1 - e^{2r})}, \end{aligned}$$

where we used part (ii) of Lemma 5.3.1. This can be also written as

$$G(z_0, z) = \frac{1}{\pi(1 - e^{2d(z_0, z)})}. \tag{5.3.11}$$

Formula (5.3.11) represents the fundamental solution of the operator $\Delta_{\mathcal{H}^3}$ in terms of the hyperbolic distance. For practical purposes we shall represent it in terms of the Euclidean distance as follows. We start by denoting

$$c = \cosh d(z_0, z) = 1 + \frac{\|z - z_0\|^2}{2y},$$

which by inversion yields $d = d(z_0, z) = \ln(c + \sqrt{c^2 - 1})$. Some elementary computation provides

$$\sqrt{c-1} = \frac{\|z - z_0\|}{\sqrt{2y}}, \qquad \sqrt{c+1} = \frac{\|z + z_0\|}{\sqrt{2y}}. \tag{5.3.12}$$

Then

$$\begin{aligned} 1 - e^d &= 1 - c - \sqrt{c^2 - 1} = -\sqrt{c-1}\big(\sqrt{c-1} + \sqrt{c+1}\big) \\ 1 + e^d &= 1 + c + \sqrt{c^2 - 1} = \sqrt{c+1}\big(\sqrt{c-1} + \sqrt{c+1}\big), \end{aligned}$$

and hence

$$\begin{aligned} \frac{1}{1 - e^{2d}} &= \frac{1}{2}\Big(\frac{1}{1 - e^d} + \frac{1}{1 + e^d}\Big) \\ &= \frac{1}{2}\frac{1}{\sqrt{c+1} + \sqrt{c-1}}\Big(\frac{1}{\sqrt{c+1}} - \frac{1}{\sqrt{c-1}}\Big) \\ &= \frac{y}{\|z + z_0\| + \|z - z_0\|}\Big(\frac{1}{\|z + z_0\|} - \frac{1}{\|z - z_0\|}\Big), \end{aligned}$$

where we used relations (5.3.12).

Proposition 5.3.3 *The fundamental solution centered at $z_0 = (0,0,1)$ of the differential operator $\Delta_{\mathcal{H}^3}$ is given by*

$$G(z_0, z) = \frac{y}{\pi}\frac{1}{\|z + z_0\| + \|z - z_0\|}\Big(\frac{1}{\|z + z_0\|} - \frac{1}{\|z - z_0\|}\Big), \tag{5.3.13}$$

where $\|\cdot\|$ denotes the Euclidean distance.

5.4 Exponential Grushin Operator

We recall the exponential Grushin operator introduced in Section 4.6

$$\Delta_{EG} = \frac{1}{2}(\partial_x^2 + e^{2x}\partial_y^2).$$

Due to the invariance with respect to the Lie group law (4.6.50), it suffices to study the fundamental solution at the origin.

Proposition 5.4.1 *The fundamental solution at the origin for the operator Δ_{EG} is given by*

$$G(x, y) = -\frac{1}{\sqrt{2}\pi^{3/2}}\int_0^\infty \int_0^\pi u^{-3/2}\, e^{-S(x,y,s,u)}\, ds du, \tag{5.4.14}$$

where

$$S(x, y, s, u) = \frac{1}{2u}[(e^x - \cos s)^2 + (\sin s)^2 + y^2].$$

Proof: Taking $\gamma \to 0$ in the Laplace transform formula provided by Corollary 4.6.3 from page 188, we obtain

$$\begin{aligned} G(x, y) &= -\int_0^\infty p_t(x, y)\, dt = -\lim_{\gamma \to 0} \int_0^\infty e^{-\frac{1}{2}\gamma^2 t} p_t(x, y)\, dt \\ &= -\frac{1}{\sqrt{2\pi}} \int_0^\infty u^{-3/2}\, I_0(e^x/u)\, e^{-\frac{1}{2u}(1+e^{2x}+y^2)}\, du. \end{aligned}$$

Using the integral representation of the modified Bessel function, see Abramowitz and Stegun [1]

$$I_0(z) = \frac{1}{\pi} \int_0^\pi e^{z \cos s}\, ds,$$

completing the square we obtain

$$\begin{aligned} G(x, y) &= -\frac{1}{\sqrt{2}\pi^{3/2}} \int_0^\infty \int_0^\pi u^{-3/2}\, e^{\frac{e^x}{u} \cos s}\, e^{-\frac{1}{2u}(1+e^{2x}+y^2)}\, ds du \\ &= -\frac{1}{\sqrt{2}\pi^{3/2}} \int_0^\infty \int_0^\pi u^{-3/2}\, e^{-\frac{1}{2u}[(e^x - \cos s)^2 + (\sin s)^2 + y^2]}\, ds du. \end{aligned}$$

■

5.5 Bessel Process with Drift in $\mathbb{R}^3$

It will be shown in Section 6.9 that the density of a Bessel process with drift in $\mathbb{R}^3$, which starts from the origin, i.e. $R_t = \sqrt{\sum_{i=1}^3 (\mu_i t + W_i(t))^2}$, is given by (6.9.150)

$$p_t(x) = \frac{2x^2}{\sqrt{2\pi} t^{3/2}} e^{-\frac{1}{2t}(x^2 + t^2|\mu|^2)} \frac{\sinh(|\mu| x)}{|\mu| x}, \qquad x > 0.$$

Proposition 6.9.3 states that $p_t(x)$ is the heat kernel of the operator

$$L_x^\mu = \frac{1}{2}\frac{d^2}{dx^2} - |\mu| \coth(|\mu| x) \frac{d}{dx} + \frac{|\mu|^2}{\sinh^2(|\mu| x)}.$$

In this section we shall find the fundamental solution of the operator L_x^μ by integrating over the time parameter in the density $p_t(x)$. We have

$$\begin{aligned}
G(x) &= -\int_0^\infty p_t(x)\,dt \\
&= -\frac{2x^2}{\sqrt{2\pi}}\frac{\sinh(|\mu|x)}{|\mu|x}\int_0^\infty \frac{1}{t^{3/2}}e^{-\frac{1}{2}(\frac{x^2}{t}+t|\mu|^2)} \\
&= -\frac{2x^2}{\sqrt{2\pi}}\frac{\sinh(|\mu|x)}{|\mu|x}\int_0^\infty \frac{|\mu|}{\tau^{3/2}}e^{-\frac{1}{2}(\frac{x^2|\mu|^2}{\tau}+\tau)}\,d\tau \\
&= -\frac{2x^2}{\sqrt{2\pi}}\frac{\sinh(|\mu|x)}{|\mu|x}\sqrt{2\pi}\frac{e^{-|\mu|x}}{|\mu|x}|\mu| \\
&= \frac{1}{|\mu|}(e^{-2|\mu|x}-1),
\end{aligned}$$

where we used substitution $\tau = |\mu|^2 t$ and Lemma 5.3.1, part (ii).

5.6 The Heisenberg Laplacian

In order to find the fundamental solution for the Heisenberg Laplacian, we shall review a few properties of modified Bessel functions of second kind, $K_\nu(x)$; for more details see also Remark 5.3.2. Their integral representation is given by Schläfli's formula, see Watson [136], pag. 183

$$K_\nu(z) = \int_0^\infty e^{-z\cosh t}\cosh(\nu t)\,dt.$$

For $\nu = 0$ we obtain

$$K_0(z) = \int_0^\infty e^{-z\cosh t}\,dt. \tag{5.6.15}$$

Another integral representation of $K_0(z)$ is provided by Mehler's formula as a cosine transform, see Abramowitz and Stegun [1], pag. 376

$$K_0(z) = \int_0^\infty \cos(z\sinh t)\,dt = \int_0^\infty \frac{\cos(zt)}{\sqrt{t^2+1}}\,dt.$$

The last integral expression implies that K_0 can be described as a Fourier transform of the reciprocal of a square root function as

$$K_0(z) = \frac{1}{2}\int_{\mathbb{R}} \frac{e^{izt}}{\sqrt{t^2+1}}\,dt, \qquad z > 0. \tag{5.6.16}$$

We recall that the heat kernel of the Heisenberg Laplacian starting from the origin is given by formula (4.10.80)

$$p_t(x_1,x_2,x_3) = \frac{1}{4\pi^2}\int_{\mathbb{R}} V(t,\xi)e^{-i\xi x_3-\xi(x_1^2+x_2^2)\coth(2\xi t)}\,d\xi,$$

where $V(t,\xi) = \frac{2\xi}{\sinh(2\xi t)}$. For the next computation we shall rewrite the kernel using some simplifying notations. We let $y = x_3$ and $|x|^2 = x_1^2 + x_2^2$. Then the heat kernel becomes

$$p_t(x,y) = \frac{1}{4\pi^2} \int_{\mathbb{R}} \frac{2\xi}{\sinh(2\xi t)} e^{-i\xi y - \xi|x|^2 \coth(2\xi t)}\, d\xi. \tag{5.6.17}$$

The next result will be useful when integrating the heat kernel with respect to the parameter t.

Lemma 5.6.1 *We have*

$$\int_0^\infty \frac{2\xi}{\sinh(2\xi t)} e^{-\xi|x|^2 \coth(2\xi t)}\, dt = K_0(|\xi|\,|x|^2).$$

Proof: We shall perform the proof in the following steps. First we let

$$J(\xi) = \int_0^\infty \frac{2\xi}{\sinh(2\xi t)} e^{-\xi|x|^2 \coth(2\xi t)}\, dt$$

and show that

$$J(\xi) = \int_0^\infty e^{-|\xi|\,|x|^2 \cosh\tau}\, d\tau. \tag{5.6.18}$$

Then applying the integral representation formula (5.6.15) with $z = |\xi|\,|x|^2$ we obtain $J(\xi) = K_0(|\xi|\,|x|^2)$.

Therefore, it suffices to show formula (5.6.18). To this end, we introduce the variable τ by

$$\cosh\tau = \text{sign}(\xi)\coth(2\xi t) = \coth(2|\xi|t). \tag{5.6.19}$$

We note that if $t \searrow 0$ then $\tau \to +\infty$ and if $t \to \infty$ then $\tau \to 0$. Then formula (5.6.18) will follow from the relation

$$d\tau = -\frac{2\xi}{\sinh(2\xi t)} dt. \tag{5.6.20}$$

To show this, applying formula

$$\cosh^{-1}(\coth x) = \ln(\coth x + \sqrt{\coth^2 x - 1}) = \ln(\cosh x + 1) - \ln(\sinh x),$$

we solve for τ from (5.6.19)

$$\tau = \ln(\cosh(2|\xi|t) + 1) - \ln(\sinh(2|\xi|t)).$$

Taking the differential yields

$$\begin{aligned} d\tau &= 2|\xi|\left(\frac{\sinh(2|\xi|t)}{\cosh(2|\xi|t)+1} - \frac{\cosh(2|\xi|t)}{\sinh(2|\xi|t)}\right)dt \\ &= 2|\xi|\frac{-1-\cosh(2|\xi|t)}{\sinh(2|\xi|t)(\cosh(2|\xi|t)+1)} = -\frac{2|\xi|}{\sinh(2|\xi|t)}dt \\ &= -\frac{2\xi}{\sinh(2\xi t)}dt, \end{aligned}$$

which recovers formula (5.6.20). ■

The next result recovers the fundamental solution constructed by Folland [57] in 1973.

Proposition 5.6.2 *The fundamental solution of the Heisenberg subLaplacian,* $\Delta_H = \frac{1}{2}(X_1^2 + X_2^2)$*, on* $\mathbb{R}^3$ *from the origin is given by*

$$G(x,y) = -\frac{1}{4\pi}\frac{1}{\sqrt{|x|^4+y^2}}, \tag{5.6.21}$$

where $x = (x_1, x_2) \in \mathbb{R}^2$ *and* $y \in \mathbb{R}$.

Proof: Integrating with respect to t in (5.6.17), interchanging integrals and using Lemma 5.6.1 we obtain

$$\begin{aligned} G(x,y) &= -\int_0^\infty p_t(x,y)\,dt \\ &= -\frac{1}{4\pi^2}\int_{\mathbb{R}}\int_0^\infty \frac{2\xi}{\sinh(2\xi t)}e^{-i\xi y-\xi|x|^2\coth(2\xi t)}\,dtd\xi \\ &= -\frac{1}{4\pi^2}\int_{\mathbb{R}} e^{-i\xi y}J(\xi)\,d\xi = -\frac{1}{4\pi^2}\int_{\mathbb{R}} e^{-i\xi y}K_0(|\xi|\,|x|^2)\,d\xi, \end{aligned} \tag{5.6.22}$$

which is the inverse Fourier transform of K_0. Since $K_0(|\xi|\,|x|^2)$ can be written as a Fourier transform, see (5.6.16)

$$K_0(|\xi|\,|x|^2) = \frac{1}{2}\int_{\mathbb{R}} e^{i\xi t}\frac{1}{\sqrt{|x|^4+t^2}}dt \tag{5.6.23}$$

we obtain

$$\int_{\mathbb{R}} e^{-i\xi y}K_0(|\xi|\,|x|^2)\,d\xi = \frac{\pi}{\sqrt{|x|^4+y^2}}.$$

Substituting in (5.6.22) we obtain the desired result. ■

The expression $d_K\big(0,(x,y)\big) = \sqrt{|x|^4+y^2} = \sqrt{(x_1^2+x_2^2)^2+y^2}$, which appears in the denominator of formula (5.6.21), is called the *Koranyi distance* from the origin. Comparing with formula (5.2.4), we note that d_K plays for the Heisenberg subLaplacian a role similar to the one the Euclidean distance plays for the Euclidean Laplacian. Sometimes we write $\|(x,y)\|_H = d_K(0,(x,y))$ and called it the *Heisenberg norm*. Using the group law invariance we may define the Koranyi distance between any two points, $(x,y),(x',y') \in \mathbb{R}^3$, by

$$d_K((x,y),(x',y')) = \|(x',y')^{-1} \circ (x,y)\|_H.$$

It is worth noting that d_K and $\|\cdot\|_H$ do not satisfy the distance and norm axioms; for instance, the homogeneity property fails, $\|\lambda(x,y)\|_H \neq |\lambda|\|(x,y)\|_H$, so the words "distance" and "norm" are used here outside their standard connotation. The connection between the Koranyi distance and stochastic processes will be done in Section 7.3.

Corollary 5.6.3 *We have*

$$\Delta_H\Big(\frac{1}{\|(x,y)\|_H}\Big) = 0, \qquad \forall (x,y) \neq (0,0,0).$$

A proof of this result by direct computation can be found in Calin et al. [38], page 41. For more details, the reader can consult Folland [58].

Remark 5.6.4 The fundamental solution from any point (x_0,y_0) can be obtained using the group invariance property. In this case we obtain

$$G\big((x_0,y_0),(x,y)\big) = -\frac{1}{4\pi}\frac{1}{d_K\big((x_0,y_0),(x,y)\big)},$$

where " $\circ$ " denotes the group law on the Heisenberg group.

5.7 The Grushin Laplacian

We shall investigate the $(n+1)$-dimensional Grushin operator with one missing direction introduced in Section 4.5.1

$$\Delta_G^{(n,1)} = \frac{1}{2}\sum_{k=1}^{n} \partial_{x_k}^2 + 2\|x\|^2\partial_y^2, \tag{5.7.24}$$

where $\|x\|^2 = x_1^2 + \cdots + x_n^2$.

Proposition 5.7.1 *Let*

$$G(x,y) = \frac{1}{(\|x\|^4 + y^2)^{n/4}}. \tag{5.7.25}$$

Then

$$\Delta_G^{(n,1)} G(x,y) = 0, \qquad \forall x,y \neq 0. \tag{5.7.26}$$

Proof: The verification is by a straightforward computation. We denote $\varphi(x,y) = \|x\|^4 + y^2$ and look for a real number p such that $G(x,y) = \varphi(x,y)^p$ satisfies relation (5.7.26). Since

$$\sum_{k=1}^{n} \partial_{x_k}^2 \varphi = 4(n+2)\|x\|^2, \qquad \sum_{k=1}^{n} (\partial_{x_k}\varphi)^2 = 16\|x\|^6$$
$$\partial_y \varphi = 2y, \qquad \partial_y^2 \varphi = 2,$$

we obtain

$$\partial_y^2 G = 2p\varphi^{p-1} + 4y^2 p(p-1)\varphi^{p-2}$$
$$\begin{aligned}\sum_{k=1}^{n} \partial_{x_k}^2 G &= p(p-1)\varphi^{p-2}\sum_{k=1}^{n}(\partial_{x_k}\varphi)^2 + p\varphi^{p-1}\sum_{k=1}^{n}\partial_{x_k}^2\varphi \\ &= 16p(p-1)\varphi^{p-2}\|x\|^6 + 4(n+2)p\varphi^{p-1}\|x\|^2,\end{aligned}$$

and therefore

$$\begin{aligned}\Delta_G^{(n,1)} G &= \frac{1}{2}\sum_{k=1}^{n} \partial_{x_k}^2 G + 2\|x\|^2 \partial_y^2 G \\ &= p\varphi^{p-2}\|x\|^2\Big(16(p-1)\|x\|^4 + 4(n+2)\varphi + 8\varphi + 16(p-1)y^2\Big) \\ &= p\varphi^{p-2}\|x\|^2\Big(16(p-1)\varphi + 4\varphi(n+4)\Big) \\ &= p\varphi^{p-2}\|x\|^2\varphi\Big(p + \frac{n}{4}\Big).\end{aligned}$$

It follows that relation (5.7.26) holds if and only if $p = -n/4$. Hence $G(x,y)$ is given by formula (5.7.25). ■

The next result deals with the classical two-dimensional Grusin operator.

Corollary 5.7.2 *Let* $\Delta_G = \frac{1}{2}(\partial_x^2 + x^2\partial_y^2)$ *and* $G(x,y) = \dfrac{1}{(x^4/4 + y^2)^{1/4}}$. *Then* $\Delta_G G(x,y) = 0$ *for* $x, y \neq 0$.

Proof: It follows from Proposition 5.7.1 by choosing $n = 1$ and then using a change of coordinates. ■

The analog of the Koranyi distance in this case is the *Grushin norm*, $\|(x,y)\|_G = (x^4/4 + y^2)^{1/4}$. This will be useful when studying the transience of Grushin diffusions in Section 7.4.

Proposition 5.7.1 implies that it is appropriate to search for a fundamental solution that is proportional to $G(x,y)$. For the sake of simplicity, we shall demonstrate this in the case $n = 1$.

Proposition 5.7.3 *The fundamental solution of the two-dimensional Grushin operator, $\Delta_G = \frac{1}{2}(\partial_x^2 + x^2\partial_y^2)$, from the origin is given by*

$$F(x,y) = -\frac{\Gamma(1/4)}{2\sqrt{2\pi}\Gamma(3/4)} \frac{1}{\left(\frac{1}{4}x^4 + y^2\right)^{1/4}}.$$

Proof: Since $G(x,y) = \left(\frac{1}{4}x^4 + y^2\right)^{-1/4}$ is not smooth at the origin, we consider its ϵ-regularization $G_\epsilon(x,y) = \left(\frac{1}{4}(x^2+\epsilon^2)^2 + y^2\right)^{-1/4}$, with $\epsilon > 0$, which is $\mathcal{C}^\infty(\mathbb{R}^2)$. A straightforward computation provides

$$\Delta_G G_\epsilon(x,y) = -\frac{\epsilon^2/\sqrt{2}}{[(x^2+\epsilon^2)^2 + 4y^2]^{5/4}}.$$

We consider $\psi_\epsilon(x,y) = \dfrac{\epsilon^2/\sqrt{2}}{[(x^2+\epsilon^2)^2 + 4y^2]^{5/4}}$ and note that $\psi_\epsilon > 0$. Assuming that $\psi_\epsilon \in L^1(\mathbb{R}^2)$, we denote

$$c = \iint_{\mathbb{R}^2} \psi_\epsilon(x,y)\,dxdy < \infty.$$

Then $\frac{1}{c}\psi_\epsilon(x,y) \to \delta(x,y)$, in distributions sense, as $\epsilon \to 0$, namely

$$\frac{1}{c}\iint_{\mathbb{R}^2} \psi_\epsilon(x,y)\varphi(x,y)\,dxdy \to \varphi(0,0), \quad \epsilon \to 0, \tag{5.7.27}$$

for any test function $\varphi \in \mathcal{C}_0^\infty$.

On the other side, using integration by parts and the dominated convergence theorem, by taking the limit $\epsilon \to 0$, we have

$$\begin{aligned}
\frac{1}{c}\iint_{\mathbb{R}^2} \psi_\epsilon(x,y)\varphi(x,y)\,dxdy &= -\frac{1}{c}\iint_{\mathbb{R}^2} \Delta_G G_\epsilon(x,y)\varphi(x,y)\,dxdy \\
&= -\frac{1}{c}\iint_{\mathbb{R}^2} G_\epsilon(x,y)\Delta_G\varphi(x,y)\,dxdy \\
&\to -\frac{1}{c}\iint_{\mathbb{R}^2} G(x,y)\Delta_G\varphi(x,y)\,dxdy \\
&= -\frac{1}{c}\iint_{\mathbb{R}^2} \Delta_G G(x,y)\varphi(x,y)\,dxdy.
\end{aligned}$$

Comparing to (5.7.27) and using the limit uniqueness yields

$$-\frac{1}{c}\iint_{\mathbb{R}^2} \Delta_G G(x,y)\varphi(x,y)\,dxdy = \varphi(0,0), \ \forall \varphi \in \mathcal{C}_0^\infty.$$

The previous relation implies that the fundamental solution is given by $F(x,y) = -\frac{1}{c}G(x,y)$. To complete the proof it suffices to compute the value of the constant c.

We shall show that the integral of ψ_ϵ is independent of ϵ. To this end, we integrate first in the variable y using Exercise 5.11.4 with $a = x^2+\epsilon^2$ and then we integrate in the variable x using Exercise 5.11.5 as follows

$$\begin{aligned} c &= \iint_{\mathbb{R}^2} \psi_\epsilon(x,y)\,dxdy = \iint_{\mathbb{R}^2} \frac{\epsilon^2}{\sqrt{2}} \frac{1}{[(x^2+\epsilon^2)^2+4y^2]^{5/4}}\,dxdy \\ &= 2\sqrt{2\pi}\frac{\Gamma(3/4)}{\Gamma(1/4)} \int_{\mathbb{R}} \frac{\epsilon^2/2}{(x^2+\epsilon^2)^{3/2}}dy = 2\sqrt{2\pi}\frac{\Gamma(3/4)}{\Gamma(1/4)}. \end{aligned}$$

■

Remark 5.7.4 Finding a fundamental solution involving the integration of $p_t(x,t)$ with respect to t is feasible if the Fourier transform formula (5.6.16) used in the proof of Proposition 5.6.2 is replaced by

$$\frac{1}{2}\int_{\mathbb{R}} e^{izt}\frac{1}{(1+t^2)^{m/2}}\,dt = \frac{\sqrt{\pi}}{\Gamma(m/2)}\left(\frac{z}{2}\right)^{\frac{m-1}{2}} K_{\frac{m-1}{2}}(z), \qquad z>0.$$

Due to the computation complexity we chose in the proof of Proposition 5.7.1 the direct verification method. The integration method using the aforementioned formula is left as a challenge for interested readers.

5.8 The Saddle Laplacian

We have seen in Section 4.13 that the operator $L = \frac{1}{2}(X^2+Y^2)$ with

$$X_1 = \partial_{x_1} + x_2\partial_{x_3}, \qquad X_2 = \partial_{x_2} + x_1\partial_{x_3},$$

has the heat kernel given by (4.13.99)

$$p_t(x_1,x_2,x_3) = \frac{1}{(2\pi)^3}\int_{\mathbb{R}^3} e^{-i\langle \mathbf{\Lambda},\mathbf{x}\rangle}\text{sech}(\lambda_3 t)e^{-\frac{1}{2\lambda_3}(\lambda_1^2+\lambda_2^2)\tanh(\lambda_3 t)}d\lambda_1 d\lambda_2 d\lambda_3. \tag{5.8.28}$$

We recall that L was called the saddle Laplacian, since X_1 and X_2 are vector fields tangent to the saddle surface $(x_1,x_2) \to (x_1,x_2,x_1x_2)$.

We shall compute next the fundamental solution from the origin, which is obtained by integrating over the t variable, $G(x) = -\int_0^\infty p_t(x)\,dt$. To accomplish this goal, we need first the following result that reduces an improper integral to a proper one.

Lemma 5.8.1 *For any $\lambda \neq 0$ and $a > 0$ we have*

$$\int_0^\infty \operatorname{sech}(\lambda t) e^{-\frac{a}{\lambda}\tanh(\lambda t)}\,dt = \frac{1}{|\lambda|}\int_0^{\pi/2} e^{-\frac{a}{|\lambda|}\sin\theta}\,d\theta.$$

Proof: Since the integrand $\operatorname{sech}(\lambda t)e^{-\frac{a}{\lambda}\tanh(\lambda t)}$ is an even function in λ, it suffices to perform the computation under the assumption $\lambda > 0$, and to replace at the end λ by its absolute value, $|\lambda|$. Using substitutions $\tau = \lambda t$, $u = \tanh\tau$, the relation $du = \frac{1}{1-u^2}du$ and the substitution $u = \sin\theta$, we have

$$\begin{aligned}
\int_0^\infty \operatorname{sech}(\lambda t)e^{-\frac{a}{\lambda}\tanh(\lambda t)}\,dt &= \frac{1}{\lambda}\int_0^\infty \operatorname{sech}(\tau)e^{-\frac{a}{\lambda}\tanh(\tau)}\,d\tau \\
&= \frac{1}{\lambda}\int_0^1 e^{-\frac{a}{\lambda}u}\frac{1}{\sqrt{1-u^2}}\,du \\
&= \frac{1}{\lambda}\int_0^{\pi/2} e^{-\frac{a}{\lambda}\sin\theta}\frac{1}{\sqrt{1-\sin^2\theta}}\cos\theta\,d\theta \\
&= \frac{1}{\lambda}\int_0^{\pi/2} e^{-\frac{a}{\lambda}\sin\theta}\,d\theta.
\end{aligned}$$

We note that the value of the integral at $\lambda = 0$ can be obtained by continuity as

$$\lim_{\lambda\to 0}\int_0^\infty \operatorname{sech}(\lambda t)e^{-\frac{a}{\lambda}\tanh(\lambda t)}\,dt = \int_0^\infty e^{-at}dt = \frac{1}{a}.$$

■

We continue the computation of the fundamental solution of L by integrating in (5.8.28) using Lemma 5.8.1 as follows

$$\begin{aligned}
G(x) &= -\int_0^\infty p_t(x)dt \\
&= \frac{-1}{(2\pi)^3}\int_{\mathbb{R}^3} e^{-i\langle \mathbf{\Lambda},\mathbf{x}\rangle}\int_0^\infty \operatorname{sech}(\lambda_3 t)e^{-\frac{1}{2\lambda_3}(\lambda_1^2+\lambda_2^2)\tanh(\lambda_3 t)}dt\,d\lambda_1 d\lambda_2 d\lambda_3 \\
&= \frac{-1}{(2\pi)^3}\int_{\mathbb{R}^3} e^{-i\langle \mathbf{\Lambda},\mathbf{x}\rangle}\frac{1}{|\lambda_3|}\int_0^{\pi/2} e^{-\frac{\lambda_1^2+\lambda_2^2}{2|\lambda_3|}\sin\theta}d\theta\,d\lambda_1 d\lambda_2 d\lambda_3.
\end{aligned}$$

Applying Fubini's theorem we integrate first in λ_1 and λ_2 as

$$G(x) = \frac{-1}{(2\pi)^3}\int_{\mathbb{R}}\frac{e^{-i\lambda_3 x_3}}{|\lambda_3|}\int_0^{\pi/2}\left(\int_{\mathbb{R}^2} e^{-i\lambda_1 x_1 - \frac{\lambda_1^2}{2|\lambda_3|}\sin\theta} e^{-i\lambda_2 x_2 - \frac{\lambda_2^2}{2|\lambda_3|}\sin\theta}d\lambda_1 d\lambda_2\right)d\theta d\lambda_3.$$

The integral with respect to λ_1 and λ_2 can be computed as a product

$$\begin{aligned}
&\int_{\mathbb{R}^2} e^{-i\lambda_1 x_1 - \frac{\lambda_1^2}{2|\lambda_3|}\sin\theta} e^{-i\lambda_2 x_2 - \frac{\lambda_2^2}{2|\lambda_3|}\sin\theta} d\lambda_1 d\lambda_2 \\
&= \int_{\mathbb{R}} e^{-i\lambda_1 x_1 - \frac{\lambda_1^2}{2|\lambda_3|}\sin\theta} d\lambda_1 \int_{\mathbb{R}} e^{-i\lambda_2 x_2 - \frac{\lambda_2^2}{2|\lambda_3|}\sin\theta} d\lambda_2 \\
&= \sqrt{\frac{2\pi|\lambda_3|}{\sin\theta}} e^{-\frac{x_1^2}{2\sin\theta}|\lambda_3|} \sqrt{\frac{2\pi|\lambda_3|}{\sin\theta}} e^{-\frac{x_2^2}{2\sin\theta}|\lambda_3|} \\
&= \frac{2\pi|\lambda_3|}{\sin\theta} e^{-\frac{x_1^2+x_2^2}{2\sin\theta}|\lambda_3|},
\end{aligned}$$

where we used formula (4.13.98). Substituting into the previous formula of $G(x)$ yields

$$\begin{aligned}
G(x) &= -\frac{1}{(2\pi)^2} \int_{\mathbb{R}} e^{-i\lambda_3 x_3} \int_0^{\pi/2} \frac{1}{\sin\theta} e^{-\frac{x_1^2+x_2^2}{2\sin\theta}|\lambda_3|} \, d\theta d\lambda_3 \\
&= -\frac{1}{(2\pi)^2} \int_0^{\pi/2} \frac{1}{\sin\theta} \int_{\mathbb{R}} e^{-i\lambda_3 x_3} e^{-\frac{x_1^2+x_2^2}{2\sin\theta}|\lambda_3|} \, d\lambda_3 d\theta, \qquad (5.8.29)
\end{aligned}$$

where we interchanged the integrals. The inner integral becomes the Fourier transform of a two-sided exponential distribution. Using the formula

$$\mathcal{F}(e^{-a|t|})(x) = \int_{\mathbb{R}} e^{-itx} e^{-a|t|} \, dt = \frac{2a}{a^2 + x^2},$$

relation (5.8.29) becomes after simplifications

$$G(x) = -\frac{x_1^2 + x_2^2}{\pi^2} \int_0^{\pi/2} \frac{d\theta}{4x_3^2 \sin^2\theta + x_1^2 + x_2^2}. \qquad (5.8.30)$$

Using the trigonometric integral formula

$$\int_0^{\pi/2} \frac{d\theta}{\sin^2\theta + b^2} = \frac{\pi}{2b\sqrt{b^2+1}}, \qquad b > 0,$$

relation (5.8.30) becomes after performing all simplifications

$$\begin{aligned}
G(x) &= -\frac{x_1^2 + x_2^2}{4x_3^2\pi^2} \int_0^{\pi/2} \frac{d\theta}{\sin^2\theta + \frac{x_1^2+x_2^2}{4x_3^2}} \qquad (5.8.31) \\
&= -\frac{1}{2\pi} \sqrt{\frac{x_1^2 + x_2^2}{x_1^2 + x_2^2 + 4x_3^2}}. \qquad (5.8.32)
\end{aligned}$$

The geometric interpretation We shall deal in the following with the geometric interpretation of the norms $|||x||| = \sqrt{x_1^2 + x_2^2 + 4x_3^2}$, and $\sqrt{x_1^2 + x_2^2}$, which come up in the fundamental solution formula (5.8.32). To accomplish this we shall employ the Hamiltonian formalism. To this end, we consider the Hamiltonian associated with the operator L, which is given as its principal symbol as

$$H(x_1, x_2, x_3, \xi_1, \xi_2, \theta) = \frac{1}{2}(\xi_1 + x_2\theta)^2 + \frac{1}{2}(\xi_2 + x_1\theta)^2.$$

In this setup the coordinates are given by (x_1, x_2, x_3), while the momenta by (ξ_1, ξ_2, θ). Then we shall solve the bicharacteristics system

$$\dot{x}_j = \frac{\partial H}{\partial \xi_j}, \quad \dot{x}_3 = \frac{\partial H}{\partial \theta}, \quad \dot{\xi}_j = -\frac{\partial H}{\partial x_j}, \quad \dot{\theta} = -\frac{\partial H}{\partial x_3}, \qquad j \in \{1, 2\}$$

subject to the boundary conditions

$$x_j(0) = 0, \qquad x_j(1) = x_j. \tag{5.8.33}$$

First, we note that since H is independent of x_3, then $\dot{\theta} = 0$, which implies that θ is constant along the solutions. Then the first derivatives of the first two coordinates are

$$\dot{x}_1 = \frac{\partial H}{\partial \xi_1} = \xi_1 + x_2\theta, \quad \dot{x}_2 = \frac{\partial H}{\partial \xi_2} = \xi_2 + x_1\theta, \tag{5.8.34}$$

while the second derivatives become

$$\ddot{x}_1 = \dot{\xi}_1 + \dot{x}_2\theta, \quad \ddot{x}_2 = \dot{\xi}_2 + \dot{x}_1\theta. \tag{5.8.35}$$

Using (5.8.34), the momenta derivatives can be written as

$$\dot{\xi}_1 = -\frac{\partial H}{\partial x_1} = -\theta(\xi_2 + x_1\theta) = -\theta\dot{x}_2$$

$$\dot{\xi}_2 = -\frac{\partial H}{\partial x_2} = -\theta(\xi_1 + x_2\theta) = -\theta\dot{x}_1.$$

Substituting back into (5.8.35) yields

$$\ddot{x}_1 = 0, \qquad \ddot{x}_2 = 0.$$

Solving subject to the boundary conditions (5.8.33) we obtain $x_1(s) = x_1 s$ and $x_2(s) = x_2 s$. Using these formulas we can find now the third coordinate as in the following

$$\begin{aligned}\dot{x}_3 &= \frac{\partial H}{\partial \theta} = (\xi_1 + x_2\theta)x_2 + (\xi_2 + x_1\theta)x_1 \\ &= \dot{x}_1 x_2 + \dot{x}_2 x_1 = x_1 x_2 s + x_2 x_1 s = 2x_1 x_2 s,\end{aligned}$$

which after integration yields $x_3(s) = x_1x_2s^2$. Therefore, the x-projection of the solution of the bicharacteristics system is given by

$$c(s) = \big(x_1(s), x_2(s), x_3(s)\big) = (x_1s, x_2s, x_1x_2s^2), \quad s \in [0,1].$$

We note that this curve belongs to the saddle $\Sigma = \{(x_1, x_2) \to (x_1, x_2, x_1x_2)\}$, while its velocity vector

$$\dot{c}(s) = \big(\dot{x}_1(s), \dot{x}_2(s), \dot{x}_3(s)\big) = (x_1, x_2, 2x_1x_2s) = (x_1, x_2, 2x_3s)$$

is tangent to the surface Σ. It turns out that

$$|||x||| = \sqrt{x_1^2 + x_2^2 + 4x_3^2} = |\dot{c}(1)|_{Eu}.$$

We shall look next at the geometric significance of the expression $\sqrt{x_1^2 + x_2^2}$. To this end, let g be the Riemannian metric on Σ such that the tangent vector fields $\{X_1, X_2\}$ are orthonormal with respect to g. Then we shall compute the length of $c(s)$ with respect to the metric g. To this end, we first decompose the velocity vector as

$$\dot{c}(s) = (x_1, 0, x_1x_2s) + (x_1, 0, x_1x_2s) = \frac{1}{s}\{x_1(s)X_{1|c(t)} + x_2(s)X_{2|c(t)}\}.$$

Then the length of c becomes

$$\ell(c) = \int_0^1 \|\dot{c}(s)\|_g\, ds = \int_0^1 \sqrt{\frac{x_1(s)^2 + x_2(s)^2}{s^2}}\, ds = \sqrt{x_1^2 + x_2^2}.$$

5.9 Green Functions and the Resolvent

The concept of the *Green function* can be applied when the integral of the heat kernel over time is not finite, such as in the case of parabolic manifolds. The Green function is defined as the Laplace transform of the heat kernel:

$$G_\alpha(x, y) = \int_0^\infty e^{-\alpha t} p_t(x, y)\, dt,$$

where $\alpha > 0$.

For instance, the Green function for the one-dimensional Brownian motion is given by

$$G_\alpha(x, y) = \int_0^\infty e^{-\alpha t} \frac{1}{\sqrt{2\pi t}} e^{-\frac{(x-y)^2}{2t}}\, dt = \frac{1}{\sqrt{2\alpha}} e^{-\sqrt{2\alpha}|x-y|}, \quad \alpha > 0.$$

More computational details regarding this example can be found at page 15.

For any continuous, bounded function f on the Riemannian manifold (M, g), the *resolvent* of $p_t(x, y)$ is defined by

$$R_\alpha f(x) = \int_M G_\alpha(x, y) f(y)\, dv(y), \tag{5.9.36}$$

where $dv(y) = \sqrt{\det g(y)}\, dy_1 \cdots dy_n$. If $\|f\| = \sup_{x\in M} |f(x)|$, then

$$\begin{aligned}|R_\alpha f(x)| &\leq \|f\| \int_M G_\alpha(x, y)\, dv(y) = \|f\| \int_0^\infty e^{-\alpha t} \int_M p_t(x, y)\, dv(y)\, dt, \\ &\leq \|f\| \int_0^\infty e^{-\alpha t}\, dt = \frac{1}{\alpha}\|f\|.\end{aligned}$$

Proposition 5.9.1 *Let f be a non-negative bounded and $\mathcal{C}^\infty$-smooth function on the Riemann manifold (M, g). Then $u = R_\alpha f$ is a bounded, $\mathcal{C}^\infty$-smooth solution of the equation*

$$\Delta_g u - \alpha u = -f. \tag{5.9.37}$$

Proof: First, we verify informally that u satisfies the equation as follows

$$\begin{aligned}\Delta_g u(x) &= \int_M \Delta_g G_\alpha(x, y) f(y)\, dv(y) = \int_M \int_0^\infty e^{-\alpha t} \Delta_g p_t(x, y)\, dt\, f(y)\, dv(y) \\ &= \int_M \int_0^\infty e^{-\alpha t} \partial_t p_t(x, y)\, dt\, f(y)\, dv(y) \\ &= \int_M \int_0^\infty \left[\partial_t\big(e^{-\alpha t} p_t(x, y)\big) + \alpha e^{-\alpha t} p_t(x, y)\right] dt\, f(y)\, dv(y) \\ &= \int_M \big(e^{-\alpha t} p_t(x, y)\big)\Big|_{t=0}^{t=\infty} + \alpha \int_0^\infty e^{-\alpha t} p_t(x, y)\, dt\, f(y)\, dv(y) \\ &= -\int_M \delta(x - y) f(y)\, dv(y) + \alpha \int_M G_\alpha(x, y) f(y)\, dv(y) \\ &= -f(x) + \alpha(R_\alpha f)(x) = -f(x) + \alpha u(x).\end{aligned}$$

Since the operator $\Delta_g - \alpha$ is elliptic, then it is hypoelliptic and hence u is a $\mathcal{C}^\infty$-smooth solution. The boundedness follows from $|u(x)| = |R_\alpha f(x)| \leq \frac{1}{\alpha}\|f\|$, which is an inequality stated previously. ■

It is worth writing the resolvent as an expectation involving the associated diffusion as follows. Interchanging integrals by Fubini's theorem, we have

$$\begin{aligned}R_\alpha f(x) &= \int_M G_\alpha(x, y) f(y)\, dy = \int_M \left(\int_0^\infty e^{-\alpha t} p_t(x, y)\, dt\right) f(y)\, dy \\ &= \int_0^\infty e^{-\alpha t} \left(\int_M p_t(x, y) f(y)\, dy\right) dt = \int_0^\infty e^{-\alpha t} \mathbb{E}^x[f(X_t)]\, dt \\ &= \mathbb{E}^x\Big[\int_0^\infty e^{-\alpha t} f(X_t)\, dt\Big],\end{aligned}$$

where X_t is the diffusion associated with the generator Δ_g, starting at $X_0 = x$. The previous formula can be also written as a Laplace transform as

$$R_\alpha f(x) = \mathcal{L}(\mu(t))(\alpha),$$

where $\mu(t) = \mathbb{E}^x[f(X_t)]$. This concept will be useful later in Section 6.14.3.

5.10 Summary

In this chapter, we explored the fundamental solutions for specific operators, including the Euclidean Laplacian, hyperbolic Laplacian, Exponential Grushin Laplacian, Heisenberg sub-Laplacian, Grushin Laplacian, and saddle Laplacian. Our primary approach involved integrating heat kernels obtained in previous chapter over the time variable to obtain explicit expressions. However, this approach is not always feasible, and in cases where the integration over time did not yield a convergent integral, we used a Laplace transform to define the concept of Green function. This concept was then used to define the concept of resolvent, which will be valuable later when studying the Cauchy problem of the heat equation on Riemannian manifolds.

5.11 Exercises

Exercise 5.11.1 Using formula (5.3.8) and Lemma 5.3.1 show that

$$K_{1/2}(x) = \sqrt{\frac{\pi}{2x}}e^{-x}.$$

Exercise 5.11.2 (Upper-half space) Let $\mathcal{H}^3 = \{(x_1, x_2, y) \in \mathbb{R}^3; y > 0\}$ be endowed with the Riemannian metric $g_{ij} = y^{-2}\delta_{ij}$. Define the Laplace-Beltrami operator

$$\Delta_{\mathcal{H}^3} = \frac{1}{2}\frac{1}{\sqrt{\det g}}\sum_{i,j=1}^{3}\Big(\sqrt{\det g}\, g^{ij}\frac{\partial}{\partial x_i}\Big),$$

where $x_3 = y$ and g^{ij} is the inverse of g_{ij}.

(i) Show that

$$\Delta_{\mathcal{H}^3} = \frac{1}{2}y^2\Big(\partial_{x_1}^2 + \partial_{x_2}^2 + \partial_y^2\Big) - \frac{1}{2}y\partial_y.$$

(ii) Is $\Delta_{\mathcal{H}^3}$ hypoelliptic?

The power of the inverse of the Heisenberg norm is met in formulas of several fundamental solutions. The local integrability about the origin is given by the following result.

Exercise 5.11.3 Let $x \in \mathbb{R}^n$ and $z \in \mathbb{R}$. We have

$$\int_0^1 \int_0^1 \frac{1}{(\|x\|^4 + z^2)^\alpha} dx dz < \infty \Leftrightarrow \alpha < \frac{n+1}{2}.$$

Exercise 5.11.4 Show that for any $a > 0$ we have

$$\int_{\mathbb{R}} \frac{1}{(a^2 + 4y^2)^{5/4}} dy = \frac{2\sqrt{\pi}}{a^{3/2}} \frac{\Gamma(3/4)}{\Gamma(1/4)}.$$

Exercise 5.11.5 Show that for any $\epsilon > 0$ we have

$$\int_{\mathbb{R}} \frac{\epsilon^2/2}{(x^2 + \epsilon^2)^{3/2}} dx = 1.$$

Exercise 5.11.6 We compute the fundamental solution for the n-dim Heisenberg Laplacian introduced in Exercise 4.17.3. More precisely, we shall show that $G(x, y) = (|x|^4 + y^2)^{-n/2}$ satisfies $\Delta_H G(x, y) = -c\delta(x, y)$, with $c = \frac{2^{3-n}\pi^{n+1}}{\Gamma(n/2)^2}$, $x \in \mathbb{R}^n$ and $y \in \mathbb{R}$.

(*a*) Verify $\Delta_H G(x, y) = 0$, for any $(x, y) \neq (0, 0)$;

(*b*) Let $G_\epsilon(x, y) = ((|x|^2 + \epsilon^2)^2 + y^2)^{-n/2}$ for $\epsilon > 0$ and denote $\psi_\epsilon(x, y) = -\Delta_H G_\epsilon(x, y)$. Show that:

(*i*) $\psi_\epsilon > 0$ for $\epsilon > 0$;

(*ii*) $\lim_{\epsilon \to 0} \psi_\epsilon(x, y) = 0$, for $(x, y) \neq (0, 0)$;

(*iii*) The integral $c(\epsilon) = \int_{\mathbb{R}^n_x \times \mathbb{R}_y} \psi_\epsilon(x, y)\, dx dy$ is independent of ϵ;

(*iv*) Find the constat $c(\epsilon)$;

(*c*) Prove that for any $\varphi \in C_0^\infty(\mathbb{R}^n_x \times \mathbb{R}_y)$ we have

$$\int_{\mathbb{R}^n_x \times \mathbb{R}_y} \Delta_H G(x, y)\varphi(x, y) dx dy = \lim_{\epsilon \to 0} \int_{\mathbb{R}^n_x \times \mathbb{R}_y} \psi_\epsilon(x, y)\varphi(x, y) dx dy = -c\varphi(0, 0).$$

Exercise 5.11.7 Compute the following expectations by using the resolvent and solving the associated differential equation (5.9.37):

(*a*) $\mathbb{E}^x \Big[\int_0^\infty e^{-\alpha t} \cos(kW_t)\, dt \Big]$

(*b*) $\mathbb{E}^x \Big[\int_0^\infty e^{-\alpha t} \sin(kW_t)\, dt \Big]$

(*c*) $\mathbb{E}^x \Big[\int_0^\infty e^{-\alpha t} e^{kW_t}\, dt \Big]$, where $\alpha > 0$, $k \in \mathbb{R}$ and W_t is a one-dimensional Brownian motion.

Chapter 6

Elliptic Diffusions

In this chapter, we focus on diffusions on a Riemannian manifold (M, g), where the diffusions are determined by the Riemannian metric g and their infinitesimal generator corresponds to the Laplace-Beltrami operator of the manifold. These diffusions can be regarded as Brownian motions associated with the manifold (M, g).

Conversely, it is possible to reverse the process, meaning that any elliptic diffusion, which can be broadly understood as a diffusion that spreads in all directions, induces a Riemannian structure. We will illustrate these concepts through various examples of elliptic diffusions in different coordinate systems.

6.1 Differential Structure Induced by a Diffusion

In this section we show how the stochastic structure of a non-degenerate diffusion on $\mathbb{R}^n$ induces naturally a Riemannian metric structure on $\mathbb{R}^n$. Thus, with each diffusion there is an associated differential geometry. The study of the associated geometric elements can help with a better understanding of the given diffusion.

We consider the Ito diffusion in $\mathbb{R}^n$

$$dX_t = b(X_t)dt + \sigma(X_t)dW_t \tag{6.1.1}$$

$$X_0 = x_0, \tag{6.1.2}$$

where $b^T = (b_1, \dots, b_n)$ is the drift vector field, $\sigma(x) \in \mathcal{M}_{n,n}(\mathbb{R})$ is the dispersion matrix and $W_t = (W_t^1, \dots, W_t^n)$ is a Brownian motion on $\mathbb{R}^n$.

We assume the diffusion X_t is everywhere *non-degenerate* in the sense that $\det \sigma(x) \neq 0$ for all $x \in \mathbb{R}$. From the physical point of view, an everywhere non-degenerate diffusion spreads in all directions from any initial point.

6.1.1 The Diffusion Metric

There is a Riemannian metric induced naturally from the diffusion structure of the Ito diffusion X_t. Since X_t is a non-degenerate diffusion, then the infinitesimal diffusion generator

$$\mathcal{A}_x = \frac{1}{2}\sum_{i,j=1}^{n}(\sigma\sigma^T)_{ij}(x)\partial_{x_i}\partial_{x_j} + \sum_{k=1}^{n} b_k(x)\partial_{x_k} \tag{6.1.3}$$

is an elliptic operator. Then we consider the matrix function $g_{ij}(\cdot)$ with the inverse given by the coefficients of the principal symbol of $\mathcal{A}_x$

$$g^{ij}(x) = (\sigma\sigma^T)_{ij}(x), \qquad x \in \mathbb{R}^n. \tag{6.1.4}$$

Since $\mathcal{A}_x$ is elliptic and $\sigma\sigma^T$ is symmetric, then g_{ij} are the coefficients of a Riemannian metric on $\mathbb{R}^n$, called *diffusion metric.* This metric, which is induced by the process X_t, will come up naturally in the study of the process X_t. The "strength" of this metric influences the diffusion tendency of X_t in $\mathbb{R}^n$ and hence it will influence the probability transition density of X_t as well as the heat kernel of the operator $\mathcal{A}_x$.

Thus, the pair $(\mathbb{R}^n, g_{ij})$ is the Riemannian manifold associated with the diffusion X_t. The question is how do some geometric properties of this manifold can shed light on the diffusion X_t?

The aforementioned diffusion metric can be generated by Euclidean scalar products of one-forms as well as vector fields as follows.

Proposition 6.1.1 *Consider the one-forms* $\sigma_i = \sum_k \sigma_{ik} dx_k$ *on* $\mathbb{R}^n$, *where* σ_{ik} *is the dispersion matrix of diffusion (6.1.1).*
(i) The inverse of the diffusion metric is generated by the Euclidean scalar products of one-forms

$$g^{ij} = \langle \sigma_i, \sigma_j \rangle.$$

(ii) The volume form on the Riemannian manifold $(\mathbb{R}^n, g_{ij})$ *is given by*

$$dv = (\det g)\sigma_1 \wedge \cdots \wedge \sigma_n,$$

where $\det g = \det(g_{ij})$.

Proof: (i) It follows from a straightforward computation using formula (6.1.4)

$$g^{ij} = (\sigma\sigma^T)_{ij} = \sum_k \sigma_{ik}\sigma^T_{kj} = \sum_k \sigma_{ik}\sigma_{jk} = \langle \sigma_i, \sigma_j \rangle.$$

(ii) Let $\eta = \sigma_1 \wedge \cdots \wedge \sigma_n$. Since any two differential n-forms on $\mathbb{R}^n$ are proportional, there is a function $f(x)$ such that $\eta = f\, dx_1 \wedge \cdots \wedge dx_n$. If

$e_i = (0, \dots, 1, \dots, 0)$, denotes the unit directional vector in the x_ith direction, using that $\{e_i\}$ is the dual base of $\{dx_i\}$, then

$$\begin{aligned} f &= \eta(e_1, \cdots, e_n) = (\sigma_1 \wedge \cdots \wedge \sigma_n)(e_1, \cdots, e_n) = \det\big(\sigma_i(e_j)\big) \\ &= \det(\sigma_{ij}) = \sqrt{\det(\sigma\sigma^T)} = \sqrt{\det(g^{ij})} = \frac{1}{\sqrt{\det g}}, \end{aligned}$$

which implies $\eta = \frac{1}{\sqrt{\det g}} dx_1 \wedge \cdots \wedge dx_n$. Therefore

$$dv = \sqrt{\det g}\, dx_1 \wedge \cdots \wedge dx_n = (\det g)\eta = (\det g)\sigma_1 \wedge \cdots \wedge \sigma_n.$$

■

Proposition 6.1.2 *Consider the vector fields $\sigma^i = \sum_k \sigma^{ik}\partial_{x_k}$ on $\mathbb{R}^n$, where $\sigma^{ij} = (\sigma^{-1})_{ij}$. Then the covariant coefficients of the metric are given by the scalar products*

$$g_{ij} = \langle \sigma^i, \sigma^j \rangle.$$

Proof: The covariant form of formula (6.1.4) writes as

$$g_{ij} = (\sigma\sigma^T)^{ij} = \sum_k \sigma^{ik}(\sigma^T)^{kj} = \sum_k \sigma^{ik}\sigma^{jk} = \langle \sigma^i, \sigma^j \rangle.$$

■

We note that the vector fields $\{\sigma^i\}$ form the dual basis of $\{\sigma_j\}$, i.e., $\sigma_j(\sigma^i) = \delta_{ij}$. Therefore, g_{ij} becomes a Gram matrix with the vector realization $\{\sigma^i\}$.

6.1.2 The Length of a Curve

We can compute now the length of a curve with respect to the diffusion metric. Let $c : [0, T] \to \mathbb{R}^n$ be a differentiable curve. The length of c on the Riemannian manifold $(\mathbb{R}^n, g_{ij})$ is given by

$$\begin{aligned} \ell_g(c) &= \int_0^T \|\dot c(t)\|_g \, dt = \int_0^T \Big(\sum_{i,j} g_{ij|c(t)} \dot c^i(t) \dot c^j(t)\Big)^{1/2} dt \\ &= \int_0^T \Big(\sum_{i,j,k} \sigma^{ik}_{|c(t)} \dot c^i(t) \sigma^{jk}_{|c(t)} \dot c^j(t)\Big)^{1/2} dt. \end{aligned}$$

If we assume now that the entries σ_{ij} are constants, denoting $\gamma(t) = \sigma \circ c(t)$, we obtain

$$\ell_g(c) = \int_0^T \Big(\sum_k (\dot\gamma^k(t))^2\Big)^{1/2} dt = \ell_{Eu}(\gamma),$$

which expresses the length of the curve c on $(\mathbb{R}^n, g_{ij})$ as the length in the Euclidean metric of the transformed curve γ.

6.1.3 Geodesics and Their Role

Between any two points, x and y, situated in a certain proximity of each other there is a unique curve with the shortest length among all the curves joining x and y. This is called the *geodesic* between x and y. The length of this geodesic, $d(x, y)$, is called the *Riemannian distance* between points x and y. This important geometric concept is related to the behavior of the diffusion X_t. For instance, the WKB method states that the transition density of X_t is sought in the form of an asymptotic series in powers of t as

$$p_t(x, y) = \frac{1}{t^{n/2}} \exp\Big\{ -\frac{d^2(x, y)}{2t} + a_1(x, y) + ta_2(x, y) + \cdots \Big\}. \tag{6.1.5}$$

The expression $S_{cls}(t, x, y) = \dfrac{d^2(x, y)}{2t}$ is called the *Riemannian action* between x and y within time t, and will come up relatively often in the study of heat kernels of elliptic operators. The functions $a_j(x, y)$ are determined from a chain of relations obtained by identifying the similar powers of t after the expression of $p_t(x, y)$ was substituted into the corresponding heat equation.

It is worthy to note that the asymptotics of $p_t(x, y)$ for t small is based on the argument that the motion of the diffusion X_t from x to y occurs mainly along the geodesic that connects x and y.

More precisely, if we apply a logarithmic function in (6.1.5) and multiply by $-2t$ yields

$$-2tp_t(x, y) = nt \ln t + d(x, y)^2 - 2ta_1(x, y) - 2t^2 a_2(x, y) + \cdots$$

Then taking the limit $t \to 0$ we obtain the asymptotic formula proved by Varadhan [133] [132] on closed Riemannian manifolds

$$\lim_{t \to 0} -2t \ln p_t(x, y) = d(x, y)^2, \tag{6.1.6}$$

under the assumption $(x, y) \in M_1 \times M_2$, where $M_1, M_2 \subset M$ are compacta.

If x and y are close enough such that there is a unique shortest geodesic joining them, then Molchanov [111] proved the asymptotic formula

$$p_t(x, y) \sim \frac{1}{t^{n/2}} e^{-\frac{d^2(x,y)}{2t}} H(x, y), \tag{6.1.7}$$

where $H(x, y)$ is given by

$$H(x, y) = d(x, y)^{\frac{n-1}{2}} \Psi^{-\frac{1}{2}}(x, y), \tag{6.1.8}$$

where $\Psi(x, y)$ can be geometrically expressed in terms of Jacobi vector fields orthogonal to the geodesic joining x and y as follows.

Let $\rho = d(x, y)$ and consider the geodesic $\gamma : [0, \rho] \to M$ connecting $x = \gamma(0)$ and $y = \gamma(\rho)$, parameterized by the arc length s. Let $\{e_1(0), \dots, e_n(0)\}$ be an orthonormal basis of T_xM, such that $e_n(0) = \dot{\gamma}(0)$. By parallel transport along γ, i.e., $\frac{D}{ds}e_k(s) = 0$, we construct the orthonormal basis $\{e_1(s), \dots, e_n(s)\}$ at $T_{\gamma(s)}M$ along γ, with $0 \le s \le \rho$. Then any Jacobi vector normal to γ can be written as $Y(s) = \sum_{k=1}^{n-1} Y_k(s)e_k(s)$. Then the matrix associated to the normal Jacobi vector fields is given by $\mathcal{Z}_{kj}(s) = e_k^j(s)$.

Under these hypotheses, the function Ψ used in (6.1.8) is given by $\Psi(x, y) = \det \mathcal{Z}(\rho)$. As long as y is not conjugate to x along γ, we have $\det \mathcal{Z}(\rho) \neq 0$.

6.1.4 Killing Vector Fields

If the metric tensor g of a Riemannian manifold is preserved along the integral curves of the vector field X, then X is called a *Killing vector field.* More precisely, let ϕ_t be the one-parameter group of diffeomorphisms generated by the vector field X, namely, for each $x \in \mathbb{R}^n$, $\phi_t(x)$ is the solution of the equation

$$\frac{d}{dt}\phi_t(x) = X_{\phi_t(x)}, \qquad \phi_0(x) = x, \qquad \forall x \in \mathbb{R}^n.$$

Equivalently, $c(t) = \phi_t(x)$ is the integral curve of X starting at $c(0) = x$. Given the point x, the previous system can be solved locally for $|t| < \epsilon$ for some small positive ϵ. The vector field X is a Killing vector field if ϕ_t are isometries for each t, namely they preserve the metric tensor, $(\phi_t)_* g = g$. In particular, ϕ_t preserves also the angle between vectors and the vector length. Using the definition of the Lie derivative as a limit

$$L_X g = \lim_{t \to 0} \frac{1}{t}(g - (\phi_t)_* g) = 0,$$

it can be shown that the condition of being a Killing vector can be written equivalently as $L_X g = 0$. Since for any two vector fields U and V we have

$$\begin{aligned}(L_X g)(U, V) &= Xg(U, V) - g(L_X U, V) - g(U, L_X V) \\ &= Xg(U, V) - g([X, U], V) - g(U, [X, V]),\end{aligned}$$

it follows that X is a Killing vector field if and only if

$$Xg(U, V) = g([X, U], V) + g(U, [X, V]), \qquad \forall U, V \in \mathcal{X}. \tag{6.1.9}$$

In particular, if $U = V = E$, with E unitary vector field, then the previous relation implies

$$g([X, E], E) = 0. \tag{6.1.10}$$

This states that if X is a Killing vector field, then the Lie bracket $[X, E]$ is perpendicular to the vector field E. This relation will be useful in the proof of the following result.

Proposition 6.1.3 *If X is a Killing vector field, then $div(X) = 0$.*

Proof: We start from Kotszul's formula of the Levi-Civita connection ∇^0 with respect to the metric g

$$\begin{aligned} 2g(\nabla^0_V X, U) &= Vg(X, U) + Xg(U, V) - Ug(V, X) - \\ &\quad - g(V, [X, U]) + g(X, [U, V]) + g(U, [V, X]). \end{aligned}$$

Making $U = V = E$, with E unitary vector field, we obtain

$$2g(\nabla^0_E X, E) = Eg(X, E) - Eg(E, X) - g(E, [X, E]) + g(E, [E, X]) = 0,$$

where we used $[X, E] = -[E, X]$, the symmetry of g and relation (6.1.10). It follows that $\nabla^0_E X$ is perpendicular on the vector field E. Taking the trace with respect to an orthonormal basis $\{E_i\}$ at p we obtain

$$(divX)(p) = \sum_{i=1}^{n} g(\nabla^0_{E_i} X, E_i) = 0,$$

which is the desired relation. ■

Remark 6.1.4 Using $\nabla^0_E X - \nabla^0_X E = [E, X]$, formula (6.1.10) and the fact that $g(\nabla^0_E X, E) = 0$, it follows that $\nabla^0_X E$ is orthogonal to E.

Remark 6.1.5 Using formula $L_X dv = div(X)dv$, see the forthcoming Proposition 6.5.9, we obtain that the volume form of the Riemannian manifold $(\mathbb{R}^n, g_{ij})$ is preserved along any Killing vector field X, i.e., $L_X dv = 0$.

The next computation produces an equation for Killing vector fields involving the Levi-Civita connection. Using the metrical and torsion-free properties of the Levi-Civita connection combined with the equation (6.1.9) we obtain

$$\begin{aligned} g(\nabla^0_U X, V) + g(U, \nabla^0_V X) &= g(\nabla^0_X U - [X, U], V) + g(U, \nabla^0_X V - [X, V]) \\ &= g(\nabla^0_X U, V) + g(U, \nabla^0_X V) - g([X, U], V) \\ &\quad - g(U, [X, V]) \\ &= Xg(U, V) - g([X, U], V) - g(U, [X, V]) \\ &= 0. \end{aligned}$$

Hence, a Killing vector field X is characterized by the equation

$$g(\nabla^0_U X, V) + g(U, \nabla^0_V X) = 0, \quad \forall U, V. \tag{6.1.11}$$

In local coordinates this becomes

$$(\nabla^0_{\partial_{x_i}} X)^j + (\nabla^0_{\partial_{x_j}} X)^i = 0.$$

As a consequence, letting $U = X$ in (6.1.11) and using that $Vg(X,X) = 2g(X, \nabla^0_V X)$, yields

$$g(\nabla^0_X X, V) + \frac{1}{2} V g(X,X) = 0, \quad \forall V. \tag{6.1.12}$$

Therefore, if X is Killing, then X has constant length if and only if X is a geodesic vector field (its integral curves are geodesics).

Remark 6.1.6 It is worthy to note that if X is a Killing vector field, there is a local system of coordinates, $(x_1, \dots, x_{n-1}, \tau)$, such that $X = \partial_\tau$ and $\partial_\tau g_{ij} = 0$, namely $g_{ij} = g_{ij}(x_1, \dots, x_{n-1})$.

Remark 6.1.7 However, in general, the determinant of the metric, $\det g$, is not preserved along Killing vector fields. One counterexample is the upper-half plane, $\mathcal{H}^2$, with $ds^2 = y^{-2}(dx^2 + dy^2)$ and $\det g = y^{-4}$. On this space there are three linearly independent Killing vector fields

$$X = \partial_x, \ \ Y = x\partial_x + y\partial_y, \ \ Z = (x^2 - y^2)\partial_x + 2xy\partial_y.$$

We notice that $X(\det g) = 0$, but $Y(\det g) \neq 0$ and $Z(\det g) \neq 0$.

The determinant $\det g = vol(\partial_{x_1}, \dots, \partial_{x_n})$ signifies the volume of the n-dimensional parallelepiped generated by the local coordinate vector fields, $\partial_{x_1}, \dots, \partial_{x_n}$. Employing the local coordinates given by Remark 6.1.6 we obtain that $\det g$ depends only on $(x_1, \dots, x_{n-1})$, while the Killing vector field is given by ∂_{x_n}. Therefore, in this local coordinates, the determinant $\det g$ becomes invariant along the given Killing vector field. Since $\det g$ is not a tensor, this relation does not neccessarily hold true in other coordinate systems.

It is worth noting that in general, a Killing vector field does not preserve the metric coefficients, g_{ij}, alone, even if it does preserve the 2-contravariant metric tensor, $g = \sum_{i,j} g_{ij}(x) dx^i \otimes dx^j$.

Remark 6.1.8 The Lie bracket $[X,Y]$ of two Killing vector fields X and Y is also a Killing vector field. This follows from the properties of the Lie derivative

$$L_{[X,Y]}g = L_X L_Y g - L_Y L_X g = L_X g - L_Y g = g - g = 0.$$

Therefore, the set of all Killing vector fields on a given Riemannian manifold of dimension n forms a Lie algebra of dimension r, with $0 \leq r \leq n(n+1)/2$. The case $r = 0$ occurs for spaces which do not admit any symmetries, while the case $r = n(n+1)/2$ corresponds to spaces with the maximum number of symmetries, such as $\mathbb{R}^n$, $\mathbb{S}^n$, or $\mathcal{H}^n$.

The existence of Killing vector fields imply the presence of symmetries in the metric, which in turn will affect the behavior of the associated diffusion. The larger the number of Killing vector fields, the more symmetric the diffusion law tends to be. These symmetries could get incorporated into the transition probability of the associated diffusion.

More specifically, if d denotes the Riemannian distance on the manifold (M, g) and ϕ_τ is the group of isometries associated with the Killing vector field X, then $d(x, y) = d\big(\phi_\tau(x), \phi_\tau(y)\big)$. Consequently, the Riemannian action, $S(t; x, y) = \dfrac{d(x, y)^2}{2t}$, is also preserved by ϕ_τ. Therefore, if the heat kernel is given by

$$p_t(x, y) \sim \frac{C}{t^{n/2}} e^{-S(t;x,y)},$$

then it is invariant by ϕ_τ.

There is a variational interpretation of a Killing vector as a momentum conservation law. Consider now the local coordinates $(x_1, \ldots, x_{n-1}, \tau)$ given in Remark 6.1.6. Then g_{ij} and its derivatives are independent of τ since $\partial_\tau \partial_{x_k} g^{ij}(x) = \partial_{x_k} \partial_\tau g^{ij}(x) = 0$. Therefore, Christoffel symbols, Γ^k_{ij}, do not dependent on τ. Hence, the principal symbol of the Laplace-Beltrami operator Δ_g

$$H(x, \tau; \xi) = \frac{1}{2} \sum_{i,j} g^{ij}(x) \xi_i \xi_j - \frac{1}{2} \sum_{k,i,j} g^{ij}(x) \Gamma^k_{ij}(x) \xi_k$$

is also independent of τ. Considering $H(x, \tau; \xi)$ as a Hamiltonian function, we associate the bicharacteristics curves $(x(s), \tau(s), \xi(s))$ and notice that $\dot{\xi}_n(s) = -\partial_\tau H = 0$, which means the moment ξ_n is preserved along the trajectory.

6.1.5 The Dispersion Induced n-form

The one-forms $\sigma_i = \sum_k \sigma_{ik} dx_k$ on $\mathbb{R}^n$ have coefficients drawn from the ith row of the dispersion matrix σ. Their wedge product

$$\eta = \sigma_1 \wedge \cdots \wedge \sigma_n$$

is the n-form associated with the diffusion X_t, which is induced by dispersion.

Remark 6.1.9 We note that in general, the form η is not parallel transported along Killing vector fields with respect to the Levi-Civita connection. To show this, we recall that the volume form dv is parallel with respect to the Levi-Civita connection, we have $\nabla^0 dv = 0$, see for instance Calin and Udriste [41], page 259. Using Proposition 6.1.1 and the linear connection properties, we have

$$\begin{aligned}\nabla^0_Y \eta &= \nabla^0_Y\Big(\frac{1}{\det g}dv\Big) = Y\Big(\frac{1}{\det g}\Big)\,dv + \frac{1}{\det g}\underbrace{\nabla^0_Y dv}_{=0}\\ &= -\frac{1}{(\det g)^2}Y(\det g)\,dv \neq 0,\end{aligned}$$

because $Y(\det g) \neq 0$, see Remark 6.1.7.

We shall characterize next all linear connections, ∇^*, such that $\nabla^*_Y \eta = 0$ for any Y vector field on $\mathbb{R}^n$. Namely, we shall look for all linear connections under which η is parallely transported. First, we recall that the coefficients of connection ∇^*, denoted by ${\Gamma^*_{ij}}^k$, are given by

$$\nabla^*_{\partial_{x_i}} \partial_{x_j} = \sum_{k=1}^n {\Gamma^*_{ij}}^k \partial_{x_k}.$$

The characterization of the connection ∇^* will be given as a constraint imposed on its coefficients.

Proposition 6.1.10 *We have* $\nabla^* \eta = 0$ *if and only if*

$$\sum_k {\Gamma^*_{ik}}^k = -\frac{1}{2}\partial_{x_i}(\ln \det g), \qquad \forall i \in \{1, \ldots, n\}. \tag{6.1.13}$$

Proof: We have $\nabla^*\eta = 0$ if and only if η is parallel transported along a basis, namely $\nabla^*_{\partial_{x_i}} \eta = 0$, for all $1 \le i \le n$. Using the relation between η and dv given by Proposition 6.1.1, as well as the properties of a linear connection we have

$$\begin{aligned}0 = \nabla^*_{\partial_{x_i}} \eta &= \nabla^*_{\partial_{x_i}}\Big(\frac{1}{\det g}\,dv\Big) = \partial_{x_i}\Big(\frac{1}{\det g}\Big)\,dv + \frac{1}{\det g}\nabla^*_{\partial_{x_i}} dv\\ &= \frac{1}{\det g}\Big(-\frac{1}{\det g}\partial_{x_i}(\det g)\,dv + \nabla^*_{\partial_{x_i}} dv\Big),\end{aligned}$$

which implies

$$\nabla^*_{\partial_{x_i}} dv = \partial_{x_i}(\ln(\det g))\,dv. \tag{6.1.14}$$

Relation (6.1.14) holds if and only if it is satisfied on a basis, i.e.

$$\nabla^*_{\partial_{x_i}} dv(\partial_{x_1}, \ldots, \partial_{x_n}) = \partial_{x_i}(\ln(\det g))\,dv(\partial_{x_1}, \ldots, \partial_{x_n}). \tag{6.1.15}$$

Using that $dv = \sqrt{\det g}\, dx_1 \wedge \cdots \wedge dx_n$, the left side of (6.1.15) can be computed as

$$\begin{aligned}
\nabla^*_{\partial_{x_i}} dv(\partial_{x_1}, \ldots, \partial_{x_n}) &= \partial_{x_i} dv(\partial_{x_1}, \ldots, \partial_{x_n}) - dv(\nabla^*_{\partial_{x_i}} \partial_{x_1}, \cdots, \partial_{x_n}) \\
&\quad \cdots - dv(\partial_{x_1}, \cdots, \nabla^*_{\partial_{x_i}} \partial_{x_n}) \\
&= \partial_{x_i} \sqrt{\det g} - dv\Big(\sum_k \Gamma^k_{i1} \partial_{x_k}, \ldots, \partial_{x_n}\Big) \\
&\quad \cdots - dv\Big(\partial_{x_1}, \cdots, \sum_k \Gamma^k_{in} \partial_{x_k}\Big) \\
&= \frac{1}{2\sqrt{\det g}} \partial_{x_i}(\det g) - dv(\partial_{x_1}, \ldots, \partial_{x_n}) \sum_k {\Gamma^*_{ik}}^k \\
&= \Big(\frac{1}{2}\partial_{x_i}(\ln \det g) - \sum_k {\Gamma^*_{ik}}^k\Big) \sqrt{\det g}, \qquad (6.1.16)
\end{aligned}$$

where we used the skew-symmetry of the volume form. The right side of (6.1.15) becomes

$$\partial_{x_i}(\ln(\det g))\, dv(\partial_{x_1}, \ldots, \partial_{x_n}) = \partial_{x_i}(\ln(\det g))\sqrt{\det g}. \qquad (6.1.17)$$

Equating (6.1.16) and (6.1.17) provides

$$\sum_k {\Gamma^*_{ik}}^k = -\frac{1}{2}\partial_{x_i}(\ln \det g),$$

which is the desired relation. ∎

If the linear connection ∇^* is replaced by the Levi-Civita connection ∇^0 of the Riemannian manifold (M, g_{ij}), then a similar relation to (6.1.13) holds, but with a flipped sign. The next result deals with this property.

Proposition 6.1.11 *On any Riemannian manifold the contraction of the Christoffel symbols is given by*

$$\sum_{j=1}^n \Gamma^j_{ij} = \frac{1}{2}\partial_{x_i}(\ln \det g). \qquad (6.1.18)$$

Proof: Summing over the index j in the formula of the Christoffel symbols and using the symmetry of the metric coefficients, we obtain

$$\sum_{j=1}^n \Gamma^j_{ij} = \sum_j \frac{1}{2} g^{jp}\Big(\partial_{x_j} g_{ip} + \partial_{x_i} g_{jp} - \partial_{x_p} g_{ij}\Big) = \sum_{j,p} \frac{1}{2} g^{jp} \partial_{x_i} g_{jp}.$$

Using the formula for the inverse

$$g^{jp} = \frac{\partial \det g}{\partial g_{jp}} \frac{1}{\det g}$$

and applying the chain rule, the previous relation becomes

$$\sum_{j=1}^{n} \Gamma_{ij}^{j} = \frac{1}{2} \frac{1}{\det g} \partial_{x_i}(\det g) = \frac{1}{2} \partial_{x_i}(\ln \det g).$$

■

Corollary 6.1.12 *The Riemannian volume form is parallel with respect to the Levi-Civita connection.*

Proof: Let ∇^0 be the the Levi-Civita connection. A computation provides a formula similar to (6.1.16)

$$\nabla^0_{\partial_{x_i}} dv(\partial_{x_i}, \ldots, \partial_{x_n}) = \Big(\frac{1}{2}\partial_{x_i}(\ln \det g) - \sum_k \Gamma_{ik}{}^k\Big)\sqrt{\det g}. \tag{6.1.19}$$

Applying (6.1.18) we obtain $\nabla^0_{\partial_{x_i}} dv = 0$. Since the volume element is parallel along a basis, then it will be parallel with respect to ∇^0 along any vector field.

■

In the following we shall exploit the sign similarity between formulae (6.1.13) and (6.1.18). We consider the linear connection $\overline{\nabla}$ as the average between connection ∇^* and Levi-Civita connection ∇^0, namely

$$\overline{\nabla} = \frac{1}{2}(\nabla^0 + \nabla^*).$$

The connection coefficients satisfy the average relation

$$\overline{\Gamma}_{ij}^k = \frac{1}{2}\Big(\Gamma_{ij}^k + \Gamma^*_{ij}{}^k\Big).$$

Contracting over $j = k$ and using (6.1.13) and (6.1.18) we obtain

$$\sum_k \overline{\Gamma}_{ik}^k = \frac{1}{2}\Big(\sum_k \Gamma_{ik}^k + \sum_k \Gamma^*_{ik}{}^k\Big) = 0. \tag{6.1.20}$$

A computation provides a formula similar to (6.1.16)

$$\begin{aligned}
\overline{\nabla}_{\partial_{x_i}} dv(\partial_{x_1}, \ldots, \partial_{x_n}) &= \Big(\frac{1}{2}\partial_{x_i}(\ln \det g) - \sum_k \overline{\Gamma}_{ik}{}^k\Big)\sqrt{\det g} \\
&= \Big(\frac{1}{2}\partial_{x_i}(\ln \det g)\Big)\sqrt{\det g},
\end{aligned} \tag{6.1.21}$$

which implies

$$\overline{\nabla}_{\partial_{x_i}} dv = \frac{1}{2}\partial_{x_i}(\ln\det g)\, dv.$$

Using the linearity of the connection in the first argument we obtain

$$\overline{\nabla}_X dv = \frac{1}{2}X(\ln\det g)\, dv,$$

for any vector field X on $\mathbb{R}^n$. This formula provides the evolution of the Riemannian volume along X with respect to the connection $\overline{\nabla}$.

6.2 Brownian Motion on Manifolds

On any Riemannian manifold we have a distinguished diffusion which is associated with the Laplace Beltrami operator. This process is commonly referred to as the Brownian motion on the manifold. The use of this term is analogous to the context of Euclidean space $\mathbb{R}^n$, where the Laplace-Beltrami operator corresponds to the conventional Laplacian, and the associated diffusion process is an n-dimensional Brownian motion.

In the upcoming sections, we will denote a connected Riemannian manifold by (M, g), where M represents the manifold and g denotes the metric tensor. We have the option to describe all the objects on the manifold M either globally or by utilizing local coordinate systems, denoted as $(x_1, \dots, x_n)$.

6.2.1 Gradient Vector Fields

The gradients taken with respect to the diffusion metric g_{ij} will describe the change of a function in a given direction taking into consideration the diffusion influence of the process X_t. For any smooth function $f : M \to \mathbb{R}$, its *g-gradient* is defined to be the vector field $\nabla_g f$ on $\mathbb{R}^n$ characterized by the relation

$$g(\nabla_g f, V) = df(V), \tag{6.2.22}$$

for any vector field V on $\mathbb{R}^n$ and, as usual, $g(\cdot,\cdot)$ denotes the two-bilinear form with coefficients g_{ij}, namely

$$g(u, v) = \sum_{i,j=1}^n g_{ij}u_i v_j, \qquad \forall u, v \in \mathbb{R}^n,$$

and the differential df is the one-form represented by $df = \sum_{i=1}^n (\partial_{x_i} f)\, dx_i$. Substituting in (6.2.22) we obtain the expression of the g-gradient of f in local coordinates

$$\nabla_g f = \sum_{j=1}^n (\nabla_g f)^j \partial_{x_j},$$

with

$$(\nabla_g f)^j = \sum_{i=1}^n g^{ij}\partial_{x_i} f = \sum_{i=1}^n (\sigma\sigma^T)_{ij}\partial_{x_i} f = \sum_{i,k=1}^n \sigma_{ik}\sigma_{jk}\partial_{x_i} f.$$

6.2.2 Divergence of Vector Fields

The divergence of a vector field X on Riemannian manifolds can be characterized equivalently as follows:

1. As the rate of change of the Riemannian volume element along the integral curves of the vector field X, via the formula $L_X dv = (divX)dv$, where L_X denotes the Lie derivative in the direction of X and dv is the volume form on (M, g_{ij}). This will be the subject of Proposition 6.5.9.

2. Given a linear connection ∇ on $(\mathbb{R}^n, g_{ij})$, the divergence of X is defined as the trace in the first argument of the connection as

$$\text{div}X = Trace(Y \to g(\nabla_Y X, Y).$$

3. Using local coordinates and the connection coefficients Γ_{ij}^k of the linear connection ∇, the trace formula from part 2 becomes

$$\text{div}(X) = \sum_i \partial_{x_i} X^i + \sum_{i,j} \Gamma_{ij}^i X^j, \tag{6.2.23}$$

where $X = \sum_k X^k \partial_{x_k}$ and $\nabla_{\partial_{x_i}} \partial_{x_j} = \sum_k \Gamma_{ij}^k \partial_{x_k}$. This can be shown as in the following computation

$$\begin{aligned}
Trace(Y \to g(\nabla_Y X, Y)) &= \sum_{i,j} g^{ij} g(\nabla_{\partial_{x_i}} X, \partial_{x_j}) \\
&= \sum_{i,j,k} g^{ij} g(\nabla_{\partial_{x_i}} X^k \partial_{x_k}, \partial_{x_j}) \\
&= \sum_{i,j,k} g^{ij} g(\partial_{x_i} X^k\, \partial_{x_k} + X^k \Gamma_{ik}^p \partial_{x_p}, \partial_{x_j}) \\
&= \sum_{i,j,k} g^{ij} g_{jk} \partial_{x_i} X^k + g^{ij} g_{pj} X^k \Gamma_{ik}^p \\
&= \sum_i \partial_{x_i} X^i + \sum_{i,k} X^k \Gamma_{ik}^i.
\end{aligned}$$

If ∇ is the Levi-Civita connection, then it is symmetric, so $\Gamma_{ij}^k = \Gamma_{ji}^k$. In this case we can write

$$\begin{aligned}
\text{div}(X) &= \sum_i \partial_{x_i} X^i + \sum_{i,j} \Gamma_{ji}^i X^j = \sum_i \partial_{x_i} X^i + \sum_j \frac{1}{2} X^j \partial_{x_j} \ln(\det g) \\
&= \sum_i \partial_{x_i} X^i + \frac{1}{2} X(\ln \det g),
\end{aligned}$$

where we used (6.1.18). The first term is the Euclidean divergence of X, while the second is a metric adjustment.

We make the remark that for parts 2 and 3 the linear connection is arbitrary, while in part 1 the divergence is defined with respect to the Levi-Civita connection on (M, g_{ij}).

6.2.3 The Laplace-Beltrami Operator

Let ∇^0 denote the Levi-Civita connection on the Riemannian manifold (M, g_{ij}) and $f : M \to \mathbb{R}$ be a twice differentiable function. The *Laplace-Beltrami operator* is defined by

$$\Delta_g f = \frac{1}{2}\mathrm{div}(\nabla_g f),$$

where the divergence is taken with respect to the connection ∇^0. A computation in local coordinates, see [36], page 24, provides

$$\Delta_g f = \frac{1}{2}\Big(\sum_{i,j} g^{ij}\partial_{x_i}\partial_{x_j} f - \sum_{i,j,k} g^{ij}\Gamma_{ij}^k \partial_{x_k} f\Big), \tag{6.2.24}$$

where

$$\Gamma_{ij}^k = \frac{1}{2}\sum_p g^{kp}(\partial_{x_j} g_{ip} - \partial_{x_p} g_{ij} + \partial_{x_i} g_{pj}).$$

are the Christoffel symbols of the second kind. Formula (6.2.24) has the advantage of explicitly showing the quadratic and linear derivatives involved in the operator Δ_g, fact that will become useful while comparing to the expression of a diffusion generator operator.

6.2.4 Brownian Motions on (M, g)

We consider a difussion process, X_t, on M, starting at $X_0 = x$, which in local coordinates resembles equation (6.1.1)

$$dX_t = b(X_t)dt + \sigma(X_t)dW_t, \tag{6.2.25}$$

with $b = \sum_k b_k \partial_{x_k}$ and $\sigma = \sigma_{ij}(x)$ such that:

1. $(\sigma\sigma^T)_{ij} = g^{ij}$, i.e. the Riemannian metric corresponds locally to the diffusion metric induced by X_t;

2. The drift vector field has the components

$$b_k = -\frac{1}{2}\sum_{i,j} g^{ij}\Gamma_{ij}^k. \tag{6.2.26}$$

Then the infinitesimal generator (6.1.3) of X_t becomes the Laplace-Beltrami operator with respect to the diffusion metric g, namely, $\mathcal{A}_x = \Delta_g$. In this case, the diffusion X_t is regarded as a Brownian motion process on the Riemannian manifold (M, g).

Even if the properties of the process X_t are geometrically invariant, the process is often expressed in local coordinates. In fact, the form of the equation (6.2.25) is coordinate dependent. Under a change of coordinates $\bar{x}^i = \bar{x}^i(x_1, \dots, x_n)$ the diffusion coefficients change tensorially as

$$\bar{g}^{ij} = \sum_{ab} g^{ab} \frac{\partial \bar{x}_i}{\partial x_a} \frac{\partial \bar{x}_j}{\partial x_b}.$$

Therefore, if a diffusion is nondegenerate in one local system of coordinates, then it will be the same in all systems.

Since Γ_{ij}^k is not a tensor, then b_k given by the contraction of the linear connection coefficients will not be a tensor either. More specifically, using that the connection coefficients change as

$$\bar{\Gamma}_{jk}^i = \Gamma_{qr}^p \frac{\partial x_q}{\partial \bar{x}_j} \frac{\partial x_r}{\partial \bar{x}_k} \frac{\partial \bar{x}_i}{\partial x_p} + \frac{\partial \bar{x}_i}{\partial x_p} \frac{\partial^2 x_p}{\partial \bar{x}_j \partial \bar{x}_k},$$

the contraction becomes

$$\bar{g}^{jk} \bar{\Gamma}_{jk}^i = \Gamma_{qr}^p g^{qr} \frac{\partial \bar{x}_i}{\partial x_p} + g^{\alpha\beta} \frac{\partial \bar{x}_k}{\partial x_\alpha} \frac{\partial \bar{x}_j}{\partial x_\beta} \frac{\partial \bar{x}_i}{\partial x_p} \frac{\partial^2 x_p}{\partial \bar{x}_j \partial \bar{x}_k},$$

fact that shows that the drift given by (6.2.26) does not change as a tensor (due to the last additive term). Thus, the drift may vanish in some local coordinates and be different than zero in others.

In fact, for any given point p on the manifold, we can always choose a local system of coordinates about p such that the drift vanishes at p, namely, $b(p) = 0$. This can be achieved by choosing a normal system of coordinates centered at p, such that $g_{ij}(p) = \delta_{ij}$, $\Gamma_{ij}^k(p) = 0$ and $\frac{\partial g_{ij}}{\partial x_k}(p) = 0$.

In the case of a diffusion on a large dimensional manifold, finding the drift components b_k using formula (6.2.26) is a tedious task due to the computation of the symbols Γ_{ij}^k. The next result avoids this difficulty by reducing the computation to finding the determinant of the metric and its inverse. We consider the summation convention over repeated indices.

Lemma 6.2.1 *On any Riemannian manifold with metric coefficients g_{ij} we have*

$$-\Gamma_{ir}^j g^{ir} = \frac{1}{\sqrt{\det g}} \partial_{x_k} \Big(\sqrt{\det g}\, g^{jk} \Big). \tag{6.2.27}$$

Proof: The proof is performed in the following steps.
Step 1. We show that

$$\partial_{x_k} g_{ij} = \Gamma^h_{ki} g_{hj} + \Gamma^h_{kj} g_{hi}. \tag{6.2.28}$$

This follows from the fact that the Levi-Civita connection ∇ is a metric connection, namely

$$0 = \left(\nabla_{\partial_{x_k}} g\right)_{ij} = \partial_{x_k} g_{ij} - \Gamma^h_{ki} g_{hj} - \Gamma^h_{kj} g_{hi}.$$

Step 2. The derivatives of the inverse metric are given by

$$\partial_{x_k} g^{pr} = -\Gamma^r_{ki} g^{ip} - \Gamma^p_{kj} g^{jr}. \tag{6.2.29}$$

To show this we differentiate with respect to x_k in $g_{ij} g^{jr} = \delta^r_i$ to obtain

$$(\partial_{x_k} g_{ij}) g^{jr} = -g_{ij}\, \partial_{x_k} g^{jr}.$$

We contract the previous relation multiplying by g^{ip} and summing over the index i

$$(\partial_{x_k} g_{ij}) g^{jr} g^{ip} = -\partial_{x_k} g^{pr}.$$

Then using formula (6.2.28) we have

$$\begin{aligned}
\partial_{x_k} g^{pr} &= -(\partial_{x_k} g_{ij}) g^{jr} g^{ip} \\
&= -\left(\Gamma^h_{ki} g_{hj} + \Gamma^h_{kj} g_{hi}\right) g^{jr} g^{ip} \\
&= -\Gamma^r_{ki} g^{ip} - \Gamma^p_{kj} g^{jr},
\end{aligned}$$

where in the last identity we used metric contraction formulas.
Step 3. The double contraction of Christoffel symbols is given by

$$-\Gamma^r_{ki} g^{ki} = \partial_{x_k} g^{kr} + \Gamma^k_{kj} g^{jr}. \tag{6.2.30}$$

This formula follows by letting $p = k$ in formula (6.2.29) and summing over the index k and then rearranging the terms.
Step 4. We show that

$$\frac{1}{\sqrt{\det g}} \partial_{x_k}\left(\sqrt{\det g}\, g^{jk}\right) = -\Gamma^j_{ip} g^{ip}. \tag{6.2.31}$$

Using the product rule and Proposition 6.1.11 we have

$$\begin{aligned}
\frac{1}{\sqrt{\det g}} \partial_{x_k}\left(\sqrt{\det g}\, g^{jk}\right) &= \partial_{x_k} g^{jk} + \frac{1}{2\sqrt{\det g}} (\partial_{x_k} g)\, g^{jk} \\
&= \partial_{x_k} g^{jk} + \frac{1}{2} \partial_{x_k}\left(\ln \det g\right) g^{jk} \\
&= \partial_{x_k} g^{jk} + \Gamma^i_{ki} g^{jk} \\
&= -\Gamma^j_{ip} g^{ip},
\end{aligned}$$

where in the last equality we used formula (6.2.30).

■

Using Lemma 6.2.1 and equation (6.2.26) we obtain the following result.

Proposition 6.2.2 *If the drift vector field of diffusion (6.1.1) has the components given by*

$$b_k = \frac{1}{2}\frac{1}{\sqrt{\det g}}\partial_{x_j}\Big(\sqrt{\det g}\, g^{kj}\Big), \tag{6.2.32}$$

then the infinitesimal generator (6.1.3) becomes a Laplace operator on $(\mathbb{R}^n, g)$ *with respect to the diffusion metric* g*, namely,* $\mathcal{A}_x = \Delta_g$.

Remark 6.2.3 If $Z_1, \dots, Z_r$ are r linearly independent Killing vector fields such that $[Z_i, Z_j] = 0$, for all $i, j \le r$, then there is a local system of coordinates, $(x_1, \dots, x_n)$, such that Z_k becomes coordinate vector fields, $Z_1 = \partial_{x_1}, \dots, Z_r = \partial_{x_r}$.Then if follows from the definition of the Lie derivative that the metric coefficients depend only on $n-r$ variables, $g_{ij} = g_{ij}(x_{r+1}, \dots, x_n)$. Consequenly, using formula (6.2.32) it follows that the drift vector field depends also on $n-r$ coordinates, $b_k = b_k(x_{r+1}, \dots, x_n)$.

In particular, if $r = n$, then g_{ij} become constants and $b_k = 0$. In this case we obtain $dX_t = \sigma dW_t$, namely

$$X_t^j = x_0^j + \sum_{k=1}^{n} \sigma_{jk} W_t^k \tag{6.2.33}$$

is a Brownian motion on $\mathbb{R}^n$ endowed with the Riemannian metric $ds^2 = \sum_{ij}(\sigma\sigma^T)_{ij}dx_i dx_j$. If σ is an orthogonal matrix, then (6.2.33) becomes a Brownian motion with respect to the Euclidean metric $ds^2 = \sum_j dx_j^2$.

Remark 6.2.4 If a Killing vector field Z is unit length, then by (6.1.12) Z is a geodesic vector field. Then we can consider a semigeodesic system of coordinates defined by considering coordinates $u = (u^1, \dots, u^{n-1})$ in some $(n-1)$-dimensional hypersurface $\mathcal{N}$ and a moving parameter τ along the geodesics orthogonal to $\mathcal{N}$ such that:

(*i*) the parameter τ is the length of the geodesic arc starting at $\mathcal{N}$;

(*ii*) Z is the velocity vector field to the geodesic flow;

(*iii*) $\mathcal{N} = \{\tau = 0\}$;

(*iv*) in the system of coordinates (u, τ) the metric takes the form

$$ds^2 = d\tau^2 + \sum_{i,j=1}^{n-1} \bar{g}_{ij}(u) du^i du^j,$$

with $\bar{g}_{ij}$ Riemannian metric on $\mathcal{N}$. We have also used that $\bar{g}_{ij}(u)$ does not depend on τ and $g(Z, Z) = g(\partial_\tau, \partial_\tau) = g_{nn} = 1$.

Under these conditions the Laplace-Beltrami operator on the manifold (M, g) can be written as

$$\begin{aligned}\Delta &= \frac{1}{2}\frac{\partial^2}{\partial \tau^2} + \frac{1}{2}\bar{g}^{ij}(u)\frac{\partial^2}{\partial u^i \partial u^j} + \frac{1}{2\sqrt{\det \bar{g}}}\Big(\frac{\partial}{\partial u^j}\bar{g}^{ij}\sqrt{\det \bar{g}}\Big)\frac{\partial}{\partial u^i}\\ &= \frac{1}{2}\frac{\partial^2}{\partial \tau^2} + \Delta_{\mathcal{N}},\end{aligned}$$

where $\Delta_{\mathcal{N}}$ is the Laplace-Beltrami operator on the hypersurface $(\mathcal{N}, \bar{g})$.

Consequently, passing to the associated diffusions, the Brownian motion on (M, g) starting at (u_0, τ_0) can be written as (X_t, Y_t), where X_t is a Brownian motion on the hypersurface $(\mathcal{N}, \bar{g})$ starting at u_0 and $Y_t = \tau_0 + W_t$, with W_t one-dimensional Brownian motion. Obviously, the diffusions X_t and Y_t are independent. For instance, Brownian motions on a torrus, $T = \mathbb{S}^1 \times \mathbb{S}^1$, or on a cylinder, $C = \mathbb{R} \times \mathbb{S}^1$, globally satisfy the aforementioned description.

6.3 One-dimensional Diffusions

We shall present a few examples of Brownian motions on Riemannian manifolds, starting with the one-dimensional case.

Example 6.3.1 Let $\sigma(x) > 0$ and we consider the one-dimensional diffusion

$$dX_t = b(X_t)dt + \sigma(X_t)dW_t.$$

We shall look for the drift $b(x)$ such that X_t becomes a Brownian motion with respect to a Riemannian metric on $\mathbb{R}$. In this case $\sigma\sigma^T = \sigma^2 = g^{11}$ and $g_{11} = \sigma^{-2}$. There is only one Christoffel symbol in this case, which is given by

$$\Gamma^1_{11} = \frac{1}{2}g^{11}\partial_x g_{11} = -\frac{1}{\sigma(x)}\sigma'(x) = -(\ln \sigma(x))'.$$

Then relation (6.2.26) becomes

$$b(x) = -\frac{1}{2}g^{11}\Gamma^1_{11} = \frac{1}{2}\sigma(x)\sigma(x)'.$$

Therefore, if X_t satisfies the diffusion equation

$$dX_t = \frac{1}{2}(\sigma\sigma')(X_t)\,dt + \sigma(X_t)dW_t, \tag{6.3.34}$$

for a smooth positive real function $\sigma(x)$, then X_t is a Brownian motion on the real line $\mathbb{R}$ endowed with the metric $ds = \frac{1}{\sigma(x)}dx$. This metric induces a distance on $\mathbb{R}$ given by

$$d(x_1, x_2) = \Big| \int_{x_1}^{x_2} \frac{1}{\sigma(x)}\, dx \Big|.$$

It is worth noting that the equation (6.3.34) can be solved explicitly as follows. We look for a solution of the form $X_t = g(W_t)$, with g twice continuous differentiable function. By Ito's formula, X_t satisfies

$$dX_t = \frac{1}{2}g''(W_t)dt + g'(W_t)dW_t.$$

Comparing to (6.3.34) we look for a function g that verifies

$$\begin{aligned}(\sigma\sigma')(X_t) &= g''(W_t)\\ \sigma(X_t) &= g'(W_t).\end{aligned}$$

This system is equivalent to

$$\begin{aligned}(\sigma\sigma')(g(W_t)) &= g''(W_t)\\ \sigma(g(W_t)) &= g'(W_t),\end{aligned}$$

which is satisfied by the function g that is the solution of the differential equation $g'(u) = \sigma(g(u))$. Substituting $x = g(u)$, the method of separation of variables implies

$$\int_{g(u_0)}^{g(u_1)} \frac{dy}{\sigma(y)} = u_1 - u_0. \tag{6.3.35}$$

This can be written as

$$d\big(g(u_0), g(u_1)\big) = |u_1 - u_0|, \qquad \forall u_0, u_1 \in \mathbb{R}.$$

Thus, $g : \mathbb{R} \to \mathbb{R}$ is an isometry between the metric spaces $(\mathbb{R}, d)$ and $(\mathbb{R}, d_{Eu})$, where $d_{Eu}(u_0, u_1) = |u_1 - u_0|$. Sometimes this isometry can be found explicitly by computing the integral on the left side of (6.3.35) and then inverting the function thus obtained. Knowing the isometry is equivalent to finding the solution $X_t = g(W_t)$.

We shall supply next a few particular cases of Brownian motions on $(\mathbb{R}, d)$.

1. Let $\sigma = c > 0$, constant. Then $X_t = cW_t$ is a Brownian motion on $\mathbb{R}$ with respect to the distance $d(x_1, x_2) = c^{-1}|x_2 - x_1|$.

2. The case $\sigma(x) = 1 + x^2$. Then a solution of

$$dX_t = X_t(1 + X_t^2)\, dt + (1 + X_t^2)dW_t, \tag{6.3.36}$$

can be regarded as a Brownian motion on $\mathbb{R}$ with respect to the distance

$$d(x_1, x_2) = |\arctan(x_2) - \arctan(x_1)|.$$

In this case we have $g(u) = \tan(u + C)$, with C constant, so a solution is $X_t = \tan(W_t + C)$, with $C = \tan^{-1}(X_0)$.

3. Consider $\sigma(x) = e^x$. Then the solution of

$$dX_t = \frac{1}{2}e^{2X_t}dt + e^{X_t}dW_t$$

is a Brownian motion on $\mathbb{R}$ endowed with the distance $d(x_1, x_2) = |e^{x_2} - e^{x_1}|$.

4. Let $\sigma(x) = \cosh(x)$. Then the solution of

$$dX_t = \frac{1}{4}\sinh(2X_t)\,dt + \cosh(X_t)\,dW_t$$

can be considered as a Brownian motion on $(\mathbb{R}, d)$, with

$$d(x_1, x_2) = \left|\tan^{-1}(\sinh(x_2)) - \tan^{-1}(\sinh(x_1))\right|.$$

6.4 Multidimensional Diffusions

We shall consider several examples of diffusions on some familiar Riemannian manifolds.

6.4.1 n-Dimensional Hyperbolic Diffusion

We consider the Ito diffusion $X_t = (X_t^1, \ldots, X_t^{n-1}, Y_t)$ in the upper half-space

$$\mathcal{H}^n = \{z = (x, y); x \in \mathbb{R}^{n-1}, y > 0\}.$$

given by the following system

$$\begin{aligned} dX_t^1 &= Y_t\,dW_t^1 + b_1(X_t)dt \\ &\cdots\cdots\cdots\cdots\cdots\cdots \\ dX_t^{n-1} &= Y_t\,dW_t^{n-1} + b_{n-1}(X_t)dt \\ dY_t &= Y_t\,dW_t^n + b_n(X_t)dt, \end{aligned}$$

with initial condition $X_0 = (0, \cdots, 0, 1)$, with W_t^k independent Brownian motions and $b_k(x)$ drift functions subject to be found later, such that X_t becomes a Brownian motion on a Riemann space endowed with the metric $g_{ij} = (\sigma\sigma^T)^{ij} = y^{-2}\delta_{ij}$. These are the coefficients of a Riemannian metric on the upper half-space $\mathcal{H}^n$. It can be shown that all Christoffel symbols are zero with the exception of the following

$$\begin{aligned} \Gamma_{ii}^n &= \frac{1}{y}, \qquad i < n \\ \Gamma_{ii}^k &= 0, \qquad k < n \\ \Gamma_{nk}^k &= \Gamma_{kn}^k = -\frac{1}{y}. \end{aligned}$$

A simple computation of the contraction of Christoffel symbols shows that

$$\sum_{i,j} g^{ij}\Gamma_{ij}^k = \begin{cases} 0, & \text{for} \quad k < n \\ (n-2)y, & \text{for} \quad k = n. \end{cases}$$

We choose the drift vector $b = (b^1, \cdots, b^n)$ given by the relation (6.2.26) by

$$b_k = \begin{cases} 0, & \text{for} \quad k < n \\ -\dfrac{n-2}{2}y, & \text{for} \quad k = n. \end{cases}$$

Therefore, for X_t to be a Brownian motion on the Riemann space $(\mathcal{H}^n, g_{ij})$ the infinitesimal generator of X_t should be

$$\mathcal{A}_x = \frac{1}{2}y^2\Big(\sum_{i=1}^{n-1} \partial_{x_i}^2 + \partial_y^2\Big) - \frac{n-2}{2}\, y\, \partial_y.$$

This is the Laplace-Beltrami operator on the upper half-space $(\mathcal{H}^n, g_{ij})$. The resulting diffusion process

$$\begin{aligned} dX_t^1 &= Y_t\, dW_t^1 \\ &\cdots\cdots\cdots\cdots\cdots\cdots \\ dX_t^{n-1} &= Y_t\, dW_t^{n-1} \\ dY_t &= Y_t\, dW_t^n - \frac{n-2}{2}Y_t dt \end{aligned}$$

is the *hyperbolic diffusion* on $\mathcal{H}^n$. We usually study the diffusion with the initial condition $(X_0^1, \cdots, X_0^{n-1}, Y_0) = (0, 0, \cdots, 0, 1)$. The previous system of stochastic differential equations can be solved starting from the last equation. Using Ito's formula we obtain

$$d(\ln Y_t) = \frac{1}{Y_t}dY_t - \frac{1}{2}\frac{1}{Y_t^2}(dY_t)^2 = -\frac{n-1}{2}dt + dW_t^n.$$

Then integrating and taking the exponential yields

$$Y_t = e^{-\frac{n-1}{2}t + W_t^n}. \tag{6.4.37}$$

It is worth noting that Y_t is a martingale just in the case $n = 2$. For $n > 2$ the component Y_t drifts away exponentially, with expectation $\mathbb{E}[Y_t] = e^{(1-\frac{n}{2})t}$. The other components of the hyperbolic diffusion can be obtained as Ito integrals

$$X_t^j = \int_0^t Y_s \, dW_s^j = \int_0^t e^{-\frac{n-1}{2}s + W_s^n} \, dW_s^j, \qquad 1 \le j \le n-1. \tag{6.4.38}$$

It is worth noting the special case $n = 2$, when the diffusion belongs to the upper half-plane and all its drift components are zero, $b_k = 0$, case in which the diffusion becomes

$$dX_t = Y_t \, dW_t^1$$
$$dY_t = Y_t \, dW_t^2,$$

with the correspondent generator $\mathcal{A}_x = \frac{1}{2} y^2 (\partial_x^2 + \partial_y^2)$. The hyperbolic diffusion in the the upper half-plane $\mathcal{H}^2$ will be studied in detail in the next section.

6.4.2 Two-Dimensional Hyperbolic Diffusion

The upper half-plane $\mathcal{H}^2 = \{(x, y); y > 0\} \subset \mathbb{R}^2$ endowed with the Riemannian metric $ds^2 = \frac{1}{y^2}(dx^2 + dy^2)$ is a Riemannian space of negative constant curvature, $K = -1$. The associated Laplace-Beltrami operator is given by the Poincaré's operator

$$\Delta_P = \frac{1}{2} y^2 (\partial_x^2 + \partial_y^2).$$

Its heat kernel has been found by McKean [108] in the integral representation

$$p_t\big((x_0, y_0); (x, y)\big) = \frac{\sqrt{2}}{(2\pi t)^{3/2}} e^{-\frac{t}{2}} \int_d^\infty r e^{-\frac{r^2}{2t}} \frac{dr}{\sqrt{\cosh r - \cosh d}},$$

where $d = d_h\big(((x_0, y_0), (x, y)\big)$ is the hyperbolic distance computed in $\mathcal{H}^2$ with respect to the hyperbolic metric ds.

In this section we shall study the two-dimensional hyperbolic diffusion and provide a formula for the heat kernel of the operator Δ_P in terms of the Hartman-Watson density function. We also calculate the characteristic functions of the components of the associated diffusion.

The two-dimensional hyperbolic diffusion, (X_t, Y_t), is the diffusion associated with the operator Δ_P and is the strong solution of the system of stochastic equations

$$dX_t = Y_t \, dW_t \tag{6.4.39}$$
$$dY_t = Y_t \, dB_t, \tag{6.4.40}$$

with W_t and B_t independent Brownian motions. Assuming that the diffusion starts at $(X_0, Y_0) = (x_0, y_0)$, we obtain

$$X_t = x_0 + \int_0^t Y_s \, dW_s \tag{6.4.41}$$

$$Y_t = y_0 + \int_0^t Y_s \, dB_s. \tag{6.4.42}$$

The equation (6.4.42) has an explicit solution as follows. We start from the equation (6.4.40), which models linear noise. To obtain the solution we substitute $Z_t = \ln(Y_t)$ and then apply Ito's formula to obtain

$$dZ_t = -\frac{1}{2}dt + dB_t.$$

Then integrating and taking the exponential yields $Y_t = y_0 e^{B_t - \frac{t}{2}}$. Then the system (6.4.41)-(6.4.42) becomes

$$X_t = x_0 + y_0 \int_0^t e^{B_s - \frac{s}{2}} \, dW_s \tag{6.4.43}$$

$$Y_t = y_0 e^{B_t - \frac{t}{2}}. \tag{6.4.44}$$

We notice that $M_t = \int_0^t e^{B_s - \frac{s}{2}} \, dW_s$ is a continuous martingale and

$$\begin{aligned} \mathbb{E}[M_t^2] &= \mathbb{E}\Big[\Big(\int_0^t e^{B_s - \frac{s}{2}} \, dW_s\Big)^2\Big] = \mathbb{E}\Big[\int_0^t e^{2B_s - s} \, ds\Big] \\ &= \int_0^t \mathbb{E}[e^{2B_s - s}] \, ds = \int_0^t e^s \, ds = e^t - 1 < \infty. \end{aligned}$$

Since $\langle M \rangle_t = \int_0^t e^{2B_s - s} \, ds \to \infty$ as $t \to \infty$, then by Theorem 1.1.11 there is a DDS Brownian motion $(\beta_t)_{t \geq 0}$, independent of $(B_t)_{t \geq 0}$, such that we can represent $M_t = \beta_{A_t^{(-1/2)}}$, where[1]

$$A_t^{(-1/2)} = \int_0^t e^{2B_s - s} \, ds = \langle M \rangle_t.$$

It suffices to study only the diffusion starting at $(x_0, y_0) = (0, 1)$, which is given by

$$X_t = \beta_{A_t^{(-1/2)}}, \qquad Y_t = e^{B_t - \frac{t}{2}}, \qquad t \geq 0. \tag{6.4.45}$$

[1]In general, we have the notation $A_t^{(\nu)} = \int_0^t e^{2(B_s + \nu s)} \, ds$.

This follows from the fact that the transformations $z \to \frac{az+b}{cz+d}$, with $ad-bc=1$, are isometries of the hyperbolic plane $\{z \in \mathbb{C}; Im(z) > 0\}$. Two distinguished types of these isometries are the translations $z \to z+b$ (obtained for $a=d=1$ and $c=0$) and the dilations $z \to a^2 z$ (obtained for $b=c=0$, $d=1/a$). Then any point $(x_0, y_0) = x_0 + iy_0$ can be mapped to $i = (0,1)$ by applying a translation with $b=-x_0$ and then a dilation with $a=\sqrt{y_0}$. Therefore, we may assume the hyperbolic diffusion starting at $(0,1)$.

The diffusion component Y_t is a continuous martingale with respect to the natural filtration generated by B_t. Since $\langle Y\rangle_t = \int_0^t Y_s^2\,ds = \int_0^t e^{2B_s - s}\,ds = A_t^{(-1/2)}$, there is a DDS Brownian motion $(\gamma_t)_{t\geq 0}$, independent of $(B_t)_{t\geq 0}$, such that $Y_t = \gamma_{A_t^{(-1/2)}}$.

To conclude, the hyperbolic diffusion starting at $(1,0)$ has both components given by time-changed Brownian motions at the same stochastic clock $A_t^{(-1/2)}$ as follows

$$X_t = \beta_{A_t^{(-1/2)}}, \qquad Y_t = \gamma_{A_t^{(-1/2)}}, \qquad t \geq 0, \tag{6.4.46}$$

where the Brownian motions β_t and γ_t are individually independent of B_t. However, it is worth to note that β_t and γ_t are not independent processes.

The characteristic functions From (6.4.46) we get that X_t and Y_t have the same law and hence their characteristic functions are equal

$$\phi_{X_t}(\lambda) = \phi_{Y_t}(\lambda) = \mathbb{E}[e^{-\frac{1}{2}\lambda^2 A_t^{(-1/2)}}].$$

This is a consequence of the tower property

$$\begin{aligned}\phi_{X_t}(\lambda) &= \mathbb{E}[e^{i\lambda X_t}] = \mathbb{E}[\,\mathbb{E}[e^{i\lambda X_t}\,|\,B[0,t]\,]\,]\\ &= \mathbb{E}[\,\mathbb{E}[e^{i\lambda\beta_{A_t^{(-1/2)}}}\,|\,B[0,t]\,]\,] = \mathbb{E}[e^{-\frac{1}{2}\lambda^2 A_t^{(-1/2)}}].\end{aligned}$$

The previous expectation can be computed as follows. If we employ notations $A_t^{(\nu)} = \int_0^t e^{2(B_s+\nu s)}\,ds$ and $A_t = A_t^{(0)}$, then by Girsanov's theorem, see Proposition 1.1.18, for any Borel function $f: R_+ \to \mathbb{R}_+$ we have the following reduction formula

$$\mathbb{E}[f(A_t^{(\nu)})] = \mathbb{E}[f(A_t)e^{\nu B_t - \frac{1}{2}\nu^2 t}].$$

Choosing $f(x) = e^{-\frac{1}{2}\lambda^2 x}$ and $\nu = -1/2$ yields

$$\mathbb{E}[e^{-\frac{1}{2}\lambda^2 A_t^{(-1/2)}}] = \mathbb{E}[e^{-\frac{1}{2}\lambda^2 A_t} e^{-\frac{1}{2}(B_t + \frac{t}{4})}] = e^{-\frac{t}{8}}\mathbb{E}[e^{-\frac{1}{2}(\lambda^2 A_t + B_t)}].$$

This expectation can be computed using the joint distribution of (A_t, B_t), which is given by

$$P(A_t \in du, B_t \in dx) = \frac{1}{u} e^{-\frac{1+e^{2x}}{2u}} \theta_{e^x/u}(t)\,du dx, \tag{6.4.47}$$

where $\theta_r(t)$ is the Hartman-Watson density

$$\theta_r(t) = \frac{r}{\sqrt{2\pi^3 t}} e^{\frac{\pi^2}{2t}} \int_0^\infty e^{-\frac{y^2}{2t}} e^{-r\cosh y} \sinh y \sin\frac{\pi y}{t}\, dy, \tag{6.4.48}$$

see Borodin and Salmilen [15], page 82.

Thus, the previous expectation has the following explicit formula

$$\begin{aligned}\mathbb{E}[e^{-\frac{1}{2}(\lambda^2 A_t + B_t)}] &= \int_{\mathbb{R}} \int_0^\infty e^{-\frac{1}{2}\lambda^2 u} e^{-\frac{1}{2}x} \frac{1}{u} e^{-\frac{1+e^{2x}}{2u}} \theta_{e^x/u}(t)\, du dx \\ &= \int_{\mathbb{R}} e^{-\frac{1}{2}x} \Big(\int_0^\infty \frac{1}{u} \theta_{e^x/u}(t) e^{-\frac{1}{2}\lambda^2 u} e^{-\frac{1+e^{2x}}{2u}}\, du \Big)\, dx.\end{aligned}$$

To conclude the computation, the characteristic function of X_t (or Y_t) is given by the integral formula

$$\phi_{X_t}(\lambda) = e^{-\frac{t}{8}} \int_{\mathbb{R}} e^{-\frac{1}{2}x} \Psi_t(x,\lambda)\, dx,$$

where

$$\Psi_t(x,\lambda) = \int_0^\infty \frac{1}{u} \theta_{e^x/u}(t) e^{-\frac{1}{2}\lambda^2 u} e^{-\frac{1+e^{2x}}{2u}}\, du.$$

The joint distribution of (X_t, Y_t) It suffices to find the joint probability density of the diffusion (X_t, Y_t) starting at $(X_0, Y_0) = (0, 1)$

$$\begin{aligned} X_t &= \int_0^t e^{B_s - \frac{s}{2}} dW_s \\ Y_t &= e^{B_t - \frac{t}{2}}, \quad t \geq 0, \end{aligned}$$

where B_t is a one-dimensional Brownian motion. To this end, we shall evaluate the expectation $\mathbb{E}[f(X_t)g(Y_t)]$ in two ways for any Borel functions $f, g : \mathbb{R}_+ \to \mathbb{R}_+$. First, by the expectation definition we have

$$\mathbb{E}[f(X_t)g(Y_t)] = \int_0^\infty \int_0^\infty f(x)g(y)p_t(x,y)dxdy, \tag{6.4.49}$$

where

$$P(X_t \in dx, Y_t \in dy) = p_t(x,y)dxdy.$$

On the other hand, we can compute the expectation using Girsanov's theorem, see Proposition 1.1.17. To this end, changing the probability measure from P to Q by

$$dQ_{|\mathcal{F}_s} = e^{-\frac{\lambda^2 t}{2} - \lambda B_t} dP_{|\mathcal{F}_s}, \quad 0 \leq s \leq t$$

with $\lambda = -1/2$, we obtain that $W_s = B_s - \frac{1}{2}s$, $0 \leq s \leq t$, becomes a Brownian motion under the measure Q. Then we can write

$$
\begin{aligned}
\mathbb{E}^P[f(X_t)g(Y_t)] &= \mathbb{E}^Q[f(X_t)g(Y_t)e^{\frac{1}{8}t-\frac{1}{2}B_t}] \\
&= e^{-\frac{t}{8}}\mathbb{E}^Q\Big[f\Big(\int_0^t e^{B_s-\frac{1}{2}s}ds\Big)g(e^{B_t-\frac{1}{2}t})e^{-\frac{1}{2}(B_t-\frac{1}{2}t)}\Big] \\
&= e^{-\frac{t}{8}}\mathbb{E}^Q\Big[f\Big(\int_0^t e^{W_s}\,ds\Big)g(e^{W_t})e^{-\frac{1}{2}W_t}\Big] \\
&= e^{-\frac{t}{8}}\mathbb{E}^Q\Big[f(\tilde{A}_t)g(e^{W_t})e^{-\frac{1}{2}W_t}\Big],
\end{aligned}
$$

where $\tilde{A}_t = \int_0^t e^{W_s}\,ds$. The joint density of $(\tilde{A}_t, W_t)$ is known and can be found by substituting $\beta = 1/2$ and $x = 0$ in the following formula, see [15], page 175

$$
P_x\Big(\int_0^t e^{2\beta W_s}\,ds \in dy,\, W_t \in dz\Big) = \frac{|\beta|}{2y}e^{\frac{-(e^{2\beta x}+e^{2\beta z})}{2\beta^2 y}}\,\mathbf{i}_{\beta^2 t/2}\Big(\frac{e^{\beta(x+z)}}{\beta^2 y}\Big)dydz,
$$

with $\mathbf{i}_y(z) = 2\theta_z(2y)$ and θ_z denoting the Hartman-Watson density, which is given under an integral representation by (6.4.48). Thus, we obtain

$$
P(\tilde{A}_t \in dy, W_t \in dz) = \frac{1}{4y}e^{-2(1+e^z)/y}\,\mathbf{i}_{t/8}\Big(\frac{4e^{z/2}}{y}\Big)\,dydz = \phi_t(y,z)\,dydz.
$$

Now we are able to transform the previous expectation into an iterated integral and then use the substitution $w = \ln y$ to obtain

$$
\begin{aligned}
\mathbb{E}^P[f(X_t)g(Y_t)] &= e^{-\frac{t}{8}}\int_0^\infty\int_0^\infty f(u)g(e^w)e^{-\frac{1}{2}w}\phi_t(u,w)\,dudw \\
&= e^{-\frac{t}{8}}\int_0^\infty\int_0^\infty f(u)g(y)y^{-3/2}\phi_t(u,\ln y)\,dudy \\
&= e^{-\frac{t}{8}}\int_0^\infty\int_0^\infty f(x)g(y)y^{-3/2}\phi_t(x,\ln y)\,dxdy. \qquad (6.4.50)
\end{aligned}
$$

Comparing relations (6.4.49) and (6.4.50) yields

$$
\begin{aligned}
p_t(x,y) &= e^{-\frac{t}{8}}y^{-3/2}\phi_t(x,\ln y) \\
&= e^{-\frac{t}{8}}y^{-3/2}\frac{1}{4x}e^{-2(1+y)/x}\,\mathbf{i}_{t/8}\Big(\frac{4\sqrt{y}}{x}\Big) \\
&= \frac{1}{2}x^{-1}y^{-3/2}e^{-\frac{t}{8}-2(1+y)/x}\,\theta_{\frac{4\sqrt{y}}{x}}\Big(\frac{t}{4}\Big),
\end{aligned}
$$

where $\theta_r(t)$ is the Hartman-Watson density given by (6.4.48). The joint probability density of the hyperbolic diffusion (X_t, Y_t) provides the heat kernel for $\Delta_P(x,y) = \frac{1}{2}y^2(\partial_x^2 + \partial_y^2)$ from $(0,1)$.

Remark 6.4.1 It is worth noting that the Poincaré's operator $\Delta_P(x, y)$ in hyperbolic coordinates (θ, φ) defined by

$$x = (\sinh\theta\cos\varphi)/(\cosh\theta - \sinh\theta\sin\varphi)$$
$$y = 1/(\cosh\theta - \sinh\theta\sin\varphi),$$

becomes

$$\Delta_P(\theta, \varphi) = \frac{1}{2}\Big(\partial_\theta^2 + \frac{1}{\sinh^2\theta}\partial_\varphi^2 + \coth\theta\,\partial_\theta\Big),$$

see Lao and Orsingher [95]. Thus, the two-dimensional hyperbolic diffusion in hyperbolic coordinates takes the form

$$d\theta_t = dW_1(t) + \frac{1}{2}\coth\theta_t\,dt$$
$$d\varphi_t = \mathrm{csch}\theta_t\,dW_2(t).$$

6.4.3 Financial Interpretation

The two-dimensional hyperbolic diffusion can be used to model the evolution of a stock with stochastic volatility following a "toy" GARCH model.

The "toy" GARCH model If S_t and V_t denote the value of a stock and its volatility, the GARCH model states that the stock evolution follows the following system

$$dS_t = \mu S_t\,dt + \sqrt{V_t}S_t\,dW_t$$
$$dV_t = a(\omega - V_t)dt + bV_t\,dB_t,$$

with W_t and B_t Brownian motions and μ, a, b and ω model parameters, see Calin [23], page 351. Substituting $X_t = \ln S_t$ and applying Ito's lemma we get

$$dX_t = \Big(\mu - \frac{1}{2}V_t\Big)dt + \sqrt{V_t}\,dW_t$$
$$dV_t = a(\omega - V_t)dt + bV_t\,dB_t,$$

Neglecting the drift term and setting $\omega = 0$, $a = -1$ and $b = 2$, we obtain the following "toy" model

$$dX_t = \sqrt{V_t}\,dW_t \tag{6.4.51}$$
$$dV_t = V_t dt + 2V_t\,dB_t, \tag{6.4.52}$$

The reduction to a "toy" GARCH The two-dimensional hyperbolic diffusion (6.4.39)-(6.4.40)

$$dX_t = Y_t\,dW_t$$
$$dY_t = Y_t\,dB_t,$$

can be reduced to the system (6.4.51)-(6.4.52) by substituting $V_t = Y_t^2$. To show this we apply Ito's formula

$$\begin{aligned} dV_t &= 2Y_t\, dY_t + (dY_t)^2 = 2Y_t(Y_t dB_t) + Y_t^2(dB_t)^2 \\ &= 2Y_t^2 dB_t + Y_t^2 dt = 2V_t dB_t + V_t dt, \end{aligned}$$

which is equation (6.4.52).

To conclude this section, a two-dimensional hyperbolic diffusion can model the evolution of a stock with stochastic volatility as follows: the x-component of the diffusion is the stock value and the y-component is the square root of the volatility. For further applications of differential geometry to finance, the reader is reffered to Labordére [92].

In the following sections we shall provide a few examples of diffusions for which the drift expression (6.2.32) can be computed explicitly. Some of the next manifolds can be covered by only one chart, while others by several charts. If a Brownian motion lives on a manifold, then the chart map projects the process in $\mathbb{R}^n$ where we study its stochastic differential equations. We note that the same Brownian motion can be projected by different maps into distinct looking systems of equations.

6.4.4 Hyperbolic Space Form

We consider the Riemannian manifold $(\mathcal{D}^n, g_{ij})$ with $\mathcal{D}^n = \{x \in \mathbb{R}^n; \|x\| \leq 1\}$ an n-dimensional unit disk and

$$g_{ij}(x) = \frac{4a^2\delta_{ij}}{(1-\|x\|^2)^2},$$

with $a > 0$ constant. Since the metric has a diagonal form, its determinant is given by

$$\det g = \frac{2^{2n}a^{2n}}{(1-\|x\|^2)^{2n}}$$

and its inverse takes the form

$$g^{ij}(x) = \frac{(1-\|x\|^2)^2}{4a^2}\delta_{ij}.$$

The drift provided by formula (6.2.32) can be computed explicitly as

$$\begin{aligned} b_k &= \frac{1}{2}\frac{1}{\sqrt{\det g}}\partial_{x_j}\left(\sqrt{\det g}\; g^{kj}\right) \\ &= \frac{n-2}{4a^2}(1-\|x\|^2)x_k, \qquad k = 1, \ldots, n. \end{aligned}$$

Also, the dispersion matrix σ satisfying

$$\sigma(x)\sigma^T(x) = g^{-1}(x) = \frac{(1-\|x\|^2)^2}{4a^2}\mathbb{I}_n$$

is given by

$$\sigma(x) = \frac{1-\|x\|^2}{2a}\mathbb{I}_n,$$

so it is diagonal and rotational invariant. The associated diffusion $X(t) = \big(X_1(t), \dots, X_n(t)\big)$ with drift $b(x)$ and dispersion $\sigma(x)$ becomes

$$dX_k(t) = \frac{n-2}{4a^2}(1-\|X(t)\|^2)X_k(t)\,dt + \frac{1-\|X(t)\|^2}{2a}dW_k(t), \qquad k = 1, \dots, n,$$

with $W_k(t)$ independent one-dimensional Brownian motions. In the vectorial notation this writes as

$$dX(t) = \frac{n-2}{4a^2}(1-\|X(t)\|^2)X(t)\,dt + \frac{1-\|X(t)\|^2}{2a}dW(t).$$

The process $X(t)$ can be considered as a Brownian motion on the Riemannian manifold $(\mathcal{D}^n, g_{ij})$.

Diffusion on Poincaré's disk The distinguished case $n = 2$ corresponds to the Poincaré disk $\mathcal{D}^2 = \{x \in \mathbb{R}^2; \|x\| \le 1\}$. Here the drift vanishes and the diffusion becomes

$$dX_1(t) = \frac{1-|X(t)|^2}{2a}\,dW_1(t) \tag{6.4.53}$$

$$dX_2(t) = \frac{1-|X(t)|^2}{2a}\,dW_2(t), \tag{6.4.54}$$

where $|X(t)|^2 = X_1(t)^2 + X_2(t)^2$ and $W_1(t)$, $W_2(t)$ are independent Brownian motions. This diffusion is rotational invariant. It makes sense to choose the initial point at the origin, $X(0) = 0$.

The law of the Brownian motion X_t on $\mathcal{D}^n$ is induced by the law of the Brownian motion on the upper half-space $\mathcal{H}^n$. To accomplish this, we note that the map $\tau : \mathcal{D}^n \to \mathcal{H}^n$, given by

$$\tau(x) = |x + e_n|^{-2}(x + e_n) - \frac{1}{2}e_n \tag{6.4.55}$$

is an isometry between the Riemannian manifolds $(\mathcal{D}^n, g_{ij})$ and $(\mathcal{H}^n, h_{ij})$, where

$$g_{ij} = \frac{4}{(1-|x|^2)^2}\delta_{ij}, \quad h_{ij}(y) = \frac{1}{y_n^2}\delta_{ij}, \tag{6.4.56}$$

and $e_n = (0, \dots, 0, 1)$. This means

$$h_{ij}(y) = g(\tau_* \partial_{x_i}, \tau_* \partial_{x_j}) = \frac{\partial \tau^k}{\partial x_i} \frac{\partial \tau^r}{\partial x_j} g_{kr}(x). \tag{6.4.57}$$

If let $Y_t = \tau(X_t)$ be the image of X_t thorough the aforementioned isometry, then Y_t is a Brownian motion on the space $(\mathcal{H}^n, h_{ij})$, which has been investigated in Section 6.4.1. Then the law of X_t can be obtained from the law of Y_t via formula

$$p_{X_t}(t; x_0, x) = p_{Y_t}(t; \tau(x_0), \tau(x)) \det J_\tau(x), \tag{6.4.58}$$

where $J_\tau(x) = \left(\frac{\partial \tau^k}{\partial x_j}\right)_{j,k}$ is the Jacobian of the map τ. Its determinant can be computed by applying a determinant in relation (6.4.57) and obtaining

$$\det h(y) = (\det J_\tau(x))^2 \det g(x),$$

or equivalently,

$$\det J_\tau(x) = \left(\frac{\det h(\tau(x))}{\det g(x)}\right)^{1/2} = \frac{(1 - |x|^2)^n}{2^n \tau_n(x)^n},$$

where

$$\tau_n(x) = \frac{x_n + 1}{|x + e_n|^2} - \frac{1}{2} = \frac{1 + x_n}{|x|^2 + 2x_n + 1}.$$

6.4.5 The Sphere $\mathbb{S}^n$

The sphere in $\mathbb{R}^{n+1}$ can be parameterized in several ways. We shall consider the corresponding diffusion equations in each of these cases. While the equations may appear distinct in various local coordinate systems, the fundamental nature of the diffusion remains geometrically invariant.

Stereographic coordinates The unit sphere $\mathbb{S}^n$ can be parameterized by the stereographic projection from the north pole $N = (0, \dots, 0, 1)$ by $\phi_N : \mathbb{S}^n \backslash \{N\} \to \mathbb{R}^n$

$$\phi_N(x_1, \dots, x_n, x_{n+1}) = \frac{1}{1 - x_{n+1}}(x_1, \dots, x_n),$$

with the inverse $\phi_N^{-1} : \mathbb{R}^n \to \mathbb{S}^n \backslash \{N\}$

$$\phi_N^{-1}(u) = \frac{(2u, |u|^2 - 1)}{|u|^2 + 1},$$

with $u = (u_1, \dots, u_n)$ and $|u|^2 = \sum_k u_k^2$. A computation shows that the induced Riemannian metric on the unit sphere $\mathbb{S}^n$ by the stereographic projection is given by

$$g_{ij}(u) = \langle \partial_{u_i} \phi_N^{-1}(u), \partial_{u_j} \phi_N^{-1}(u) \rangle = \frac{4}{(1 + |u|^2)^2} \delta_{ij}.$$

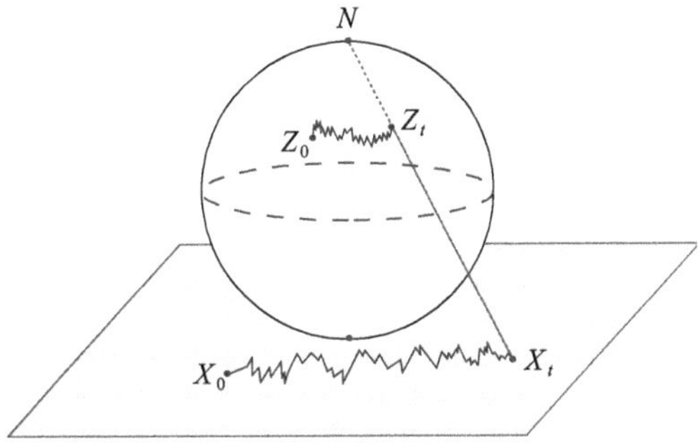

Figure 6.1: *The Brownian motion Z_t on the sphere $\mathbb{S}^n$ is projected via the stereographic projection into the diffusion $X_t = \phi_N(Z_t)$ on $\mathbb{R}^n$.*

Since the determinant and the inverse matrix are given by

$$\det g = \frac{2^{2n}}{(1+|u|^2)^{2n}}, \quad g^{ij}(u) = \frac{1}{4}(1+|u|^2)^2\delta_{ij},$$

then the drift given by formula (6.2.32) takes the form

$$\begin{aligned} b_k(u) &= \frac{1}{2}\frac{1}{\sqrt{\det g}}\partial_{x_j}\Big(\sqrt{\det g}\, g^{kj}\Big) \\ &= \frac{2-n}{4}(1+|u|^2)u_k, \qquad k = 1,\dots,n. \end{aligned}$$

Since $\sigma(u)\sigma^T(u) = g^{-1} = \dfrac{(1+|u|^2)^2}{2^2}\mathbb{I}_n$, we choose the dispersion

$$\sigma(u) = \frac{1+|u|^2}{2}\mathbb{I}_n.$$

Then the associated diffusion $X(t)$ on $\mathbb{S}^n$ in stereographic coordinates satisfies the following stochastic differential equations

$$dX_k(t) = \frac{2-n}{4}(1+|X(t)|^2)X_k(t)dt + \frac{1+|X(t)|^2}{2}dW_k(t), \qquad k = 1,\dots,n$$

with $W_k(t)$ independent Brownian motions. In fact, $X(t)$ represents the stereographic projection of the Brownian motion Z_t on $\mathbb{S}^n$ onto $\mathbb{R}^n$, see Fig. 6.1.

In the case $n = 2$ the drift cancels and we obtain a diffusion $X(t)$ on the unit sphere $\mathbb{S}^2$ given by

$$dX_1(t) = \frac{1+|X(t)|^2}{2}\, dW_1(t) \tag{6.4.59}$$

$$dX_2(t) = \frac{1+|X(t)|^2}{2}\, dW_2(t), \tag{6.4.60}$$

where $|X(t)|^2 = X_1(t)^2 + X_2(t)^2$ and $W_1(t)$, $W_2(t)$ are independent Brownian motions. It is worth noting the similarity with formulas (6.4.53)-(6.4.54) of the diffusion on Poincaré's disk.

Spherical coordinates for $\mathbb{S}^2$ The unit sphere $\mathbb{S}^2$ can be parameterized by spherical coordinates as $\phi : [0, \pi] \times [0, 2\pi) \to \mathbb{R}^3$

$$\phi(\theta, \varphi) = (\cos\varphi \sin\theta, \sin\varphi \sin\theta, \cos\theta).$$

Then the induced Riemannian metric on $\mathbb{S}^2$ has the coefficients

$$\begin{aligned} g_{\theta\theta} &= \langle \partial_\theta \phi, \partial_\theta \phi \rangle = 1 \\ g_{\theta\varphi} &= g_{\varphi\theta} = \langle \partial_\theta \phi, \partial_\varphi \phi \rangle = 0 \\ g_{\varphi\varphi} &= \langle \partial_\varphi \phi, \partial_\varphi \phi \rangle = \sin^2\theta. \end{aligned}$$

The dispersion satisfying $\sigma\sigma^T = g^{-1}$ is given by

$$\sigma = \begin{pmatrix} 1 & 0 \\ 0 & \csc\theta \end{pmatrix}.$$

On the other side, the drift vector $b = (b_\theta, b_\varphi)$ given by formula (6.2.32) has the following components

$$\begin{aligned} b_\theta &= \frac{1}{2}\frac{1}{\sin\theta}\Big\{\partial_\theta(\sin\theta\, g^{\theta\theta} + \partial_\varphi(\sin\theta\, g^{\varphi\theta})\Big\} = \frac{1}{2}\cot\theta \\ b_\varphi &= \frac{1}{2}\frac{1}{\sin\theta}\Big\{\partial_\theta(\sin\theta\, g^{\theta\varphi} + \partial_\varphi(\sin\theta\, g^{\varphi\varphi})\Big\} = 0. \end{aligned}$$

It follows that the Brownian motion on the unit sphere $\mathbb{S}^2$ in polar coordinates is associated to the stochastic process (θ_t, φ_t) that satisfies the system

$$\begin{aligned} d\theta_t &= dW_t^1 + \frac{1}{2}\cot\theta_t\, dt \\ d\varphi_t &= \csc\theta_t dW_t^2, \end{aligned}$$

with W_t^1, W_t^2 independent Brownian motions. We note that in spherical coordinates the diffusion has a nonzero drift, $b = (\frac{1}{2}\cot\theta_t, 0)^T$, while in stereographic coordinates the drift is zero, see the system (6.4.59)-(6.4.60). Its associated generator is given by the spherical Laplacian on $\mathbb{S}^2$

$$\begin{aligned} \mathcal{A} &= \frac{1}{2}(\partial_\theta^2 + (\csc\theta)^2\,\partial_\varphi^2) + \frac{1}{2}\cot\theta\,\partial_\theta \\ &= \frac{1}{\sin\theta}\partial_\theta(\sin\theta\,\partial_\theta) + \frac{1}{\sin^2\theta}\partial_\varphi^2. \end{aligned}$$

Orthogonal projection coordinates Consider the hypersphere of radius r and dimension n centered at the origin

$$\mathbb{S}^n(r) = \{x \in \mathbb{R}^{n+1}; x_1^2 + \cdots + x_{n+1}^2 = r^2\}.$$

Its upper-half, $\mathbb{S}^n_+(r) = \{x \in \mathbb{S}^n(r); x_{n+1} > 0\}$, is parameterized by the map $\phi : D^n(r) \to \mathbb{R}^{n+1}$

$$\phi(u_1, \dots, u_n) = (u_1, \dots, u_n, \sqrt{r^2 - |u|^2}),$$

where $|u|^2 = \sum_{k=1}^n u_k^2$ and $D^n(r) = \{u \in \mathbb{R}^n; |u| < r\}$. A computation shows that the Riemannian metric induced on $\mathbb{S}^n_+(r)$ by ϕ is given by

$$g_{ij}(u) = \langle \partial_{u_i}\phi, \partial_{u_j}\phi \rangle = \delta_{ij} + \frac{u_i u_j}{r^2 - |u|^2},$$

where $\{\partial_{u_i}\phi\}$ are the coordinate vector fields on the sphere and $\langle\, , \rangle$ denotes the Euclidean scalar product on $\mathbb{R}^{n+1}$. In matrix notation this can be written as $g = \mathbb{I}_n + \dfrac{1}{r^2 - |u|^2} uu^T$. To find the inverse matrix g^{-1}, we look for an inverse of the type

$$g^{-1} = \mathbb{I}_n - a^2 uu^T$$

where a is a constant to be determined from the multiplication relation $gg^{-1} = \mathbb{I}_n$. Using $uu^T uu^T = u(u^T u)u^T = |u|^2 uu^T$, we show that

$$gg^{-1} = \mathbb{I}_n + \frac{1 - a^2 r^2}{r^2 - |u|^2} uu^T.$$

Therefore, choosing $a = 1/r$, or $a = -1/r$, we obtain the inverse $g^{-1} = \mathbb{I}_n - \dfrac{1}{r^2} uu^T$, or in its equivalent components version

$$g^{ij}(u) = \delta_{ij} - \frac{1}{r^2} u_i u_j. \tag{6.4.61}$$

It is easier to compute the determinant of g^{ij} rather than the determinant of g_{ij}. For this we need the following result.

Lemma 6.4.2 *Let $u \in \mathbb{R}^n$ and consider the matrix $M = uu^T - k\mathbb{I}_n$ with $k \in \mathbb{R}$. Then* $\det M = (-k)^{n-1}(|u|^2 - k)$.

Proof: A direct computation shows that the matrix M has n linear independent eigenvectors given by

$$\begin{aligned}
\xi^{(1)} &= \Big(-\frac{x_n}{x_1}, 0, \dots, 0, 1\Big)^T \\
\xi^{(2)} &= \Big(-\frac{x_{n-1}}{x_1}, 0, \dots 1, 0\Big)^T \\
&\dots\dots\dots\dots\dots\dots\dots\dots \\
\xi^{(n-1)} &= \Big(-\frac{x_2}{x_1}, 1, \dots 0, 0\Big)^T \\
\xi^{(n)} &= \Big(\frac{x_1}{x_n}, \frac{x_2}{x_n}, \dots \frac{x_{n-1}}{x_n}, 1\Big)^T,
\end{aligned}$$

with corresponding eigenvalues

$$\lambda_1 = \cdots = \lambda_{n-1} = -k, \qquad \lambda_n = |u|^2 - k.$$

Then the determinant is obtained by taking the product of the aforementioned eigenvalues. ■

Writing $g^{ij} = -\frac{1}{r^2}(u_i u_j - r^2\delta_{ij})$ and applying Lemma 6.4.2 yields

$$\det g^{ij} = \frac{(-1)^n}{r^{2n}} \det(u_i u_j - r^2\delta_{ij}) = 1 - \frac{|u|^2}{r^2}.$$

Then inverting

$$\det g_{ij} = \frac{1}{\det g^{ij}} = \frac{r^2}{r^2 - |u|^2}.$$

Using formula (6.2.32) the drift becomes

$$b_k(u) = \frac{1}{2}\sqrt{r^2 - |u|^2}\Bigg(\underbrace{\partial_{x_k}\frac{1}{\sqrt{r^2-|u|^2}}}_{A} - \frac{1}{r^2}\underbrace{\sum_{p=1}^n \partial_{x_p}\frac{u_p u_k}{\sqrt{r^2-|u|^2}}}_{B}\Bigg). \tag{6.4.62}$$

We compute the terms A and B separately as follows

$$\begin{aligned}
A &= \frac{u_k}{(r^2 - |u|^2)^{3/2}} \\
B &= u_k \sum_{p\neq k} \partial_{x_p}\bigg(\frac{u_p}{\sqrt{r^2-|u|^2}}\bigg) + \partial_{x_k}\bigg(\frac{u_k^2}{\sqrt{r^2-|u|^2}}\bigg) \\
&= \frac{u_k}{(r^2-|u|^2)^{3/2}}\Big(n(r^2 - |u|^2) + r^2\Big).
\end{aligned}$$

Substituting back into (6.4.62) yields

$$b_k(u) = \frac{u_k}{2(r^2-|u|^2)}\left(1 - \frac{n(r^2-|u|^2)+r^2}{r^2}\right) = -\frac{n}{2r^2}u_k,$$

which implies $b(u) = -\frac{n}{2r^2}u$, namely the drift is proportional to the position vector u. Then the generator of the diffusion takes the form

$$\begin{aligned}\mathcal{A} &= \frac{1}{2}\Delta_g = \frac{1}{2}(\sigma(u)\sigma(u)^T)_{ij}\partial_{u_i}\partial_{u_j} + b_k(u)\partial_{u_k}\\ &= \frac{1}{2}g^{ij}(u)\partial_{u_i}\partial_{u_j} + b_k(u)\partial_{u_k}\\ &= \frac{1}{2}\Big(\Delta_n - \frac{1}{r^2}\sum_{i,j=1}^{n} u_iu_j\partial_{u_i}\partial_{u_j}\Big) - \frac{n}{2r^2}\sum_{k=1}^{n} u_k\partial_{u_k},\end{aligned}$$

where $\Delta_n = \sum_{k=1}^n \partial^2_{u_k}$. To construct the associated diffusion on $\mathbb{S}^n_+(r)$ we need to solve the matrix equation

$$\sigma(u)\sigma(u)^T = \mathbb{I}_n - \frac{1}{r^2}uu^T. \tag{6.4.63}$$

We look for a solution of the form $\sigma(u) = \mathbb{I}_n + cuu^T$ where c is a function of u subject to be determined from computation. Using that $u^Tu = |u|^2$, we have

$$\begin{aligned}\sigma(u)\sigma^T(u) &= (\mathbb{I}_n + cuu^T)(\mathbb{I}_n + cuu^T)\\ &= \mathbb{I}_n + 2cuu^T + c^2u(u^Tu)u^T\\ &= \mathbb{I}_n + (2c + c^2|u|^2)uu^T.\end{aligned}$$

Comparing to (6.4.63) we obtain $2c+c^2|u|^2 = -\frac{1}{r^2}$, which implies $c = \frac{\sqrt{r^2-|u|^2}-r}{r|u|^2}$. Hence, the dispersion matrix becomes

$$\sigma(u) = \mathbb{I}_n - \frac{r-\sqrt{r^2-|u|^2}}{r|u|^2}uu^T.$$

Then the associated diffusion X_t on $\mathbb{S}^n_+(r)$ satisfies

$$dX_t = -\frac{n}{2r^2}X_t dt + \Big(\mathbb{I}_n - \frac{r-\sqrt{r^2-|X_t|^2}}{r|X_t|^2}X_tX_t^T\Big)dW_t, \tag{6.4.64}$$

with W_t Brownian motion on $\mathbb{R}^n$. We assume the initial condition is $X_0 = 0 \in \mathbb{R}^n$, i.e., the diffusion starts from the center of the disk $D^n(r)$ and gets killed when it reaches the boundary $\partial D^n(r)$. It is worth noting that the drift term of (6.4.64) depends on the sphere curvature. As a limit situation, when

$r \to \infty$, the upper half-sphere becomes a plane. In this case $c = 0$, $\sigma(u) = \mathbb{I}_n$ and then the equation (6.4.64) becomes $dX_t = dW_t$, i.e., $X_t = W_t$, becomes an n-dimensional Brownian motion on $\mathbb{R}^n$.

The diffusion X_t lives in the n-dimensional disk $D^n(r)$. If we consider the $(n+1)$th component defined by $Y_t = \sqrt{r^2 - |X_t|^2}$, then the $(n+1)$-dimensional process (X_t, Y_t) is a Brownian motion on the sphere $\mathbb{S}^n(r)$.

6.4.6 Conformally Flat Manifolds

A Riemannian manifold (M, g) of dimension n is called *conformally flat* if each point has a neighborhood $\mathcal{U}$ and there is a system of coordinates $u = (u_1, \dots, u_n)$ on $\mathcal{U}$ such that

$$g_{ij}(u) = e^{\rho(u)}\delta_{ij}, \qquad \forall u \in \mathcal{U},$$

where $\rho : \mathcal{U} \to \mathbb{R}$ is a smooth function. We shall consider in the following $\Delta_n = \frac{1}{2}\sum_{j=1}^n \partial_{u_j}^2$ and denote the gradient with respect to the metric g by ∇_g.

Proposition 6.4.3 (*i*) *The Laplace-Beltrami operator on the manifold* $(\mathcal{U}, g)$ *is given by*

$$\Delta_g = e^{-\rho(u)}\Delta_n + \frac{1}{2}\Big(\frac{n}{2} - 1\Big)\nabla_g \rho.$$

(*ii*) *The associated diffusion is given by*

$$dX_t = e^{-\frac{\rho(X_t)}{2}}dW_t + \frac{1}{2}\Big(\frac{n}{2} - 1\Big)\nabla_g \rho\, dt.$$

Proof: Using formula (6.2.24) we have

$$\begin{aligned} \Delta_g &= \frac{1}{2}\sum_{i,j} g^{ij}(u)\partial_{u_i}\partial_{u_j} - \frac{1}{2}\sum_{i,j,k} g^{ij}\Gamma_{ij}^k \partial_{u_k} \\ &= e^{-\rho(u)}\Delta_n + \sum_k b_k(u)\partial_{u_k}. \end{aligned} \tag{6.4.65}$$

In the following we shall compute the drift vector using formula (6.2.32). Since $\det g = e^{n\rho}$, $g^{ij} = e^{-\rho}\delta_{ij}$, we have $\sqrt{\det g}g^{kj} = e^{(\frac{n}{2}-1)\rho}\delta_{kj}$, we obtain

$$\begin{aligned} b_k &= \frac{1}{2}\frac{1}{\sqrt{\det g}}\sum_j \partial_{u_j}\big(\sqrt{\det g}g^{kj}\big) \\ &= \frac{1}{2}e^{-\frac{n}{2}\rho}\partial_{u_k}e^{(\frac{n}{2}-1)\rho} = \frac{1}{2}\Big(\frac{n}{2} - 1\Big)e^{-\rho}\partial_{u_k}\rho. \end{aligned}$$

Since the kth component of the g-gradient of ρ is given by

$$(\nabla_g \rho)^k = \sum_i g^{ik}\partial_{u_i}\rho = e^{-\rho}\partial_{u_k}\rho,$$

we obtain $b_k = \frac{1}{2}\Big(\frac{n}{2} - 1\Big)(\nabla_g \rho)^k$, or

$$b = \sum_k b_k \partial_{u_k} = \frac{1}{2}\Big(\frac{n}{2} - 1\Big)\nabla_g \rho. \tag{6.4.66}$$

Substituting in (6.4.65) we obtain the desired relation of part (i). To obtain part (ii), we write the diffusion as $dX_t = \sigma(X_t)dW_t + b(X_t)dt$ with the dispersion $\sigma_{ij} = e^{-\rho/2}\delta_{ij}$ and drift given by (6.4.66). ∎

Driftless diffusions There are two cases when the diffusion X_t becomes driftless in the system of coordinates $(u_1, \dots, u_n)$.

1. The case when $\rho =$constant, which implies $dX_t = e^{\rho/2}dW_t$ and then $X_t = W_{e^{-\rho}t}$, i.e. X_t is a n-dimensional Brownian motion under a time change.

2. The case when $n = 2$, which corresponds to Brownian motions on surfaces. This will be treated in the next section as a distinguished case of a Riemannian manifold which is conformally flat.

6.4.7 Brownian Motions on Surfaces

Let (M, g) be a two-dimensional surface. It is known that on any surface, locally, about each point there is a system of *isothermal coordinates*, (u, v), such that the metric becomes in these coordinates conformal to the Euclidean metric, namely $ds^2 = e^{\rho(u,v)}(du^2 + dv^2)$. Then, using that $n = 2$, Proposition 6.4.3 can be applied to obtain the Laplace-Beltrami operator $\Delta_g = e^{-\rho(u)}\Delta_n$ and the two-dimensional driftless diffusion $dX_t = e^{-\frac{\rho(X_t)}{2}}dW_t$. Furthermore, in a system of isothermal coordinates, the Gaussian curvature K of the surface is related to the function ρ by the relation

$$K = -\frac{1}{2}e^{-\rho}(\partial_u^2 + \partial_v^2)\rho = -\Delta_g \rho. \tag{6.4.67}$$

The diffusion

$$dX_t = e^{-\frac{\rho(X_t)}{2}}dW_t$$

represents a Brownian motion on the surface (M, g) in isothermal coordinates, as being associated with the Laplace-Beltrami operator Δ_g. Here, $W_t = (W_t^1, W_t^2)$ denotes a two-dimensional Brownian motion on $\mathbb{R}^2$.

We shall show that the diffusion X_t can be obtained from a two-dimensional Brownian motion by a random change of time. To this end, in Theorem 1.1.14 we set

$$u(t, \omega) = 0, \quad v(t, \omega) = 1, \quad b = 0, \quad \sigma = e^{-\rho/2}.$$

Thus, the Ito process $dY_t = dW_t$ becomes a two-dimensional Brownian motion, $Y_t = W_t$. Then the condition $u(t, \omega) = c(t, \omega)b(Y_t)$ is obviously satisfied, while the relation $vv^T(t,\omega) = c(t,\omega)\,\sigma\sigma^T(Y_t)$ is satisfied for $c(t,\omega) = e^{\rho(W_t)}$. The time change $\beta_t(\omega) = \int_0^t e^{\rho(W_s)}\,ds$ is increasing and depends on the Gaussian curvature K by relation (6.4.67). Let $\alpha_t(\omega)$ denote its inverse. Then

$$X_t = W_{\alpha_t}, \qquad X_{\beta_t} = W_t.$$

The concluding remark is that Brownian motions on Riemannian surfaces can be regarded as two-dimmensional Brownian motions under a time change, which depends on the Gaussian curvature of the surface.

Example 6.4.4 We consider the particular case of the upper half-plane, $\mathcal{H}^2 = \{(u,v); v > 0\}$, with the metric $ds^2 = \frac{1}{v^2}(du^2 + dv^2)$. Then we have $\rho = -\ln(v^2)$, and the diffusion $dX_t = e^{-\frac{\rho(X_t)}{2}}dW_t$ becomes the system

$$dX_t^1 = X_t^2\,dW_t^1, \quad dX_t^2 = X_t^2\,dW_t^2.$$

The time change in this case is $\beta_t = \int_0^t e^{-\ln(W_s^2)^2}\,ds = \int_0^t \frac{1}{B_s^2}\,ds$, where we denoted $B_t = W_t^2$.

Remark 6.4.5 If the surface M is compact (with empty boundary) and allows a global system of isothermal coordinates, then integrating the equation $K = -\Delta_g\rho$ over M and using Gauss-Bonett's theorem we obtain that the Euler characteristic vanishes, $\chi(M) = 0$, namely the surface M has genus 1.

6.4.8 Diffusions and Brownian Motions

A diffusion process on a Riemannian manifold (M, g) can be seen as a generalization of a Brownian motion, but not all diffusions are necessarily Brownian motions. However, the deviation from a Brownian motion can be characterized as a drift, as we will explain shortly.

Let's consider a diffusion process denoted by X_t on the manifold M. In a local coordinate system, it can be expressed as:

$$dX_t = \sigma(X_t)dW_t + h(X_t)dt$$

where $\sigma\sigma^T = g$ and h is a vector field on the manifold M. Now, let's define a vector field $V = h - b$, where b is given by the expression (6.2.32). By introducing this vector field, we can rewrite the diffusion process as:

$$dX_t = d\widehat{X}_t + V(X_t)dt,$$

where $\widehat{X}_t$ represents the Brownian motion on the manifold (M, g). Therefore, we can recover the Brownian motion from the diffusion process X_t using the following formula:

$$\widehat{X}_t = X_t - \int_0^t V(X_s)ds.$$

It is important to mention that the transition densities of X_t and $\widehat{X}_t$ are connected through a formula established by Molchanov [111]. To explain this relationship, let's consider two points, x and y, in close proximity on the closed manifold M. Suppose there exists a unique geodesic, denoted by γ, that connects these points. Specifically, $\gamma(0) = x$ and $\gamma(\rho) = y$, where $\rho = d_g(x, y)$ represents the geodesic distance. In the limit as t approaches 0, we have the following expression:

$$p_{\widehat{X}_t}(t, x, y) \sim p_{X_t}(t, x, y) e^{-\int_0^\rho g(V_{\gamma(s)}, \dot{\gamma}(s))ds}.$$

It is worth noting that the integral $\int_0^\rho g(V_{\gamma(s)}, \dot{\gamma}(s))ds$ denotes the work done by the vector field V along the minimizing geodesic γ between x and y.

6.5 Stokes' Formula for Brownian Motions

In this section we shall develop a Stokes' type formula for Brownian motions, which relates the rate of change of the likelihood that a Brownian motion belongs to a region at time t with its flux across the region's border. This will be obtained as a consequence of Kolmogorov's equation, Fourier's law and the Divergence Theorem.

6.5.1 Escape Probability for Brownian Motions

We start by considering a planar Brownian motion $B_t = (B_t^1, B_t^2)$ starting at $B_0 = x^0$. Let Ω be an open bounded domain of class C^1 that contains x^0. We are concerned with the following question:

What is the probability that the Brownian motion B_t is outside of Ω at time t?

Denoting the probability density of B_t by $p_t(x)$, we start by evaluating the chance that the Brownian motion belongs to the region Ω at time t by

$$P(B_t \in \Omega) = \int_\Omega p_t(x)\, dx.$$

The change in probability between instances t and $t + \Delta t$ can be evaluated as the difference

$$P(B_{t+\Delta t} \in \Omega) - P(B_t \in \Omega) = \int_\Omega \big(p_{t+\Delta t}(x) - p_t(x)\big)\, dx.$$

Dividing by Δt, taking $\Delta t \to 0$, and using that $p_t(x)$ satisfies the heat equation, we obtain the following expression for the rate of change

$$\frac{d}{dt} P(B_t \in \Omega) = \int_\Omega \partial_t p_t(x)\, dx = \frac{1}{2} \int_\Omega \Delta_x p_t(x)\, dx, \qquad (6.5.68)$$

where here $\Delta_x = \partial^2_{x_1} + \partial^2_{x_2}$. Using Green's formula

$$\int_\Omega \Delta_x u(x)\, dx = \int_{\partial\Omega} \frac{\partial u}{\partial \nu}\, ds,$$

with $u \in C^1(\Omega)$, ν being the outward unit normal to the boundary curve $\partial\Omega$ and $\frac{\partial u}{\partial \nu} = \langle \nabla u, \nu \rangle$ being the normal derivative, relation (6.5.68) becomes

$$\frac{d}{dt} P(B_t \in \Omega) = \frac{1}{2} \int_{\partial\Omega} \frac{\partial p_t}{\partial \nu}\, ds. \tag{6.5.69}$$

This formula has the following physical significance. The left side is the rate of change of probability that the Brownian motion belongs to the region Ω. On the right side, the normal derivative $\frac{\partial p_t}{\partial \nu}$ represents the *heat flux*, which is roughly equal to the number of Brownian particles that cross the border $\partial\Omega$ per unit of time at location x. More precisely, the flux $Q(x, t)$ is given by the *Fourier law* as a normal derivative

$$Q(x, t) = -\frac{1}{2} \frac{\partial p_t(x)}{\partial \nu},$$

where the minus sign is consistent with the fact that the heat moves from a higher density to a lower density region. The coefficient $1/2$ has been chosen for simplicity reasons and can be achieved by a change in the units of measure. The *total flux* at time t across the border $\partial\Omega$ is obtained by integrating the flux over the entire border

$$F_t = \int_{\partial\Omega} Q(x, t)\, ds = -\frac{1}{2} \int_{\partial\Omega} \frac{\partial p_t(x)}{\partial \nu}\, ds,$$

where ds denotes the length measure on $\partial\Omega$. F_t represents roughly the number of Brownian particles crossing the entire border $\partial\Omega$ at time t. Therefore, formula (6.5.69) can be written equivalently as

$$\frac{d}{dt} P(B_t \in \Omega) = -F_t. \tag{6.5.70}$$

This provides the interpretation of the total flux across the border $\partial\Omega$ as the negative rate of change of the probability inside the domain Ω. Applying the Fundamental Theorem of Calculus implies

$$P(B_t \in \Omega) = 1 - \int_0^t F_u\, du,$$

where we used $P(B_0 \in \Omega) = 1$, since B_t starts from a point situated inside Ω. This is equivalent to

$$\int_0^t F_u\, du = P(B_t \notin \Omega).$$

The left side of this formula is the *cumulative total flux*, which roughly means the number of Brownian particles that cross the border $\partial\Omega$ during the time interval $[0, t]$. The right side is the chance that the Brownian motion is outside of the region Ω at time t. This represents a Stokes' type relation since it relates the probability of being into a domain with the dynamics of the Brownian motion through the border of that domain.

Remark 6.5.1 To provide a clearer interpretation of the previous equation as a relation of Stokes' type, we shall revisit Stokes' formula $\int_\Omega d\omega = \int_{\partial\Omega} \omega$. This formula relates the integral of the exterior derivative of a one-form ω over a domain Ω to the integral of the one-form itself over the boundary $\partial\Omega$. In the context of this discussion, we recall that a one-form can be interpreted as representing a force field.

Hence, the left-hand side integral in the equation represents the net change in energy within the domain Ω. On the other hand, the right-hand side integral represents the work done across the boundary $\partial\Omega$.

Consequently, in essence, Stokes' formula asserts that the net change in the internal energy of the domain Ω is equal to the work done along its boundary. In the context of our case, the concept of energy is replaced by probability, while the energy density is represented by probability density. Similarly, the notion of work is substituted by the concept of flux.

Definition 6.5.2 *Let B_t be a Brownian motion that starts at $B_0 \in \Omega$. The escape probability of the Brownian motion B_t from the domain Ω is the chance that B_t is outside of Ω at time t, namely*

$$\epsilon_t(\Omega) = P(B_t \notin \Omega).$$

We conclude this section with the following result, which shows how the flux across the boundary $\partial\Omega$ controls the probability inside the domain Ω.

Proposition 6.5.3 *The escape probability of a Brownian motion from Ω is given by the cumulative flux through the boundary $\partial\Omega$*

$$\epsilon_t(\Omega) = \int_0^t F_u \, du.$$

It is worth noting that this can be written also as the integral of the heat flux over the cylinder $\partial\Omega \times [0, t]$ as

$$\epsilon_t(\Omega) = \iint_{\partial\Omega \times [0,t]} Q(x, u) \, du ds.$$

Example 6.5.4 In the following we shall compute the flux of a planar Brownian motion for the case when Ω is the disk of radius r

$$\Omega = \{x \in \mathbb{R}^2; \|x\| < r\}$$

and the initial point for the Brownin motion is the center of the circle, $B_0 = (0,0)$. Using that $p_t(x_1, x_2) = \frac{1}{2\pi t} e^{-\frac{x_1^2+x_2^2}{2t}}$, then the gradient of the density is given by $\nabla p_t(x) = -\dfrac{p_t(x)}{t}(x_1, x_2)$. Using that the outward unit normal to the circle $\partial\Omega$ is $\nu(x) = x/r$, then the normal derivative of p_t along $\partial\Omega$ is given by

$$\frac{\partial p_t}{\partial \nu}_{|\partial\Omega} = -\frac{p_t(x)}{rt}\langle (x_1, x_2), (x_1, x_2)\rangle_{|\partial\Omega} = -\frac{r}{t} p_t(x)_{|\partial\Omega}.$$

Then the total flux becomes

$$F_t = -\frac{1}{2}\int_{\partial\Omega} \frac{\partial p_t}{\partial \nu}\, ds = \frac{r}{2t}\int_{\partial\Omega} \frac{1}{2\pi t} e^{-\frac{r^2}{2t}}\, ds = \frac{r^2}{2t^2} e^{-\frac{r^2}{2t}}.$$

Then the rate of change of probability given by (6.5.70) becomes

$$\frac{d}{dt} P(B_t \in \Omega) = -F_t = -\frac{r^2}{2t^2} e^{-\frac{r^2}{2t}}. \tag{6.5.71}$$

The negative sign reflects that the probability $P(B_t \in \Omega)$ decreases over time.

Using Proposition 6.5.3 we obtain the following closed form expression for the escape probability from the ball Ω

$$\epsilon_t(\Omega) = P(B_t \notin \Omega) = \int_0^t \frac{r^2}{2u^2} e^{-\frac{r^2}{2u}}\, du = e^{-\frac{r^2}{2t}}.$$

We also note that

$$\lim_{t\to\infty} \epsilon_t(\Omega) = \lim_{t\to\infty} e^{-\frac{r^2}{2t}} = 1,$$

which means that B_t escapes from Ω almost surely as t gets large.

It is worth noting that the total flux F_t has a global maximum reached at $t = r^2/4$. Therefore, the probability $P(B_t \in \Omega)$ has a global minimum rate of change realized also for $t = r^2/4$, which is equal to $-\frac{8}{e^2 r^2}$. Thus, we obtain the following upper bound for the escape probability from the disk Ω

$$P(B_t \notin \Omega) \le \frac{8t}{e^2 r^2}.$$

Remark 6.5.5 (i) We can arrive to the same result by using that the two-dimensional Bessel process $R_t^{(2)}$ has the probability density given by the Wald's distribution

$$f_t(x) = \frac{1}{t} x e^{-\frac{x^2}{2t}}, \qquad x > 0,\ t > 0,$$

as follows

$$\epsilon_t(\Omega) = P(B_t \notin \Omega) = P(R_t^{(2)} > r) = \int_r^\infty \frac{1}{t} x e^{-\frac{x^2}{2t}}\, dx = e^{-\frac{r^2}{2t}}.$$

(*ii*) We can also invert the reasoning: starting from the formula of the escape probability obtained via Proposition 6.5.3 we can retrieve the distribution law of the Bessel process $R_t^{(2)}$ by taking a derivative

$$\begin{aligned} f_{R_t^{(2)}}(r) &= -\frac{d}{dr} P(R_t^{(2)} > r) = -\frac{d}{dr}\epsilon_t(\Omega) \\ &= -\frac{d}{dr} e^{-\frac{r^2}{2t}} = \frac{1}{t} r e^{-\frac{r^2}{2t}}. \end{aligned}$$

This idea will be used again later in Section 6.6 to find the law of a Bessel process on a Riemannian manifold.

If the Brownian motion does not start from the center of the disk, then the computation of the escape probability will involve some special functions, as the next example shows.

Example 6.5.6 We consider now a two-dimensional Brownian motion, $X_t = x^0 + B_t$, starting from $x^0 = (x_1^0, x_2^0) \in \Omega = \{x \in \mathbb{R}^2; \|x\| < r\}$, with $x^0 \neq (0,0)$. We let $r_0 = \|x^0\|$ and note that $r_0 < r$. The transition probability from x^0 to x within time t is given by the gaussian

$$p_t(x; x^0) = \frac{1}{2\pi t} e^{-\frac{1}{2t}\left((x_1 - x_1^0)^2 + (x_2 - x_2^0)^2\right)}.$$

The gradient of p_t is expressed as

$$\nabla p_t = (\partial_{x_1} p_t, \partial_{x_2} p_t) = -\frac{1}{t} p_t(x; x^0)\,(x - x^0).$$

Taking the scalar product between the gradient of p_t and the outward unit normal to the circle, $\nu(x) = \dfrac{x}{r}$, we obtain

$$\langle \nabla p_t, \nu \rangle = -\frac{1}{rt} p_t(x; x^0)\big(\|x\|^2 - \langle x, x^0\rangle\big).$$

Denoting $\theta = arg(x, x^0)$ and using $ds = r d\theta$, the total flux can be computed as

$$\begin{aligned} F_t &= -\frac{1}{2}\int_{\partial\Omega} \langle \nabla p_t, \nu\rangle\, ds = \frac{1}{2rt}\frac{1}{2\pi t}\int_0^{2\pi} e^{-\frac{1}{2t}(r^2 + r_0^2 - 2rr_0\cos\theta)}(r^2 - rr_0\cos\theta) r d\theta \\ &= \frac{r}{2t}\frac{1}{2\pi t} e^{-\frac{r^2 + r_0^2}{2t}} \int_0^{2\pi} e^{\frac{rr_0}{t}\cos\theta}(r - r_0\cos\theta)\, d\theta \\ &= \frac{r}{2t^2} e^{-\frac{r^2 + r_0^2}{2t}} \Big\{ r I_0\Big(\frac{rr_0}{t}\Big) - r_0 I_1\Big(\frac{rr_0}{t}\Big)\Big\}, \end{aligned}$$

where we denoted by $I_0(\cdot)$ and $I_1(\cdot)$ the first two modified Bessel functions of the first type and we used their integral representations

$$\int_0^{2\pi} e^{z\cos\theta}\, d\theta = 2\pi I_0(z) \tag{6.5.72}$$

$$\int_0^{2\pi} e^{z\cos\theta}\cos\theta\, d\theta = 2\pi I_1(z). \tag{6.5.73}$$

The escape probability of X_t from the disk Ω is now computed as a cumulative flux using Proposition 6.5.3. Assuming $r_0 \neq 0$, using the substitution $v = \frac{rr_0}{u}$, we obtain the following integral expression for the escape probability

$$\epsilon_t(\Omega) = \int_0^t F_u\, du = \frac{1}{2}\int_{\frac{rr_0}{t}}^{\infty} e^{-\frac{r^2+r_0^2}{2rr_0}v}\Big(\frac{r}{r_0}I_0(v) - I_1(v)\Big)\, dv. \tag{6.5.74}$$

It is worth noting that since $r_0 < r$ we have $\frac{r}{r_0}I_0(v) > I_0(v) > I_1(v)$ for any $v > 0$, which implies the positivity of $\epsilon_t(\Omega)$.

6.5.2 Entrance Probability of a Brownian Motion

If the process X_t starts outside of domain Ω, the *entrance probability* of X_t in Ω at time t is given by $P(X_t \in \Omega)$. We shall compute explicitly this probability in the case when $X_t = B_t$ is a two-dimensional Brownian motion (starting at the origin) and

$$\Omega = \{x \in \mathbb{R}^2; \|x - c\| \leq r\}$$

is a disk of radius r centered at $c \in \mathbb{R}^2$, with $\|c\| > r$. The entrance probability can be evaluated as

$$P(B_t \in \Omega) = \int_\Omega p_t(x)\, dx = \frac{1}{2\pi t}\int_\Omega e^{-\frac{\|x\|^2}{2t}}\, dx.$$

Instead of computing the integral over Ω directly, we shall reduce it to an integral over a circle. To this end we start to compute the rate of change of the entrance probability as

$$\begin{aligned}\frac{d}{dt}P(B_t \in \Omega) &= \int_\Omega \partial_t p_t(x)\, dx = \frac{1}{2}\int_\Omega (\partial_{x_1}^2 + \partial_{x_2}^2)p_t(x)\, dx \\ &= \frac{1}{2}\int_\Omega div(\nabla p_t(x))\, dx = \frac{1}{2}\int_{\partial\Omega} \frac{\partial p_t}{\partial \nu}(x)\, ds,\end{aligned} \tag{6.5.75}$$

where we used Kolmogorov's equation and the Divergence Theorem. Since the gradient of probability density has the expression $\nabla p_t(x) = -\frac{1}{t}p_t(x)\, x$

(the Brownian motion starts at the origin) and the outward unit normal to $\partial\Omega$ is given by $\nu(x) = (x-c)/r$, then the normal derivative becomes

$$\begin{aligned}\frac{\partial p_t}{\partial \nu}(x)\Big|_{\partial\Omega} &= \langle \nabla p_t(x), \nu(x)\rangle = -\frac{1}{tr} p_t(x)\langle x, x-c\rangle \\ &= -\frac{1}{tr} p_t(x)\langle x-c+c, x-c\rangle = -\frac{r}{t} p_t(x) - \frac{\|c\|}{t} p_t(x)\cos\theta,\end{aligned} \tag{6.5.76}$$

where $\theta = \arg(c, x-c)$ is a central angle on the circle $\partial\Omega$, which parameterizes the vector x with respect to the circle center c. Substituting (6.5.76) into (6.5.75) yields

$$\frac{d}{dt} P(B_t \in \Omega) = -\frac{r}{2t} A - \frac{\|c\|}{2t} B, \tag{6.5.77}$$

where

$$A = \int_{\partial\Omega} p_t(x)\, ds, \qquad B = \int_{\partial\Omega} p_t(x)\cos\theta\, ds$$

are two integrals that can be computed explicitly as follows. The first integral can be computed using polar coordinates

$$\begin{aligned} A &= \frac{1}{2\pi t}\int_{\partial\Omega} e^{-\frac{\|x-c+c\|^2}{2t}} = \frac{1}{2\pi t} e^{-\frac{r^2+\|c\|^2}{2t}} \int_0^{2\pi} e^{-\frac{r\|c\|}{t}\cos\theta}\, r d\theta \\ &= \frac{r}{t} e^{-\frac{r^2+\|c\|^2}{2t}} I_0\Big(\frac{r\|c\|}{t}\Big), \end{aligned} \tag{6.5.78}$$

where we used the integral representation of the modified Bessel function I_0 given by (6.5.72) and the fact that I_0 is even.[2] To compute the second integral we use a similar approach using the integral representation (6.5.73)

$$\begin{aligned} B &= \frac{1}{2\pi t}\int_{\partial\Omega} e^{-\frac{1}{2t}\|x\|^2}\, ds = \frac{1}{2\pi t}\int_{\partial\Omega} e^{-\frac{1}{2t}(r^2+\|c\|^2+2r\|c\|\cos\theta)}\cos\theta\, ds \\ &= \frac{1}{2\pi t} e^{-\frac{r^2+\|c\|^2}{2t}} \int_0^{2\pi} e^{-\frac{r\|c\|}{t}\cos\theta}\cos\theta\, r d\theta = -\frac{r}{t} e^{-\frac{r^2+\|c\|^2}{2t}} I_1\Big(\frac{r\|c\|}{t}\Big), \end{aligned} \tag{6.5.79}$$

where we used that I_1 is an odd function.

Then substituting (6.5.78) and (6.5.79) into (6.5.77) provides the rate of change of the entrance probability

$$\frac{d}{dt} P(B_t \in \Omega) = \frac{r}{2t^2} e^{-\frac{r^2+\|c\|^2}{2t}} \Big[\|c\| I_1\Big(\frac{r\|c\|}{t}\Big) - r I_0\Big(\frac{r\|c\|}{t}\Big)\Big]. \tag{6.5.80}$$

[2]The fact that I_0 is even follows from its series expansion $I_0(z) = \sum_{m\geq 0} \frac{(z/2)^{2m}}{(m!)^2}$.

It is worth noting that the rate is maximum at the time τ, which is the solution of the equation

$$\frac{I_1}{I_0}\Big(\frac{r\|c\|}{\tau}\Big) = \frac{r}{\|c\|}. \tag{6.5.81}$$

The existence and uniqueness of the solution follows from the fact that the function $f(x) = \frac{I_1}{I_0}(x)$ is an increasing sigmoid-shaped function with $f(0) = 0$ and $\lim_{x\to\pm\infty} = \pm 1$. Then the graph of $f(x)$ intersects the horizontal line $y = \frac{r}{\|c\|}$ at exactly one point, which corresponds to the unique solution of (6.5.81).

Using that $B_0 = 0 \notin \Omega$, the entrance probability in Ω at time T can be constructed by integration from formula (6.5.80) as

$$P(B_T \in \Omega) = \int_0^T \frac{d}{dt}P(B_t \in \Omega)\,dt = J - I, \tag{6.5.82}$$

where the integrals I and J can be written and evaluated using a change of variables as

$$I = \int_0^T \frac{r^2}{2t^2} e^{-\frac{r^2+\|c\|^2}{2t}} I_0\Big(\frac{r\|c\|}{t}\Big)\,dt = \frac{r}{2\|c\|}\int_{r\|c\|/T}^{\infty} e^{-\frac{r^2+\|c\|^2}{2r\|c\|}v} I_0(v)\,dv,$$

$$J = \int_0^T \frac{r\|c\|}{2t^2} e^{-\frac{r^2+\|c\|^2}{2t}} I_1\Big(\frac{r\|c\|}{t}\Big)\,dt = \frac{1}{2}\int_{r\|c\|/T}^{\infty} e^{-\frac{r^2+\|c\|^2}{2r\|c\|}v} I_1(v)\,dv.$$

Then relation (6.5.82) becomes

$$P(B_T \in \Omega) = \frac{1}{2}\int_{r\|c\|/T}^{\infty} e^{-\frac{r^2+\|c\|^2}{2r\|c\|}v}\Big(I_1(v) - \frac{r}{\|c\|}I_0(v)\Big)\,dv, \tag{6.5.83}$$

which provides the entrance probability in the disk Ω. We note that equation (6.5.81) implies $\dfrac{I_1}{I_0}(v) > \dfrac{r}{\|c\|}$ for $v > \frac{r\|c\|}{T}$, fact that checks the positivity of the entrance probability.

Now we can prove Lemma 2.19.3, which was quoted and used at page 119.

Lemma 6.5.7 *Let $\Omega = \{x \in \mathbb{R}^2; \|x\| \le r\}$ be a disk of radius r and $x_0 \in \partial\Omega = \{x \in \mathbb{R}^2; \|x\| = r\}$ be a point on the boundary of Ω. Consider a planar Brownian motion B_t starting from the point x_0. The exit probability from the disk is given by*

$$P(B_t \notin \Omega) = 1 - \frac{1}{2}\int_0^{\frac{r^2}{t}} \phi(v)\,dv,$$

with ϕ given by (2.19.55).

Proof: Integrating between 0 and t in formula (6.5.70)

$$\frac{d}{dt}P(B_t \in \Omega) = -F_t,$$

we obtain

$$P(B_t \in \Omega) - P(B_0 \in \Omega) = -\int_0^t F_u \, du.$$

Since $B_0 = x_0 \in \partial\Omega$, then $P(B_0 \in \Omega) = \int_\Omega \delta(x - x_0)\, dx = \frac{1}{2}$. Heuristically, Dirac's function $\delta(x - x_0)$ sitting at the border point x_0 is approximated by bell curves that are situated half inside and half outside Ω. Hence

$$P(B_t \in \Omega) = \frac{1}{2} - \int_0^t F_u \, du. \tag{6.5.84}$$

The integral of the flux can be obtained from formula (6.5.74) by letting $r = r_0$

$$\int_0^t F_u \, du = \frac{1}{2}\int_{\frac{r^2}{t}}^{\infty} e^{-v}\big(I_0(v) - I_1(v)\big)\, dv. \tag{6.5.85}$$

Then the escape probability becomes

$$\begin{aligned} P(B_t \notin \Omega) &= 1 - \frac{1}{2} + \int_0^t F_u \, du = \frac{1}{2}\Big(1 + \int_{\frac{r^2}{t}}^{\infty} e^{-v}\big(I_0(v) - I_1(v)\big)\, dv\Big) \\ &= \frac{1}{2}\Big(1 + \int_{\frac{r^2}{t}}^{\infty} \phi(v)\, dv\Big) = 1 - \frac{1}{2}\int_0^{\frac{r^2}{t}} \phi(v)\, dv. \end{aligned}$$

■

6.5.3 Stokes' Formula for Ito Diffusions

In this section we shall develop similar formulas to the ones provided in Section 6.5.1 for Brownian motions on Riemannian manifolds. To this end, we consider the Ito diffusion in $\mathbb{R}^n$

$$dX_t = b(X_t)dt + \sigma(X_t)dW_t \tag{6.5.86}$$

$$X_0 = x_0, \tag{6.5.87}$$

where $b^T = (b_1, \cdots, b_n)$, $\sigma(x) \in \mathcal{M}_{n,n}(\mathbb{R})$ and $W_t = (W_t^1, \cdots, W_t^n)$ is a Brownian motion on $\mathbb{R}^n$. We shall further assume that the drift b is related to dispersion σ by the relation (6.2.26). This means that X_t can be considered as a Brownian motion on the Riemannian space $(\mathbb{R}^n, g_{ij})$, with $g_{ij} = (\sigma\sigma^T)^{ij}$.

For this type of diffusion, the infinitesimal generator is the Laplace-Beltrami operator on $(\mathbb{R}^n, g_{ij})$, namely

$$\mathcal{A}_x = \frac{1}{2}\mathrm{div}\nabla_g.$$

Let Ω be an open and bounded domain of class C^1 in $\mathbb{R}^n$, such that $x_0 \in \Omega$. We shall denote by $p_t(x) = p(x_0, x; 0, t)$ the transition probability density for the process X_t from x_0 to x within time t, so we have

$$P(X_t \in \Omega) = \int_\Omega p_t(x)\, dx.$$

Taking the derivative and using the Kolmogorov's equation and the Divergence Theorem (Theorem 6.5.8) we obtain

$$\begin{aligned}\frac{d}{dt}P(X_t \in \Omega) &= \int_\Omega \partial_t p_t(x)\, dx = \int_\Omega \mathcal{A}_x p_t(x)\, dx \\ &= \frac{1}{2}\int_\Omega \mathrm{div}(\nabla_g p_t(x))\, dx = \frac{1}{2}\int_{\partial\Omega} g\big(\nabla_g p_t, \nu\big)\, d\mathbf{s},\end{aligned} \tag{6.5.88}$$

where $d\mathbf{s}$ denotes the surface area element. The negative g-gradient, $-\nabla_g p_t$, represents a vector field that points into the direction of minimum variation of the probability density p_t. Defining the flux at the point $x \in \partial\Omega$ and time t as the negative half of the normal component of the g-gradient of density p_t, i.e.,

$$Q(x,t) = -\frac{1}{2}g\big(\nabla_g p_t, \nu\big)$$

and the total flux across the border $\partial\Omega$ by

$$F_t = \int_{\partial\Omega} Q(x,t)\, d\mathbf{s},$$

then substituting in (6.5.88) we obtain the formula for the escape probability from Ω in terms of the cumulative flux as

$$\frac{d}{dt}P(X_t \in \Omega) = -F_t.$$

Then integrating and using that $X_0 = x_0 \in \Omega$ we obtain

$$P(X_t \notin \Omega) = \int_0^t F_u\, du. \tag{6.5.89}$$

This states that the chances that the Brownian motion X_t has left the domain Ω at time t is given by the cummulative total flux through the boundary $\partial\Omega$.

6.5.4 The Divergence Theorem

This section states and proves the Divergence Theorem on Riemannian manifolds. The proof will be derived from Stokes' integral formula and it is included here for the sake of completeness.

We shall work in the following setup: the space $\mathbb{R}^n$ is endowed with a Riemannian metric with coefficients g_{ij}; we consider a compact domain Ω in $\mathbb{R}^n$ with an orientable boundary $\partial\Omega$ of class C^1. This means that for any point $x_0 \in \partial\Omega$ there is a neighborhood $\mathcal{V}_{x_0}$ of x_0 in $\mathbb{R}^n$ and a neighborhood $\mathcal{U}$ of the origin, $0 \in \mathbb{R}^n$, such that there is a homeomorphism $\phi : \mathcal{V}_{x_0} \to \mathcal{U}$ with ϕ, ϕ^{-1} of class C^1 satisfying

$$\phi(\mathcal{V}_{x_0} \cap \Omega) = \mathcal{U} \cap \{y \in \mathbb{R}^n; y_n > 0\}$$

$$\phi(\mathcal{V}_{x_0} \cap \partial\Omega) = \mathcal{U} \cap \{y \in \mathbb{R}^n; y_n = 0\}$$

$$\phi(\mathcal{V}_{x_0} \cap \Omega^c) = \mathcal{U} \cap \{y \in \mathbb{R}^n; y_n < 0\},$$

where $\Omega^c = \mathbb{R}^n \backslash \Omega$. We let $\psi = \phi_n$, where $\phi = (\phi_1, \cdots, \phi_n)$. Then locally about a point $x_0 \in \partial\Omega$ the boundary can be written as $\{x \in \mathcal{V}_{x_0}; \psi(x) = 0\}$. Assuming further that $\nabla\psi(x_0) \neq 0$, we define the unit normal to $\partial\Omega$ at x_0 by $\nu(x_0) = \nabla\psi(x_0)/g(\nabla\psi(x_0), \nabla\psi(x_0))$ and the tangent plane to $\partial\Omega$ at x_0 by

$$T_{x_0}\partial\Omega = \{\xi \in \mathbb{R}^n; g(\xi, \nu(x_0)) = 0\}.$$

We also note that the Riemannian metric g of the ambient space $\mathbb{R}^n$ induces a scalar product on $\partial\Omega$, which is also a Riemannian metric on $\partial\Omega$.

Theorem 6.5.8 *If the aforementioned conditions are satisfied, then*

$$\int_\Omega \text{div} X \, dv = \int_{\partial\Omega} g(X, \nu) \, ds,$$

where dv is the volume measure on $(\mathbb{R}^n, g_{ij})$, ds is the area measure on the hypersurface $\partial\Omega$ and div$X$ *represents the divergence of the vector field X taken with respect to the metric g_{ij}.*

Before providing the proof, we shall review some prerequisites.

The Lie derivative The rate of change of a geometric object (such as a function, a vector field, a differential form, a metric, or in general, a tensor) along the integral curves of a given vector field X is called the *Lie derivative* with respect to X. In this respect, the regular derivative of a real function, $f'(x)$, can be regarded as a Lie derivative with respect to the vector field $X = \frac{d}{dx}$.

We shall define and use in this section only the Lie derivative of p-forms. Let U be an open subset of $\mathbb{R}^n$ and ω a p-form on U. Let X be a vector field on U. Then the *Lie derivative* of ω with respect to X is the p-form denoted by $L_X\omega$ and defined by

$$(L_X\omega)(Y_1,\dots,Y_p) = X\omega(Y_1,\dots,Y_p) - \sum_{i=1}^{p} \omega(Y_1,\dots,[X,Y_i],\dots,Y_p), \quad (6.5.90)$$

where Y_i are vector fields on U and $[X,Y_i] = XY_i - Y_iX$ denotes the Lie bracket of vector fields X and Y_i. For one-forms the previous definition becomes

$$(L_X\omega)(Y) = X\omega(Y) - \omega([X,Y]),$$

while for zero-forms (i.e. functions) the definition takes the simple form $L_X f = X(f)$. We note that the definition of Lie derivative, L_X, does not depend on any metric structure.[3]

We recall that the volume form on the Riemannian manifold $(\mathbb{R}^n, g_{ij})$ is given in local coordinates by $dv = \sqrt{\det g}\, dx_1 \wedge \cdots \wedge dx_n$. This form can be used as a measure of integration; thus, the volume of the compact set Ω is given by $\mathrm{vol}(\Omega) = \int_\Omega dv$.

The next result relates the divergence of a vector field X with the Lie derivative of the volume element dv. Its physical interpretation is that the volume element expands (contracts) along positive (negative) divergence vector fields. Also, a zero divergence vector field corresponds to a conservation of the volume form along the integral curves of the vector field X.

The next result considers the divergence taken with respect to the Levi-Civita connection.

Proposition 6.5.9 *Let X be a vector field on $\mathbb{R}^n$ and dv be the volume element on the Riemannian manifold $(\mathbb{R}^n, g_{ij})$. Then*

$$L_X\, dv = (\mathrm{div} X)\, dv. \quad (6.5.91)$$

Proof: It suffices to show that relation (6.5.91) holds on a basis, i.e.

$$\big(L_X\, dv\big)(\partial_{x_1},\dots,\partial_{x_n}) = (\mathrm{div} X)\, dv(\partial_{x_1},\dots,\partial_{x_n}). \quad (6.5.92)$$

[3]Concurently, the covariant derivative with respect to the Levi-Civita connection, ∇^0_X, does depend on the metric structure of the manifold.

A computation using formula (6.5.90) yields

$$
\begin{aligned}
(L_X\, dv)(\partial_{x_1},\dots,\partial_{x_n}) &= X\, dv(\partial_{x_1},\dots,\partial_{x_n}) - \sum_{i=1}^n dv(\partial_{x_1},\dots,[X,\partial_{x_i}],\dots,\partial_{x_n})\\
&= X(\sqrt{\det g}) - \sum_{i=1}^n dv(\partial_{x_1},\dots,-\sum_k \frac{\partial X^k}{\partial x_i}\partial_{x_k},\dots,\partial_{x_n})\\
&= X(\sqrt{\det g}) + \sum_{i=1}^n \frac{\partial X^i}{\partial x_i} dv(\partial_{x_1},\dots,\partial_{x_n})\\
&= X(\sqrt{\det g}) + \sum_{i=1}^n \frac{\partial X^i}{\partial x_i}\sqrt{\det g}\\
&= \Big(\sum_{i=1}^n \frac{\partial X^i}{\partial x_i} + \frac{1}{\sqrt{\det g}} X(\sqrt{\det g})\Big)\sqrt{\det g}. \qquad (6.5.93)
\end{aligned}
$$

The second term inside the parenthesis in (6.5.93) can be computed as

$$
\begin{aligned}
\frac{1}{\sqrt{\det g}} X(\sqrt{\det g}) &= \frac{1}{\sqrt{\det g}}\sum_k X^k(\partial_{x_k}\sqrt{\det g})\\
&= \sum_k X^k \frac{1}{2\det g}\partial_{x_k} g = \sum_k X^k \partial_{x_k}(\ln\sqrt{\det g})\\
&= \sum_{k,i} X^k \Gamma^i_{ki}, \qquad (6.5.94)
\end{aligned}
$$

where we used the contraction of Christoffel symbols formula (6.1.18)

$$
\sum_i \Gamma^i_{ki} = \partial_{x_k}(\ln \det \sqrt{g}).
$$

Substituting (6.5.94) into (6.5.93) yields

$$
\begin{aligned}
\frac{1}{\sqrt{\det g}} X(\sqrt{\det g}) &= \Big(\sum_{i=1}^n \frac{\partial X^i}{\partial x_i} + \sum_{k,i} X^k\Gamma^i_{ki}\Big)\sqrt{\det g}\\
&= (\mathrm{div} X)\sqrt{\det g} = (\mathrm{div} X)\, dv(\partial_{x_1},\dots,\partial_{x_n}),
\end{aligned}
$$

where we used the divergence formula (6.2.23). Hence, relation (6.5.92) is verified. ■

Cartan's decomposition formula The *exterior derivative* d is an operator that acts on p-forms and yields $(p+1)$-forms as follows

$$d\omega(Y_1, \ldots, Y_{p+1}) = \sum_{i=1}^{p+1} (-1)^{i+1} Y_i\, \omega(Y_1, \ldots, \hat{Y_i}, \ldots, Y_{p+1}) + \sum_{i<j} (-1)^{i+j} \omega([Y_i, Y_j], Y_1, \ldots, \hat{Y_i}, \ldots, \hat{Y_j}, \ldots, Y_{p+1}),$$

where $\hat{Y_i}$ means that Y_i is missing from the argument of the form ω.

Next we define the *interior multiplication* operator, i_X, that acts on p-forms and yields $(p-1)$-forms as follows. Let ω be a p-form and X be a vector field on $\mathbb{R}^n$. We define the $(p-1)$-form $i_X\omega$ by

$$(i_X\omega)(X_1, \ldots, X_{p-1}) = \begin{cases} 0, & \text{if } p = 0 \\ \omega(X, X_1, \ldots, X_{p-1}), & \text{if } p \geq 1 \end{cases}$$

and call it the interior multiplication of X with the form ω.

Cartan's decomposition formula provides a relation among the Lie derivative and the exterior differentiation and interior multiplication operators.

Proposition 6.5.10 *For any p-form ω and vector field X we have*

$$L_X\omega = i_X(d\omega) + d(i_X\omega).$$

For a proof the reader is referred to the book [26], page 23.

The Proof of the Divergence Theorem First we note that the exterior derivative applied to the volume form dv is equal to zero

$$d(dv) = 0.$$

It is worth noting that the inner d is not an exterior derivative; it stands for the volume measure. This identity follows from the general fact that the exterior derivative of any n-form on $\mathbb{R}^n$ vanishes.

Applying Cartan's decomposition formula with $\omega = dv$ and using the aforementioned remark, we obtain

$$L_X dv = i_X\big(d(dv)\big) + d(i_X dv) = d(i_X dv). \tag{6.5.95}$$

Then Stokes' formula

$$\int_\Omega d\eta = \int_{\partial\Omega} \eta$$

applied for the $(n-1)$-form $\eta = i_X dv$ becomes

$$\int_\Omega \text{div} X\, dv = \int_\Omega L_X\, dv = \int_\Omega d(i_X dv) = \int_{\partial\Omega} i_X dv,$$

where in the first equality we used Proposition 6.5.9 and for the second equality we applied relation (6.5.95).
In order to finish the proof, it suffices to show that

$$i_X dv = g(X, \nu)\, ds \tag{6.5.96}$$

along $\partial\Omega$. We shall verify this relation at an arbitrary point $x_0 \in \partial\Omega$ using an orthonormal basis. Let $\{E_1, \dots, E_{n-1}\}$ be $(n-1)$ vector fields that form an orthonormal basis in the tangent space $T_{x_0}\partial\Omega$. Using the properties of the area element we have

$$g(X, \nu)\, ds(E_1, \dots, E_{n-1}) = g(X, \nu).$$

Since $\{\nu, E_1, \dots, E_{n-1}\}$ is an orthonormal basis at x_0 in $(\mathbb{R}^n, g_{ij})$, we obtain

$$dv(\nu, E_1, \dots, E_{n-1}) = 1.$$

Expanding at the point x_0 as $X = g(X, \nu)\nu + \sum_{k=1}^{n-1} X^k E_k$, using the definition of the interior multiplication operator and linearity property of forms, we obtain

$$\begin{aligned}
(i_X dv)(E_1, \dots, E_{n-1}) &= dv(X, E_1, \dots, E_{n-1}) \\
&= \sum_{k=1}^{n-1} X^k \underbrace{dv(E_k, E_1, \dots, E_{n-1})}_{=0} \\
&\quad + g(X, \nu) \underbrace{dv(\nu, E_1, \dots, E_{n-1})}_{=1} \\
&= g(X, \nu) = g(X, \nu)\, ds(E_1, \dots, E_{n-1}).
\end{aligned}$$

We hence verified relation (6.5.96) on a basis, which suffices for its validity. Hence, the Divergence Theorem is totally proven.

Before attempting to compute the flux across the boundary of a Riemannian hypersphere we shall state a few preliminary results which serves this purpose.

6.5.5 The Riemannian Action

The geodesics on a Riemannian manifold (M, g) can be constructed variationally from the energy Lagrangian

$$L(\phi, \dot{\phi}) = \frac{1}{2}|\dot{\phi}|_g^2 = \frac{1}{2} g_{ij}(\phi)\dot{\phi}^i \dot{\phi}^j.$$

More specifically, the Euler-Lagrange equations

$$\frac{d}{du}\frac{\partial L}{\partial \dot\phi^k} = \frac{\partial L}{\partial \phi^k}, \qquad 1 \le k \le n,$$

become

$$\ddot\phi^k(u) + \Gamma^k_{ij}(\phi(u))\dot\phi^i(u)\dot\phi^j(u) = 0, \qquad 1 \le k \le n.$$

which are the geodesics equations in local coordinates. They can be written invariantly as $\nabla^0_{\dot\phi}\dot\phi = 0$, where ∇^0 denotes the Levi-Civita connection.

The *Riemannian action* between x_0 and x within time τ is defined by

$$S(x_0, x; \tau) = \inf_{\phi \in \mathcal{P}_{x_0,x;\tau}} \int_0^\tau L(\phi, \dot\phi)\, du,$$

where $\mathcal{P}_{x_0,x;\tau} = \{\phi : [0,\tau] \to M, \phi(0) = x_0, \phi(\tau) = x\}$ is the space of smooth paths joining the given points and parametrized by the interval $[0,\tau]$. Consequently, the action can be written in terms of the geodesic ϕ connecting x_0 and x as

$$S(x_0, x; \tau) = \int_0^\tau L(\phi, \dot\phi)\, du = \frac{1}{2}\int_0^\tau |\dot\phi(u)|^2 du = \frac{1}{2}\int_0^\tau g_{ij}(\phi(u))\dot\phi^i(u)\dot\phi^j(u).$$

The points x_0 and x are considered close enough, so there is a unique geodesic joining them. Then the Riemannian distance between x_0 and x is given by the length of this geodesic as

$$d(x_0, x) = \int_0^\tau |\dot\phi(u)|_g du = \int_0^\tau \sqrt{g_{ij}(\phi(u))\dot\phi^i(u)\dot\phi^j(u)}\, du.$$

Proposition 6.5.11 *The relation between the Riemannian action and the Riemannian distance is given by*

$$S(x_0, x; \tau) = \frac{d(x_0, x)^2}{2\tau}, \qquad \forall \tau > 0.$$

Proof: First we note that the magnitude of the tangent vector to a geodesic is constant along the geodesic, i.e., $|\dot\phi(u)|$ =constant. This follows from the metrical property of the Levi-Civita connection combined with the zero acceleration property of geodesics

$$\dot\phi\, g(\dot\phi, \dot\phi) = 2g(\nabla^0_{\dot\phi}\dot\phi, \dot\phi) = 0.$$

In this case the Cauchy inequality

$$\int_0^\tau |\dot\phi(u)|_g \cdot 1\, du \le \Big(\int_0^\tau |\dot\phi(u)|_g^2\, du\Big)^{1/2}\Big(\int_0^\tau du\Big)^{1/2}$$

becomes identity

$$\int_0^\tau |\dot{\phi}(u)|_g \, du = \Big(\int_0^\tau |\dot{\phi}(u)|_g^2 \, du \Big)^{1/2} \tau^{1/2},$$

which becomes $d(x_0, x)^2 = (2S(x_0, x; \tau))^{1/2} \tau^{1/2}$, or equivalently, $S(x_0, x; \tau) = \dfrac{d(x_0, x)^2}{2\tau}$.

■

6.5.6 The Eiconal Equation

The equation $\|\nabla_g f\|_g = 1$ is called the *eiconal equation* on the Riemannian manifold (M, g). Its solutions are smooth functions f on M whose gradient magnitude is equal to 1. The following result makes the connection with the Riemannian distance.

Proposition 6.5.12 *The Riemannian distance from x_0 to x on (M, g)*

$$f(x) = d(x_0, x)$$

satisfies the eiconal equation $\|\nabla_g f\|_g = 1$, with the initial condition $f(x_0) = 0$.

Proof: We start from the expression of the classical action between x_0 and x within time τ on the Riemannian manifold (M, g)

$$S = S(x_0, x; \tau) = \frac{1}{2} \int_0^\tau g_{ij}(x(u)) \dot{x}^i(u) \dot{x}^j(u) \, du = \frac{d^2(x_0, x)}{2\tau}, \tag{6.5.97}$$

where $x(u)$ denotes the geodesic between $x(0) = x_0$ and $x(t) = x$. We write next that S satisfies the Hamilton-Jacobi equation

$$\frac{\partial S}{\partial \tau} + H(x, \nabla_g S) = 0, \tag{6.5.98}$$

with the classical Hamiltonian

$$H(x, p) = \frac{1}{2} \|p\|_g^2 = \frac{1}{2} \sum_{i,j} g_{ij}(x) p^i p^j.$$

Substituting (6.5.97) into (6.5.98) we obtain

$$-\frac{1}{2\tau^2} d^2(x_0, x) + \frac{1}{2} \frac{d^2(x_0, x)}{\tau^2} \|\nabla_g d\|^2 = 0,$$

which implies $\|\nabla_g d\|^2 = 1$. ■

Remark 6.5.13 A function $\varphi(x)$ is called a potential for the vector field F if $F = \nabla_g \varphi$. If the potential is considered to be the distance $\varphi(x) = d(x_0, x)$ then the associated vector field, $\nabla_g d(x_0, x)$, is the geodesic vector field, whose integral lines are geodesics emanating from x_0. This geodesic flow is normal to the geodesic spheres centered at x_0. More details are given by the next result.

Proposition 6.5.14 *Let $r > 0$ and consider $\mathcal{S}_r = \{x \in \mathbb{R}^n; d(x_0, x) = r\}$ be the hypersphere of radius r centered at x_0 on the Riemannian manifold $(\mathbb{R}^n, g_{ij})$. Then the outward unit normal to $\mathcal{S}_r$ is given by*

$$\nu(x) = \nabla_g d(x_0, x), \qquad \forall x \in \mathcal{S}_r.$$

Proof: We assume $r > 0$, small enough, such that there are no conjugate points to the center inside the hypershpere $\mathcal{S}_r$. We let $f(x) = d(x_0, x)$ and write $\mathcal{S}_r = f^{-1}(\{r\})$. Under the hypothesis that $\mathcal{S}_r$ is a smooth surface, the Euclidean gradient $\nabla_{Eu} f(x)$ is normal to the tangent space $T_x\mathcal{S}_r$, i.e.,

$$\langle \nabla_{Eu} f(x), w \rangle = 0, \qquad \forall w \in T_x\mathcal{S}_r. \tag{6.5.99}$$

Now, if W is a tangent vector field on $\mathcal{S}_r$ in a neighborhood of x we can express $W(f)$ in the following two equivalent ways:

(i) Using the Euclidean scalar product as: $W(f) = \langle \nabla_{Eu} f(x), W \rangle$;
(ii) Using the metric g as: $W(f) = g(\nabla_g f(x), W)$.

Therefore, if $w = W_x$, we obtain

$$\langle \nabla_{Eu} f(x), w \rangle = g(\nabla_g f(x), w). \tag{6.5.100}$$

From (6.5.99) and (6.5.100) yields

$$g(\nabla_g f(x), w) = 0, \quad \forall w \in T_x\mathcal{S}_r.$$

Hence, $\nabla_g f$ is a normal vector to the hypersphere $\mathcal{S}_r$. Using Proposition 6.5.12

$$\|\nabla_g f\|_g^2 = \|\nabla_g d(x_0, x)\|^2 = 1, \tag{6.5.101}$$

so $\nu(x) = \nabla_g f(x)$ is the unit normal. The normal vector points outward because $f(x)$ increases as x gets further from x_0. ∎

6.5.7 Computing the Flux

We consider the Riemannian manifold $(\mathbb{R}^n, g_{ij})$, where g_{ij} is the diffusion metric, and $\Omega = \{x \in \mathbb{R}^n; d(x_0, x) \leq r\}$ is the Riemannian ball of radius r centered at x_0. We also assume that the radius of injectivity of the manifold at x_0 is larger than r. The boundary of Ω is the hypersphere $\partial\Omega = \mathcal{S}_r$, with the outward unit normal given by Proposition 6.5.14

$$\nu(x) = \nabla_g d(x_0, x).$$

Let X_t be a Brownian motion on $(\mathbb{R}^n, g_{ij})$, which starts at x_0. In order to compute the flux of X_t across $\partial\Omega$ we need to be able to find the normal component of the gradient of the density $p_t(x_0, x)$, which is $g(\nabla_g p_t, \nu(x))$. We shall perform the computation under the following simplifying *ansatz*:

The probability density p_t is rotationally invariant, i.e., it depends on the Riemannian distance between x_0 and x

$$p_t(x_0, x) = \phi_t\big(d(x_0, x)\big), \tag{6.5.102}$$

with ϕ_t positive smooth function.

Then $\nabla_g p_t = \phi_t'(d(x_0, x))\nabla_g d(x_0, x)$ and using the eiconal equation we obtain

$$g(\nabla_g p_t, \nu(x)) = g(\nabla_g p_t, \nabla_g d(x_0, x)) = \phi_t'\big(d(x_0, x)\big)\|\nabla_g d(x_0, x)\|_g^2 = \phi_t'\big(d(x_0, x)\big).$$

Then the total flux across the boundary $\partial\Omega$ becomes

$$\begin{aligned} F_t &= -\frac{1}{2}\int_{\partial\Omega} g(\nabla_g p_t, \nu(x))\, ds = -\frac{1}{2}\int_{\partial\Omega} \phi_t'\big(d(x_0, x)\big)\, ds \\ &= -\frac{1}{2}\int_{\partial\Omega} \phi_t'(r)\, ds = -\frac{1}{2}\phi_t'(r) vol(\mathcal{S}_r), \end{aligned} \tag{6.5.103}$$

where we applied that the distance is independent of s and we used the notation $\phi_t'(r) = \frac{d}{dr}\phi_t(r)$.

Therefore, the escape probability of X_t from the Riemannian ball Ω is obtained using formula (6.5.89)

$$\begin{aligned} P(X_t \notin \Omega) &= \int_0^t F_u\, du = -\frac{1}{2} vol(\mathcal{S}_r)\int_0^t \phi_u'(r)\, du \\ &= -\frac{1}{2} vol(\mathcal{S}_r)\frac{d}{dr}\int_0^t \phi_u(r)\, du. \end{aligned} \tag{6.5.104}$$

We provide next an example of Riemannian manifold where these computations can be carried out explicitly.

Example 6.5.1 (Poincaré's upper half-space) The set

$$\mathcal{H}^3 = \{z = (x_1, x_2, y) \in \mathbb{R}^3; (x_1, x_2) \in \mathbb{R}^2, y > 0\}$$

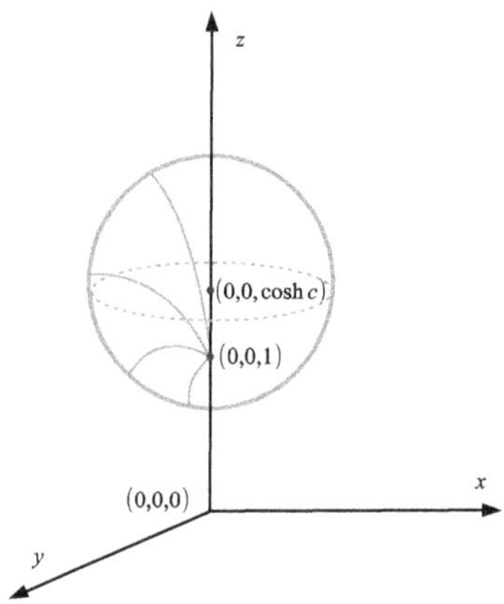

Figure 6.2: *The hyperbolic sphere centered at* $(0,0,1)$ *is a Euclidean sphere with the center located at* $(0,0,\cosh c)$. *The geodesics are arcs of circle or vertical lines.*

endowed with the metric $g_{ij} = y^{-2}\delta_{ij}$ forms a Riemannian manifold, called the *upper half-space*. It can be shown that the distance $d(z,z')$ between any two points $z, z' \in \mathcal{H}^3$ satisfies

$$\cosh d(z,z') = 1 + \frac{\|z-z'\|^2}{2yy'}, \tag{6.5.105}$$

where $\|\cdot\|$ stands for the Euclidean norm in $\mathbb{R}^3$. This formula is useful to show that the hyperbolic sphere centered at $z_0 = (0,0,1)$ of radius c, i.e.,

$$\mathcal{S}_c = \{z \in \mathbb{R}^3; d(z_0, z) = c\}$$

is an offcentered Euclidean sphere. This follows from the computation

$$\begin{aligned} d(z_0, z) = c \Leftrightarrow \|z - z_0\|^2 = 2y(\cosh c - 1) \Leftrightarrow \\ |x|^2 + y^2 - 2y\cos c + 1 = 0 \Leftrightarrow \\ |x|^2 + (y - \cosh c)^2 = \sinh^2 c, \end{aligned}$$

which is the equation of a sphere centered at $(0,0,\cosh c)$ with radius $\sinh c$, see Fig. 6.2. The fact that the point $(0,0,1)$ is contained inside the sphere follows from the fact that $\cosh c - 1 < \sinh c$.

The hyperbolic diffusion $Z(t) = \big(X_1(t), X_2(t), Y(t)\big)$ on $\mathcal{H}^3$ satisfies

$$\begin{aligned} dX_1(t) &= Y(t)dW_t^1 \\ dX_2(t) &= Y(t)dW_t^2 \\ dY(t) &= Y(t)dW_t^3 - \frac{1}{2}Y(t)\,dt, \end{aligned}$$

and starts from the point $z_0 = (x_0^1, x_0^2, y_0) = (0, 0, 1)$. As usual, we assumed that W_t^k are independent Brownian motions. The associated generator with the diffusion $Z(t)$ is given by

$$\mathcal{A} = \frac{1}{2}y^2(\partial_{x_1}^2 + \partial_{x_2}^2 + \partial_y^2) - \frac{1}{2}y\partial_y.$$

The transition probability of diffusion $Z(t)$ depends on the hyperbolic distance $r = d(z_0, z)$ and is given by

$$p_t(z_0, z) = \phi_t(r) = \frac{1}{(2\pi t)^{3/2}}e^{-\frac{t}{2}}\frac{r}{\sinh r}e^{-\frac{r^2}{2t}},$$

and hence it is of form (6.5.102). This formula can be found, for instance, in Avramidi [6], page 279. Therefore, the flux of the hyperbolic diffusion across the hyperbolic sphere $\mathcal{S}_r$ is given by formula (6.5.103). Using that the area of the hyperbolic sphere is $vol(\mathcal{S}_r) = 4\pi\sinh^2 r$, by direct computation we obtain

$$\begin{aligned} F_t &= -\frac{1}{2}\phi_t'(r)vol(\mathcal{S}_r) \\ &= \frac{\sinh r}{\sqrt{2\pi t^3}}e^{-(\frac{t}{2}+\frac{r^2}{2t})}\Big(\frac{r^2}{t} + r\coth r - 1\Big). \end{aligned}$$

The escape probability of diffusion $Z(t)$ from the hyperbolic ball $\Omega_r = \{z \in \mathbb{R}^3; d(z, z_0) \le r\}$ is given by formula (6.5.104)

$$P(Z(t) \notin \Omega_r) = \frac{\sinh r}{\sqrt{2\pi}}\int_0^t u^{-3/2}e^{-(\frac{u}{2}+\frac{r^2}{2u})}\Big(\frac{r^2}{u} + r\coth r - 1\Big)\,du. \quad (6.5.106)$$

We note that the integral (6.5.106) cannot be computed in closed form for $t < \infty$. However, for $t \to \infty$ the integral can be written as the sum of two integrals that can be expressed in terms of modified Bessel functions via formula (5.3.8).

6.6 Bessel Processes on Riemannian Manifolds

The usual Bessel process of order n can be represented geometrically as the Euclidean distance between an n-dimensional Brownian motion W_t and the origin of $\mathbb{R}^n$, namely

$$R_t = \|W_t\| = \Big(\sum_{k=1}^{n}(W_t^k)^2\Big)^{1/2}.$$

In a similar way, one can introduce the Bessel process R_t^g on a Riemannian manifold (M, g). To this end, we consider the process X_t on the Riemannian manifold (M, g), which is a Brownian motion with respect to the metric g, starting at $X_0 = x_0$. This means that the infinitesimal generator of X_t is equal to the Laplace-Beltrami operator Δ_g on (M, g). Then the Bessel process R_t^g is defined by

$$R_t^g = d(x_0, X_t),$$

i.e., it is equal to the Riemannian distance between x_0 and X_t. This can be also characterized in terms of geodesic curves as follows.

We assume the manifold has the property that any point can be connected to x_0 by a unique geodesic. All Riemannian manifolds have this property locally about x_0, i.e., for x close enough to x_0 there is only one minimizing geodesic joining x_0 and x. However, hyperbolic manifolds have this property satisfied globally.

Since we parameterize the process X_t in a local chart, it makes sense to assume that the manifold M is the entire $\mathbb{R}^n$, or an open subset of $\mathbb{R}^n$. Let $\gamma_t : [0,1] \to \mathbb{R}^n$, $\gamma_t(s) = \big(\gamma_t^1(s), \cdots, \gamma_t^n(s)\big)$ be the geodesic on $(\mathbb{R}^n, g)$, connecting the points $\gamma_t(0) = x_0$ and $X_t = \gamma_t(1)$. Then

$$R_t^g = \int_0^1 \|\dot\gamma_t(\tau)\|_g \, d\tau.$$

Using that geodesics are curves of acceleration zero, we have

$$\frac{d}{d\tau} g(\dot\gamma_t(\tau), \dot\gamma_t(\tau)) = 2g(\nabla_{\dot\gamma_t(\tau)}\dot\gamma_t(\tau), \dot\gamma_t(\tau)) = 0,$$

fact that implies that the magnitude of the tangent vector $\|\dot\gamma_t(\tau)\|$ is constant along the geodesic. It follows that the Bessel process can be expressed in terms of the initial geodesic velocity as

$$R_t^g = \|\dot\gamma_t(0)\|_g.$$

We shall describe in the following the law of the process R_t^g, which is

$$f_{R_t^g}(r) = \frac{d}{dr} P(R_t^g \le r), \tag{6.6.107}$$

using the cumulative flux as described in Remark 6.5.5 for the case of the usual Bessel process. Let $\Omega = \{x \in \mathbb{R}^n; d(x_0, x) \le r\}$ be the Riemannian ball of radius r on $(\mathbb{R}^n, g)$. Then

$$P(R_t^g \le r) = 1 - P(X_t \notin \Omega) = 1 - \int_0^t F_u\, du,$$

where we used the escape probability formula (6.5.89). Substituting in (6.6.107) yields

$$\begin{aligned} f_{R_t^g}(r) &= -\frac{d}{dr}\int_0^t F_u\, du \\ &= \frac{1}{2}\int_0^t \frac{d}{dr}\int_{\mathcal{S}_r} g\big(\nabla_g p_u(x_0, x), \nabla_g d(x_0, x)\big)\, ds\, du, \end{aligned} \tag{6.6.108}$$

where $p_t(x_0, x)$ denotes the transition density of X_t and we used that $\nu(x) = \nabla_g d(x_0, x)$ is the outward unit normal to the Riemannian sphere $\mathcal{S}_r = \partial\Omega$ of radius r and centered at x_0.

We can further compute the expression (6.6.108) if some additional assumptions are made on the metric g. If the metric g is rotational invariant about the point x_0, then the Laplace-Beltrami operator has the same property. Therefore, its heat kernel $p_t(x_0, x)$, which is also the density of X_t, will depend on the Riemannian distance between x_0 and x, namely

$$p_t(x_0, x) = \phi_t\big(d(x_0, x)\big),$$

with ϕ_t positive and smooth function. Under these hypotheses, using the eiconal equation (see Proposition 6.5.12), the integrand of (6.6.108) becomes

$$g\big(\nabla_g p_t(x_0, x), \nabla_g d(x_0, x)\big) = \phi_t'\big(d(x_0, x)\big)\|\nabla_g d(x_0, x)\|_g^2 = \phi_t'\big(d(x_0, x)\big).$$

Then relation (6.6.108) can be computed using Fubini's theorem as

$$\begin{aligned} f_{R_t^g}(r) &= \frac{1}{2}\int_0^t \frac{d}{dr}\int_{\mathcal{S}_r} \phi_u'\big(d(x_0, x)\big)\, ds\, du \\ &= \frac{1}{2}\int_0^t \frac{d}{dr}\Big(\phi_u'(r) vol(\mathcal{S}_r)\Big)\, du \\ &= \frac{1}{2} vol(\mathcal{S}_r)\int_0^t \phi_u''(r)\, du + \frac{1}{2}\frac{d}{dr}\Big(vol(\mathcal{S}_r)\Big)\int_0^t \phi_u'(r)\, du \\ &= \frac{1}{2} vol(\mathcal{S}_r)\frac{d^2}{dr^2}\int_0^t \phi_u(r)\, du + \frac{1}{2}\frac{d}{dr}\Big(vol(\mathcal{S}_r)\Big)\frac{d}{dr}\int_0^t \phi_u(r)\, du. \end{aligned} \tag{6.6.109}$$

Hence, formula (6.6.109) provides the probability density of the Bessel process R_t^g, which starts at x_0 on a Riemannian manifold, whose metric g has rotational symmetry about x_0.

We shall consider two examples of such manifolds: the Euclidean space and the hyperbolic space.

Example 6.6.1 We consider the Euclidean space $\mathbb{R}^n$ with the metric $g_{ij} = \delta_{ij}$ and a Brownian motion B_t starting at the origin. The geodesics in this case are straight lines starting from the origin. The radial function $\phi_t(r)$ is given by

$$\phi_t(r) = \frac{1}{(2\pi t)^{n/2}} e^{-\frac{r^2}{2t}},$$

where $r = \|x\|$. Since the area of the $(n-1)$-dimensional Euclidean hypersphere of radius r is $vol(\mathcal{S}_r) = \frac{2\pi^{n/2}}{\Gamma(n/2)} r^{n-1}$, after using the substitution $x = \frac{r^2}{2u}$, formula (6.6.109) provides the density of the n-dimensional Bessel process $R_t^{(n)} = \|B_t\|$ as

$$f_{R_t^{(n)}}(r) = \frac{r^{n-1}}{2\Gamma(n/2)} \frac{d^2}{dr^2}\left(\frac{1}{r^{n-2}} \int_{\frac{r^2}{2t}}^{\infty} x^{\frac{n}{2}-2}\, e^{-x}\, dx \right) + \frac{(n-1)r^{n-2}}{2\Gamma(n/2)} \frac{d}{dr}\left(\frac{1}{r^{n-2}} \int_{\frac{r^2}{2t}}^{\infty} x^{\frac{n}{2}-2}\, e^{-x}\, dx \right). \tag{6.6.110}$$

It is worth noting the special case $n = 2$, when the expression in the parenthesis is more simple and this formula can be computed using Fundamental Theorem of Calculus as

$$\begin{aligned} f_{R_t^{(2)}}(r) &= \frac{r}{2} \frac{d^2}{dr^2} \int_{\frac{r^2}{2t}}^{\infty} x^{-1}\, e^{-x}\, dx + \frac{1}{2} \frac{d}{dr} \int_{\frac{r^2}{2t}}^{\infty} x^{-1}\, e^{-x}\, dx \\ &= \frac{r}{t} e^{-\frac{r^2}{2t}}, \end{aligned} \tag{6.6.111}$$

which retrieves Wald's distribution of a two-dimensional Bessel process in $\mathbb{R}^2$.

Example 6.6.2 (Bessel process on the Poincaré upper half-space) We consider the Bessel process on the Poincaré upper half-space $\mathcal{H}^3$ starting from $z_0 = (0,0,1)$, see Example 6.5.1. The transition probability of the hyperbolic diffusion $Z(t)$ can be represented in terms of the hyperbolic distance $r = d(z_0, z)$ as

$$p_t(z_0, z) = \phi_t(r) = \frac{1}{(2\pi t)^{3/2}} e^{-\frac{t}{2}} \frac{r}{\sinh r} e^{-\frac{r^2}{2t}}. \tag{6.6.112}$$

The 3-dimensional hyperbolic Bessel process is equal to the hyperbolic distance between z_0 and diffusion $Z(t)$, i.e., $R_t^{(h)} = d\big(z_0, Z(t)\big)$. Substituting the relation (6.6.112) into (6.6.109) yields the density of $R_t^{(h)}$

$$f_{R_t^{(h)}}(r) = \frac{1}{\sqrt{2\pi}}(\sinh r)^2 \frac{d^2}{dr^2}\left(\frac{r}{\sinh r}\int_0^t u^{-3/2}e^{-\frac{1}{2}(u+\frac{r^2}{u})}\,du\right) + \frac{1}{\sqrt{2\pi}}\sinh(2r)\frac{d}{dr}\left(\frac{r}{\sinh r}\int_0^t u^{-3/2}e^{-\frac{1}{2}(u+\frac{r^2}{u})}\,du\right). \quad (6.6.113)$$

Denoting

$$U(r,t) = \frac{r}{\sinh r}\int_0^t u^{-3/2}e^{-\frac{1}{2}(u+\frac{r^2}{u})}\,du, \quad (6.6.114)$$

then (6.6.113) becomes

$$f_{R_t^{(h)}}(r) = \sqrt{\frac{2}{\pi}}(\sinh r)^2\Big(\frac{1}{2}\frac{d^2}{dr^2} + \coth r\frac{d}{dr}\Big)U(r,t). \quad (6.6.115)$$

An explicit formula for the process $R_t^{(h)}$ can be found as follows. Using the formula for the hyperbolic distance (6.5.105) we have

$$\begin{aligned} R_t^{(h)} &= \cosh^{-1}\left(1 + \frac{\|Z(t)-z_0\|^2}{2Y_t}\right) \\ &= \cosh^{-1}\left(\frac{1+(X_t^1)^2+(X_t^2)^2+(Y_t)^2}{2Y_t}\right), \end{aligned}$$

with X_t^j and Y_t given by (6.4.37) and (6.4.38), i.e.,

$$\begin{aligned} X_t^j &= \int_0^t e^{-s+W_s^3}\,dW_s^j, \qquad j=1,2 \\ Y_t &= e^{-t+W_t^3}, \end{aligned}$$

with W_t^1, W_t^2, W_t^3 independent Brownian motions. The aforementioned explicit formula for $R_t^{(h)}$ leads to the following expression for the distribution function

$$\begin{aligned} F_{R_t}(\rho) &= P(R_t^{(h)} \le \rho) = P\big(1+(X_t^1)^2+(X_t^2)^2+(Y_t)^2 \le 2Y_t\cosh(\rho)\big) \\ &= P\big((X_t^1)^2+(X_t^2)^2+(Y_t-\cosh\rho)^2 \le \sinh^2\rho\big) \\ &= P(Z(t)\in\Omega), \end{aligned}$$

where $\Omega = \{|x|^2+(y-\cosh\rho)^2 \le \sinh^2\rho\}$ is the Euclidean ball of radius $\sinh\rho$ centered at $(0,0,\cosh\rho)$.

Remark 6.6.3 The integral (6.6.114) cannot be evaluated exactly. However, an explicit formula is obtained for $t \to \infty$ using Lemma 5.3.1 as follows

$$\lim_{t\to\infty} U(r,t) = \frac{r}{\sinh r}\int_0^\infty u^{-3/2} e^{-\frac{1}{2}(u+\frac{r^2}{u})}\, du$$
$$= \sqrt{2\pi}\frac{e^{-r}}{\sinh r} = \frac{2\sqrt{2\pi}}{e^{2r}-1}.$$

6.7 Radial Processes on Model Manifolds

Let $(r,\theta) \in (0,\infty) \times \mathbb{S}^{n-1}$ be the polar coordinates in $\mathbb{R}^n$ and consider the Riemannian metric

$$ds^2 = dr^2 + \psi^2(r) d\theta^2,$$

where $d\theta^2$ is the standard metric on $\mathbb{S}^{n-1}$ and ψ is a smooth function such that $\phi(0) = 0$ and $\psi(r) > 0$ for $r > 0$. Then the Riemannian manifold $M_\psi = (\mathbb{R}^n, ds^2)$ is called a *model manifold.*

The Laplacian on M_ψ The Laplacian in polar coordinates is given by formula (6.2.24)

$$\Delta = \frac{1}{2}\Big(\sum_{i,j} g^{ij}\partial_{x_i}\partial_{x_j} - \sum_{i,j,k} g^{ij}\Gamma_{ij}^k\partial_{x_k}\Big) \tag{6.7.116}$$

with the inverse metric matrix

$$g^{ij} = \begin{pmatrix} 1 & 0 \\ 0 & \psi^{-2}(r)\mathbb{I}_{n-1} \end{pmatrix}. \tag{6.7.117}$$

The coefficient of the linear part of Δ is given by Lemma 6.2.1

$$-\sum_{i,j} g^{ij}\Gamma_{ij}^k = \frac{1}{\sqrt{\det g}}\partial_{x_p}(\sqrt{\det g}\, g^{kp}).$$

This can be computed by letting $x_1 = r$, $x_2 = \theta_1, \ldots, x_n = \theta_{n-1}$ and $\det g = \psi^{2(n-1)}(r)$ in the aforementioned formula, which leads to

$$-\sum_{i,j} g^{ij}\Gamma_{ij}^1 = \frac{1}{\psi^{n-1}(r)}[\partial_r(\psi^{n-1}(r)g^{11}) + \partial_\theta(\psi^{n-1}(r)g^{12})] = (n-1)\frac{\psi'(r)}{\psi(r)} \tag{6.7.118}$$

$$-\sum_{i,j} g^{ij}\Gamma_{ij}^2 = \frac{1}{\psi^{n-1}(r)}[\partial_r(\psi^{n-1}(r)g^{21}) + \partial_\theta(\psi^{n-1}(r)g^{22})] = 0. \tag{6.7.119}$$

Substituting (6.7.117), (6.7.118) and (6.7.119) into (6.7.116) yields the following formula for the Laplacian on the model manifold M_ψ

$$\Delta_\psi = \frac{1}{2}\Big\{\partial_r^2 + (n-1)\frac{\psi'(r)}{\psi(r)}\partial_r + \frac{1}{\psi^2(r)}\Delta_{\mathbb{S}^{n-1}}\Big\}, \tag{6.7.120}$$

where $\Delta_{\mathbb{S}^{n-1}}$ denotes the Laplacian on the sphere $\mathbb{S}^{n-1}$. The operator

$$\Delta_{rad} = \frac{1}{2}\Big\{\partial_r^2 + (n-1)\frac{\psi'(r)}{\psi(r)}\partial_r\Big\}$$

denotes the *radial Laplacian* on M_ψ and will play a role in the study of the Bessel processes on M_ψ.

There are three prototypes of model manifolds as follows:

1. If $\psi(r) = r$, then $M_\psi = \mathbb{R}^n$ is the Euclidean space with the Euclidean Laplacian

$$\Delta_{\mathbb{R}^n} = \frac{1}{2}\Big\{\partial_r^2 + \frac{n-1}{r}\partial_r + \frac{1}{\psi^2(r)}\Delta_{\mathbb{S}^{n-1}}\Big\}. \tag{6.7.121}$$

2. If $\psi(r) = \sinh r$, then $M_\psi = \mathcal{H}^n$ is the hyperbolic space with the hyperbolic Laplacian

$$\Delta_{\mathcal{H}^n} = \frac{1}{2}\Big\{\partial_r^2 + (n-1)\coth r\,\partial_r + \frac{1}{\psi^2(r)}\Delta_{\mathbb{S}^{n-1}}\Big\}. \tag{6.7.122}$$

3. If $\psi(r) = \sin r$, with $0 < r < \pi$, then M_ψ is the punctured sphere $\mathbb{S}^n$, with the spherical Laplacian

$$\Delta_{\mathbb{S}^n} = \frac{1}{2}\Big\{\partial_r^2 + (n-1)\cot r\,\partial_r + \frac{1}{\psi^2(r)}\Delta_{\mathbb{S}^{n-1}}\Big\}. \tag{6.7.123}$$

The Brownian motion on M_ψ The Brownian motion (r_t, θ_t) on the model manifold M_ψ is the diffusion having the infinitesimal generator given by the Laplacian (6.7.120). This process satisfies the following stochastic differential equations

$$\begin{aligned} dr_t &= \frac{n-1}{2}\frac{\psi'(r_t)}{\psi(r_t)}dt + dW_t \\ d\theta_t &= \frac{1}{\psi(r_t)}d\beta_t. \end{aligned}$$

The process W_t is a standard Brownian motion on $\mathbb{R}$, and β_t is a Brownian motion on the sphere $\mathbb{S}^{n-1}$, which is the diffusion process associated with the spherical Laplacian operator $\Delta_{\mathbb{S}^{n-1}}$.

The process r_t provides the radial component of the Brownian motion, while θ_t describes its angular part. We will proceed to separately study the process r_t, independent of θ_t.

The radial process on M_ψ The diffusion process associated with the operator Δ_{rad} is the radial process on the manifold M_ψ. This satisfies the following stochastic differential equation

$$dr_t = \frac{n-1}{2}\frac{\psi'(r_t)}{\psi(r_t)}dt + dW_t.$$

1. If $\psi(r) = r$, then we obtain an n-dimensional *Bessel process* on the Euclidean space $\mathbb{R}^n$

$$dr_t = \frac{n-1}{2r_t} dt + dW_t.$$

2. If $\psi(r) = \sinh r$, then we obtain an n-dimensional *hyperbolic Bessel process*

$$dr_t = \frac{n-1}{2} \coth r_t \, dt + dW_t,$$

which is associated with the operator

$$\Delta_{rad}^{\mathcal{H}^n} = \frac{1}{2}\Big\{\frac{d^2}{dr^2} + (n-1)\coth r \frac{d}{dr}\Big\}.$$

Setting $x = \cosh r$ and using that

$$\frac{d}{dr} = \sqrt{x^2-1}\frac{d}{dx}, \qquad \frac{d^2}{dr^2} = (x^2-1)\frac{d^2}{dx^2} + x\frac{d}{dx}$$

then $\Delta_{rad}^{\mathcal{H}^n}$ becomes

$$J_n = \frac{1}{2}\Big((x^2-1)\frac{d^2}{dx^2} + nx\frac{d}{dx}\Big), \qquad x \geq 1.$$

It is worth noting the significance of the operator $\frac{1}{2}\frac{d^2}{dx^2} + \mu\coth(\mu x)\frac{d}{dx}$ in the context of the hyperbolic space $\mathcal{H}^3$. This operator is the generator of a process known as the non-returning Brownian motion with positive drift rate μ, which is constructed by conditioning a Brownian motion with drift, $W_t^{(\mu)} = W_t + \mu t$, not to hit the point 0 for any positive t. This process, called X_t, lives on the space $(0, \infty)$. When $\mu = 1$, the generator of this process is the same as the operator $\Delta_{rad}^{\mathcal{H}^3}$, which is the generator of the hyperbolic Bessel process on $\mathcal{H}^3$. This means that the hyperbolic Bessel process on $\mathcal{H}^3$ has the same law as the non-returning Brownian motion with unit drift rate, $X_t = \{W_t + t, |, W_t + t > 0, \forall t > 0\}$. This result is shown in Rogers and Williams [122], page 436, using excessive transforms.

3. If $\psi(r) = \sin r$, with $0 < r < \pi$, then the diffusion process associated with Δ_{rad} is the n-dimensional process

$$dr_t = \frac{n-1}{2} \cot r_t \, dt + dW_t. \tag{6.7.124}$$

If we make the change of variable $x = \cos r$, then the associated generator of r_t

$$\Delta_{rad}^{\mathbb{S}^n} = \frac{1}{2}\Big\{\frac{d^2}{dr^2} + (n-1)\cot r \frac{d}{dr}\Big\}$$

becomes

$$L_n = \frac{1}{2}\Big\{(1-x^2)\frac{d^2}{dx^2} - nx\frac{d}{dx}\Big\}, \qquad -1 < x < 1. \tag{6.7.125}$$

The process $X_t = \cos r_t$ is a diffusion process on the interval $[-1,1]$. Its generator, L_n, is related to Gegenbauer polynomials or ultraspherical polynomials, $C_n^{(\alpha)}(x)$, which are a set of orthogonal polynomials on the interval [1,1] with respect to the weight function $w(x) = (1-x^2)^{\alpha-1/2}$ and can be defined by Rodrigues' formula

$$C_n^{(\alpha)}(x) = \frac{(-1)^n}{2^n n!}\frac{\Gamma(\alpha+\frac{1}{2})\Gamma(n+2\alpha)}{\Gamma(2\alpha)\Gamma(\alpha+n+\frac{1}{2})}(1-x^2)^{-\alpha+1/2}\frac{d^n}{dx^n}[(1-x^2)^{n+\alpha-1/2}],$$

for $\alpha,\beta > -1$. The orthogonality condition writes, see Abramowitz and Stegun [1], page 744

$$\int_{-1}^{1} C_m^{(\alpha)}(x)C_k^{(\alpha)}(x)w(x)dx = \frac{\pi 2^{1-2\alpha}\Gamma(k+2\alpha)}{k!(k+\alpha)\Gamma(\alpha)^2}\delta_{km}.$$

Furthermore, the Gegenbauer polynomial $C_m^{(\alpha)}$ satisfies the differential equation

$$(1-x^2)y'' - (2\alpha+1)xy' + m(m+2\alpha)y = 0.$$

Choosing $\alpha = \frac{n-1}{2}$, this becomes

$$L_n C_m^{(\frac{n-1}{2})}(x) = -\frac{1}{2}m(m+n-1)C_m^{(\frac{n-1}{2})}(x),$$

with L_n given by (6.7.125). Then

$$f_k(x) = 2^{\frac{n}{2}-1}\Gamma\Big(\frac{n-1}{2}\Big)\sqrt{\frac{k!(k+\frac{n-1}{2})}{\pi\Gamma(k+n-1)}}\,C_k^{(\frac{n-1}{2})}(x)$$

is an eigenfunction of the operator L_n with the corresponding eigenvalue $\lambda_k = -\frac{1}{2}k(k+n-1)$. Moreover, the set $\{f_k(x)\}$ forms a complete orthonormal system in the interval [-1,1] with respect to the weighting function $w(x) = (1-x^2)^{\frac{n}{2}-1}$.

Then the heat kernel of L_n can be constructed from the aforementioned system of eigenfunctions by Mercer's formula $K(x,y;t) = \sum_{k\geq 0} e^{\lambda_k t} f_k(x)f_k(y)$. Therefore, the transition density of the radial process r_t given by (6.7.124) becomes

$$\begin{aligned}p_t(r_0,r) &= K(\cos r_0,\cos r;t) = \sum_{k\geq 0} e^{\lambda_k t} f_k(\cos r_0)f_k(\cos r)\\ &= \frac{2^{n-2}}{\pi}\Gamma\Big(\frac{n-1}{2}\Big)^2\sum_{k\geq 0} e^{-\frac{1}{2}k(k+n-1)}\frac{k!\big(k+\frac{n-1}{2}\big)}{\Gamma(k+n-1)}C_k^{(\frac{n-1}{2})}(\cos r_0)C_k^{(\frac{n-1}{2})}(\cos r).\end{aligned}$$

For an approach using the Hamiltonian formalism and for the role of L_n in finding the heat kernel on the odd dimensional unit sphere, the reader is referred to Greiner [65], [66].

6.8 Potential Induced Diffusions

This section focuses on diffusions that are induced by a smooth positive potential function ρ. Depending on the specific context, this potential can be interpreted as resistivity, gravitational potential, electric potential, or other similar notions. Our main interest lies in understanding how these potentials impact the behavior of a two-dimensional Brownian motion.

To illustrate, let's consider the scenario of charged ink particles diffusing under the influence of an electric field. Mathematically, this can be represented by a hyperbolic diffusion process, where the horizon is determined by the configuration of the electrical wire.

In another example, let's examine the diffusion of water molecules above the surface of a swamp, where the primary driving force is the gravitational field. As a result, a fog layer forms, with higher density closer to the surface of the lake.

By exploring these examples and studying the effects of different potentials, we aim to deepen our understanding of how various factors impact the dynamics of diffusion processes.

Curves of minimum potential We begin by revisiting a physics problem that explores the concept of the minimum potential curve connecting two specified points in $\mathbb{R}^n$. This problem investigates the curve that requires the least amount of potential energy to travel between the given points.

To illustrate this idea, let's consider different scenarios. In the case of gravitational potential in $\mathbb{R}^2$, the minimum potential curve takes the shape of a catenary curve. On the other hand, for electrical potential, the minimum potential curve can be an arc of a circle or a straight line, depending on the specific configuration. Similarly, in the case of resistivity potential, the minimum potential curve represents the path of least resistance between the points A and B.

The curve of the lowest potential between the points A and B in $\mathbb{R}^n$ is a curve $c : [0, 1] \to \mathbb{R}^n$ with $c(0) = A$, $c(1) = B$ such that the functional

$$c \to \int_c \rho\, ds = \int_0^1 \rho(c(t))\|\dot{c}(t)\|\, dt$$

is minimum. This problem is equivalent to finding the geodesics between the points A and B on the Riemannian space $(\mathbb{R}^2, g)$, with the metric coefficients

$g_{ij} = \rho^2 \delta_{ij}$. Since the metric is conformally flat, the results of Section 6.4.6 apply, and hence the associated diffusion will be driftless.

Potential induced diffusion The aforementioned curves of minimum potential have a dynamics described by the Hamiltonian function

$$H(x, y; p_1, p_2) = \frac{1}{2} g(p, p) = \frac{1}{2} \frac{1}{\rho^2(x, y)} (p_1^2 + p_2^2). \tag{6.8.126}$$

This means that they are obtained as the (x, y)-component of the solution of the Hamiltonian system

$$\dot{q} = \frac{\partial H}{\partial p}, \qquad \dot{p} = -\frac{\partial H}{\partial q},$$

with $q = (x, y)$ and $p = (p_1, p_2)$. The associated Laplacian is obtained by constructing a second order operator whose principal symbol is given by (6.8.126)

$$\mathcal{A} = \frac{1}{2} \frac{1}{\rho^2(x, y)} (\partial_x^2 + \partial_y^2). \tag{6.8.127}$$

This is the infinitesimal generator of the plane diffusion (X_t, Y_t) with equations

$$dX_t = \frac{1}{\rho(X_t, Y_t)} dW_1(t) \tag{6.8.128}$$

$$dY_t = \frac{1}{\rho(X_t, Y_t)} dW_2(t). \tag{6.8.129}$$

If we further assume that the potential function ρ satisfies:

(i) the boundedness condition: $0 < m < \rho(x, y) < M < \infty$, for all $x, y \in \mathbb{R}$, with m and M finite lower and upper bounds;

(ii) the Lipschits condition: $|\rho(x', y') - \rho(x, y)| \leq \lambda \|(x' - x, y' - y)\|$, for all $(x, x'), (y, y') \in \mathbb{R}^2$ for some $\lambda > 0$,

then the system (6.8.128)-(6.8.129) has a unique strong solution. This follows from the fact that the aforementioned conditions implies the inequality

$$\left| \frac{1}{\rho(x', y')} - \frac{1}{\rho(x, y)} \right| \leq \frac{\lambda}{m^2} \|(x' - x, y' - y)\|$$

and from the application of the existence and uniqueness theorem of stochastic differential equations.

Sometimes, the system (6.8.128)-(6.8.129) can be solved explicitly for some particular cases of potentials ρ. A few of these cases will be discussed in the following.

1. The electric potential In this case we consider an electrical wire that passes through the x-axis with the potential given by the inverse of the distance to the wire, i.e., $\rho(x, y) = \frac{1}{y}$. The curve that minimizes the electric potential between any two points A and B in the half-plane $\mathcal{H}^2 = \{(x, y); y > 0\}$ are geodesics with respect to the hyperbolic metric $d\sigma^2 = \frac{1}{y^2}(dx^2 + dy^2)$. They are either arcs of circle or line segments with direction normal to the wire. The associated diffusion (6.8.128)-(6.8.129) becomes

$$dX_t = Y_t \, dW_1(t) \tag{6.8.130}$$

$$dY_t = Y_t \, dW_2(t), \tag{6.8.131}$$

which is a two-dimensional hyperbolic diffusion, which is approached in section 6.4.2. This model describes the diffusion dynamics of charged particles in the neighborhood of an electrical rectilinear source.

2. The gravitational potential This case considers the potential $\rho(x, y) = y$, with $y > 0$, which represents the distance to the ground. The curve that minimizes this potential between two given points is a *catenary* curve and can be described mathematically using a hyperbolic cosine. This curve is represented physically by the shape of a chain hanging between points A and B and being left free under the gravitational influence. These curves are geodesics with respect to the following x-translation invariant metric

$$d\sigma^2 = y^2(dx^2 + dy^2) = y^2(1 + y'^2)dx^2.$$

They can be obtained by solving the Euler-Lagrange equation associated with the Lagrangian function

$$L(y, y') = y\sqrt{1 + y'^2},$$

which is $yy'' = 1 + y'^2$. Substitution $p = y'$ and using the method of separation of variables we obtain $y(x) = a\cosh(x/a + b)$, with a and b constants such that the boundary conditions are satisfied.

In this case the associated diffusion (6.8.128)-(6.8.129) takes the form

$$dX_t = \frac{1}{Y_t} \, dW_1(t) \tag{6.8.132}$$

$$dY_t = \frac{1}{Y_t} \, dW_2(t). \tag{6.8.133}$$

We solve first equation (6.8.133). To this end we let $Z_t = (Y_t)^2$. By Ito's lemma

$$dZ_t = 2Y_t \, dY_t + \frac{1}{(Y_t)^2} dt = 2dW_2(t) + \frac{1}{Z_t} dt.$$

If we further substitute $R_t = \frac{1}{2}Z_t$ then we arrive at the Bessel process equation

$$dR_t = dW_2(t) + \frac{1/2}{2R_t}dt,$$

which implies that $R_t = R_t^{(3/2)}$. Hence, the second component of the diffusion is $Y_t = \sqrt{2R_t^{(3/2)}}$. To find the first component, we notice that the process $X_t = x_0 + \int_0^t \frac{1}{Y_s}\, dW_1(s)$ is a continuous martingale with quadratic variation

$$\langle X \rangle_t = \int_0^t \frac{1}{2R_s^{(3/2)}} ds.$$

By Theorem 1.1.11 there is a DDS Brownian motion β_t such that

$$X_t = \beta_{\int_0^t \frac{1}{2R_s^{(3/2)}} ds}.$$

3. The inverse square potential If we consider $\rho(x, y) = \dfrac{1}{y^2}$, with $y > 0$, the associated diffusion is given by

$$dX_t = (Y_t)^2\, dW_1(t) \tag{6.8.134}$$
$$dY_t = (Y_t)^2\, dW_2(t). \tag{6.8.135}$$

To solve (6.8.135) we consider $U_t = \dfrac{1}{Y_t}$ and apply Ito's lemma

$$dU_t = -\frac{1}{Y_t^2}dY_t + Y_t\, dt = -dW_2(t) + \frac{1}{U_t}dt.$$

If consider the reflected Brownian motion $\beta_t = -W_2(t)$, then we obtain

$$dU_t = d\beta_t + \frac{1}{U_t}dt,$$

which shows that U_t has the distribution of a 3-dimensional Bessel process, $U_t = R_t^{(3)}$. Therefore, the y-component of the diffusion has the same law as the reciprocal of the distance to the origin of a 3-dimensional Brownian motion, i.e., $Y_t = \dfrac{1}{R_t^{(3)}}$. To solve (6.8.134) we note that $X_t = x_0 + \int_0^t Y_s^2 dW_1(s)$ is a continuous martingale with quadratic variation $\langle X \rangle_t = \int_0^t \frac{1}{(R_s^{(3)})^4}\, ds$. Hence, there is a DDS Brownian motion γ_t such that $X_t = \gamma_{\int_0^t \frac{1}{(R_s^{(3)})^4} ds}$.

4. The inverse square root potential Let the potential be $\rho(x, y) = \dfrac{1}{\sqrt{y}}$. Then the action

$$\int \frac{1}{\sqrt{y}}\, ds = \int \frac{1}{\sqrt{y}} \sqrt{1 + y'^2}\, dx$$

represents the travelling time along the curve $y = y(x)$ from points $A = (x_0, y_0)$ and $B = (x_1, y_1)$ under gravitational influence. The time-minimizing curve satisfies the following associated Euler-Lagrange equation

$$1 + y'^2 + 2yy'' = 0.$$

This can be transformed into an exact equation multiplying by y'. Integrating yields $y' = \pm(C/y - 1)^{1/2}$. Substituting $y = C\sin^2\theta$ we obtain

$$x(t) = \pm\frac{C}{2}(t - \sin t) + x_0, \qquad y(t) = \frac{C}{2}(1 - \cos t) + y_0, \qquad t = 2\theta,$$

which are the parametric equations of a the cycloid with endpoints A and B. The constant C depends on the distance between the end points by

$$d_E(A, B)^2 = \frac{C^2}{4}\big(3 - 2(\sin 1 + \cos 1)\big).$$

The diffusion associated with the previous inverse square root potential is given by

$$dX_t = \sqrt{Y_t}\, dW_1(t) \tag{6.8.136}$$

$$dY_t = \sqrt{Y_t}\, dW_2(t). \tag{6.8.137}$$

To find the law of Y_t, we set $R_t = 2\sqrt{Y_t}$ and apply Ito's lemma

$$dR_t = dW_2(t) - \frac{1}{4}\frac{1}{\sqrt{Y_t}}dt = dW_2(t) - \frac{1}{2R_t}dt.$$

This is a Bessel process[4] of index $\nu = -1$. Then the vertical component of the diffusion becomes $Y_t = y_0 + R_t^2/4$. Integrating in (6.8.136) we get

$$X_t = x_0 + \frac{1}{2}\int_0^t R_s\, dW_1(s),$$

[4]We recall that $dX_t = \frac{a}{X_t}dt + dW_t$ is the stochastic differential equation of a Bessel process of index $\nu = a - \frac{1}{2}$ and dimension $n = 2a + 1$.

which is a continuous martingale with quadratic variation $\langle X\rangle_t = \frac{1}{4}\int_0^t R_s^2\,ds$. Therefore, there is a DDS Brownian motion B_t such that

$$X_t = B_{\frac{1}{4}\int_0^t R_s^2\,ds} = \frac{1}{2}B_{\int_0^t R_s^2\,ds}.$$

Hence, the solution of (6.8.136)-(6.8.137) is given by

$$(X_t, Y_t) = \frac{1}{4}(2B_{\int_0^t R_s^2\,ds}, R_t^2) + (x_0, y_0).$$

It is worth noting that in all previous cases when the potential $\rho(x, y)$ depends only on y, the x-component of diffusion is a time-changed Brownian motion and the y-component depends on a Bessel process. In the next example we consider a potential which depends on both variables x and y.

5. A rotational invariant potential We assume $\rho(x, y) = h(x^2 + y^2)$ with h smooth function. This is a potential depending only on the distance to the origin. The associated diffusion, which is given by

$$dX_t = \frac{1}{h(X_t^2 + Y_t^2)}\,dW_1(t) \tag{6.8.138}$$

$$dY_t = \frac{1}{h(X_t^2 + Y_t^2)}\,dW_2(t), \tag{6.8.139}$$

is also rotational invariant. A few distinguished cases are obtained for some particular expressions of the function h and are given in the following:

5.a Diffusion on the Poincaré disk Choosing $h(u) = \dfrac{2a}{1-u}$, we retrieve the system of stochastic differential equations (6.4.53)-(6.4.54)

$$dX_t = \frac{1 - (X_t^2 + Y_t^2)}{2a}\,dW_1(t) \tag{6.8.140}$$

$$dY_t = \frac{1 - (X_t^2 + Y_t^2)}{2a}\,dW_2(t), \tag{6.8.141}$$

which represents a hyperbolic diffusion on Poincaré's disk $\mathcal{D}^2 = \{(x, y) \in \mathbb{R}^2; x^2 + y^2 \leq 1\}$.

5.b Diffusion on the unit sphere Choosing $h(u) = \dfrac{2}{1+u}$, we retrieve the system of stochastic differential equations (6.4.59)-(6.4.60)

$$dX_t = \frac{1 + (X_t^2 + Y_t^2)}{2a}\,dW_1(t) \tag{6.8.142}$$

$$dY_t = \frac{1 + (X_t^2 + Y_t^2)}{2a}\,dW_2(t), \tag{6.8.143}$$

which represents a diffusion on the unit sphere $\mathbb{S}^2$ in stereographic coordinates, see page 277.

5.c The inverse distance potential Choosing $h(u) = \dfrac{1}{\sqrt{u}}$, we get the diffusion

$$dX_t = \sqrt{X_t^2 + Y_t^2}\, dW_1(t) \tag{6.8.144}$$

$$dY_t = \sqrt{X_t^2 + Y_t^2}\, dW_2(t). \tag{6.8.145}$$

The name comes from the fact that the diffusion evolves under the influence of the potential $\rho(x,y) = \frac{1}{\sqrt{x^2+y^2}}$, which is the inverse of the distance to the origin. This potential is associated to a source of gravitation or static electricity located at the origin, the gravitational and Coulombian forces being derived by applying a gradient to ρ.

The probability density of diffusion (X_t, Y_t) is the heat kernel of the associated generator operator (6.8.127)

$$\mathcal{A} = \frac{1}{2}(x^2 + y^2)(\partial_x^2 + \partial_y^2).$$

It is convenient to rewrite the operator in polar coordinates (r, θ) as

$$\mathcal{A} = \frac{1}{2}(r^2\partial_r^2 + r\partial_r) + \frac{1}{2}\partial_\theta^2 = \mathcal{E} + \Delta_{\mathbb{S}^1},$$

which is the sum between an Euler operator in variable r and the Laplacian on the unit circle $\mathbb{S}^1$ in variable θ. Since $[\mathcal{E}, \Delta_{\mathbb{S}^1}] = \mathcal{E}\Delta_{\mathbb{S}^1} - \Delta_{\mathbb{S}^1}\mathcal{E} = 0$, using the property of the semigroup generated by the sum of two commuting operators, we have

$$e^{t\mathcal{A}} = e^{t(\mathcal{E}+\Delta_{\mathbb{S}^1})} = e^{t\mathcal{E}}e^{t\Delta_{\mathbb{S}^1}}.$$

The heat kernel of $\mathcal{A}$ is obtained by applying the previous semigroup on Dirac's distribution and using the separation of variables r and θ

$$e^{t\mathcal{A}}\delta(r,\theta) = e^{t\mathcal{A}}\delta(r) \otimes \delta(\theta) = e^{t\mathcal{E}}\delta(r)\, e^{t\Delta_{\mathbb{S}^1}}\delta(\theta), \tag{6.8.146}$$

which is the product between the heat kernel of the Euler operator, $e^{t\mathcal{E}}\delta(r)$, and the heat kernel of the circle Laplacian, $e^{t\Delta_{\mathbb{S}^1}}\delta(\theta)$. Each of these two kernels will be computed separately as follows.

(i) The heat kernel of $\mathcal{E} = \frac{1}{2}(r^2\partial_r^2 + r\partial_r)$ is the transition probability of the associated geometric Brownian motion

$$dG_t = \frac{1}{2}G_t dt + G_t dW_t.$$

Setting $F_t = \ln G_t$ and applying Ito's lemma we obtain $dF_t = dW_t$, which implies $G_t = G_0 e^{W_t}$, which is log-normally distributed, with

$$p_{G_t}(r_0, r; t) = \frac{1}{r}\frac{1}{\sqrt{2\pi t}} e^{-\frac{1}{2t}(\ln r - \ln r_0)^2}$$

(*ii*) The heat kernel of $\Delta_{\mathbb{S}^1}$ is given by (2.18.45)

$$p_{\mathbb{S}^1}(\theta_0, \theta; t) = \frac{1}{\sqrt{2\pi t}} e^{-\frac{(\theta-\theta_0)^2}{2t}} \theta_3\Big(\frac{\pi(\theta-\theta_0)i}{t} \mid \frac{2\pi i}{t}\Big),$$

where

$$\theta_3(z|i\tau) = 1 + 2\sum_{n\geq 1} e^{-\pi n^2 \tau}\cos(2nz).$$

The heat kernel of $\mathcal{A}$ is obtained by taking the product of kernels via formula (6.8.146) as

$$\begin{aligned} e^{t\mathcal{A}}\delta(r,\theta) &= p_{G_t}(r_0, r; t) p_{\mathbb{S}^1}(\theta_0, \theta; t) \\ &= \frac{1}{2\pi r t} e^{-\frac{1}{2t}[(\ln r - \ln r_0)^2 + (\theta-\theta_0)^2]} \theta_3\Big(\frac{\pi(\theta-\theta_0)i}{t} \mid \frac{2\pi i}{t}\Big). \end{aligned}$$

Therefore, if we write the diffusion in polar coordinates as $(X_t, Y_t) = r_t e^{i\theta_t}$, then its transition density is given by

$$p_t(r_0, \theta_0; r, \theta) = \frac{1}{2\pi r t} e^{-\frac{1}{2t}[(\ln r - \ln r_0)^2 + (\theta-\theta_0)^2]} \theta_3\Big(\frac{\pi(\theta-\theta_0)i}{t} \mid \frac{2\pi i}{t}\Big),$$

for $r, r_0 > 0$ and $\theta, \theta_0 \in \mathbb{R}$. It is worth noting that the small time asymptotics takes the form

$$p_t(r_0, \theta_0; r, \theta) \sim \frac{1}{2\pi r t} e^{-\frac{1}{2t}[(\ln r - \ln r_0)^2 + (\theta-\theta_0)^2]}, \qquad t \sim 0.$$

6.9 Bessel Process with Drift in $\mathbb{R}^3$

Consider the process

$$X_t = \big(\mu_1 t + W_1(t), \mu_2 t + W_2(t), \mu_3 t + W_3(t)\big), \tag{6.9.147}$$

where $\mu = (\mu_1, \mu_2, \mu_3) \in \mathbb{R}^3$ is a given vector of length $|\mu| = \sqrt{\mu_1^2 + \mu_2^2 + \mu_3^2}$ and $W_1(t)$, $W_2(t)$ and $W_3(t)$ are three independent one-dimensional Brownian motions. The process X_t can be regarded as a 3-dimensional Brownian motion with drift rate μ. The radial part of X_t

$$R_t = R_t^{\mu} = \|X_t\| = \sqrt{\sum_{i=1}^{3} (\mu_i t + W_i(t))^2} \tag{6.9.148}$$

is called the *3-dimensional Bessel process with drift rate* μ starting at the origin.

The probability density In the following we shall compute the probability density of R_t. Since processes $X_i(t) = \mu_i t + W_i(t)$ are independent and have normal densities

$$f_{X_i}(x) = \frac{1}{\sqrt{2\pi t}} e^{-\frac{(x-\mu_i t)^2}{2t}}, \qquad t > 0$$

then the distribution function of R_t is given for any $\rho > 0$ by

$$\begin{aligned}
F(\rho) = P(R_t \le \rho) &= \int_{|x|\le\rho} f_{X_1}(x_1) f_{X_2}(x_2) f_{X_3}(x_3)\, dx_1 dx_2 dx_3 \\
&= \frac{1}{(2\pi t)^{3/2}} \int_{|x|\le\rho} e^{-\frac{1}{2t}\sum_i (x_i-\mu_i t)^2}\, dx \\
&= \frac{1}{(2\pi t)^{3/2}} \int_{|x|\le\rho} e^{-\frac{1}{2t}|x|^2}\, e^{-\frac{1}{2t}t^2|\mu|^2}\, e^{\sum_i x_i\mu_i} dx \\
&= \frac{1}{(2\pi t)^{3/2}} e^{-\frac{1}{2t}t^2|\mu|^2} \int_{|x|\le\rho} e^{-\frac{1}{2t}|x|^2+\langle x,\mu\rangle}\, dx, \qquad (6.9.149)
\end{aligned}$$

where $dx = dx_1 dx_2 dx_3$. The previous integral is rotational invariant, in the sense that if M is an arbitrary orthogonal matrix, $\det M = 1$, we have

$$I = \int_{|x|\le\rho} e^{-\frac{1}{2t}|x|^2+\langle x,\mu\rangle}\, dx = \int_{|x|\le\rho} e^{-\frac{1}{2t}|Mx|^2+\langle Mx, M\mu\rangle}\, dx.$$

In order to compute the integral, we consider a matrix M that rotates the vector μ towards $e_3 = (0,0,1)$, namely $M\mu = |\mu| e_3$. Substituting $y = Mx$, the integral becomes

$$\begin{aligned}
I &= \int_{|y|\le\rho} e^{-\frac{1}{2t}|y|^2+\langle y, |\mu| e_3\rangle} (\det M)^{-1}\, dy \\
&= \int_{|y|\le\rho} e^{-\frac{1}{2t}|y|^2+|y|\,|\mu|\cos\phi}\, dy,
\end{aligned}$$

where ϕ is the angle between the vector y and direction e_3. Employing spherical coordinates

$$y_1 = r\sin\phi\cos\theta, \quad y_2 = r\sin\phi\sin\theta, \quad y_3 = r\cos\phi,$$

with $0 \le r \le \rho$, $0 \le \phi \le \pi$ and $0 \le \theta \le 2\pi$, and using the volume element

$dy = r^2 \sin\phi \, dr d\theta d\phi$, the integral becomes

$$\begin{aligned} I &= \int_0^{\pi} \int_0^{2\pi} \int_0^{\rho} e^{-\frac{1}{2t}r^2 + r|\mu|\cos\phi} r^2 \sin\phi \, dr d\theta d\phi \\ &= 2\pi \int_0^{\rho} e^{-\frac{1}{2t}r^2} r^2 \left(\int_0^{\pi} e^{r|\mu|\cos\phi} \sin\phi \, d\phi \right) dr \\ &= 2\pi \int_0^{\rho} e^{-\frac{1}{2t}r^2} r^2 \left(\int_{-1}^{1} e^{r|\mu|v} dv \right) dr \\ &= 4\pi \int_0^{\rho} e^{-\frac{1}{2t}r^2} r^2 \frac{1}{r|\mu|} \sinh(r|\mu|) \, dr. \end{aligned}$$

Substitute in (6.9.149) yields

$$F(\rho) = \frac{4\pi}{(2\pi t)^{3/2}} e^{-\frac{1}{2t}t^2|\mu|^2} \int_0^{\rho} e^{-\frac{1}{2t}r^2} r^2 \frac{\sinh(r|\mu|)}{r|\mu|} \, dr, \qquad \rho \geq 0.$$

The probability density $p_t(\rho)$ defined by

$$P(R_t \in d\rho) = p_t(\rho) \, d\rho$$

is now obtained using the formula $p_t(\rho) = F'(\rho)$. Differentiating, we get the following probability density of R_t

$$p_t(\rho) = \frac{2\rho^2}{\sqrt{2\pi} t^{3/2}} e^{-\frac{1}{2t}(\rho^2 + t^2|\mu|^2)} \frac{\sinh(\rho|\mu|)}{\rho|\mu|}, \qquad \rho > 0. \tag{6.9.150}$$

We also have $p_t(0) = \lim_{\rho \to 0} p_t(\rho) = 0$ for $t > 0$. The graph of $p_t(\rho)$ is skewed to the right. When $|\mu|$ increases, the peak of $p_t(x)$ moves to the right.

Remark 6.9.1 It is worth noting that the density of the process R_t^{μ} depends on $|\mu|$, not on μ directly. It follows that any two Bessel processes with equal magnitude of the drift rate have the same density. Therefore, for any orthogonal matrix M, the processes R_t^{μ} and $R_t^{M\mu}$ have the same law, given by formula (6.9.150).

Remark 6.9.2 The Bessel process on drifting Brownian motions has been generalized to any dimension $k \geq 3$, see Theorem 3 in Rogers and Pitman [121]. Thus, the radial part of $W_t + \mu t$, with $\mu \in \mathbb{R}^k$ and W_t Brownian motion on $\mathbb{R}^k$ is a diffusion process on $[0, \infty)$ with the transition density

$$q_{k,\mu}(t, x, y) = e^{-\frac{t|\mu|^2}{2}} h_k(|\mu| x)^{-1} q_k(t, x, y) h_k(|\mu| y),$$

where $h_k(y) = (y/2)^{-\nu}\Gamma(\nu+1)I_\nu(y)$, $\nu = (k/2)-1$ and $q_k(t, x, y)$ the transition density of the k-dimensional Bessel process $|W_t|$ on $\mathbb{R}^k$. It is worth noting that in particular for $k = 3$ we have $h_3(y) = \dfrac{\sinh y}{y}$; then it can be shown that for $x = 0$ we retrieve formula (6.9.150).

The infinitesimal generator We are looking for a second order differential operator whose heat kernel is the aforementioned probability density.

Proposition 6.9.3 *The heat kernel for the differential operator*

$$L_x^\mu = \frac{1}{2}\frac{d^2}{dx^2} - |\mu| \coth(|\mu|x)\frac{d}{dx} + \frac{|\mu|^2}{\sinh^2(|\mu|x)}, \qquad x > 0$$

is given by

$$p_t(x) = \frac{2x^2}{\sqrt{2\pi}t^{3/2}} e^{-\frac{1}{2t}(x^2+t^2|\mu|^2)} \frac{\sinh(|\mu|x)}{|\mu|x}, \qquad x > 0.$$

Proof: We need to show

$$\partial_t p_t(x) = L_x^\mu p_t(x) \tag{6.9.151}$$

$$\lim_{t \searrow 0} p_t(x) = \delta_x. \tag{6.9.152}$$

This will be accomplished by a direct computation. First we note that we can write $p_t(x) = \varphi_t(x)\psi_t(x)$, where

$$\varphi_t(x) = \frac{1}{\sqrt{2\pi t}} e^{-\frac{x^2}{2t}}, \qquad \psi_t(x) = \frac{2}{t} e^{-\frac{t|\mu|^2}{2}} \frac{x \sinh(|\mu|x)}{|\mu|}.$$

By a straightforward computation we obtain

$$\begin{aligned}
\partial_t \varphi_t &= \frac{1}{2}\partial_x^2 \varphi_t, \qquad \partial_t \psi_t = -\Big(\frac{|\mu|^2}{2} + \frac{1}{t}\Big)\psi_t \\
\partial_x \varphi_t &= -\frac{x}{t}\varphi_t, \qquad \partial_x \psi_t = \Big(\frac{1}{x} + |\mu| \coth(|\mu|x)\Big)\psi_t \\
\partial_x^2 \psi_t &= \Big(|\mu|^2 + \frac{2|\mu|}{x} \coth(|\mu|x)\Big)\psi_t.
\end{aligned}$$

We shall find two functions, $a(x)$ and $b(x)$, such that

$$\partial_t p_t(x) = \frac{1}{2}\partial_x^2 p_t(x) + a(x)\partial_x p_t(x) + b(x)p_t(x). \tag{6.9.153}$$

Using the previous derivatives computation, the left side of (6.9.153) is given by

$$\begin{aligned}
\partial_t p_t &= \partial_t \varphi_t\, \psi_t + \varphi_t\, \partial_t \psi_t \\
&= \partial_t \varphi_t\, \psi_t - \Big(\frac{|\mu|^2}{2} + \frac{1}{t}\Big)p_t.
\end{aligned} \tag{6.9.154}$$

The derivatives of the right side of (6.9.153) can be computed as

$$\begin{aligned}\partial_x p_t &= \partial_x(\varphi_t\psi_t) = \partial_x\varphi_t\,\psi_t + \varphi_t\,\partial_x\psi_t\\ &= -\frac{x}{t}p_t + \Big(\frac{1}{x} + |\mu|\coth(|\mu|x)\Big)p_t,\end{aligned} \tag{6.9.155}$$

$$\begin{aligned}\partial_x^2 p_t &= \partial_x^2(\varphi_t\psi_t) = \partial_x^2\varphi_t\,\psi_t + 2\partial_x\varphi_t\,\partial_x\psi_t + \varphi_t\partial_x^2\psi_t\\ &= \partial_x^2\varphi_t\,\psi_t - \frac{2x}{t}\Big(\frac{1}{x} + |\mu|\coth(|\mu|x)\Big)p_t\\ &\quad + \Big(|\mu|^2 + \frac{2|\mu|}{x}\coth(|\mu|x)\Big)p_t.\end{aligned} \tag{6.9.156}$$

Substituting relations (6.9.154), (6.9.155) and (6.9.156) into equation (6.9.153) and using that $\partial_t\varphi_t = \frac{1}{2}\partial_x^2\varphi_t$, after cancelations and regrouping, we obtain

$$\begin{aligned}0 &= |\mu|^2 + \Big(-\frac{x}{t} + \frac{1}{x}\Big)\Big(a(x) + |\mu|\coth(|\mu|x)\Big)\\ &\quad + |\mu|a(x)\coth(|\mu|x) + b(x),\end{aligned}$$

which is valid for any $x > 0$ and $t > 0$. Keeping $x > 0$ fixed and taking $t \to 0$ we infer that $a(x) + |\mu|\coth(|\mu|x) = 0$. Since x was chosen arbitrary, it follows that

$$a(x) = -|\mu|\coth(|\mu|x).$$

Substituting back into the previous expression we obtain

$$b(x) = -|\mu|^2(1 - \coth^2(|\mu|x)) = \frac{|\mu|^2}{\sinh^2(|\mu|x)}.$$

Hence, relation (6.9.151) holds. To show (6.9.152) we notice first that since $p_t(x)$ is the probability density of the process (6.9.148), then $p_t \geq 0$ and $\int_0^\infty p_t(x), dx = 1$, see Exercise 6.19.1. To show the spike behavior of p_t as $t \to 0$, we notice that the mode of p_t solves the equation $\partial_x p_t(x) = 0$, which can be written as

$$|\mu|\coth(|\mu|x) = \frac{x}{t} - 1.$$

Substituting $\xi = |\mu|x$, the aforementioned equation writes equivalently as

$$1 + \xi\coth\xi = \frac{\xi^2}{|\mu|^2 t}, \qquad \xi \geq 0.$$

This equation has a unique solution, $\xi_t > 0$. For values $\xi > \xi_t$ the quadratic part on the right exceeds the left side, which increases at most linearly in ξ. We also note that $\xi_t \to 0$ as $t \to 0$. Then $\partial_x p_t(x) > 0$ for $x \in (0, \xi_t/|\mu|)$ and $\partial_x p_t(x) < 0$ for $x > \xi_t/|\mu|$. The mode of p_t occurs at $x = \xi_t/|\mu|$, which approaches zeros as $t \to 0$. ∎

Remark 6.9.4 It is worth noting the behavior when $|\mu| \to 0$. In this case we obtain that the heat kernel of the operator

$$L_x^0 = \frac{1}{2}\frac{d^2}{dx^2} - \frac{1}{x}\frac{d}{dx}$$

is given by

$$p_t(x) = \frac{2x^2}{\sqrt{2\pi}t^{3/2}} e^{-\frac{1}{2t}x^2}, \qquad x > 0.$$

The next section tries to generalize the radial part of a Brownian motion with drift to the norm of n processes.

6.10 The Norm of n Processes

Consider n stochastic processes $X_j(\mathrm{t})$ given by

$$dX_j(t) = b_j(X_1, \cdots, X_n)dt + dW_j(t), \qquad j = 1, \ldots, n, \tag{6.10.157}$$

with $W_j(t)$ independent Brownian motions and $b_j(\cdot)$ linear functions

$$b_j(x_1, \cdots, x_n) = \sum_{i=1}^{n} a_{ji}x_i, \qquad j = 1, \ldots, n$$

and $a_{ij} \in \mathbb{R}$. It is worth noting that under these hypotheses the system (6.10.157) together with the initial condition $X_j(0) = x_j$, $j = 1, \ldots, n$ has a unique strong solution.

The norm of the n-dimensional process $X(t) = \big(X_1(t), \cdots, X_n(t)\big)$ is given by the one-dimensional process

$$R_t = \|X(t)\| = \sqrt{X_1^2(t) + \cdots + X_n^2(t)}.$$

In particular, if $b_j = 0$, then R_t becomes the usual n-dimensional Bessel process.

Applying Ito's lemma we have

$$\begin{aligned} d(R_t^2) &= 2\sum_j X_j(t)dX_j(t) + ndt \\ &= 2\sum_j X_j(t)\, dW_j(t) + \Big(2\sum_j X_j(t)b_j(X(t)) + n\Big)dt. \end{aligned}$$

Consider the martingale

$$\beta_t = \int_0^t \sum_j \frac{X_j(s)dW_j(s)}{R_s}.$$

Since $(d\beta_t)^2 = \sum_j \frac{X_j^2(t)}{R_t^2} dt = dt$, then by Levy's theorem (see Proposition 1.1.9) β_t is a Brownian motion. Then we can write

$$d(R_t^2) = 2R_t\, d\beta_t + \Big(2\sum_j X_j(t) b_j(X(t)) + n\Big) dt.$$

Substituting $Z_t = R_t^2$, the previous relation becomes

$$dZ_t = 2Z_t^{1/2}\, d\beta_t + \Big(2\sum_j X_j(t) b_j(X(t)) + n\Big) dt,$$

which implies $(dZ_t)^2 = 4Z_t dt$. Applying Ito's lemma again yields

$$\begin{aligned} dR_t = d(Z_t^{1/2}) &= \frac{1}{2} Z_t^{-1/2}\, dZ_t - \frac{1}{8} Z_t^{-3/2} (dZ_t)^2 \\ &= d\beta_t + \frac{1}{2R_t}\Big(2\sum_j X_j(t) b_j(X(t)) + n - 1\Big) dt. \end{aligned}$$

Denoting $A = (a_{ji})$, we obtain

$$dR_t = d\beta_t + \left(\frac{\langle AX(t), X(t)\rangle}{R_t} + \frac{n-1}{2R_t} \right) dt, \qquad (6.10.158)$$

where $\langle\, ,\, \rangle$ denotes the Euclidean inner product. If the matrix A is of a particular form the previous expression can be simplified even further.

1. The case $A = \lambda \mathbb{I}_n$ This case recovers the n-dimensional Ornstein-Uhlenbeck process defined by the system of stochastic differential equations

$$dX_j(t) = \lambda X_j(t) dt + dW_j(t), \qquad j = 1, \ldots, n, \qquad (6.10.159)$$

with $\lambda \in \mathbb{R}$ and associate generator

$$\mathcal{A} = \frac{1}{2}\sum_{j=1}^n \partial_{x_j}^2 + \lambda \sum_{j=1}^n x_j \partial_{x_j}. \qquad (6.10.160)$$

In this case (6.10.158) becomes

$$dR_t = d\beta_t + \Big(\lambda R_t + \frac{n-1}{2R_t}\Big) dt,$$

which is the equation satisfied by the norm of an n-dimensional Ornstein-Uhlenbeck process. From here we also infer that the infinitesimal generator associated to R_t is given by

$$L = \frac{1}{2}\partial_r^2 + \Big(\lambda r + \frac{n-1}{2r}\Big)\partial_r. \qquad (6.10.161)$$

It turns out that L is the radial part of the operator $\mathcal{A}$ given by (6.10.160). Its heat kernel can be obtained by a similar computation with the one employed at page 37, see Exercise 6.19.6. We also note that for $n = 2$ we retrieve equation (1.10.60) (with λ replaced by $-\lambda$).

2. The case when A is a rotation Let $n = 2$ and assume

$$A = \begin{pmatrix} \cos u & \sin u \\ -\sin u & \cos u \end{pmatrix}$$

for some $u \in \mathbb{R}$. This corresponds to the system

$$\begin{aligned} dX_1(t) &= \big(\cos u\, X_1(t) + \sin u\, X_2(t)\big)dt + dW_1(t) \\ dX_2(t) &= \big(-\sin u\, X_1(t) + \cos u\, X_2(t)\big)dt + dW_2(t). \end{aligned}$$

In this case equation (6.10.158) becomes

$$dR_t = d\beta_t + \Big(R_t \cos u + \frac{1}{2R_t}\Big)dt,$$

where we used that $\langle AX(t), X(t)\rangle = |AX(t)|\,|X(t)| \cos\theta = R_t^2 \cos\theta$. The infinitesimal generator associated to R_t is given by the operator

$$L = \frac{1}{2}\partial_r^2 + \Big(r\cos u + \frac{1}{2r}\Big)\partial_r. \tag{6.10.162}$$

We note that (6.10.162) is a particular case of (6.10.161) with $\lambda = \cos u$.

For $u = \pi/2$ we obtain $A = \mathcal{J}$, where

$$\mathcal{J} = \begin{pmatrix} 0 & 1 \\ -1 & 0 \end{pmatrix}, \qquad \mathcal{J}^2 = -\mathbb{I}_2$$

is the complex structure on $\mathbb{R}^2$. In this case the system becomes

$$dX_1(t) = X_2(t)dt + dW_1(t) \tag{6.10.163}$$

$$dX_2(t) = -X_1(t)dt + dW_2(t). \tag{6.10.164}$$

The associated norm satisfies

$$dR_t = d\beta_t + \frac{1}{2R_t}dt,$$

while the infinitesimal generator is given by

$$L = \frac{1}{2}\partial_r^2 + \frac{1}{2r}\partial_r.$$

This implies that $R_t = \sqrt{X_1(t)^2 + X_2(t)^2}$ is a two-dimensional Bessel process.

Remark 6.10.1 The infinitesimal generator of the process given by the system (6.10.163)-(6.10.164) is given in Cartesian coordinates by

$$P = \frac{1}{2}(\partial_{x_1}^2 + \partial_{x_2}^2) - (x_1\partial_{x_2} - x_2\partial_{x_1}).$$

This is the difference between a two-dimensional Laplacian and a momentum operator, both being rotational invariant. The operator P can be rewritten using polar coordinates, $x = r\cos\theta$ and $y = r\sin\theta$, as a sum of two parts: the radial part, P_{rad}, which depends solely on the radius variable r and the angular part, P_{ang}, which depends also on the angle variable θ. This is

$$P = P_{rad} + P_{ang},$$

where

$$P_{rad} = \frac{1}{2}\partial_r^2 + \frac{1}{2r}\partial_r, \qquad P_{ang} = \frac{1}{2r^2}\partial_\theta^2 - \partial_\theta.$$

The radial part of $X(t) = \big(X_1(t), X_2(t)\big)$, that is $R_r = \|X_t\|$, has the infinitesimal generator given by the radial part of the infinitesimal generator of $X(t) = \big(X_1(t), X_2(t)\big)$, that is $P_{rad} = \frac{1}{2}\partial_r^2 + \frac{1}{2r}\partial_r$, which corresponds to a two-dimensional Bessel process.

6.11 Central Projection on $\mathbb{S}^n$

In this section we shall deal with the central projection of an $(n+1)$-dimensional Brownian motion on the n-dimensional unit sphere, which under a certain time change reduces to a diffusion living on the sphere.

Let $W_t = \big(W_1(t), \cdots, W_{n+1}(t)\big)$ be a Brownian motion in $\mathbb{R}^{n+1}$ and consider the intersection of the vectorial radius of W_t with the unit sphere $\mathbb{S}^n$. This intersection is modeled by a process on the n-dimensional unit sphere given by $Y_t = W_t/|W_t|$. We shall compute the stochastic differential equation of Y_t following Øksendal [114], page 151.

Setting $Y_t = \phi(W_t)$ with $\phi : \mathbb{R}^{n+1}\backslash\{0\} \to \mathbb{S}^n$, $\phi(y) = \dfrac{y}{|y|}$, by applying the multidimensional Ito's formula, we obtain

$$dY_t = \frac{1}{|W_t|}\sigma(Y_t)dW_t + \frac{1}{|W_t|^2}b(Y_t)dt, \tag{6.11.165}$$

where

$$\sigma(y) = \mathbb{I}_{n+1} - yy^T, \qquad b(y) = -\frac{n}{2}y, \tag{6.11.166}$$

where $y = (y_1, \cdots, y_{n+1})^T$.

For the sake of simplicity, we shall show how to get to this formula in the case $n = 1$. The general case is similar, but the work is more laborious. An application of Ito's formula yields

$$dY_1(t) = \frac{W_2^2}{|W|^3} dW_1 - \frac{W_1 W_2}{|W|^3} dW_2 - \frac{W_1}{2|W|^3} dt$$
$$dY_2(t) = -\frac{W_1 W_2}{|W|^3} dW_1 + \frac{W_1^2}{|W|^3} dW_1 - \frac{W_2}{2|W|^3} dt.$$

Substituting $W_1 = |W_1|Y_1$ and $W_2 = |W_2|Y_2$, the previous relations become

$$dY_t = \frac{1}{|W_t|} \begin{pmatrix} Y_2^2 & -Y_1Y_2 \\ -Y_1Y_2 & Y_1^2 \end{pmatrix} dW_t - \frac{1}{2|W_t|^2} \begin{pmatrix} Y_1 \\ Y_2 \end{pmatrix} dt, \tag{6.11.167}$$

which agrees with (6.11.165).

The process Y_t given by (6.11.165) is an Ito process living on $\mathbb{S}^n$. For convenience, we write it as

$$dY_t = v(t, \omega) dW_t + u(t, \omega) dt,$$

with $u(t, \omega) = \frac{1}{|W_t|^2} b(Y_t)$ and $v(t, \omega) = \frac{1}{|W_t|} \sigma(Y_t)$.

We shall see that under a suitable time change the process Y_t becomes the diffusion

$$dX_t = (\mathbb{I}_{n+1} - X_t X_t^T) dW_t - \frac{n}{2} X_t dt. \tag{6.11.168}$$

To this end, we note first that this diffusion can be written equivalently as

$$dX_t = \sigma(X_t) dW_t + b(X_t) dt,$$

with σ and b given by (6.11.166). Since the hypotheses of Theorem 1.1.14

$$u(t, \omega) = c(t, \omega) b(Y_t)$$
$$vv^T(t, \omega) = c(t, \omega) \sigma\sigma^T(Y_t)$$

hold with $c(t, \omega) = \frac{1}{|W_t(\omega)|^2}$, then it follows that

$$\beta_t = \int_0^t \frac{1}{|W_s|^2} ds, \qquad \alpha_t = \beta_t^{-1}$$

are time changes with properties

$$Y_{\alpha_t} = X_t, \qquad Y_t = X_{\beta_t}.$$

Therefore, the central projection of an $(n+1)$-dimensional Brownian motion, W_t, onto the unit sphere, $\mathbb{S}^n$, given by the process $Y_t = W_t/|W_t|$, is a diffusion process, X_t, on the unit sphere under the stochastic time β_t. We note that X_t is not a Brownian motion on the sphere.

Remark 6.11.1 The above result works for the case $n \geq 2$, since the Brownian motion W_t on $\mathbb{R}^{n+1}$ does not return to the origin almost surely. For the case $n = 1$ we should consider additional requirements, such as the Brownian motion starting outside the origin.

6.12 The Skew-product Representation

We shall develop further the analysis for the case $n = 1$ to obtain a formula for the planar Brownian motion in terms of its radial and angular parts, called the *skew-product representation*. For another proof of this formula the reader can consult Pitman and Yor [118], or Ito and McKean [83].

Let $W_t = \big(W_1(t), W_2(t)\big)$ denote a two-dimensional Brownian motion starting at $W_0 = w_0 = \rho_0 e^{\theta_0} \neq 0$, and

$$Y_t = \big(W_1(t), W_2(t)\big)/|W_t|$$

be the central projection of W_t on the unit sphere, $\mathbb{S}^1$, which satisfies equation (6.11.167). In this case the diffusion X_t on the circle $\mathbb{S}^1$, which is given by (6.11.168), takes the form

$$dX_1(t) = -\frac{1}{2}X_1(t)\,dt + X_2(t)^2 dW_1(t) - X_1(t)X_2(t)dW_2(t) \qquad (6.12.169)$$

$$dX_2(t) = -\frac{1}{2}X_2(t)\,dt - X_1(t)X_2(t)dW_1(t) + X_1(t)^2 dW_2(t). \qquad (6.12.170)$$

This diffusion depends on two independent Brownian motions, $W_1(t)$ and $W_2(t)$. We shall reduce it to a diffusion that depends on only one Brownian motion and show that X_t becomes the Brownian motion on the unit circle given by equations (2.18.42)-(2.18.43).

More precisely, we shall show the existence of a one-dimensional Brownian motion, B_t, such that

$$dX_1(t) = -\frac{1}{2}X_1(t)\,dt - X_2(t)\,dB_t \qquad (6.12.171)$$

$$dX_2(t) = -\frac{1}{2}X_2(t)\,dt + X_1(t)\,dB_t. \qquad (6.12.172)$$

To achieve this goal, we consider the process

$$B_t = \int_0^t X_1(s)dW_2(s) - \int_0^t X_2(s)dW_1(s),$$

which is a continuous martingale, with differential

$$dB_t = X_1(t)dW_2(t) - X_2(t)dW_1(t)$$

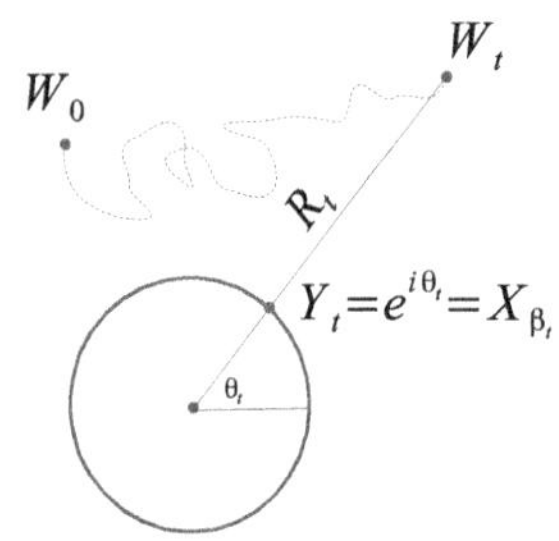

Figure 6.3: *The central projection Y_t of a Brownian motion W_t on the unit circle.*

and quadratic variation

$$\langle B, B\rangle_t = \int_0^t (dB_t)^2 = \int_0^t \left(X_1^2(s) + X_2^2(s)\right) ds = t.$$

Then, by Lévy's theorem, it follows that B_t is a Brownian motion. Then substituting the expression of dB_t into equations (6.12.169)-(6.12.170) we obtain (6.12.171)-(6.12.172).

The Brownian motion B_t represents the angular stochastic parameter of the process X_t. By the result of the previous section, there is a time change, $\beta_t = \int_0^t \frac{1}{R_s^2} ds$, such that $Y_t = X_{\beta_t}$, where X_t is a Brownian motion on the unit circle and $R_s = |W_s|$ is a Bessel process of dimension 2. This means

$$Y_1(t) = X_1(\beta_t) = \cos(\theta_0 + B_{\beta_t}) \tag{6.12.173}$$
$$Y_2(t) = X_2(\beta_t) = \sin(\theta_0 + B_{\beta_t}). \tag{6.12.174}$$

Therefore, the process Y_t can be represented in polar coordinates as

$$Y_t = e^{i\theta_t} = e^{i(\theta_0 + B_{\beta_t})}.$$

Since $Y_t = W_t/|W_t|$, we obtain the following representation of a two-dimensional Brownian motion in polar coordinates

$$W_t = |W_t|e^{i\theta_t} = R_t e^{i(\theta_0 + B_{\beta_t})}, \tag{6.12.175}$$

with $\beta_t = \int_0^t \frac{1}{R_s^2} ds$ and R_s a two-dimensional Bessel process starting at $R_0 = \rho_0 \neq 0$, see Fig. 6.3. The angular process θ_t is called *the winding number* of W_t about zero. It is worth noting that the Brownian motion B_t is independent of the process R_t. Since the skew-product representation (6.12.175) reduces some problems involving planar Brownian motions to the study of its radial and angular parts, it is worth investigating their laws.

The radial part The radial part of the Brownian motion is a two-dimensional Bessel process that starts at $\rho_0 \neq 0$

$$R_t = R_t^{(2)} = |W_t|$$

with the probability density given by

$$P_{\rho_0}(R_t^{(2)} \in d\rho) = \frac{\rho}{t} e^{-\frac{\rho_0^2+\rho^2}{2t}} I_0\Big(\frac{\rho_0 \rho}{t}\Big) d\rho, \tag{6.12.176}$$

where I_0 denotes the modified Bessel function of the first kind of order zero, see for instance Calin et al. [29], page 180.

The angular part The angular process θ_t associated to a planar Brownian motion, W_t, has a relatively complex expression, which depends on several special functions. We shall compute its characteristic function under the assumptions $\rho_0 \neq 0$ and $\theta_0 = 0$. We have by the tower property

$$\begin{aligned}\mathbb{E}[e^{i\lambda\theta_t}] &= \mathbb{E}[e^{i\lambda B_{\beta_t}}] = \mathbb{E}[\,\mathbb{E}[e^{i\lambda B_{\beta_t}} | W[0,t]]\\ &= \mathbb{E}[e^{-\frac{1}{2}\lambda^2 \beta_t}] = \mathbb{E}[e^{-\frac{1}{2}\lambda^2 \int_0^t \frac{1}{R_s^2} ds}].\end{aligned}$$

The last expression is the Laplace transform of the time change β_t evaluated at $s = \frac{1}{2}\lambda^2$. An exact formula for this [5] can be found in Borodin and Salminen [15], page 392, formula 1.20.3, in terms of a Whittaker function as

$$\mathbb{E}[e^{-\frac{1}{2}\lambda^2 \int_0^t \frac{1}{R_s^2} ds}] = \frac{\sqrt{2t}\,\Gamma(1+\frac{\lambda}{2})}{\rho_0\;\Gamma(1+\lambda)} e^{-\frac{\rho_0^2}{4t}} M_{-\frac{1}{2},\frac{\lambda}{2}}\Big(\frac{\rho_0^2}{2t}\Big). \tag{6.12.177}$$

The previous Whittaker function with parameters $(-\frac{1}{2}, \frac{\lambda}{2})$ can be written in terms of a Kummer function as

$$M_{-\frac{1}{2},\frac{\lambda}{2}}(x) = x^{\frac{\lambda+1}{2}} e^{-\frac{x}{2}} M\Big(\frac{\lambda}{2} + 1, \lambda + 1, x\Big).$$

[5] This is obtained by choosing $n = 2$, $\nu = 0$, $x = \rho_0$ and $\lambda = \gamma$ in the aforementioned formula 1.20.3

$$\mathbb{E}_x e^{\left(-\frac{\gamma^2}{2}\int_0^t \frac{ds}{(R_s^{(n)})^2}\right)} = \frac{(2t)^{(\nu+1)/2}\Gamma(1+\nu/2+\sqrt{\nu^2+\gamma^2}/2)}{x^{\nu+1}\Gamma(1+\sqrt{\nu^2+\gamma^2})} e^{-\frac{x^2}{4t}} M_{-\nu/2-1/2,\sqrt{\nu^2+\gamma^2}/2}\Big(\frac{x^2}{2t}\Big),$$

where

$$M_{n,m}(x) = x^{m+1/2} e^{-x/2} M(m-n+1/2, 2m+1, x)$$

is a Whittaker function that depends on the Kummer function, see also page 647 in [15].

Substituting in (6.12.177) we obtain

$$\mathbb{E}[e^{-\frac{1}{2}\lambda^2\int_0^t \frac{1}{R_s^2}ds}] = \Big(\frac{\rho_0^2}{2t}\Big)^{\frac{\lambda}{2}}\frac{\Gamma(1+\frac{\lambda}{2})}{\Gamma(1+\lambda)}e^{-\frac{\rho_0^2}{2t}}M\Big(\frac{\lambda}{2}+1,\lambda+1,\frac{\rho_0^2}{2t}\Big). \qquad (6.12.178)$$

The Kummer function on the right side has the following integral representation[6]

$$M\Big(\frac{\lambda}{2}+1,\lambda+1,\frac{\rho_0^2}{2t}\Big) = \frac{\Gamma(\lambda+1)}{\Gamma\Big(\frac{\lambda}{2}+1\Big)\Gamma\Big(\frac{\lambda}{2}\Big)}\int_0^1 e^{\frac{\rho_0^2}{2t}u}u^{\frac{\lambda}{2}}(1-u)^{\frac{\lambda}{2}-1}du.$$

Therefore, (6.12.178) becomes

$$\begin{aligned}\mathbb{E}[e^{-\frac{1}{2}\lambda^2\int_0^t \frac{1}{R_s^2}ds}] &= \Big(\frac{\rho_0^2}{2t}\Big)^{\frac{\lambda}{2}}\frac{e^{-\frac{\rho_0^2}{2t}}}{\Gamma\Big(\frac{\lambda}{2}\Big)}\int_0^1 e^{\frac{\rho_0^2}{2t}u}u^{\frac{\lambda}{2}}(1-u)^{\frac{\lambda}{2}-1}du \\ &= \Big(\frac{\rho_0^2}{2t}\Big)^{\frac{\lambda}{2}}\frac{1}{\Gamma\Big(\frac{\lambda}{2}\Big)}\int_0^1 e^{\frac{\rho_0^2}{2t}v}(1-v)^{\frac{\lambda}{2}}v^{\frac{\lambda}{2}-1}dv, \end{aligned} \qquad (6.12.179)$$

where $v = 1-u$. We set

$$G_\lambda(x) = \frac{x^{\frac{\lambda}{2}}}{\Gamma\Big(\frac{\lambda}{2}\Big)}\int_0^1 e^{xv}(1-v)^{\frac{\lambda}{2}}v^{\frac{\lambda}{2}-1}dv.$$

Then the characteristic function of the argument θ_t is given by

$$\mathbb{E}[e^{i\lambda\theta_t}|\theta_0 = 0] = G_\lambda\Big(\frac{\rho_0^2}{2t}\Big).$$

Remark 6.12.1 Using the integral formula for confluent hypergeometric functions, see Lebedev [98], page 266

$${}_1F_1(a,b;x) = \frac{\Gamma(b)}{\gamma(a)\Gamma(b-a)}\int_0^1 e^{xt}t^{a-1}(1-t)^{b-a-1}dt, \quad Re\, b > Re\, a > 0,$$

the function $G_\lambda(x)$ can be also written as

$$G_\lambda(x) = x^{\frac{\lambda}{2}}\frac{\Gamma\Big(1+\frac{\lambda}{2}\Big)}{\Gamma(\lambda+1)}\,{}_1F_1(\lambda/2,\lambda+1;x).$$

---

[6]The integral representation of a Kummer function is

$$M(a,b,x) = \frac{\Gamma(b)}{\Gamma(a)\Gamma(b-a)}\int_0^1 e^{xu}u^{a-1}(1-u)^{b-a-1}\,du, \qquad Re\ b > Re\ a > 0.$$

We can also compute more complex expressions, such as the expectation $\mathbb{E}[f(R_t)e^{i\lambda\theta_t}]$, with $f:\mathbb{R}_+\to\mathbb{R}_+$ Borel function. To accomplish this, conditioning over the history $W[0,t]$ and applying the tower property, we obtain

$$\begin{aligned}\mathbb{E}[f(R_t)e^{i\lambda\theta_t}] &= \mathbb{E}[f(R_t)e^{i\lambda B_{\beta_t}}] = \mathbb{E}[f(R_t)\mathbb{E}[e^{i\lambda B_{\beta_t}}|W[0,t]]\\ &= \mathbb{E}[f(R_t)e^{-\frac{1}{2}\lambda^2\beta_t}] = \mathbb{E}[f(R_t)e^{-\frac{1}{2}\lambda^2\int_0^t\frac{1}{R_s^2}ds}]. \end{aligned}\tag{6.12.180}$$

To continue the computation we shall use the formula for the joint probability between R_t and β_t, which can be obtained from formula 1.20.8 in Borodin and Salminen [15], page 392

$$P_{\rho_0}(\beta_t\in dy, R_t\in dz) = \frac{z}{2t}e^{-\frac{\rho_0^2+z^2}{2t}}\,\mathbf{i}_{y/2}\Big(\frac{\rho_0 z}{t}\Big)dydz,$$

where

$$\mathbf{i}_y(z) = \mathcal{L}_\gamma^{-1}(I_{\sqrt{\gamma}}(z)) = \frac{ze^{\pi^2/4y}}{\pi\sqrt{\pi y}}\int_0^\infty e^{-z\cosh u-u^2/4y}\sinh u\sin(\pi u/2y)du,$$

for $Re z>0$. We shall actually use this relation in its integral form

$$\int_0^\infty e^{-\gamma y}\mathbf{i}_y(z)\,dy = I_{\sqrt{\gamma}}(z).\tag{6.12.181}$$

Substituting in (6.12.180) we have

$$\begin{aligned}\mathbb{E}[f(R_t)e^{-\frac{1}{2}\lambda^2\beta_t}] &= \int_0^\infty\!\!\int_0^\infty f(z)e^{-\frac{1}{2}\lambda^2 y}\frac{z}{2t}e^{-\frac{\rho_0^2+z^2}{2t}}\,\mathbf{i}_{y/2}\Big(\frac{\rho_0 z}{t}\Big)dydz\\ &= \frac{1}{2t}e^{-\frac{\rho_0^2}{2t}}\int_0^\infty zf(z)e^{-\frac{z^2}{2t}}\int_0^\infty e^{-\frac{1}{2}\lambda^2 y}\mathbf{i}_{y/2}\Big(\frac{\rho_0 z}{t}\Big)\,dy\,dz\\ &= \frac{1}{t}e^{-\frac{\rho_0^2}{2t}}\int_0^\infty zf(z)e^{-\frac{z^2}{2t}}\int_0^\infty e^{-\lambda^2 y}\mathbf{i}_y\Big(\frac{\rho_0 z}{t}\Big)\,dy\,dz\\ &= \frac{1}{t}e^{-\frac{\rho_0^2}{2t}}\int_0^\infty zf(z)e^{-\frac{z^2}{2t}}I_\lambda\Big(\frac{\rho_0 z}{t}\Big)\,dz,\end{aligned}$$

where the integral with respect to y was computed in terms of the modified Bessel function I_λ using formula (6.12.181). Substituting in (6.12.180) we arrive to the formula

$$\mathbb{E}[f(R_t)e^{i\lambda\theta_t}] = \frac{1}{t}e^{-\frac{\rho_0^2}{2t}}\int_0^\infty zf(z)e^{-\frac{z^2}{2t}}I_\lambda\Big(\frac{\rho_0 z}{t}\Big)\,dz.\tag{6.12.182}$$

There are two distinguished particular cases:

(*i*) If $f(z)=1$, we obtain another formula for the characteristic function of the angular part

$$\mathbb{E}[e^{i\lambda\theta_t}] = \frac{1}{t}e^{-\frac{\rho_0^2}{2t}} \int_0^\infty z e^{-\frac{z^2}{2t}} I_\lambda\Big(\frac{\rho_0 z}{t}\Big)\, dz$$
$$= \frac{1}{2t}e^{-\frac{\rho_0^2}{2t}} \int_0^\infty e^{-\frac{u}{2t}} I_\lambda\Big(\frac{\rho_0 \sqrt{u}}{t}\Big)\, du, \tag{6.12.183}$$

where we used the substitution $u = z^2$.

(*ii*) If $f(z) = 1/z$, we obtain

$$\mathbb{E}\Big[\frac{e^{i\lambda\theta_t}}{R_t}\Big] = \frac{1}{t}e^{-\frac{\rho_0^2}{2t}} \int_0^\infty e^{-\frac{z^2}{2t}} I_\lambda\Big(\frac{\rho_0 z}{t}\Big)\, dz. \tag{6.12.184}$$

Remark 6.12.2 The Hartman-Watson probability measure on $\mathbb{R}_+$ is denoted by $\eta_r(du)$, for $r > 0$, and is defined by

$$\Big(\frac{I_{|\lambda|}}{I_0}\Big)(r) = \int_0^\infty e^{-\frac{1}{2}\lambda^2 u}\eta_r(du), \qquad \lambda \in \mathbb{R}.$$

This density has been used also in Section 4.6 for computing the heat kernel of the exponential-Grushin operator. The probabilistic interpretation of the Hartman-Watson probability measure in terms of the angular process of a Brownian motion is given by the following conditional expectation relation, see Yor [142] or Revuz and Yor [120], page 355

$$\mathbb{E}[e^{i\lambda(\theta_t - \theta_0)}|R_t = \rho] = \mathbb{E}[e^{-\frac{1}{2}\lambda^2\int_0^t \frac{ds}{R_s^2}}] = \Big(\frac{I_{|\lambda|}}{I_0}\Big)\Big(\frac{\rho_0\rho}{t}\Big).$$

For more information regarding this probability measure, the reader is referred to Hartman and Watson [73], Yor [140], or Gulisashvili [69], page 99.

Remark 6.12.3 The skew product representation of a planar Brownian motion can be extended to the $(n+1)$-dimensional case as follows. Thus, a Brownian motion W_t in $\mathbb{R}^{n+1}$, for $n \geq 1$, can be written as

$$W_t = |W_t|\gamma_{\int_0^t \frac{ds}{|W_s|^2}}, \tag{6.12.185}$$

where the radial process $(|W_t|; t \geq 0)$ is an $(n+1)$-dimensional Bessel process in $\mathbb{R}^{n+1}$ and the angular process $(\gamma_t; t \geq 0)$ is a Brownian motion on the unit sphere $\mathbb{S}^n$. It is worth noting that the radial process $|W_t|$ and the angular motion γ_t are independent processes. Also, the angular process satisfies the equation (6.11.168).

6.13 The h-Laplacian

Let h be a smooth positive function defined on the Riemannian manifold (M, g). We define the elliptic operator $L^{(h)}$ by

$$L^{(h)}u = \frac{1}{2}\text{div}\,(h\nabla_g u),$$

where $u : M \to \mathbb{R}$ is a smooth function, $\nabla_g u$ is the gradient of u with respect to the metric g and div stands for the divergence taken with respect to the Levi-Civita connection. We note that in the particular case when $h = c$, constant, then $L^{(c)} = c\Delta_g$, where Δ_g is the Laplace-Beltrami operator on M. We shall study how the function h is related to the drift of the Brownian motion on (M, g).

Using Exercise 6.19.11, part (a), we can write $L^{(h)}$ is terms of the Laplace-Beltrami operator on M as

$$L^{(h)}u = h\Delta_g u + \frac{1}{2}g(\nabla_g u, \nabla_g h).$$

Using the local coordinates expression

$$\Delta_g u = \frac{1}{2}g^{ij}\Big(\frac{\partial^2 u}{\partial x_i \partial x_j} - \Gamma^k_{ji}\frac{\partial u}{\partial x_k}\Big),$$

and $\nabla_g u = g^{ik}\frac{\partial u}{\partial x_i}\frac{\partial}{\partial x_k}$ we obtain the expression of $L^{(h)}u$ in local coordinates as

$$L^{(h)}u = \frac{1}{2}hg^{ij}\frac{\partial^2 u}{\partial x_i \partial x_j} + \frac{1}{2}\Big(g^{kj}\frac{\partial h}{\partial x_j} - hg^{ij}\Gamma^k_{ij}\Big)\frac{\partial u}{\partial x_k}.$$

Then the diffusion on M associated to $L^{(h)}$ satisfies

$$dX_t^{(h)} = \sigma dW_t + \big(\frac{1}{2}\nabla_g h + hb\big)dt,$$

where W_t is an n-dimensional Brownian motion on $\mathbb{R}^n$, $(\sigma\sigma^T)_{ij} = hg^{ij}$ and $b = b_k\partial_{x_k}$ is the drift vector field of the Brownian motion on (M, g) given by (6.2.26) at page 260.

We shall investigate the existence of a local system of coordinates in which the drift of the diffusion $X_t^{(h)}$ vanishes. To this end, we notice that the equation $\frac{1}{2}\nabla_g h + hb = 0$ can be written equivalently as $b = \nabla_g(-\frac{1}{2}\ln h)$, i.e. b is a gradient vector field with the potential $\phi = -\frac{1}{2}\ln h$. Then the equation $\nabla_g\phi = b$ is equivalent to $g^{rk}\frac{\partial\phi}{\partial x_s} = g^{ij}\Gamma^k_{ij}$, or $\frac{\partial\phi}{\partial x_s} = g^{ij}\Gamma_{ijs}$ where $\Gamma_{ijs} = g_{ks}\Gamma^k_{ij}$. The closeness conditions that need to be satisfied by the metric are

$$\frac{\partial}{\partial x_r}\Big(g^{ij}\Gamma_{ijs}\Big) = \frac{\partial}{\partial x_s}\Big(g^{ij}\Gamma_{ijr}\Big).$$

6.14 Stochastically Complete Manifolds

In this section we deal with two concepts and their interrelationships: the geodesic completeness and the stochastic completeness. The former concept represents a geometrical property involving geodesics and it is equivalent to the concept of metrically completeness of the manifold as a metric space. The latter concept involves the behavior of Brownian motions on the manifold, whose life can be extended indefinitely; it is shown that this concept is equivalent to the uniqueness of the Cauchy problem of the heat equation on the manifold M.

6.14.1 Geodesic Completeness

Let (M, g) be a connected Riemannian manifold. The Riemannnian metric g defines a natural distance on M as follows

$$d(p, q) = \inf\{\ell(\phi); \phi : [0, 1] \to M, \phi(0) = p, \phi(1) = q\},$$

where $\ell(\phi) = \int_0^1 ||\dot{\phi}(s)||_g \, ds$ is the length of the continuous curve ϕ considered with respect to g. We note that $||\dot{\phi}(s)||_g$ denotes the magnitude of the velocity to the curve and it is given by $||\dot{\phi}(s)||_g^2 = \sum_{i,j} g_{ij}(\phi(s))\dot{\phi}^i(s)\dot{\phi}^j(s)$. Since M is a connected manifold, there is at least one continuous curve ϕ between any two given points on the manifold. Thus, the pair (M, d) becomes a metric space. This space is called *complete* if any Cauchy sequence $(x_n)_n \subset M$ is convergent.

A Riemannian manifold (M, g) is called *geodesically complete* if any geodesic curve on the manifold can be extended indefinitely. This means that for any given $x_0 \in M$ and $v \in T_{x_0}M$ the initial value problem

$$\begin{aligned} \ddot{\phi}^k(s) + \Gamma^k_{ij}(\phi(s))\dot{\phi}^i(s)\dot{\phi}^j(s) &= 0, \qquad 1 \le k \le n \\ \phi(0) &= x_0 \\ \dot{\phi}(0) &= v \end{aligned}$$

has a global solution $\phi : [0, +\infty) \to M$. As stated by standard ODEs results, the aforementioned system has a local solution, $\phi : [0, \epsilon) \to M$, for some $\epsilon > 0$. However, there is a possibility of extension to a global solution if extra conditions on the rate of growth of the Christoffel symbols Γ^k_{ij} are satisfied. These conditions can be seen as constraints on the variation of the Riemannian metric coefficients.

If the solution of the aforementioned system is denoted by ϕ_v, then the application that maps a vector of small enough magnitude, $v \in T_pM$, into $\phi_v(1)$ is called the *exponential map*. In the case of a geodesically complete Riemannian manifold the exponential map is defined globally, $\exp_p : T_pM \to M$. We note that the Riemannian distance between the image of the exponential map and the initial point p is given by

$$d(p, \exp_p(v)) = d(p, \phi_v(1)) = \int_0^1 \|\dot{\phi}_v(t)\|_g \, dt = \|\dot{\phi}_v(0)\|_g = |v|,$$

where we used that the velocity of a geodesic has a constant magnitude, as inferred by the relation

$$\frac{d}{dt}\|\dot{\phi}(t)\|^2 = 2g_{k\ell}\dot{\phi}^\ell(\ddot{\phi}^k + \Gamma_{ij}^k(\phi)\dot{\phi}^i\dot{\phi}^j) = 0,$$

see Calin and Chang [36], page 49.

Remark 6.14.1 Some examples of geodesically complete manifolds include: the Euclidean space $\mathbb{R}^n$, the hyperbolic space $\mathcal{H}^n$, the Euclidean sphere $\mathbb{S}^n$, the real projective space $\mathbb{P}_n(\mathbb{R})$. In general, Riemannian manifolds with non-positive sectional curvature are geodesically complete. Any compact Riemannian manifold is also geodesically complete.

Hopf and Rinow's result shows that the concepts of geodesical completeness and metrical completeness can be unified as follows, see do Carmo [49]

Theorem 6.14.2 *On a connected and smooth Riemannian manifold* (M, g) *the following conditions are equivalent:*
(i) Any closed and bounded set in M *is compact;*
(ii) M *is a complete metric space;*
(iii) M *is geodesically complete.*

It is worth noting that on a geodesically complete manifold any two points $p, q \in M$ can be joined by a length minimizing geodesic, i.e., a property of global connectedness by Riemannian geodesics holds. This result is a consequence of properties of the exponential map.

The stochastic analog of the concept of geodesically completeness, which refers to geodesics, is the stochastically completeness, which refers to the underlying Brownian motion. We shall study this concept in the next section.

6.14.2 Stochastic Completeness

Stochastic completeness of a Riemannian manifold, (M, g), refers to the property that any Brownian motion, B_t, starting at any point x in the manifold M has infinite life time, i.e., it is defined for all time $t \geq 0$. Since the Brownian particle does not escape to infinity in finite time, we can write the stochastic completeness of a Riemannian manifold (M, g) equivalently as

$$P(B_t \in M) = \int_M p_t(x, y) dv(y) = 1, \qquad \forall x \in M, \; t > 0,$$

where $p_t(x,y)$ is the transition probability of the Brownian motion B_t and $dv(y) = \sqrt{\det g}\, dy_1 \dots dy_n$ is the volume measure on (M, g). We note that this definition does not assume the manifold (M, g) to be geodesically complete.

Equivalently, the manifold (M, g) is stochastically incomplete if there is an initial point x from where the Brownian motion B_t starts and a time $t > 0$ such that

$$P(B_t \in M) = \int_M p_t(x,y)dv(y) < 1.$$

Example 6.14.3 Since the Brownian motion on $\mathbb{R}^n$ does not escape to infinity in finite time, it follows that the Euclidean space $\mathbb{R}^n$ is stochastically complete. This can be also checked directly as

$$\int_{\mathbb{R}^n} \frac{1}{(2\pi t)^{n/2}} e^{\frac{\|x-y\|^2}{2t}} dy = \prod_{j=1}^{n} \int_{\mathbb{R}} \frac{1}{(2\pi t)^{1/2}} e^{\frac{(x_j-y_j)^2}{2t}} dy_j = 1.$$

Example 6.14.4 The unit sphere $\mathbb{S}^n$ with the natural metric endowed from $\mathbb{R}^{n+1}$ is stochastically complete. This is heuristically obvious, since the Brownian motion cannot escape to infinity on this compact manifold. Similar considerations hold on an ellipsoid. In general, any complete manifold with the Ricci curvature bounded from below is stochastically complete, see Yau [139].

The fact that stochastic completeness and geodesic completeness are two independent concepts is infered from the following examples:

Example 6.14.1 Let $D = \{x \in \mathbb{R}^m; |x| \le r\}$, $m \ge 3$, and consider the manifold $M = \mathbb{R}^m \backslash D$ endowed with the Euclidean metric. Then M is neither stochastically complete nor geodesically complete. The former follows from the transience property of the Brownian motion on $\mathbb{R}^m$, $m \ge 3$. This means that a Brownian motion starting at $x_0 \in M$, i.e., $|x_0| > r$, hits the ball D in finite time with probability $r/|x_0| < 1$. Therefore, the Brownian motion starting at x_0 leaves the manifold M with a positive probability. The latter follows from the fact that the lines intersecting the ball D cannot be extended indefinitely.

Example 6.14.2 The manifold $M = \mathbb{R}^m \backslash \{0\}$, $m \ge 3$, endowed with the Euclidean metric is stochastically complete but not geodesically complete. The former follows from Exercise 6.14.1 by taking $r \to 0$, while the latter is implied by the fact that lines cannot be extended through the origin.

Example 6.14.3 Azencott [7] provided an example of a non-compact manifold that is geodesically complete but not stochastically complete. In this counterexample the negative curvature grows quickly with distance acting as a drift that pulls the Brownian particle towards infinity in a finite time.

Geodesically complete manifolds are not necessarily stochastically complete, unless extra hypothesis are assumed. Sufficient conditions in terms of lower bounded Ricci curvature was provided by S.-T. Yau [139], while a general sufficient condition in terms of the volume growth was proved in Grigor'yan [67]. A relation with the maxima principles can be found in Pigola et.al [117].

6.14.3 The Cauchy Problem

Heat Kernel for the Laplace-Beltrami Operator

We reacall that the Laplace-Beltrami operator on the Riemannian manifold (M, g), denoted by Δ_g, in local coordinates $(x_1, \dots, x_n)$ is expressed as

$$\Delta_g = \frac{1}{2}\frac{1}{\sqrt{\det g}}\sum_{i=1}^{n}\frac{\partial}{\partial x_i}\Big(\sqrt{\det g}\, g^{ij}\frac{\partial}{\partial x_j}\Big). \tag{6.14.186}$$

Since g is a Riemannian metric, the operator Δ_g is elliptic. It is known that for any point q in the Riemannian manifold (M, g), there exists a small enough neighborhood U_q of q such that every two points within U_q can be connected by a unique geodesic. This fact is a consequence of the fact that the exponential map, $\exp_q$, is a local diffeomorphism.

The function $p_t(x, y) = p(t; x, y)$ is a *local heat kernel* for Δ_g in the open neighborhood U_q if it is a solution to the heat equation

$$\begin{aligned} \frac{\partial p_t}{\partial t} &= \Delta_g p_t, \qquad t > 0 \\ \lim_{t\to 0} p_t(x, y) &= \delta(x - y), \qquad \forall x, y \in U_q. \end{aligned}$$

Since Δ_g is elliptic, the heat kernel $p_t(x, y)$ is C^∞-smooth for any $t > 0$.

In particular, if (M, g) is geodesically complete, the neighborhood U_q can be expanded to cover the entire manifold M. In this scenario, the heat kernel $p_t(x, y)$ becomes a *global heat kernel* for Δ_t. This heat kernel depends in a certain extent on the Riemannian distance between the points x and y, see Chapter 4.

Remark 6.14.5 It is important to note that the expressions for the local and global heat kernels may differ. For example, the Laplacian on the unit circle parameterized by the arc length θ is represented by $\Delta = \frac{1}{2}\frac{d^2}{d\theta^2}$. The local heat kernel for the interval $U_0 = (-\pi/2, \pi/2)$ is expressed as $p_{\text{loc}}(t; \theta_0, \theta) = \frac{1}{\sqrt{2\pi t}}e^{-\frac{(\theta-\theta_0)^2}{2t}}$, while the global heat kernel is expressed as $p_{\text{glob}}(t; \theta_0, \theta) = \frac{1}{\sqrt{2\pi t}}\sum_{k=-\infty}^{\infty} e^{-\frac{2k\pi+(\theta-\theta_0)^2}{2t}}$.

The difference between these expressions becomes more apparent when considering the heat kernel as a representation of heat distribution, starting from a concentrated spike at the point y at time $t = 0$. In the case of compact, connected manifolds (such as the unit circle), heat propagation primarily occurs along geodesics and will eventually return to its starting point, leading to a formula involving a sum of an infinite number of terms.

This phenomenon is linked to the nature of the underlying diffusion process, which in this case is a Brownian motion, that is recurrent. This means that the process returns an infinite number of times to its starting point.

The Cauchy Problem Formulation

For $T > 0$ we consider the *Cauchy problem* for the Laplace-Beltrami operator on (M, g)

$$\begin{aligned} \frac{\partial u}{\partial t} &= \Delta_g u, \qquad (t,x) \in (0,T) \times M \\ u_{|t=0} &= f, \end{aligned}$$

where f is a continuous bounded function on M. The solution u is considered in the classical sense, i.e. u meets the requirements of being $\mathcal{C}^2$ in terms of its variables x_i and $\mathcal{C}^1$ in terms of t.

Existence of a Classical Solution

We can verify by a direct computation that the function $u(t, x)$ defined by

$$u(t,x) = \int_M p_t(x,y) f(y)\, dv(y) \tag{6.14.187}$$

is a smooth solution of the Cauchy problem by performing the following calculations:

$$\begin{aligned} \frac{\partial}{\partial t} u(t,x) &= \frac{\partial}{\partial t} \int_M p_t(x,y) f(y)\, dv(y) = \int_M \frac{\partial}{\partial t} p_t(x,y) f(y)\, dv(y) \\ &= \int_M \Delta_g p_t(x,y) f(y)\, dv(y) = \Delta_g \int_M p_t(x,y) f(y)\, dv(y) \\ &= \Delta_g u(t,x), \end{aligned}$$

$$\begin{aligned} \lim_{t\to 0} u(t,x) &= \lim_{t\to 0} \int_M p_t(x,y) f(y)\, dv(y) = \int_M \lim_{t\to 0} p_t(x,y) f(y)\, dv(y) \\ &= \int_M \delta(x-y) f(y)\, dv(y) = f(x). \end{aligned}$$

Furthermore, the aforementioned solution is bounded since f is bounded

$$|u(t,x)| \leq \sup_{y\in M} |f(y)| \int_M p_t(x,y)\, dv(y) \leq \sup_{y\in M} |f(y)| = \|f\|.$$

We are interested in conditions under which the solution (6.14.187) is unique.

Solution Uniqueness

In general, the solution to the aforementioned Cauchy problem is not unique. For instance, the problem

$$\frac{\partial u}{\partial t} = \frac{1}{2}\frac{\partial^2 u}{\partial x^2}, \qquad (t,x) \in (0,\infty) \times \mathbb{R}$$
$$u_{|t=0} = 0$$

has, besides the trivial solution, $u \equiv 0$, the nonzero C^∞-smooth solution

$$u(t,x) = \sum_{k\geq 0} \frac{x^{2k}}{(2k)!} h^{(k)}(t),$$

where

$$h(t) = \begin{cases} e^{-\frac{1}{t^2}}, & \text{if } t > 0 \\ 0, & \text{if } t \leq 0. \end{cases}$$

The reason for non-uniqueness is the fact that the previous solution increases too fast as $|x| \to \infty$.

The solution to the aforementioned Cauchy problem is unique only if the solutions involved belong to specific classes of functions that do not grow too rapidly at infinity. Some of these classes include:

1. The *class* $\mathcal{M}$, which is composed of continuous functions defined on $\mathbb{R}^n \times [0,\infty)$ that are bounded on any set $\mathbb{R}^n \times [0,T]$ for any positive T.

2. The *Tikhonov class*, consisting of solutions that increase not faster than an exponential of a quadratic function, i.e. $|u(t,x)| \leq Ae^{a|x|^2}$, for all $t > 0$ and $x \in \mathbb{R}^n$, with A and a being positive constants.

3. The *Täcklind class*, consisting of solutions satisfying the bound $|u(t,x)| \leq Ce^{f(|x|)}$ for all $t > 0$ and $x \in \mathbb{R}^n$, where $f(r)$ is a positive, convex, and increasing function on $(0,\infty)$ satisfying $\displaystyle\int^\infty \frac{r}{f(r)} dr = \infty$.

4. Another class of solutions for which uniqueness holds in the case of Riemannian manifolds can be found in Grigor'yan (see Theorem 2 in [67]). This general result can be stated as follows:

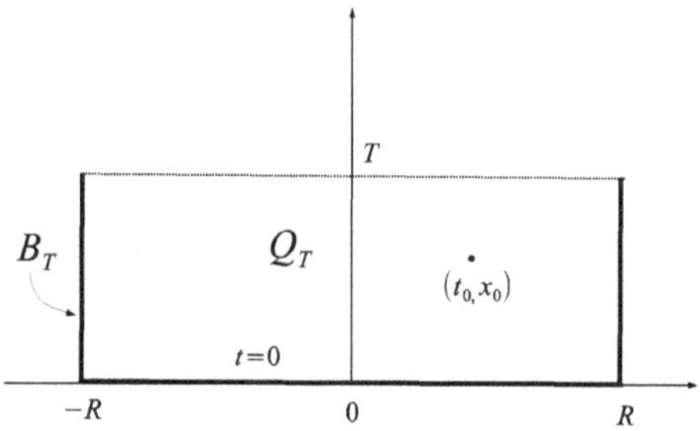

Figure 6.4: *The minimum principle for operator L applied to $v_\pm$: Since $v_\pm > 0$ on B_T, then $v_\pm > 0$ on Q_T.*

Theorem 6.14.6 *Let (M, g) be a complete connected Riemannian manifold, and let $u(x,t)$ be a solution to the Cauchy problem*

$$\frac{\partial u}{\partial t} = \Delta u, \qquad (t,x) \in (0,T) \times M$$
$$u_{|t=0} = 0.$$

Assume that, for some $x_0 \in M$ and for all $R > 0$ we have

$$\int_0^T \int_{B(x_0,R)} u^2(t,x)\, dv(x)\, dt \le e^{f(R)}, \tag{6.14.188}$$

where $f(r)$ is a positive increasing function on $(0, +\infty)$ such that $\int^\infty \frac{r}{f(r)}\, dr = \infty$. Then $u \equiv 0$ in $(0,T) \times M$.

It is important to note that in the case of the Euclidean space, $M = \mathbb{R}^n$, class 4 implies class 3; furthermore, class 3 with the choice $f(r) = Cr^2$ implies class 2. Obviously, class 1 is a particular case of class 2. The proof of these results requires the utilization of cleverly constructed auxiliary functions.

To get the general flavor, we shall include next the proof for the uniqueness in the case of bounded solutions for the particular case $M = \mathbb{R}$. This property is a consequence of the minimum principle for parabolic operators.

Theorem 6.14.7 *Let $u \in \mathcal{C}([0,T] \times \mathbb{R})$ such that $\partial_t u, \partial_x u, \partial_x^2 u \in \mathcal{C}((0,T] \times \mathbb{R})$ be a solution for the Cauchy problem*

$$\partial_t u - \frac{1}{2}\partial_x^2 u = 0, \qquad (t,x) \in (0,T] \times \mathbb{R}$$
$$u_{|t=0} = 0.$$

If there is a constant $M > 0$ such that $|u(t, x)| \le M$ for any $(t, x) \in [0, T] \times \mathbb{R}$, then $u \equiv 0$.

Proof: Let $a = \sqrt{T/2}$ and consider the function $w(t, x) = t\cosh(x/a) + 1$. We notice that $w \in \mathcal{C}^2([0, T] \times \mathbb{R})$. The following relations can be easily verified

$$\begin{aligned}
Lw &\ge 0, \quad (t, x) \in [0, T] \times \mathbb{R} \\
w &> 0, \quad (t, x) \in [0, T] \times \mathbb{R} \\
\lim_{|x| \to \infty} w(t, x) &= \infty, \text{ uniformly in } t \in [0, T],
\end{aligned}$$

where $L = \partial_t + \frac{1}{2}\partial_x^2$. Let $(t_0, x_0) \in [0, T] \times \mathbb{R}$ be an arbitrary point and we shall show that $u(t_0, x_0) = 0$. To this end, it suffices to show that $|u(t_0, x_0)| < \epsilon$ for an arbitrary fixed $\epsilon > 0$.

We let $\eta = \dfrac{\epsilon}{w(t_0, x_0)} > 0$ and consider the functions $v_\pm(t, x) = \eta w(t, x) \pm u(t, x)$. These functions satisfy the following properties

$$\begin{aligned}
Lv_\pm &= \eta Lw \ge 0 \\
v_{\pm|t=0} &= \eta w_{|t=0} \pm u_{|t=0} = \eta w_{|t=0} > 0.
\end{aligned}$$

Since u is bounded, then $\lim_{|x| \to \infty} v_\pm(t, x) = \eta \lim_{|x| \to \infty} w(t, x) = +\infty$. Therefore, there is $R > 0$ such that $|x_0| < R$ and $v_\pm(t, x) > 0$ for $|x| = R$ and any $t \in [0, T]$.

Denoting $Q_T = [0, T] \times [-R, R]$ and $B_T = \big([-R, R] \times \{0\}\big) \cup \big(\{\pm R\} \times [0, T)\big)$, then the minimum principle for the hyperbolic operator L applied to $v_\pm$ provides

$$\inf_{(t,x) \in Q_T} v_\pm(t, x) = \inf_{(t,x) \in B_T} v_\pm(t, x) > 0,$$

see Fig. 6.4. Therefore, $v_\pm(t, x) > 0$ for any $(t, x) \in Q_T$, fact that implies $|u(t, x)| < \eta w(t, x)$ on Q_T. Evaluating at (t_0, x_0) yields

$$|u(t_0, x_0)| < \eta w(t_0, x_0) = \epsilon.$$

Since this inequality holds for an arbitrary $\epsilon > 0$, it follows that $u(t_0, x_0) = 0$. Since the point (t_0, x_0) was chosen arbitrary in $[0, T] \times \mathbb{R}$, it follows that $u \equiv 0$. ■

Minimality of the heat semigroup The minimality property of the heat semigroup states that if u_0 is an initial temperature distribution and

$$u(x, t) = \int_M p_t(x, y) u_0(y)\, dv(y)$$

is the corresponding temperature distribution at time t, then for any other temperature distribution $v(x,t)$ that satisfy the same Cauchy problem

$$\partial_t v = \Delta v$$
$$v_{|t=0} = u_0,$$

we have $u(x,t) \leq v(x,t)$ for all $t > 0$ and $x \in M$.

This property follows from the maximum principle for parabolic equations applied to the difference $w(x,t) = v(x,t) - u(x,t)$. Since we have $\partial_t w - \Delta w = 0$ and $w(0,x) = 0$, then the maximum principle implies $w(x,t) \geq 0$ for all $t > 0$ and $x \in M$. This completes the proof of the minimality property.

Remark 6.14.8 It is worthy to note that the the probabilistic analogue of the minimality property of the heat semigroup is the "strong Markov property" of the Brownian motion. The strong Markov property states that the future evolution of a Brownian motion process is conditionally independent of its past given its present position. Both of these properties describe the evolution of a process over time and its dependence on initial conditions. The difference is that the minimality property of the heat semigroup is a deterministic result, while the strong Markov property of Brownian motion is a probabilistic one.

There is a proof variant of Theorem 6.14.7 using the minimality property of the heat semigroup. This proof variant is given in the following. We consider a bounded solution u of the Cauchy problem

$$\partial_t u - \Delta u - 0, \qquad (t,x) \in (0,T] \times \mathbb{R}^n$$
$$u_{|t=0} = 0.$$

Dividing by M in the boundedness condition $|u| \leq M$, we may assume that $|u| \leq 1$. Then we construct the functions $w_{\pm} = 1 \pm u$, which satisfies the Cauchy problem

$$\partial_t w_{\pm} - \Delta w_{\pm} = 0, \qquad (t,x) \in (0,T] \times \mathbb{R}^n$$
$$w_{\pm}|t = 0 = 1.$$

By the aforementioned minimality property of the heat semigroup we obtain

$$w_{\pm} \geq \int_{\mathbb{R}^n} p_t(x,y) \cdot 1\, dy = 1,$$

which implies $1 \pm u \geq 1$. This imply both $u \geq 0$ and $u \leq 0$, which leads to $u = 0$.

Remark 6.14.9 The previous result holds on spaces more general than $\mathbb{R}^n$, namely on Riemannian manifolds that satisfy the stochastic completenes condition $\int_M p_t(x,y)\, dv(y) = 1$, see the forthcomming Theorem 6.14.11.

The next result provides a strict inequality regarding the resolvent of the heat kernel (see also Section 5.9, page 242).

Lemma 6.14.10 *(a) Let $\alpha > 0$ and $\psi : [0, \infty) \to \mathbb{R}$ be a continuous function such that $0 \leq \psi(t) \leq 1$, for every $t \geq 0$. If there is $t_0 > 0$ such that $\psi(t_0) < 1$, then*

$$\int_0^\infty e^{-\alpha t}\psi(t)\,dt < \frac{1}{\alpha}.$$

(b) Let M be a stochastically incomplete manifold. Then there is $x_0 \in M$ such that $(R_\alpha 1)(x_0) < \frac{1}{\alpha}$, where R_α denotes the resolvent of the heat kernel.

Proof: (a) Integrating in the inequality $\psi(t) \leq 1$, we get $\int_0^\infty e^{-\alpha t}\psi(t)\,dt \leq \frac{1}{\alpha}$. By contradiction, we assume $\int_0^\infty e^{-\alpha t}\psi(t)\,dt = \frac{1}{\alpha}$. This is equivalent to $\int_0^\infty e^{-\alpha t}(1 - \psi(t))\,dt = 0$. Since the integrand $e^{-\alpha t}(1 - \psi(t)) \geq 0$, it follows that $e^{-\alpha t}(1 - \psi(t)) = 0$, i.e. $\psi(t) = 1$ for all t, which is a contradiction.
(b) Let $\psi(t) = \int_M p_t(x_0, y)\,dv(y)$. We have $0 \leq \psi(t) \leq 1$. Since M is stochastically incomplete, there is $x_0 \in M$ and $t_0 > 0$ such that

$$\psi(t_0) = \int_M p_{t_0}(x_0, y)\,dv(y) < 1.$$

Applying part (a) we obtain

$$(R_\alpha 1)(x_0) = \int_0^\infty e^{-\alpha t}\int_M p_t(x_0, y)\,dv(y)\,dt = \int_0^\infty e^{-\alpha t}\psi(t)\,dt < \frac{1}{\alpha}.$$

■

A characterization of stochastic completeness in terms of the unique existence of a temperature density on a manifold is obtained from the following result, see Grigor'yan [68]:

Theorem 6.14.11 *Let $0 < T < \infty$ and (M, g) be a Riemannian manifold and f an arbitrary continuous bounded function on M. The following conditions are equivalent:*
(a) M is stochastically complete;
(b) The Cauchy problem

$$\frac{\partial u}{\partial t} = \Delta_g u, \qquad (t, x) \in (0, T) \times M$$
$$u_{|t=0} = f$$

has at most one bounded classical solution.

Proof: "(*a*) $\Longrightarrow$ (*b*)" We consider M stochastically complete, so we have

$$\int_M p_t(x,y)\,dv(y) = 1. \tag{6.14.189}$$

By contradiction, we assume that the Cauchy problem has two distinct bounded solutions, u_1 and u_2. Let $w = u_1 - u_2$. Then w is bounded and satisfies a Cauchy problem with zero initial condition. Without loss of generality we may assume $|w| \leq 1$. Then we construct the functions $u_\pm = 1 \pm w$, which satisfy the problem

$$\frac{\partial u}{\partial t} = \Delta_g u, \qquad (t,x) \in (0,T) \times M \tag{6.14.190}$$

$$u_{|t=0} = 1. \tag{6.14.191}$$

Using (6.14.189) the minimal solution of (6.14.190)-(6.14.191) is given by

$$z(t,x) = \int_M p_t(x,y) \cdot 1\,dv(y) = 1.$$

From the minimality property of the solution of the heat equation we have

$$u_\pm(t,x) \geq z(t,x) = 1,$$

which imply both $w \geq 0$ and $w \leq 0$. Hence w vanishes and hence $u_1 = u_2$, which is a contradiction. Therefore, the Cauchy problem has at most one bounded solution.

"(*b*) $\Longrightarrow$ (*a*)" By contradiction, we assume that M is stochastically incomplete. Then by Lemma 6.14.10 part (*b*) there is an $x_0 \in M$ such that

$$0 < (R_\alpha 1)(x_0) < \frac{1}{\alpha}.$$

Set $w(x) = (R_\alpha 1)(x)$. Then $0 < w(x) \leq \frac{1}{\alpha}$ and $w(x_0) < \frac{1}{\alpha}$. By Proposition 5.9.1 the function $w(x)$ is $\mathcal{C}^\infty$-smooth and satisfies the equation $\Delta_g w - \alpha w = -1$. Let $z(x) = \frac{1}{\alpha} - w(x)$. Then z is bounded with $0 \leq |z(x)| \leq \frac{1}{\alpha} + |w(x)| \leq \frac{2}{\alpha}$. We note that $z \neq 0$, because $z(x_0) = \frac{1}{\alpha} - w(x_0) > 0$. Since

$$\Delta_g z = -\Delta_g w = -\alpha w + 1 = \alpha\Big(z - \frac{1}{\alpha}\Big) + 1 = \alpha z,$$

then $z(x)$ is a bounded eigenfunction of the Laplacian with the corresponding eigenvalue α, namely $\Delta_g w = \alpha z$. Then we can check that the function $\xi(t,x) = e^{\alpha t} z(x)$ satisfies the Cauchy problem

$$\frac{\partial \xi}{\partial t} = \Delta_g \xi, \qquad (t,x) \in (0,T) \times M$$

$$\xi_{|t=0} = z.$$

Besides $\xi(t,x)$, another classical solution the the previous problem is obtained convolving the heat kernel with the initial condition, namely

$$u(t,x) = \int_M p_t(x,y)z(y)\,dv(x).$$

Both solutions, $\xi(t,x)$ and $u(t,x)$, are bounded as it can be inferred from the following estimations

$$|\xi(t,x)| \le e^{\alpha t}|z(x)| \le e^{\alpha T}\frac{2}{\alpha}$$
$$|u(t,x)| \le \|z\| \int_M p_t(x,y)\,dv(x) \le \|z\| \le \frac{2}{\alpha},$$

where we used $T < \infty$. Using part (b), the aforementioned Cauchy problem has at most one bounded solution, fact that implies $u(t,x) = \xi(t,x)$. We shall show that this identity leads to a contradiction. To this end, we estimate

$$\sup_x |u(t,x)| \le \|z\| \int_M p_t(x,y)z(y)\,dv(y) = \|z\|$$
$$\sup_x |u(t,\xi)| = e^{\alpha t}\sup_x |z(x)| = e^{\alpha t}\|z\| > \|z\|, \quad t > 0,$$

which leads to the contradictory relation

$$\|z\| < \sup_x |u(t,\xi)| = \sup_x |u(t,x)| \le \|z\|.$$

Hence, (M,g) is stochastically complete. ■

Corollary 6.14.12 *Let M be stochastically complete and f a continuous bounded function on M. Then there is only one bounded solution for the Cauchy problem for the heat equation with initial condition f. This solution is given by*

$$u(t,x) = \int_M p_t(x,y)f(y)\,dv(y). \tag{6.14.192}$$

Proof: We can verify that u is a bounded solution of the Cauchy problem, with

$$u(t,x) \le \|f\| \int_M p_t(x,y)\,dv(y) = \|f\|.$$

By Theorem 6.14.11, there are no other bounded solutions besides (6.14.192). ■

Remark 6.14.13 Stochastic completeness signifies that, in almost all cases, the Brownian motion particles will never escape the manifold. This means that the manifold operates as an isolated system, devoid of heat exchange with the exterior. Theorem 6.14.11 shows that this condition is equivalent to the unique determination of heat distribution on the manifold, given a bounded initial temperature.

Theorem 6.14.11 provides a way to show that a manifold is not stochastically complete, by finding two bounded solutions to the associated Cauchy problem with a bounded initial condition.

Example 6.14.14 For instance, we consider $M = \{x \in \mathbb{R}^2; \|x\| \leq 1\}$ to be the unit disk endowed with the Euclidean metric. Obviously, a two-dimensional Brownian motion starting at the origin, $W_t = (W_t^1, W_t^2)$, can escape the disk in finite time, so the probability $P(W_t \in M) < 1$, for t large, which implies that M is stochastically incomplete.

This problem is equivalent to the non-uniqueness of solutions of the associated Cauchy problem

$$\begin{aligned} \partial_t u &= \Delta_2\, u, \qquad (t,x) \in (0,T) \times M \\ u_{|t=0} &= f, \end{aligned}$$

for any continuous bounded function f on the disk M. For the sake of simplicity we assume that the initial distribution of temperature is rotational invariant, i.e. $f(x) = \rho(r)$, where $r = |x|$. Then the Cauchy problem has infinitely many solutions

$$u(t,x) = \sum_{n\geq 1} c_n e^{-\frac{1}{2}(\frac{z_{0n}}{a})^2 t} J_0\Big(\frac{r}{a} z_{0n}\Big),$$

which depend on the parameter $a > 0$. Here, J_0 denotes the Bessel function of the first kind of order zero, z_{0n} is the n-th zero of J_0 and the coefficients c_n are determined from the initial condition

$$\rho(r) = \sum_{n\geq 1} c_n J_0\Big(\frac{r}{a} z_{0n}\Big),$$

with the Fourier-Bessel coefficients given by

$$c_n = \frac{\int_0^a \rho(r) J_0\Big(\frac{r}{a} z_{0n}\Big) r\, dr}{\int_0^a J_0^2\Big(\frac{r}{a} z_{0n}\Big) r\, dr}.$$

It is worth noting that the solution $u(t,x)$ vanishes on the circle of radius a. These solutions are obtained by the method of separation of variables t and ρ

and using that J_0 is the bounded solution of

$$z^2\frac{d^2f}{dz^2} + z\frac{df}{dz} + z^2 f = 0.$$

For details see Haberman, [70], page 301.

The connection between the stochastic completeness of the manifold (M, g) and the Laplace-Beltrami operator Δ_g can be seen in the following result.

Proposition 6.14.15 *Let (M, g) be a stochastically complete Riemannian manifold. Then the fundamental solution $G(x, \cdot) \notin L^1(M)$, for any $x \in M$.*

Proof: The condition of stochastic completeness, $\int_M p_t(x, y), dv(y) = 1$, after integration and using Fubini's theorem leads to the following conclusion

$$\begin{aligned}-\int_M G(x, y)\, dv(y) &= \int_M \int_0^\infty p_t(x, y) dt\, dv(y) = \int_0^\infty \int_M p_t(x, y) dv(y)\, dt\\ &= \int_0^\infty dt = \infty,\end{aligned}$$

so $G(x, \cdot) \notin L^1(M)$. ■

The long run behavior of the temperature on a compact manifold as the average value of the initial temperature is given as follows.

Proposition 6.14.16 *Consider the Cauchy problem*

$$\begin{aligned}\frac{\partial u}{\partial t} &= \Delta_g u, \qquad (t, x) \in (0, \infty) \times M\\ u_{|t=0} &= f\end{aligned}$$

on a stochastically complete, compact (without boundary) Riemannian manifold (M, g), where f is continuous on M. Then the bounded solution $u(t, x)$ satisfies

$$\lim_{t\to\infty} u(t, x) = f_{\text{ave}}(x),$$

where $f_{\text{ave}} = \frac{1}{vol(M)} \int_M f(x)\, dv(x)$.

Proof: The function f is bounded as a continuous function on a compact set M. Since (M, g) is stochastically complete, then the bounded solution u of the aforementioned Cauchy problem is unique and is given by the integral formula

$$u(t, x) = \int_M p_t(x, y)\, f(y)\, dv(y).$$

Let $w(x) = \lim_{t\to\infty} u(t, x)$. Then w is harmonic on M, i.e., $\Delta_g w = 0$. Since w is continuous, by Weiestrass theorem it will reach the global minima and maxima on the compact set M at points x_0 and x_1, respectively. By the maximum principle of elliptic operators, the global extremes of the function w are reached on the boundary, $x_0, x_1 \in \partial M = \emptyset$. This implies that w is a constant function on M, $w(x) = c$.

Let $\mathcal{E}(t) = \int_M u(t,x)\, dv(x)$ be the total heat in M at time t. By the divergence theorem we obtain

$$\frac{d\mathcal{E}(t)}{dt} = \int_M \frac{\partial u(t,x)}{\partial t}\, dv(x) = \int_M \Delta_g u(t,x)\, dv(x) = \int_M \text{div}(\nabla_g u(t,x))\, dv(x) = 0,$$

which implies the conservation law of energy $\mathcal{E}(t) = \mathcal{E}(0) = \int_M f(x)\, dv(x)$. This also implies

$$\lim_{t\to\infty} \mathcal{E}(t) = \int_M f(x)\, dv(x).$$

On the other hand, using the definition of $\mathcal{E}(t)$ we have

$$\lim_{t\to\infty} \mathcal{E}(t) = \int_M w(x)\, dv(x) = c\,\text{vol}(M).$$

Equating the right sides of the last two relations yields $c = f_{\text{ave}}$, which is equivalent to the desired relation. ■

Remark 6.14.17 From the kinetic molecular theory of gases it is known that the average kinetic energy of the particles in a gas is proportional to the temperature of the gas. Since the total heat energy is constant, then the total average kinetic energy of the gas in M is preserved. In the long run the gas temperature tends to the steady state, which is constant on M.

Proposition 6.14.18 *Any compact Riemannian manifold without boundary, (M, g), is stochastically complete.*

Proof: We need to show that that for any point x in the manifold M and any positive value of t, the integral of the heat kernel $p_t(x,y)$ over the entire manifold is equal to 1. The proof involves showing that the time derivative of this integral vanishes, which implies that the integral is constant in time. To this end, applying the divergence theorem, we have

$$\partial_t \int_M p_t(x,y)\, dv(y) = \int_M \partial_t p_t(x,y)\, dv(y) = \int_M \Delta_g p_t(x,y)\, dv(y) = 0,$$

and hence, $\int_M p_t(x,y)\, dv(y) = C$, constant. Taking the limit as t approaches 0 and using properties of the Dirac delta function, we can determine the value of this constant to be 1 as follows

$$C = \lim_{t\to 0+} \int_M p_t(x,y)\, dv(y) = \int_M \lim_{t\to 0+} p_t(x,y)\, dv(y) = \int_M \delta_x(y)\, dv(y) = 1,$$

which completes the proof. ■

Corollary 6.14.19 *The Cauchy problem*

$$\partial_t v = \Delta_g v, \quad v(0,x) = 1$$

on a compact Riemannian manifold, without boundary, has only one bounded solution, $v = 1$.

Proof: Since M is stochastically complete, by Proposition 6.14.11 the Cauchy problem has a unique bounded solution and this is given by

$$u(t,x) = \int_M p_t(x,y) \cdot 1 \, dv(y) = 1.$$

■

Using a system of eigenfunctions, a complete characterization of stochastically complete manifolds can be provided as follows.

Proposition 6.14.20 *Let* $\phi_k(x)$ *be the* k*-th eigenfunction of the Laplacian* Δ_g *on the Riemannian manifold* (M,g)*. Then* M *is stochastically complete if and only if*

$$\int_M \phi_k(x) \, dv(x) = 0, \qquad \forall k \geq 1. \tag{6.14.193}$$

Proof: Without restricting the generality we may assume the normality condition $\int_M \phi_k(x)^2 \, dv(x) = 1$. Then expanding the heat kernel as

$$p_t(x,y) = \frac{1}{vol(M)} + \sum_{k\geq 1} e^{\lambda_k t} \phi_k(x)\phi_k(y)$$

and then integrate over M, yields

$$\int_M p_t(x,y) \, dv(y) = 1 + \sum_{k\geq 1} e^{\lambda_k t} \phi_k(x) \int_M \phi_k(y) \, dv(y),$$

for any $x \in M$ and $t > 0$. Then the stochastically completeness condition $\displaystyle\int_M p_t(x,y) \, dv(y) = 1$ is equivalent to

$$\sum_{k\geq 1} e^{\lambda_k t} \phi_k(x) \int_M \phi_k(y) \, dv(y) = 0, \qquad \forall x \in M, t > 0.$$

Since the eigenfunctions $\{\phi_k(x)\}_{k\geq 1}$ are independent, the previous condition is further equivalent to the vanishing coefficients conditions

$$e^{\lambda_k t} \int_M \phi_k(y) \, dv(y) = 0, \qquad \forall k \geq 1,$$

which is equivalent to equation (6.14.193). ■

Example 6.14.4 Let $M = \mathbb{S}^1$, with the Riemannian metric $ds^2 = d\theta^2$, where θ is the central angle. The eigenfunctions of the Laplacian $\Delta = \frac{1}{2}\frac{d^2}{d\theta^2}$ are $\phi_k(\theta) = \frac{1}{\sqrt{2\pi}}e^{ik\theta}$, $k \geq 1$. Since $\int_0^{2\pi} e^{ik\theta}\,d\theta = 0$, it follows by Proposition 6.14.20 that $\mathbb{S}^1$ is stochastically complete.

More consequences of Proposition 6.14.20 can be found in Exercises 6.19.7 and 6.19.8.

6.14.4 The Volume Test

The volume test for determining stochastic completeness was introduced by Grigor'yan [67] in 1986. In simpler terms, if the volume of the geodesic ball grows at a slow enough rate as its radius approaches infinity, then the manifold is stochastically complete.

We start with a geodesically complete Riemannian manifold, (M, g), so that geodesics can be extended indefinitely. As a result, we may consider geodesic balls of any size. The volume of a geodesic ball centered at x and having a radius of r will be indicated by $V(x, r)$.

Theorem 6.14.21 *Let (M, g) be a complete connected manifold. If there is a point $x_0 \in M$ such that*

$$\int^{\infty} \frac{r}{V(x_0, r)}\,dr = \infty,$$

then M is stochastically complete.

Proof: In order to prove the stochastic completeness, it suffices to verify part (b) of Theorem 6.14.11. By contradiction, we consider that there are two distinct bounded solutions, u_1 and u_2 of the Cauchy problem. Then the difference $u = u_1 - u_2$ is a bounded solution of the problem

$$\frac{\partial u}{\partial t} = \Delta u, \qquad (t, x) \in (0, T) \times M$$
$$u_{|t=0} = 0.$$

If we let $|u| \leq M$, then

$$\int_0^T \int_{B(x_0,r)} u^2(t, x)\,dv(x)\,dt \leq TM^2 V(x_0, r).$$

Then inequality (6.14.188) holds for $f(r) = \ln V(x_0, r) + C$, with $C > 2\ln M + \ln T$. Applying Theorem 6.14.6 it follows that $u \equiv 0$, and hence $u_1 = u_2$ and hence part (b) of Theorem 6.14.11 holds.

∎

Remark 6.14.22 In particular, the condition of a manifold to be stochastically complete is fulfilled if the volume of geodesic balls, $V(x_0, r)$, grows at most as Ce^{ar^2}, where C and a are positive constants. This condition applies to two well-known examples of stochastically complete manifolds, the Euclidean space and the hyperbolic space.

Since the volume of an n-dimensional sphere in the Euclidean space with radius r is given by $V_{Eu}(r) = \frac{2\pi^{n/2}}{\Gamma(n/2)} r^n$, then it can be shown to satisfy the aforementioned exponential growth condition. It can be also verified that the volume of an n-dimensional sphere of radius r in the hyperbolic space $\mathcal{H}^n$ is given by

$$V_{hyp}(r) = \frac{2\pi^{n/2}}{\Gamma(n/2)} \int_0^r \sinh^{n-1}(u)\, du,$$

see Chavel [45]. This expression also satisfies the aforementioned growth volume condition.

6.15 Maximum Integral Principles

This section introduces several maximum principles that have an integral nature, extending the classical maximum principle for the operator $L = \partial_t - \Delta_g$ to an integral form.

Let (M, g) be a compact Riemannian manifold, without boundary. We start from the integral expression

$$Q_\alpha(t) = \frac{1}{\alpha} \int_M u(t,x)^\alpha \, dv(x), \qquad \alpha > 0,$$

where $u(t,x)$ satisfies the heat equation $Lu = 0$ on M, having the initial condition $u(0,x) = f(x)$, with f continuous function.

The case $\alpha = 1$ retrieves the total heat, $Q_1(t) = \mathcal{E}(t)$, which has been shown to be preserved over time at page 352. Our next case of concern is $\alpha \geq 2$, integer.

Taking the derivative with respect to time and using Exercise 6.19.11 and the divergence theorem, we get

$$\begin{aligned}\frac{d}{dt} Q_\alpha(t) &= \int_M u(t,x)^{\alpha-1} \partial_t u(t,x)\, dv(x) = \int_M u(t,x)^{\alpha-1}\, \Delta_g u \, dv(x) \\ &= \frac{1}{2}\int_M \operatorname{div}(u^{\alpha-1}\nabla u)\, dv(x) - \frac{1}{2}\int_M g(\nabla u^{\alpha-1}, \nabla u)\, dv(x) \\ &= -\frac{\alpha-1}{2}\int_M u^{\alpha-2}\|\nabla u\|_g^2\, dv(x). \qquad (6.15.194)\end{aligned}$$

Therefore, if u is a solution of $Lu = 0$ with $u \geq 0$, and $\alpha \geq 2$, then the previous relation implies that $Q_\alpha(t)$ is non-increasing. Therefore, we have

$$Q_\alpha(t) \leq Q_\alpha(0) = \frac{1}{\alpha} \int_M f(x)^\alpha \, dv(x), \qquad \forall t \geq 0.$$

This inequality can be written in terms of the norm $\| \cdot \|_\alpha$ as

$$\|u(t, \cdot)\|_\alpha \leq \|f(\cdot)\|_\alpha, \tag{6.15.195}$$

where $\|f\|_\alpha = \Big(\int_M |f(x)|^\alpha \, dv(x) \Big)^\alpha$, with $\alpha \geq 2$.

The nonnegativity condition $u \geq 0$ implies also $f \geq 0$. Using the result

$$\lim_{n \to \infty} \Big(\int_M |f(x)|^n \, dv(x) \Big)^{1/n} = \max_{x \in M} |f(x)|$$

and taking $\alpha \to \infty$ in relation (6.15.195) yields $\|u(t, \cdot)\|_\infty \leq \|f(\cdot)\|_\infty$, or

$$\max_{x \in M} u(t, x) \leq \max_{x \in M} f(x), \qquad \forall x \in M, \ t \geq 0,$$

which retrieves a maximum principle for L on compact manifolds.

A distinguished case of the previous inequality occurs in the case $\alpha = 2$. First, substituting $\alpha = 2$ into inequality (6.15.194), we obtain

$$\frac{d}{dt} Q_2(t) \leq -\frac{1}{2} \int_M \|\nabla u\|_g^2 \, dv(x) \leq 0,$$

which holds for any solution u of $Lu = 0$. By *any* solution we mean not necessarily non-negative solutions like in the case $\alpha \neq 2$. This implies that $Q_2(t)$ is non-increasing, namely we have

$$\int_M u(t, x)^2 \, dv(x) \leq \int_M f(x)^2 \, dv(x), \qquad \forall t \geq 0. \tag{6.15.196}$$

Remark 6.15.1 Since $Q_2(t)$ is non-increasing, we have $Q_2(t) \leq Q_2(0) = \frac{1}{2} \int_M f(x)^2 \, dv(x)$. As a consequence, we obtain

$$\lim_{t \to \infty} Q(t) \leq \frac{1}{2} \int_M f(x)^2 \, dv(x).$$

Using the result of Proposition 6.14.16, we can compute the limit on the left side as

$$\lim_{t \to \infty} Q(t) = \frac{1}{2} \int_M \lim_{t \to \infty} u(t, x)^2 \, dv(x) = \frac{1}{2} \int_M (f_{\text{ave}})^2 \, dv(x) = \frac{1}{2} vol(M) (f_{\text{ave}})^2.$$

Then the previous inequality becomes

$$(f_{\text{ave}})^2 \leq (f^2)_{\text{ave}},$$

which recovers the result of Exercise 6.19.11.

In the following we shall attempt to get to the inequality (6.15.196) in a direct way. We start from the classical solution

$$u(t,x) = \int_M p_t(x,y) f(y)\, dv(y),$$

which is the only bounded solution, since M is stochastically complete, see Proposition 6.14.18. In this case, the Cauchy integral inequality implies

$$u(t,x)^2 = \Big(\int_M p_t(x,y) f(y)\, dv(y)\Big)^2 \leq \int_M p_t(x,y)^2 dv(y) \int_M f(y)^2\, dv(y),$$

and hence the left side of (6.15.196) can be estimated as

$$\int_M u(t,x)^2\, dv(x) \leq \int_M f(y)^2\, dv(y) \int_M \int_M p_t(x,y)^2\, dv(y) dv(x). \quad (6.15.197)$$

The double integral can be evaluated by employing the expansion of the heat kernel in terms of the orthonormal system of eigenfunctions $\{\phi_i(x)\}$ as

$$p_t(x,y) = \sum_{i\geq 0} e^{\lambda_i t} \phi_i(x)\phi_j(x), \quad (6.15.198)$$

where $\Delta_g \phi_i = \lambda_i \phi_i$, with $\lambda_0 = 0$ and $\lambda_i < 0$ for $i \geq 1$. Using Fubini's theorem and the orthonormality relations

$$\int_M \phi_i(x)\phi_j(x)\, dv(x) = \delta_{ij},$$

we have

$$\begin{aligned}\int_M \int_M p_t(x,y)^2\, dv(y)dv(x) &= \sum_{i,j\geq 0} e^{(\lambda_i+\lambda_j)t} \int_M \phi_i(x)\phi_j(x) dv(x) \int_M \phi_i(y)\phi_j(y) dv(y) \\ &= \sum_{i\geq 0} e^{2\lambda_i t} = Z(2t), \quad (6.15.199)\end{aligned}$$

where we used the *partition function* $Z(t) = \sum_{i\geq 0} e^{\lambda_i t}$. Since using (6.15.198) it can be checked that

$$Z(t) = \int_M p_t(x,x) dv(x), \quad (6.15.200)$$

then $Z(t)$ is also called the *trace of the heat kernel.*

Substituting (6.15.199) back in (6.15.197) yields the inequality

$$\int_M u(t,x)^2\, dv(x) \leq Z(2t) \int_M f(y)^2\, dv(y). \quad (6.15.201)$$

It is worth noting that the inequality (6.15.201) is weaker than (6.15.196) for all values of t. This follows from the following properties of the partition function:

Proposition 6.15.2 *Let $Z(t)$ be the partition function on the compact Riemannian manifold (M, g). Then*
(i) $Z(t) > 1$, for all $t > 0$;
(ii) $Z(0+) = +\infty$;
(iii) $\lim_{t\to\infty} Z(t) = 1$;
(iv) $Z(t)$ is strictly decreasing in t.

Proof: (i) Separating the eigenvalue $\lambda_0 = 0$ from the rest of the sum, yields

$$Z(t) = 1 + \sum_{k\geq 1} e^{\lambda_k t} > 1.$$

(iii) Since $\partial_t p_t(x, y) = \Delta_y p_t(x, y)$, then when t tends to infinity the limit is the equilibrium density, $\lim_{t\to\infty} p_t(x, y) = \dfrac{1}{vol(M)}$. Therefore

$$\lim_{t\to\infty} Z(t) = \lim_{t\to\infty} \int_M p_t(x, x)\, dv(x) = 1.$$

(iv) It follows from $Z'(t) = \sum_{k\geq 1} \lambda_k e^{\lambda_k t} < 0$ and $\lambda_k < 0$.

■

Remark 6.15.3 Using (6.15.199) and (6.15.200) we obtain a reduction of the double integral to the single integral along the diagonal of the heat kernel as

$$\int_M \int_M p_t(x, y)^2\, dv(x) dv(y) = \int_M p_{2t}(x, x)\, dv(x).$$

Remark 6.15.4 It is worth speculating on the physical interpretation of the previous computations. If the manifold M represents the states of a thermodynamic system, then $p_t(x, y)$ represents the probability that the state x can make the transition to state y in time t. Similarly, $p_t(x, x)$ describes the probability that the state x remains unchanged (or returns to itself) in time t. The integral of these probabilities gives the partition function $Z(t)$. This is a fundamental concept in statistical mechanics used to calculate various thermodynamic properties of a system.

The partition function can also be expressed as a sum of exponentials of the negative energy levels of the system, described by the eigenvalues $E_i = -\lambda_i > 0$. If the partition function is equal to 1, it indicates that all energy levels are equally likely to be occupied by the particles or molecules in the system. This is only true for systems that are in thermal equilibrium, as suggested by the fact that $\lim_{t\to\infty} p_t(x, y) = \dfrac{1}{vol(M)}$ is a uniform probability density on M.

For smaller values of time t, most physical systems have partition functions that are much larger than 1, indicating that some energy levels are more likely to be occupied than others.

The partition function is related to the underlying Brownian motion by relation (6.15.200). Its geometric interpretation is given by the following example.

Example 6.15.1 The geometric interpretation of the heat kernel is based on the fact[7] that the heat flows between two points x and y on a manifold M mainly along the geodesics joining x and y, see Calin et al. [29]. Then the heat kernel is given by the contribution of each geodesic weighted by a volume element $V_k(t)$, which describes the density of geodesics as

$$p_t(x,y) = \sum_{k\geq 1} V_k(t) e^{-\frac{d_k(x,y)^2}{2t}},$$

where $d_k(x,y)$ is the length of the k-th geodesic joining the points x and y. Then the partition function on the compact manifold M becomes

$$Z(t) = \int_M p_t(x,x)\,dv(x) = \sum_{k\geq 1} V_k(t) \int_M e^{-\frac{\ell_k(x)^2}{2t}}\,dv(x), \tag{6.15.202}$$

where $\ell_k(x)$ is the length of the k-th geodesic loop at x. For a computation in the case of the unit circle, $\mathbb{S}^1$, see Exercise 6.19.13.

6.15.1 The Boltzmann Distribution

The *Boltzmann distribution* associated with a compact Riemannian manifold (M,g) is given by the discrete distribution

$$p_k = \frac{e^{\lambda_k t}}{Z(t)}, \quad k \geq 0 \tag{6.15.203}$$

where λ_k is the k-th eigenvalue of the Laplacian Δ_g on M and $Z(t) = \sum_{k\geq 0} e^{\lambda_k t}$ is the partition function. The parameter t, which stands for time is physically associated with the inverse of temperature.

[7]This idea is also behind the van Vleck's formula in $\mathbb{R}^n$, $p_t(x,y) = V(t)e^{-\frac{d(x,y)^2}{2t}}$, where $V(t) = (2\pi t)^{-n/2}$ stands for the van Vleck determinant.

This distribution captures global geometric properties of the manifold. It is known that the coefficients of the Minakshisundaram-Pleijel asymptotic expansion of the partition function

$$Z(t) \sim \frac{1}{(2\pi t)^{n/2}}(a_0 + a_1 t + a_2 t^2 + \cdots), \quad t \searrow 0$$

encodes the global geometry of M. For instance, $a_0 = vol(M)$ is the manifold's volume and $a_1 = \frac{1}{12}\int_M R(x)dv(x)$ is proportional to the total scalar curvature of M. We shall discuss next about the expectation of the Boltzmann distribution.

The *entropy* of the manifold (M, g) is defined by

$$S(t) = -\sum_{k\geq 0} p_k(t)\ln p_k(t). \tag{6.15.204}$$

Substituting (6.15.203) into the previous formula, after some algebraic manipulations, yields

$$S(t) = \ln Z(t) - t\frac{d}{dt}\ln Z(t). \tag{6.15.205}$$

It is worthy to note that $S(t)$ depends on $\ln Z(t)$, which is called *configurational entropy*[8]. The previous relation can be inverted as follows.

Proposition 6.15.5 *For any $t > 0$ the partition function can be expressed in terms of the entropy as*

$$Z(t) = e^{t\int_t^\infty \frac{1}{u^2}S(u)du}. \tag{6.15.206}$$

Proof: Substituting $U(t) = \ln Z(t)$, we obtain the differential equation

$$-tU'(t) + U(t) = S(t),$$

which can be solved by multiplying with the integrating factor $\mu = -t^{-2}$. Integrating between t_0 and t yields

$$U(t) = t\Big(\frac{1}{t_0}U(t_0) - \int_{t_0}^t \frac{1}{u^2}S(u)du\Big), \tag{6.15.207}$$

where $0 < t_0 < t$. By Proposition 6.15.2, part (*iii*), we get $\lim_{t\to\infty} U(t) = 0$. Then taking the limit in (6.15.207) yields

$$0 = \lim_{t\to\infty} t\Big(\frac{1}{t_0}U(t_0) - \int_{t_0}^t \frac{1}{u^2}S(u)du\Big).$$

[8]In Physics this describes the number of ways that the particles in the system can be arranged while still maintaining the same energy.

This implies

$$\int_{t_0}^{\infty} \frac{1}{u^2} S(u)du = \frac{1}{t_0} U(t_0), \quad \forall t_0 > 0.$$

Replacing t_0 by t and then solving for $U(t)$ and finally taking the exponential we arrive to formula (6.15.206). ■

It is interesting to note that formula (6.15.206) shows that the partition function at time t encodes the entropy values at the present and all future time instances of t. Consequently, the transition density of the Brownian motion on (M, g) at time t contains information about all future entropy values.

Using the definition of the partition function, we can express the sum of a series of exponentials as an exponential of a cumulative quadratic average of the entropy as follows.

Corollary 6.15.6 *We have*

$$\sum_{k \geq 0} e^{\lambda_k} = e^{\int_1^{\infty} \frac{1}{u^2} S(u)du}.$$

Using the entropy one can find an upper bound for the partition function as $Z(t) \leq Z(1)^t$, for all $t \geq 1$, see Exercise 6.19.15.

6.15.2 Action Weighted Integrals

Let x_0 be a fixed point on the Riemannian manifold M and let

$$S(t, x) = S_{cl}(x, y; t) = \frac{d(x_0, x)^2}{2t}$$

be the Riemannian action on a Riemannian manifold between points x and y within time t, see page 299 for details. We also recall the Hamilton-Jacobi equation satisfied by the action

$$\partial_t S(t, x) + \frac{1}{2} \|\nabla_g S\|_g^2 = 0,$$

which will be useful shortly.

The following statement is a specific case of a result originally stated in Aronson [5].

Proposition 6.15.7 *Let (M, g) be a compact Riemannian manifold and u be a $\mathcal{C}^2$-solution of the heat equation $\partial_t u = \Delta_g u$ on M. Then the function*

$$J(t) = \int_M u^2(t, x) e^{S(t,x)} \, dv(x)$$

is non-decreasing in variable t.

Proof: Differentiating under the integral sign, using the chain rule, the heat equation and the Hamilton-Jacobi equation, we have

$$\begin{aligned}J'(t) &= \int_M (2ue^S\partial_t e^S + u^2e^S\partial_t S)\,dv(x) = \int_M (2ue^S\Delta_g u - \frac{1}{2}u^2e^S\|\nabla S\|^2)\,dv\\&= -\frac{1}{2}\int_M u^2e^S\|\nabla S\|^2 - 2\int_M g(\nabla(ue^S),\nabla u)\,dv,\end{aligned}$$

where for the last integral we have used Exercise 6.19.11 (a) together with the divergence theorem. Applying chain rule, $\nabla(ue^S) = e^S\nabla u + e^S u\nabla S$, we can complete the integrand to a perfect square as follows

$$\begin{aligned}J'(t) &= -\frac{1}{2}\int_M e^S\Big(u^2\|\nabla S\|^2 + 4\|\nabla S\|^2 + 4u\,g(\nabla u,\nabla S)\Big)dv\\&= -\frac{1}{2}\int_M e^S\|u\nabla S + 2\nabla u\|^2\,dv \le 0,\end{aligned}$$

which implies that $J(t)$ is non-decreasing. ■

6.16 The Role of the Exponential Map

Let (M, g) be a geodesically complete Riemannian manifold and p be a fixed point in M. The exponential map, $\exp : T_pM \to M$, maps a tangent vector v at p to the point obtained by following the geodesic γ_v on M, starting at p with initial velocity v, by an unit arc. Specifically, we have $\exp(v) = \gamma_v(1) = \gamma_{\frac{v}{|v|}}(|v|)$ for all $v \in T_pM\backslash 0$, and $\exp(0) = p$.

By considering a Brownian motion B_t on the tangent space $T_pM \approx \mathbb{R}^n$ with $B_0 = 0$, we can define a process on M as $X_t = \exp(B_t)$, which satisfies $X_0 = p$. The problem at hand is to determine the stochastic differential equation governing the behavior of this process. This setting is motivated by the interest in studying the behavior of X_t under the influence of Brownian motion in the tangent space.

Exponential map on $\mathbb{S}^1$ To begin, let us consider the case of the unit circle, $M = \mathbb{S}^1$, and fix the point $p = (1, 0)$. In this setting, B_t represents a Brownian motion on the real line, which corresponds to the tangent space of $\mathbb{S}^1$ at p, i.e., $T_p\mathbb{S}^1 \approx \mathbb{R}$. The exponential map in this case takes the form

$$X_t = \exp(B_t) = \gamma_{\epsilon_t}(|B_t|) = (\cos(|B_t|), \epsilon_t\sin(|B_t|)) = (\cos(B_t), \sin(B_t)),$$

where $\epsilon_t = B_t/|B_t| = \text{sign}(B_t)$. As a result, we obtain a process which satisfies (2.18.42)-(2.18.43), see page 109, which represents a Brownian motion on the circle $\mathbb{S}^1$. Thus, the exponential map takes a Brownian motion on the tangent line of $\mathbb{S}^1$ and maps it to a Brownian motion on the unit circle, see Fig. 6.5 a.

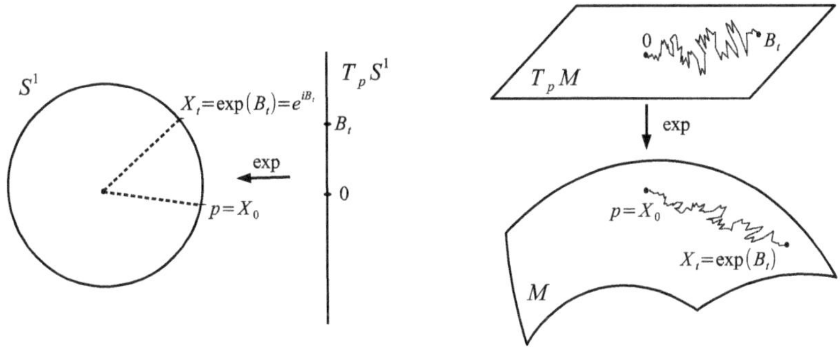

(a) (b)

Figure 6.5: *A Brownian motion B_t through the exponential map:* (a) The case of the circle $\mathbb{S}^1$; (b) *The case of a surface M.*

Exponential map on a surface We consider the case of a surface M, so $n = 2$. Let $p \in M$ be a fixed point and $T_pM \approx \mathbb{R}^2$ be the tangent space at p. We define a Brownian motion B_t on T_pM in polar coordinates as $B_t = (B_t^1, B_t^2) = (R_t \cos\theta_t, R_t \sin\theta_t)$, where R_t and θ_t are the radial and angular processes respectively. The process R_t is a two-dimensional Bessel process given by

$$dR_t = d\beta_t + \frac{1}{R_t}dt,$$

where β_t is a one-dimensional Brownian motion. The angular process is given by

$$\theta_t = \alpha_{\int_0^t R_s^{-2} ds},$$

where α_t and β_t denote two independent one-dimensional Brownian motions. Moreover, we have[9]

$$d\theta_t = \frac{1}{R_t}dW_t,$$

where W_t is a one-dimensional Brownian motion independent of β_t. We also

[9]We use the property $d\alpha_{\phi(t)} = \sqrt{\phi'(t)}\,dW_t$, or equivalently, $\alpha_{\phi(t)} = \int_0^t \sqrt{\phi'(s)}\,dW_s$, with $\phi(t)$ increasing and differentiable satisfying $\phi(0) = 0$.

have the following relations from standard stochastic calculus

$$(dR_t)^2 = (d\beta_t)^2 = dt \tag{6.16.208}$$

$$(d\theta_t)^2 = \frac{1}{R_t^2}(dW_t)^2 = \frac{1}{R_t^2}dt, \tag{6.16.209}$$

which will be used shortly in Ito's lemma.

Jacobi vectors Since M is geodesically complete, any point on the surface M can be obtained by following the geodesic $\gamma_\theta(t)$ with initial direction $(\cos\theta, \sin\theta) \in T_pM$ starting at p on M. Then M has a global parametrization $f : [0, 2\pi) \times [0, \infty) \to \mathbb{R}^3$, given by $f(\theta, t) = \gamma_\theta(t)$. Differentiating with respect to parameters we obtain two coordinate vector fields tangent to the surface

$$\frac{\partial f}{\partial t} = \dot{\gamma}_\theta(t), \quad \frac{\partial f}{\partial \theta} = Y(\theta, t). \tag{6.16.210}$$

The first one represents the velocity along the geodesic flow, while the second is the *Jacobi field* along the geodesic $t \to \gamma_\theta(t)$. The variation field Y satisfies a differential equation involving the curvature of the surface, as follows.

Let $\dfrac{D}{dt}$ and $\dfrac{D}{d\theta}$ denote the covariant derivatives with respect to the coordinate curves $t \to \gamma(\theta, t)$ and $\theta \to \gamma(\theta, t)$, respectively. This means

$$\frac{D}{dt} = \nabla_{\dot{\gamma}_\theta(t)} = \nabla_{\partial_t}, \qquad \frac{D}{d\theta} = \nabla_{Y(\theta,t)} = \nabla_{\partial_\theta},$$

where ∇ denotes the Levi-Civita connection on (M, g).

Since the coordinate vector fields commute, $[Y, \dot{\gamma}] = [\partial_\theta, \partial_t] = 0$, then using that ∇ is symmetric, we have $\nabla_Y\dot{\gamma} - \nabla_{\dot{\gamma}}Y = [Y, \dot{\gamma}] = 0$. Therefore $\nabla_Y\dot{\gamma} = \nabla_{\dot{\gamma}}Y$, or equivalently

$$\frac{D}{d\theta}\frac{\partial f}{\partial t} = \frac{D}{dt}\frac{\partial f}{\partial \theta}. \tag{6.16.211}$$

Using the facts that geodesics have zero covariant acceleration, $\nabla_{\dot{\gamma}}\dot{\gamma} = 0$, the commutation of the vector fields $\dot{\gamma}$ and Y, and the formula (6.16.211), then the definition of the Riemannian tensor field provides

$$\begin{aligned}
\mathbf{R}(\dot{\gamma}, Y)\dot{\gamma} &= \nabla_{\dot{\gamma}}\nabla_Y\dot{\gamma} - \nabla_Y\nabla_{\dot{\gamma}}\dot{\gamma} - \nabla_{[\dot{\gamma},Y]}\dot{\gamma} \\
&= \nabla_{\dot{\gamma}}\nabla_Y\dot{\gamma} \\
&= \frac{D}{dt}\frac{D}{d\theta}\frac{\partial f}{\partial t} = \frac{D}{dt}\frac{D}{dt}\frac{\partial f}{\partial \theta} \\
&= \frac{D^2}{dt^2}Y.
\end{aligned}$$

We have obtained an equation for the Jacobi vector field,

$$\frac{D^2}{dt^2}Y - \mathbf{R}(\dot\gamma, Y)\dot\gamma = 0, \tag{6.16.212}$$

called the *Jacobi equation.* In the previous equation the parameter θ is considered fixed, while $t \in [0, +\infty)$.

The exponential of B_t We shall denote the surface parametrization by $f = (f^1, f^2, f^3)$. Now, we are interested in the process $X_t = \exp(B_t) = \gamma_{\theta_t}(R_t) = f(\theta_t, R_t)$, which is a process on M starting at p, see Fig. 6.5 b. Then $X_t^k = f^k(\theta_t, R_t)$, $k = 1, 2, 3$. To find the stochastic differential equation satisfied by X_t^k, we can apply Ito's lemma, which provides

$$\begin{aligned} dX_t^k =& \frac{\partial f^k}{\partial \theta}(\theta_t, R_t)\, d\theta_t + \frac{1}{2}\frac{\partial^2 f^k}{\partial \theta^2}(\theta_t, R_t)\, (d\theta_t)^2 + \frac{\partial f^k}{\partial t}(\theta_t, R_t)\, dR_t \\ &+ \frac{1}{2}\frac{\partial^2 f^k}{\partial t^2}(\theta_t, R_t)\, (dR_t)^2 \\ =& \Big(\frac{1}{2R_t^2}\frac{\partial^2 f^k}{\partial \theta^2}(\theta_t, R_t) + \frac{1}{R_t}\frac{\partial f^k}{\partial t}(\theta_t, R_t) + \frac{1}{2}\frac{\partial^2 f^k}{\partial t^2}(f_t, R_t)\Big)\, dt \\ &+ \frac{1}{R_t}\frac{\partial f^k}{\partial \theta}(\theta_t, R_t)\, dW_t + \frac{\partial f^k}{\partial t}(\theta_t, R_t)\, d\beta_t, \end{aligned}$$

where we used relations (6.16.208)-(6.16.209) and that $dW_t\, d\beta_t = 0$, which follows from the fact that W_t and β_t are independent Brownian motions. Employing the vector field interpretation of partial derivatives as in relation (6.16.210) we obtain

$$\begin{aligned} dX_t^k =& \Big(\frac{1}{2R_t^2}(Y^k)'(\theta_t, R_t) + \frac{1}{R_t}\dot\gamma_{\theta_t}^k(R_t) + \frac{1}{2}\ddot\gamma_{\theta_t}^k(R_t)\Big)\, dt \\ &+ \frac{1}{R_t}Y^k(\theta_t, R_t)\, dW_t + \dot\gamma_{\theta_t}^k(R_t)\, d\beta_t. \end{aligned}$$

We note that the derivation with respect to θ was denoted by prime, " $'$ ", while the derivative with respect t by dot, " $\cdot$ ". Integrating and dropping the index k yields

$$\begin{aligned} X_t = \exp(B_t) =& p + \int_0^t \Big(\frac{1}{2R_s^2}Y'(\theta_s, R_s) + \frac{1}{R_s}\dot\gamma(\theta_s, R_s) + \frac{1}{2}\ddot\gamma(\theta_s, R_s)\Big)\, ds \\ &+ \int_0^t \frac{1}{R_s}Y(\theta_s, R_s)\, dW_s + \int_0^t \dot\gamma(\theta_s, R_s)\, d\beta_s, \end{aligned} \tag{6.16.213}$$

where we used the two-argument notation $\gamma_\theta(t) = \gamma(\theta, t)$.

The drift term in the stochastic process X_t is represented by the first integral. The other two Ito integrals correspond to sources of uncertainty, where the first integral accounts for angular spread and the second integral accounts for the radial dispersion.

The geometry of the drift rate Given the variation $\gamma(\theta, t)$ of the geodesic $\gamma(t)$, let us consider the vector function $\mu : [0, 2\pi) \times [0, \infty) \to \mathbb{R}^3$ given by

$$\mu(\theta, t) = \frac{1}{2t^2} Y'(\theta, t) + \frac{1}{t}\dot{\gamma}(\theta, t) + \frac{1}{2}\ddot{\gamma}(\theta, t). \tag{6.16.214}$$

Then taking the expectation in (6.16.213) and using the properties of Ito integrals, yields

$$\mathbb{E}[X_t] = p + \int_0^t \mathbb{E}[\mu(\theta_s, R_s)]\, ds.$$

The function $\mu(\theta, t)$ represents a vector field in $\mathbb{R}^3$. To understand it better we shall compute its tangent and normal parts with respect to the surface M. Since for any $p \in M$ we have the orthonormal split $\mathbb{R}^3 = T_pM \bigoplus \mathcal{N}_p$ with respect to the Euclidean scalar product, then

$$\mu(\theta, t) = \mu(\theta, t)^{\tan} + \mu(\theta, t)^{\text{nor}},$$

with $\mu(\theta, t)^{\tan} \in T_{f(\theta,t)}M$ and $\mu(\theta, t)^{\text{nor}} \in \mathcal{N}_{f(\theta,t)}$.

We shall continue by finding $\mu(\theta, t)^{\tan}$. To this end, we notice that $\ddot{\gamma}(\theta, t)^{\tan} = 0$ since γ is a geodesic on M and then $\ddot{\gamma}(\theta, t)$ is normal on M. The velocity $\dot{\gamma}(\theta, t)$ does not have any normal part, since it belongs to the tangent space. To compute the tangent component of $Y'(\theta, t)$, we claim that

$$Y'(\theta, t)^{\tan} = \frac{D}{d\theta} Y(\theta, t).$$

This is a consequence of Gauss' formula as follows. We recall first the expression of the Levi-Civita connection on $\mathbb{R}^3$

$$\overline{\nabla}_U V = \sum_i U(V^i) e_i$$

for any two vector fields U and V on $\mathbb{R}^3$. Choosing U and V vector fields on M, then Gauss' formula states

$$\begin{aligned}\overline{\nabla}_U V &= (\overline{\nabla}_U V)^{\tan} + (\overline{\nabla}_U V)^{\text{nor}} \\ &= \nabla_U V + L(U, V),\end{aligned}$$

where ∇ is the Levi-Civita connection on M and L denotes the second fundamental form of M. Taking $U = V = \frac{\partial}{\partial\theta} f(\theta, t) = Y(\theta, t)$ we obtain $\overline{\nabla}_Y Y = \nabla_Y Y + L(Y, Y)$, which can be written equivalently as

$$Y'(\theta, t) = \frac{\partial^2}{\partial\theta^2} f(\theta, t) = \frac{\overline{D}}{d\theta} Y = \frac{D}{d\theta} Y + L(Y, Y),$$

where we used that the covariant differentiation in $\mathbb{R}^3$ along parameter θ is $\frac{\overline{D}}{d\theta} = \frac{\partial}{\partial\theta}$. Taking the tangent and the normal parts yields

$$Y'(\theta,t)^{\tan} = \frac{D}{d\theta}Y, \qquad Y'(\theta,t)^{\text{nor}} = L(Y,Y).$$

Therefore, we arrived at

$$\mu(\theta,t)^{\tan} = \frac{1}{2t^2}\frac{D}{d\theta}Y + \frac{1}{t}\dot{\gamma}(\theta,t) \tag{6.16.215}$$

$$\mu(\theta,t)^{\text{nor}} = \frac{1}{2t^2}L(Y,Y) + \frac{1}{2}\ddot{\gamma}(\theta,t), \tag{6.16.216}$$

where L is the second fundamental form of the surface M in $\mathbb{R}^3$. If ξ denotes the unit normal vector field to M, then there is a symmetric function h such that $L(U,V) = h(U,V)\xi$, for any two vector fields U and V on M. Since $\ddot{\gamma}(\theta,t) = \kappa_n(\theta,t)\xi$, where κ_n is the normal curvature along geodesic γ, we have

$$\mu(\theta,t)^{\text{nor}} = \Big(\frac{1}{2t^2}h(Y,Y) + \frac{1}{2}\kappa_n(\theta,t)\Big)\xi.$$

To analyze the behavior of the tangent vector field $\mu(\theta,t)^{\tan}$ on the surface M, we need to decompose it into two directions. The first direction is the radial direction, which points in the direction of the geodesic velocity, and the second direction is the transversal direction, which points in the direction of the Jacobi vector field. Hence

$$\mu(\theta,t)^{\tan} = \mu(\theta,t)^{\tan}_{\text{rad}} + \mu(\theta,t)^{\tan}_{\text{tran}}, \tag{6.16.217}$$

with

$$\mu(\theta,t)^{\tan}_{\text{rad}} = \langle \mu(\theta,t)^{\tan}, \dot{\gamma}(\theta,t)\rangle, \qquad \mu(\theta,t)^{\tan}_{\text{tran}} = \langle \mu(\theta,t)^{\tan}, Y(\theta,t)\rangle.$$

The fact that the aforementioned directions are orthogonal is a consequence of Gauss' lemma as follows. Let $S_p(t) = \{\gamma(\theta,t); \theta \in [0,2\pi)\}$ be the one-dimensional sphere centered at p of radius t on the surface M. By Gauss' lemma, the radial direction, $\dot{\gamma}(\theta,t)$ is normal to the sphere, namely on the tangent direction $\frac{\partial}{\partial\theta}\gamma(\theta,t) = Y(\theta,t)$, which is the Jacobi field. Therefore, $\langle Y,\dot{\gamma}\rangle = 0$. We shall study in the following the radial and transversal components.

The radial component Since γ is unit speed, using the metric properties of $\dfrac{D}{d\theta}$ and the commutation relation (6.16.211) we can compute the radial part as follows:

$$\begin{aligned}
\mu(\theta,t)^{\tan}_{\rm rad} &= \langle \mu(\theta,t)^{\tan}, \dot\gamma(\theta,t)\rangle = \frac{1}{2t^2}\langle \frac{D}{d\theta}Y, \dot\gamma\rangle + \frac{1}{t}\langle \dot\gamma, \dot\gamma\rangle \\
&= \frac{1}{2t^2}\Big\{ \frac{D}{d\theta}\langle Y, \dot\gamma\rangle - \langle Y, \frac{D}{d\theta}\dot\gamma\rangle \Big\} + \frac{1}{t} \\
&= -\frac{1}{2t^2}\langle Y, \frac{D}{d\theta}\dot\gamma\rangle + \frac{1}{t} \\
&= -\frac{1}{2t^2}\langle Y, \frac{D}{d\theta}\frac{\partial}{\partial t} f(\theta,t)\rangle + \frac{1}{t} \\
&= -\frac{1}{2t^2}\langle Y, \frac{D}{dt}\frac{\partial}{\partial \theta} f(\theta,t)\rangle + \frac{1}{t} \\
&= -\frac{1}{2t^2}\langle Y, \frac{D}{dt}Y\rangle + \frac{1}{t} = -\frac{1}{4t^2}\frac{D}{dt}\langle Y, Y\rangle + \frac{1}{t} \\
&= -\frac{1}{4t^2}\frac{d}{dt}\|Y(\theta,t)\|^2 + \frac{1}{t}\cdot \qquad (6.16.218)
\end{aligned}$$

It can be observed that the magnitude of the Jacobi field influences the radial component, indicating that it is dependent on the radial rate at which geodesics spread apart.

Remark 6.16.1 The magnitude of the Jacobi field is determined by the Gaussian curvature K of the surface M, which can be seen from the following asymptotic formula as follows:

$$\|Y(\theta,t)\|^2 = t^2 - \frac{1}{3}K_{|\gamma(\theta,t)}\, t^4 + \mathcal{R}(\theta,t),$$

with remainder $\mathcal{R}(\theta,t) = o(t^4)$. The effect of the curvature on the geodesic spread can be observed from this formula: positively curved surfaces tend to have a smaller spread, while negatively curved surfaces have a larger spread. This formula follows from expanding the function $h(\theta,t) = \langle Y(\theta,t), Y(\theta,t)\rangle$ in a Taylor series as a function of t and computing the first few coefficients and using the Jacobi equation.

The following result illustrates the radial evolution of the tangent component $Y'(\theta,t)^{\tan}$.

Proposition 6.16.2 *Let $V = Y'(\theta,t)^{\tan}$ and denote by $\mathbf{R}$ the Riemannian curvature tensor. Then V solves the following initial value problem*

$$\begin{aligned}
\frac{D}{dt}V(\theta,t) &= \frac{D^2}{d\theta^2}\dot\gamma(\theta,t) + \mathbf{R}(\dot\gamma(\theta,t), Y(\theta,t))Y(\theta,t), \quad t>0 \\
V(\theta,0) &= 0.
\end{aligned}$$

Proof: An analogous approach to the one employed in proving the Jacobi equation (6.16.212) gives us

$$\begin{aligned}\frac{D}{dt}V(\theta,t) &= \frac{D}{dt}\frac{D}{d\theta}Y(\theta,t) = \frac{D}{d\theta}\frac{D}{dt}Y(\theta,t) + \mathbf{R}\big(\dot\gamma(\theta,t),Y(\theta,t)\big)Y(\theta,t)\\ &= \frac{D}{d\theta}\frac{D}{dt}\frac{\partial f}{\partial\theta}(\theta,t) + \mathbf{R}\big(\dot\gamma(\theta,t),Y(\theta,t)\big)Y(\theta,t)\\ &= \frac{D}{d\theta}\frac{D}{d\theta}\frac{\partial f}{\partial t}(\theta,t) + \mathbf{R}\big(\dot\gamma(\theta,t),Y(\theta,t)\big)Y(\theta,t)\\ &= \frac{D^2}{d\theta^2}\dot\gamma(\theta,t) + \mathbf{R}\big(\dot\gamma(\theta,t),Y(\theta,t)\big)Y(\theta,t).\end{aligned}$$

Also, since $f(\theta,0)=\gamma(\theta,0)=p$, then

$$V(\theta,0) = Y'(\theta,0)^{\tan} = \Big(\frac{\partial^2}{\partial\theta^2}f(\theta,0)\Big)^{\tan} = 0.$$

■

The transversal component Using that $\dot\gamma$ and Y are orthogonal and that $\frac{D}{d\theta}$ is metrical, we have

$$\begin{aligned}\mu(\theta,t)^{\tan}_{\text{tran}} &= \langle \mu(\theta,t)^{\tan}, Y(\theta,t)\rangle = \frac{1}{2t^2}\langle\frac{D}{d\theta}Y,Y\rangle + \frac{1}{t}\langle\dot\gamma,Y\rangle\\ &= \frac{1}{4t^2}\frac{D}{d\theta}\langle\, Y,Y\rangle = \frac{1}{4t^2}\frac{\partial}{\partial\theta}\|Y(\theta,t)\|^2.\end{aligned} \tag{6.16.219}$$

This relation shows how the radial component is affected by the magnitude of the Jacobi field, indicating its dependence on the rate at which geodesics spread apart in the transversal direction.

Proposition 6.16.3 *The tangent component of the drift rate is given by*

$$\mu(\theta,t)^{\tan} = \frac{1}{4t^2}\Big(\frac{\partial}{\partial\theta}+\frac{\partial}{\partial t}\Big)\|Y(\theta,t)\|^2 + \frac{1}{t}, \qquad t>0. \tag{6.16.220}$$

Proof: The proof can be obtained by substituting equations (6.16.218) and (6.16.219) into expression (6.16.217).

■

6.17 Brownian Motion in Geodesic Coordinates

We will now focus on the inverse problem: finding the expression of a Brownian motion on a surface M, equipped with the natural metric from $\mathbb{R}^3$, in geodesic coordinates. Suppose M is geodesically complete. Then, any point on M can be represented as $\gamma_\theta(t)$, where γ_θ is a unit speed geodesic starting at p on M with initial direction $(\cos\theta, \sin\theta) \in T_pM$. A global parametrization of M can be given by $f : [0, 2\pi) \times [0, \infty) \to \mathbb{R}^3$, where $f(\theta, t) = \gamma_\theta(t)$. The coordinate vector fields on the surface are expressed by (6.16.210). Then the coefficients of the metric induced on M from $\mathbb{R}^3$ in coordinates (θ, t) are computed as

$$\begin{aligned}
g_{tt} &= \langle \frac{\partial f}{\partial t}, \frac{\partial f}{\partial t} \rangle = \langle \dot\gamma_\theta(t), \dot\gamma_\theta(t) \rangle = 1 \\
g_{t\theta} = g_{\theta t} &= \langle \frac{\partial f}{\partial t}, \frac{\partial f}{\partial \theta} \rangle = \langle \dot\gamma_\theta(t), V(\theta, t) \rangle = 0 \\
g_{\theta\theta} &= \langle \frac{\partial f}{\partial \theta}, \frac{\partial f}{\partial \theta} \rangle = \|V(\theta, t)\|^2.
\end{aligned}$$

We employ the notation $h(\theta, t) = \|V(\theta, t)\|$ to denote the magnitude of the variation field. Then

$$g_{ij} = \begin{pmatrix} 1 & 0 \\ 0 & h^2 \end{pmatrix}, \qquad g^{ij} = \begin{pmatrix} 1 & 0 \\ 0 & h^{-2} \end{pmatrix} \tag{6.17.221}$$

and $\det g = h^2$. The Laplace-Beltrami operator in geodesic coordinates is expressed as follows

$$\begin{aligned}
\Delta &= \frac{1}{2} \frac{1}{\sqrt{\det g}} \sum_{j=1}^{2} \frac{\partial}{\partial x_j} \left(\sqrt{\det g}\, g^{ij} \frac{\partial}{\partial x_i} \right) \\
&= \frac{1}{2} \frac{1}{h} \Big(\frac{\partial}{\partial_t} \big(h \frac{\partial}{\partial t} \big) + \partial_\theta \big(h^{-1} \frac{\partial}{\partial \theta} \big) \Big) \\
&= \frac{1}{2} \Big(\frac{\partial^2}{\partial t^2} + \frac{1}{h^2} \frac{\partial^2}{\partial \theta^2} \Big) + \frac{1}{2h} \frac{\partial h}{\partial t} \frac{\partial}{\partial t} - \frac{1}{2h^3} \frac{\partial h}{\partial \theta} \frac{\partial}{\partial \theta}.
\end{aligned}$$

The dispersion and drift vectors are given by

$$\sigma = \begin{pmatrix} 1 & 0 \\ 0 & h^{-1} \end{pmatrix}, \qquad b = \Big(\frac{1}{2h} \frac{\partial h}{\partial t}, -\frac{1}{2h^3} \frac{\partial h}{\partial \theta} \Big)^T.$$

Thus, the Brownian motion on M in geodesic coordinates, denoted by (Θ_t, T_t), can be written as follows:

$$dT_t = dW_t^1 + \frac{1}{2h(\Theta_t, T_t)} \frac{\partial h}{\partial t}(\Theta_t, T_t)\, dt \tag{6.17.222}$$

$$d\Theta_t = \frac{1}{h(\Theta_t, T_t)} dW_t^2 - \frac{1}{2h^3(\Theta_t, T_t)} \frac{\partial h}{\partial \theta}(\Theta_t, T_t)\, dt. \tag{6.17.223}$$

where W_t^1, W_t^2 are two independent one-dimensional Brownian motions. Thus, the Brownian motion on M is described by two processes: radial process, T_t, and the angular process, Θ_t. As it can seen from the system (6.17.222)-(6.17.223), the aforementioned processes depend on each other and also on the magnitude of h. We shall investigate these equations in the particular case of surfaces of constant curvature. We shall handle first flat surfaces, then positive curvature and finally negative curvature surfaces.

Zero curvature surfaces: We consider the case when the surface M is flat, namely, $K = 0$, where K denotes the Gaussian curvature. By the *Theorema Egregium*, the curvature K can be written only in terms of the coefficients of the metric given by (6.17.221). A computation provides

$$K = -\frac{1}{h}\frac{\partial^2 h}{\partial t^2}.$$

The vanishing curvature condition implies the existence of two smooth function $a(\cdot)$ and $b(\cdot)$ such that $h(\theta, t) = a(\theta)t + b(\theta)$. Since $h(\theta, 0) = \|V(\theta, 0)\| = 0$, then it follows that $b(\theta) = 0$. Then

$$\frac{1}{2h}\frac{\partial h}{\partial t} = \frac{a(\theta)}{2a(\theta)t} = \frac{1}{2t}.$$

Then equation (6.17.222) becomes

$$dT_t = dW_t^1 + \frac{1}{2T_t}\,dt, \tag{6.17.224}$$

which is exactly the equation of a two-dimensional Bessel process.

Using $h(\theta, t) = a(\theta)t$, the angular process (6.17.223) becomes

$$d\Theta_t = \frac{1}{a(\Theta_t)T_t}dW_t^2 - \frac{a'(\Theta_t)}{2a^3(\Theta_t)T_t^2}\,dt. \tag{6.17.225}$$

(*i*) Assuming spherical symmetry about point p, the equation can be simplified by imposing the invariance of h with respect to the angle θ. Let us assume $a(\theta) = 1$ without loss of generality. With this assumption, the drift term disappears, leaving us with the equation

$$d\Theta_t = \frac{1}{T_t}dW_t^2. \tag{6.17.226}$$

It should be noted that the drift term is an indication of the non-spherical symmetry of the magnitude of $V(\theta, t)$. The previous equation implies that Θ_t is a martingale and by Theorem 1.1.11 there is a DDS Brownian motion β_t such that $\Theta_t = \beta_{\int_0^t \frac{1}{R_s^2}ds}$.

(*ii*) Another special case is when the drift of Θ_t only depends on T_t and not on Θ_t. This can be achieved by setting $a(\theta)$ to satisfy $\frac{a'(\theta)}{2a^3(\theta)} = c$, where

c is a real constant. The solution is given by $a(\theta) = \frac{a(0)}{\sqrt{1-4ca(0)^2\theta}}$. Since it makes sense to have solutions $a(\theta)$ that do not blow up for finite values of θ, we consider $c = -\delta^2 < 0$. Then the drift rate of Θ_t is positive and we have

$$d\Theta_t = \frac{\sqrt{1+4\delta^2\Theta_t}}{T_t} dW_t^2 + \frac{\delta^2}{T_t^2}\, dt,$$

where we assumed $a(0) = 1$.

Constant positive curvature surfaces: Let us consider surfaces with a constant Gaussian curvature $K = a^2 > 0$ for $a > 0$. Then the equation

$$\frac{\partial^2}{\partial t^2} h(\theta, t) + Kh(\theta, t) = 0$$

has a general solution of the form $h(\theta, t) = A_1(\theta)\cos(at) + A_2(\theta)\sin(at)$. By using the initial condition $h(\theta, 0) = 0$, we get $A_1(\theta) = 0$ and hence the solution becomes $h(\theta, t) = A_2(\theta)\sin(at)$. Since

$$\frac{1}{2h(\theta,t)} \frac{\partial}{\partial t} h(\theta, t) = \frac{a}{2}\cot(at),$$

the drift rate of T_t becomes independent of θ. Therefore, the relation (6.17.222) simplifies to

$$dT_t = dW_t^1 + \frac{a}{2}\cot(aT_t)dt. \tag{6.17.227}$$

The angular process given by (6.17.223) becomes

$$d\Theta_t = \frac{1}{A_2(\theta)\sin(aT_t)} dW_t^2 - \frac{A_2'(\theta)}{2A_2^3(\theta)\sin^2(aT_t)}\, dt. \tag{6.17.228}$$

This process becomes driftless under the assumption of spherical symmetry about p with the normalization condition $A(\theta) = 1/a$. Hence, we have

$$d\Theta_t = \frac{a}{\sin(aT_t)} dW_t^2.$$

It is worth noting that as the curvature a^2 approaches zero, we retrieve equation (6.17.226).

Constant negative curvature surfaces: Assuming a negative constant Gaussian curvature $K = -a^2 < 0$, the equation $\frac{\partial^2}{\partial t^2} h(\theta, t) + Kh(\theta, t) = 0$ has the general solution $h(\theta, t) = A_1(\theta)\cosh(at) + A_2(\theta)\sinh(at)$. Using the initial condition $h(\theta, 0) = 0$, we obtain $h(\theta, t) = A_2(\theta)\sinh(at)$. Since $\frac{1}{2h(\theta,t)} \frac{\partial}{\partial t} h(\theta, t) = \frac{a}{2}\coth(at)$, the drift rate of T_t is independent of θ, and the relation (6.17.222) becomes

$$dT_t = dW_t^1 + \frac{a}{2}\coth(aT_t)\, dt. \tag{6.17.229}$$

The angular process given by (6.17.223) becomes

$$d\Theta_t = \frac{1}{A_2(\theta)\sinh(aT_t)}dW_t^2 - \frac{A_2'(\theta)}{2A_2^3(\theta)\sinh^2(aT_t)}dt,$$

which simplifies to a driftless process

$$d\Theta_t = \frac{a}{\sinh(aT_t)}dW_t^2$$

if we assume spherical symmetry about p with the choice $A_2(\theta) = 1/a$.

It is a well-established fact that two surfaces with the same constant curvature are locally isometric, meaning that they have the same intrinsic geometry, see for instance Millman and Parker [110], page 149. Therefore, regardless of their extrinsic differences, the diffusion behavior on these surfaces is identical, provided that they belong to the same class of constant curvature surface (flat, positive, or negative). However, diffusions corresponding to different types of curvatures are expected to exhibit distinct behaviors. The comparisons of their drifts are provided below.

The drift of the radial process T_t is affected by the sign of the curvature. Equations (6.17.224), (6.17.227) and (6.17.229) describe the radial process T_t for the flat, positive, or negative constant curvature surfaces, respectively. By the inequality $a\cot(ax) < \frac{1}{x} < a\coth(ax)$, for any $a \neq 0$ and $x > 0$, we can establish the following inequalities for the expected value of the radial process:

$$\frac{a}{2}\int_0^t \mathbb{E}[\cot(aT_s)]\,ds < \int_0^t \mathbb{E}\Big[\frac{1}{T_s}\Big]\,ds < \frac{a}{2}\int_0^t \mathbb{E}[\coth(aT_s)]\,ds.$$

These inequalities represent the expected value of the radial process, $\mathbb{E}[T_t]$, for the three types of constant curvature surfaces. The lowest expected value (and therefore the smallest drift) is observed for the positive curvature case, while the highest expected value is observed for the negative curvature case.

6.18 Summary

This chapter explored the study of Brownian motions on Riemannian manifolds, which are diffusions associated with the Laplace-Beltrami operator under the Riemannian metric. Various examples are examined, including hyperbolic diffusions, Brownian motions on spheres, and surfaces in isothermal and geodesic coordinates. Bessel processes and their generalizations to Riemannian manifolds, known as radial processes, are also investigated.

The chapter covered the skew-product representation of Brownian motions obtained through the central projection onto $\mathbb{S}^n$ in $\mathbb{R}^{n+1}$. Furthermore, the modeling of Brownian motions escaping a given region was addressed using a Stokes' type formula involving the flux over the region's boundary. Explicit computations were provided for two-dimensional balls.

The chapter also highlighted the connection between the stochastic structure of a non-degenerate diffusion on $\mathbb{R}^n$ and its induced Riemannian metric structure. The study of the associated geometric elements aids in better understanding the diffusion process.

Geodesic completeness and stochastic completeness were discussed, with the former relating to geometrical properties involving geodesics and metric completeness of the manifold as a metric space. The latter pertains to the behavior of Brownian motions on the manifold, extending indefinitely, and is equivalent to the uniqueness of the Cauchy problem for the heat equation on M. Conditions for stochastic completeness were explored, including those based on volume growth.

The Boltzmann distribution, partition function, and their interrelations on Riemannian manifolds were examined, along with maximum integral principles. An analysis of the Brownian motion via exponential map as well as under geodesic coordinates is done on surfaces and a discussion of its behavior depending on the Gaussian curvature is performed.

6.19 Exercises

Exercise 6.19.1 Let $p_t(\rho) = \dfrac{2\rho^2}{\sqrt{2\pi}t^{3/2}} e^{-\frac{1}{2t}(\rho^2+t^2|\mu|^2)} \dfrac{\sinh(\rho|\mu|)}{\rho|\mu|}$, $\rho > 0$. Prove that $\displaystyle\int_0^\infty p_t(\rho)\, d\rho = 1$.

Exercise 6.19.2 Let $\sigma(x)$ be a positive and differentiable function on $\mathbb{R}$ and consider the one-dimensional manifold $M = (\mathbb{R}, ds)$, with the metric $ds = \frac{1}{\sigma}dx$.
(a) Use formula (6.14.186) to show that the Laplacian on M is given by

$$\Delta_\sigma = \frac{1}{2}\sigma^2 \frac{d^2}{dx^2} + \frac{1}{2}\sigma(x)\sigma'(x)\frac{d}{dx}.$$

(b) Show that the Brownian motion on M starting at zero satisfies

$$dX_t = \sigma(X_t)dW_t + \frac{1}{2}\sigma(X_t)\sigma'(X_t)dt, \qquad X_0 = 0.$$

Exercise 6.19.3 (a) Find a strong solution for the equation

$$dX_t = \frac{1}{2}e^{2X_t}dt + e^{X_t}dW_t$$
$$X_0 = 0.$$

(b) Specify the metric ds on $\mathbb{R}$ such that X_t is a Brownian motion on $(\mathbb{R}, ds)$.
(c) Find the isometry g described at page 265.

Exercise 6.19.4 (a) Show that the differential operator

$$L = \frac{1}{2}(\cos x)^2 \frac{d^2}{dx^2} - \frac{1}{4}\sin(2x)\frac{d}{dx}$$

is hypoelliptic on $\mathbb{R}$.
(b) Show that the diffusion X_t associated to the operator L is a Brownian motion on $M = (\mathbb{R}, ds)$, with the metric $ds = \sec x\, dx$).
(c) Assuming $X_0 = 0$, show that $-\frac{\pi}{2} < X_t < \frac{\pi}{2}$. We notice that the diffusion cannot pass over the singularities of the metric.

Exercise 6.19.5 (a) Show that the map $\tau : \mathcal{D}^n \to \mathcal{H}^n$ given by (6.4.55) is an isometry of Riemannian spaces, with the metrics are given by (6.4.56).
(b) Using formula (6.4.58), find the law of the Brownian motion on the Poincaré's ball $\mathcal{D}^3$.

Exercise 6.19.6 Find the heat kernel of the operator

$$L = \frac{1}{2}\partial_r^2 + \Big(\lambda r + \frac{n-1}{2r}\Big)\partial_r, \qquad r > 0$$

by solving equations (6.10.159) and then finding the transition probability of the norm process R_t.

Exercise 6.19.7 If there exists a positive integer k such that

$$\int_M \phi_k(x)\, dv(x) \neq 0,$$

where $\phi_k(x)$ is the k-th eigenfunction of the Laplace-Beltrami operator Δ_g, show that the Riemannian manifold (M, g) is stochastically incomplete.

Exercise 6.19.8 Prove in at least two different ways that $\mathbb{S}^2$ endowed with the natural metric induced from $\mathbb{R}^3$ is stochastically complete.

Exercise 6.19.9 Let (M, g) be a connected, compact Riemannian manifold, without boundary.
(a) Use formula (6.14.186) to show that for any smooth function $\phi : M \to \mathbb{R}$ we have

$$\Delta_g(\phi^2) = 2\phi\Delta\phi + \|\nabla\phi\|_g^2,$$

where $\nabla\phi$ is the gradient of ϕ taken with respect to the metric g.

(b)(Hopf) Prove that harmonic functions are constants, i.e. if $\Delta\phi = 0$, then ϕ is a constant.

(c) Let ϕ be an eigenvector satisfying $\Delta\phi = \lambda\phi$ on M. Show that

$$\lambda = -\frac{\frac{1}{2}\int_M \|\nabla\phi\|^2\, dv}{\int_M \phi^2 dv}.$$

Exercise 6.19.10 Let (M, g) be a connected, compact Riemannian manifold. Use Cauchy's integral inequality to prove that for any continuous function $u : M \to \mathbb{R}$ we have

$$(u_{\text{ave}})^2 \leq (u^2)_{\text{ave}},$$

where $u_{\text{ave}} = \dfrac{1}{vol(M)}\displaystyle\int_M u(x)\, dv(x)$ and $(u^2)_{\text{ave}} = \dfrac{1}{vol(M)}\displaystyle\int_M u(x)^2\, dv(x)$.

Exercise 6.19.11 Let X be a vector field on the Riemannian manifold (M, g) and denote by $\text{div}\,(X)$ the divergence of X given by formula (6.2.23).

(a) Prove that for any smooth function $f : M \to \mathbb{R}$ we have

$$\text{div}\,(fX) = f\,\text{div}(X) + g(\nabla f, X),$$

where ∇f denotes the gradient of f with respect to the metric g.

(b) If Δ_g is the Laplace-Beltrami operator on (M, g), show that

$$\Delta_g(f^\alpha) = \alpha f^{\alpha-1}\Big(\Delta_g f + \frac{\alpha-1}{2}\|\nabla f\|_g^2\Big), \qquad \forall\alpha > 0.$$

Exercise 6.19.12 Consider the differential operator L on $\mathbb{R}^n$ given by

$$Lu = \frac{1}{2}\text{div}\,(\phi^2\nabla_{Eu}u),$$

where ϕ is a given smooth function on $\mathbb{R}^n$. Show that the associated diffusion X_t on $\mathbb{R}^n$ satisfies

$$dX_t = \phi(X_t)dW_t + \phi(X_t)(\nabla_{Eu}\phi)(X_t)\, dt,$$

where W_t is an n-dimensional Brownian motion.

Exercise 6.19.13 (a) Show that the eigenvalues of the Laplacian $\Delta = \frac{1}{2}\frac{d}{ds^2}$ on $\mathbb{S}^1$ are $\lambda_k = -k^2/2$ and verify that the partition function is $Z(t) = \sum_{k\geq 0} e^{-\frac{k^2}{2}t}$.

(b) Starting from the heat kernel formula of $\mathbb{S}^1$

$$p_t(s_0, s) = \frac{1}{\sqrt{2\pi t}}\sum_{n\in\mathbb{Z}} e^{-\frac{1}{2t}(2n\pi+s-s_0)^2},$$

use the trace formula (6.15.200) to find the partition function $Z(t)$.

(c) Find the entropy associated with the Boltzmann distribution on $\mathbb{S}^1$.

Exercise 6.19.14 This is a continuation of Exercise 6.19.13, which computes an integral representation of the partition function $Z(t) = \sum_{k\geq 0} e^{-\frac{k^2}{2}t}$ on the unit circle.

(a) Show that

$$\sum_{k=0}^{n-1} e^{-\frac{k^2}{2}t} = \lim_{n\to\infty} \mathbb{E}\Big(\frac{1-e^{inW_t}}{1-e^{iW_t}}\Big),$$

where W_t is a one-dimensional Brownian motion.

(b) Use trigonometry to verify the identity

$$\frac{1-e^{inx}}{1-e^{ix}} = \frac{1}{2}(1-\cos(nx)) + \frac{1}{2}\cot\frac{x}{2}\sin(nx) + \frac{i}{2}\Big[\cot\frac{x}{2}(1-\cos(nx)) - \sin(nx)\Big]$$

for any $x \in \mathbb{R}$.

(c) Using parts (a) and (b) show that $Z(t) = \frac{1}{2} + \frac{1}{\sqrt{2\pi t}}F(t)$, where

$$F(t) = \frac{1}{2}\lim_{n\to\infty}\int_{\mathbb{R}} \cot\frac{x}{2}\sin(nx)e^{-\frac{x^2}{2t}}\,dx.$$

(d) Verify that $F(1) = \pi$ and deduce that $\sum_{k\geq 0} e^{-\frac{k^2}{2}} = \frac{1}{2} + \sqrt{\frac{\pi}{2}}$.

(e) Use Exercise 6.19.15 to infer the inequality $Z(t) < 2^t$ for $t > 1$.

Exercise 6.19.15 Let $Z(t)$ be the partition function of the compact manifold (M, g). Show that $Z(t) \leq Z(1)^t$, for all $t \geq 1$ and $Z(t) > Z(1)^t$, for all $0 < t < 1$.

Exercise 6.19.16 Let $\phi : M \to \mathbb{R}$ be a smooth function on the compact Riemannian manifold (M, g) and consider a solution u of the equation $\partial_t u = \Delta_g u$ on M. Let

$$J(t) = \int_M u(t,x)^2 e^{\phi(x)}\,dv(x).$$

(a) Use the Cauchy inequality to prove

$$-2ug(\nabla u, \nabla\phi) \leq 2\|\nabla u\|^2 + \frac{1}{2}u^2\|\nabla\psi\|^2.$$

(b) Show that

$$J'(t) \leq \frac{1}{2}\int_M u^2 e^{\phi}\|\nabla\phi\|^2\,dv(x).$$

(c) Assume $\|\nabla\phi\| \leq a$ on M. Then show the inequality

$$J(t) \leq J(0)e^{\frac{1}{2}a^2 t}, \qquad t \geq 0.$$

(d) Deduce the uniqueness property: If $u_{t=0} = 0$ then $u = 0$ for all $t > 0$.

Exercise 6.19.17 Let (M, g) be a compact Riemannian manifold, without boundary. Let $x_0 \in M$ be a fixed point and let $d(x_0, x)$ be the Riemannian distance between x_0 and x. Consider the integral

$$J(t) = \int_M u^2(t, x)e^{\alpha d(x_0, x)}\, dv(x),$$

where $\alpha \in \mathbb{R}$ and $u(t, x)$ satisfies the Cauchy problem

$$\partial_t u = \Delta_g u, \qquad t > 0$$
$$u(0, x) = e^{-\frac{\alpha}{2}d(x_0, x)}, \; x \in M.$$

(a) Prove the inequality

$$J(t) \leq vol(M)e^{\frac{\alpha^2 t}{2}}.$$

(b) Can you find a stochastic interpretation for the inequality

$$\Big(u^2(t, x)e^{\alpha d(x_0, x)}\Big)_{\text{ave}} \leq \mathbb{E}[e^{\alpha W_t}],$$

where W_t is a one-dimensional Brownian motion?

Exercise 6.19.18 In order to investigate whether there is a solution u of the heat equation $\partial_t u = \Delta_g u$ on the connected, compact Riemannian manifold (M, g) such that the function

$$J(t) = \int_M u^2(t, x)e^{S(t,x)}\, dv(x)$$

becomes an invariant over time, we need to solve the equation

$$u\nabla S + 2\nabla u = 0. \tag{6.19.230}$$

(a) Show that equation (6.19.230) has two solutions: either $u_1(t, x) = 0$, or $u_2(t, x) = e^{-S(t,x)/2}$.

(b) Does the function $u_2(t, x) = e^{-S(t,x)/2}$ solve the heat equation on M? What conclusion do you infer?

Exercise 6.19.19 (a) Verify that the Laplacian on $\mathbb{R}^2$ can be written in polar coordinates $(r, \theta) \in (0, \infty) \times (0, 2\pi)$ as

$$\frac{1}{2}(\partial_x^2 + \partial_y^2) = \frac{1}{2}\partial_r^2 + \frac{1}{2r}\partial_r + \frac{1}{2r^2}\partial_\theta^2.$$

(b) Show that the associated diffusion on $\mathbb{R}^2$ in polar coordinates, (R_t, θ_t), becomes

$$dR_t = dB_t + \frac{1}{2R_t}dt$$
$$d\theta_t = \frac{1}{R_t^2}dW_t,$$

with B_t and W_t independent one-dimensional Brownian motions.

Exercise 6.19.20 (a) Verify that the Laplacian on $\mathbb{R}^3$ can be written in spherical coordinates $(r, \theta, \varphi) \in (0, \infty) \times (0, 2\pi) \times (0, \pi)$ as

$$\frac{1}{2}(\partial_x^2 + \partial_y^2 + \partial_z^2) = \frac{1}{2}\partial_r^2 + \frac{1}{r}\partial_r + \frac{1}{2r^2}\partial_\theta^2 + \frac{1}{2r^2}\cot\theta\partial_\theta + \frac{1}{2r^2\sin^2\theta}\partial_\varphi^2.$$

(b) Show that the associated diffusion in spherical coordinates, $(R_t, \theta_t, \varphi_t)$, becomes

$$dR_t = dW_t^1 + \frac{1}{R_t}dt$$
$$d\theta_t = \frac{1}{R_t}dW_t^2 + \frac{1}{2R_t^2}\cot\theta_t dt$$
$$d\varphi_t = \frac{1}{R_t\sin\theta_t}dW_t^3,$$

with W_t^j independent one-dimensional Brownian motions.

Exercise 6.19.21 Consider the vector fields $X = x_1\partial_{x_2}$ and $Y = x_2\partial_{x_1}$ on $\mathbb{R}^2$.
(a) Write the equations of the diffusion $(X_1(t), X_2(t))$ associated to the operator $L = \frac{1}{2}(X^2 + Y^2)$;
(b) Let $Z(t) = X_1(t)^2 + X_2(t)^2$. Show that the expectation $\mathbb{E}[Z(t)]$ increases exponentially, provided $Z(0) \neq 0$.

Chapter 7

Sub-Elliptic Diffusions

This chapter deals with particular types of degenerate diffusions, called *sub-elliptic diffusions*. These are diffusions whose generators are sub-elliptic operators that can be written as a sum of squares of vector fields, which satisfy the bracket-generating condition introduced in Chapter 3. In this case the associated sub-Riemannian geometry satisfies the global connectivity property by curves tangent to the vector fields.

We are interested in sub-elliptic diffusions that are horizontal processes, namely their infinitesimal increments satisfy some non-holonomic constraints. It is expected that these processes diffuse in the entire space satisfying a similar property to the connectivity property by horizontal curves. Therefore, we shall mostly address their transience properties.

The first few sections of this chapter are based on the results of Calin [21] and they address some prototype examples of sub-elliptic diffusions.

7.1 Heisenberg Diffusion

This section deals with a diffusion on the Heisenberg group, $(\mathbb{R}^3, \circ)$. This group is defined by the law

$$(x_1, x_2, x_3)\circ(x_1', x_2', x_3') = \big(x_1+x_1', x_2+x_2', x_3+x_3'+2(x_2x_1'-x_2'x_1)\big), \quad x, x' \in \mathbb{R}^3.$$

$(\mathbb{R}^3, \circ)$ is a non-commutative Lie group with the Lie algebra generated by the left invariant vector fields

$$X = \partial_{x_1} + 2x_2\partial_{x_3}, \qquad Y = \partial_{x_2} - 2x_1\partial_{x_3}, \qquad X = \partial_{x_3}. \tag{7.1.1}$$

Since the following commutation relations holds

$$[X, Y] = -4Z, \quad [X, Z] = [Y, Z] = 0,$$

the horizontal distribution $p \to \mathcal{D}_p$, with

$$\mathcal{D}_p = \{aX_p + bY_p; a, b \in \mathbb{R}\}$$

is non-integrable (as implied by the Frobenius theorem, since $Z_p \notin \mathcal{D}_p$), and satisfies the bracket-generating condition with non-holonomy of second degree (or, equivalently, of step 2). Consequently, any two points in $\mathbb{R}^3$ can be joined by a horizontal curve, namely a curve whose tangent vector belongs to the horizontal distribution $\mathcal{D}$. It can be shown that these curves are actually of polynomial type (of 3rd degree), as shown in Calin et al. [38], page 12.

The sum of squares operator, $\Delta_H = \frac{1}{2}(X^2 + Y^2)$, is a sub-elliptic operator, called the Heisenberg Laplacian. Since the bracket-generating condition holds, this operator is hypoelliptic by Hörmander's theorem. Its smooth heat kernel has been found in Section 4.10 and its fundamental solution was computed in Section 5.6. Expanding in coordinates, we have

$$\Delta_H = \frac{1}{2}\sum_{i,j} a_{ij}\partial_{x_i}\partial_{x_j},$$

where

$$a = (a_{ij}) = \begin{pmatrix} 1 & 0 & 2x_2 \\ 0 & 1 & -2x_1 \\ 2x_2 & -2x_1 & 4(x_1^2 + x_2^2) \end{pmatrix}$$

is the coefficients matrix. This is degenerate and can be also written as the product $a = \sigma\sigma^T$, with dispersion matrix

$$\sigma = (\sigma_{ij}) = \begin{pmatrix} 1 & 0 & 0 \\ 0 & 1 & 0 \\ 2x_2 & -2x_1 & 0 \end{pmatrix}.$$

Definition 7.1.1 *The Heisenberg diffusion on the group* $(\mathbb{R}^3, \circ)$ *is a continuous stochastic process* $H_t = \big(H_1(t), H_2(t), H_3(t)\big)$, *whose generator is the Heisenberg Laplacian* Δ_H.

Consequently, the Heisenberg diffusion starting at x^0 is the solution of the stochastic differential system

$$\begin{aligned} dH_1(t) &= dW_1(t) \\ dH_2(t) &= dW_2(t) \\ dH_3(t) &= 2H_2(t)dW_1(t) - 2H_1(t)dW_2(t), \qquad H_0 = x^0, \end{aligned}$$

where $W_t = \big(W_1(t), W_2(t), W_3(t)\big)$ is a 3-dimensional Brownian motion. Integrating, we can write the Heisenberg diffusion explicitly as

$$\begin{aligned} H_1(t) &= x_1^0 + W_1(t) \\ H_2(t) &= x_2^0 + W_2(t) \\ H_3(t) &= x_3^0 + 2x_2^0 W_1(t) - 2x_1^0 W_1(t) + 2\int_0^t \Big(W_2(s)dW_1(s) - W_1(s)dW_2(s)\Big). \end{aligned}$$

These equations simplify a lot if we assume $x^0 = (0,0,0)$. We can always assume the Heisenberg diffusion starting at the origin due to the group law invariance. More precisely, if $L_x : \mathbb{R}^3 \to \mathbb{R}^3$ denotes the left translation on the Heisenberg group, i.e., $L_x x' = x \circ x'$, then both X and Y are left invariant and hence Δ_H is also left invariant. Consequently, if H_t starts at x^0, then $H_t' = L_x H_t$ is a Heisenberg diffusion starting at $x' = L_x x^0 = x \circ x^0$. Hence, up to a left translation, any Heisenberg diffusion can be considered as starting at the origin $(0,0,0)$.

Under this simplifying assumption, the Heisenberg diffusion looks like a two-dimensional Brownian motion in the first two components and as a Lévi area process in the third, namely

$$\begin{aligned} H_1(t) &= W_1(t) \\ H_2(t) &= W_2(t) \\ H_3(t) &= 2S_t, \end{aligned}$$

where $S_t = \displaystyle\int_0^t \Big(W_2(s)dW_1(s) - W_1(s)dW_2(s)\Big)$, see Section 1.11. The marginal probability densities are given by

$$p_{H_1(t)}(x) = p_{H_2(t)}(x) = \frac{1}{\sqrt{2\pi t}} e^{-\frac{x^2}{2t}}, \qquad p_{H_3(t)}(x) = \frac{1}{4t}\operatorname{sech}\Big(\frac{\pi x}{4t}\Big), \qquad x \in \mathbb{R},$$

where we used (1.12.77). The joint probability density of difussion $H_t = \big(H_1(t), H_2(t), H_3(t)\big)$ is given by the heat kernel of Δ_H as found in Section 4.10

$$p_t(x_1, x_2, x_3) = \frac{1}{4\pi^2}\int_{\mathbb{R}} V(t,\xi) e^{-i\xi x_3 - \xi(x_1^2 + x_2^2)\coth(2\xi t)}\, d\xi, \tag{7.1.2}$$

where $V(t,\xi) = \dfrac{2\xi}{\sinh(2\xi t)}$.

Remark 7.1.2 We notify here the similarity between the Heisenberg diffusion and the diffusion on a saddle, which was introduced in Section 4.13. The later is a 3-dimensional process $X_t = \big(W_1(t), W_2(t), 2W_1(t)W_2(t)\big)$, with $W_1(t)$ and

$W_2(t)$ independent one-dimensional Brownian motions. By Proposition 2.11.1 it follows that the marginal distributions of the Heisenberg diffusion, H_t, and the saddle diffusion, X_t, are the same. However, the joint distributions are not the same.

Another similarity consists of the form of their generating vector fields. In the case of the aforementioned diffusion on a saddle they are $X = \partial_{x_1} + 2x_2\partial_{x_3}$, $Y = \partial_{x_2} + 2x_1\partial_{x_3}$, while in the Heisenberg case they are given by (7.1.1). However, the former vector fields satisfy the commutation relation, $[X, Y] = 0$, while the latter do not, which marks the difference between the aforementioned examples of elliptic and sub-elliptic diffusions.

We shall show that the Heisenberg diffusion evolves along the horizontal distribution. For this it suffices to show that $dH_t \in \mathcal{D}_{H_t}$, i.e. the infinitesimal increments are contained in the horizontal distribution. Using

$$dH_3(t) = 2H_2(t)dW_1(t) - 2H_1(t)dW_2(t) = 2H_2(t)dH_1(t) - 2H_1(t)dH_2(t),$$

we can write

$$\begin{aligned} dH_t &= \big(dH_1(t), dH_2(t), dH_3(t)\big) \\ &= \big(dH_1(t), dH_2(t), 2H_2(t)dH_1(t) - 2H_1(t)dH_2(t)\big) \\ &= \big(1, 0, 2H_2(t)\big)dH_1(t) + \big(0, 1, -2H_1(t)\big)dH_2(t) \\ &= X_{H_t}dH_1(t) + Y_{H_t}dH_2(t) \in \mathcal{D}_{H_t}. \end{aligned}$$

Hence, during the time step dt the Heisenberg diffusion H_t moves the amount $dW_1(t)$ into the X direction and $dW_2(t)$ into the Y direction. These movements are independent and normally distributed.

Now we assume that the process H_t models the diffusion of a drop of ink into a certain crystal, with molecules oriented in the direction of the horizontal distribution, so that each ink particle trajectory $t \to H_t(\omega)$ represents a piece-wise continuous horizontal curve, for each state ω. We have seen that via the global connectivity theorem, any point of $\mathbb{R}^3$ can be accessed from the origin by a piece-wise continuous horizontal curve. If this curve is an ink particle trajectory, then the diffusion H_t will spread in the long run to the entire $\mathbb{R}^3$. This is equivalent to saying that the set of all paths fills the entire space, i.e. $\{H_t(\omega); t \geq 0, \omega \in \Omega\} = \mathbb{R}^3$. In this case, any ball outside the origin will be visited by some paths, i.e. the probability that H_t hits to the ball is strictly greater than zero. We shall deal with the study of these probabilities next.

Exit probabilities We recall the definition of the Heisenberg norm (or Koranyi distance) introduced in Section 5.6

$$\|x\|_H = \big((x_1^2 + x_2^2)^2 + x_3^2\big)^{1/2}.$$

The property which will be useful here is the one given by Proposition 5.6.2, namely the inverse of the Heisenberg norm is proportional to the fundamental solution of the Heisenberg Laplacian, i.e.

$$\Delta_H\Big(\frac{1}{\|x\|_H}\Big) = 0, \qquad \forall x \neq (0,0,0). \tag{7.1.3}$$

We note that this relation is very similar to the one that holds for the Euclidean Laplacian on $\mathbb{R}^3$

$$\Delta\Big(\frac{1}{\|x\|_{Eu}}\Big) = 0, \qquad \forall x \neq (0,0,0),$$

where $\Delta = \frac{1}{2}(\partial^2_{x_1} + \partial^2_{x_2} + \partial^2_{x_3})$; from this perspective, the Heisenberg norm, $\|x\|_H$, plays a role for the Heisenberg group that is analog to role of the Euclidean norm, $\|x\|_{Eu}$, for the Euclidean space. Therefore, the transitivity and recursivity of the Heisenberg diffusion, H_t will be approached in a way similar to the one applied for the 3-dimensional Brownian motion on $\mathbb{R}^3$.

A few more relations between the Heisenberg Laplacian and the Koranyi distance, which will be useful later, are given in the following, see Calin et al. [38], page 40:

Lemma 7.1.3 *We have*

$$(i) \quad \Delta_H(\|x\|_H^2) = 12(x_1^2 + x_2^2);$$

$$(ii) \quad \Delta_H(\|x\|_H) = \frac{2(x_1^2 + x_2^2)}{\|x\|_H}.$$

Proof: (i) Since $X(\|x\|_H^2) = 4\big(x_1(x_1^2 + x_2^2) + x_2x_3\big)$, then by differentiation we get

$$X^2(\|x\|_H^2) = 4X\big(x_1(x_1^2 + x_2^2) + x_2x_3\big) = 12(x_1^2 + x_2^2).$$

Similarly, we obtain $X^2(\|x\|_H^2) = 12(x_1^2 + x_2^2)$. Taking the average, yields

$$\Delta_H(\|x\|_H^2) = \frac{1}{2}(X^2 + Y^2)(\|x\|_H^2) = 12(x_1^2 + x_2^2).$$

(ii) Since $X(\|x\|_H^2) = 2\|x\|_H X(\|x\|_H)$, using the previous formula for $X(\|x\|_H^2)$ implies

$$X(\|x\|_H) = \frac{2}{\|x\|_H}\big(x_1(x_1^2 + x_2^2) + x_2x_3\big).$$

Similarly, we have

$$Y(\|x\|_H) = \frac{2}{\|x\|_H}\big(x_2(x_1^2 + x_2^2) - x_1x_3\big).$$

Summing the squares of the previous two formulas, yields

$$X(\|x\|_H)^2+Y(\|x\|_H)^2 = \frac{4}{\|x\|_H^2}\Big((x_1^2+x_2^2)^3+x_3^2(x_1^2+x_2^2)\Big) = 4(x_1^2+x_2^2). \quad (7.1.4)$$

Substituting $f(x) = g(x) = \|x\|_H$, into the formula

$$\Delta_H(fg) = f\Delta_H(g) + g\Delta_H(f) + 2\big(X(f)X(g) + Y(f)Y(g)\big),$$

we obtain

$$\Delta_H(\|x\|_H^2) = 2\|x\|_H\Delta_H(\|x\|_H) + 2\big(X(\|x\|_H)^2 + Y(\|x\|_H)^2\big).$$

Using part (i) and formula (7.1.4) the previous expression becomes

$$12(x_1^2 + x_2^2) = 2\|x\|_H\Delta_H(\|x\|_H) + 8(x_1^2 + x_2^2).$$

Solving for $\Delta_H(\|x\|_H)$, we arrive at the desired result.

■

The *Heisenberg ball* centered at $a \in \mathbb{R}^3$ and having the radius r is defined by $B_H(a,r) = \{x \in \mathbb{R}^3; \|x-a\|_H \leq r\}$. It looks like a bulky ball, or like a beveled cylinder. If $B_{Eu}(a,r) = \{x \in \mathbb{R}^3; \|x-a\|_{Eu} \leq r\}$ denotes the Euclidean ball centered at $a \in \mathbb{R}^3$ and having the radius r, it is worth asking the inclusion relationship between the Euclidean and the Heisenberg balls. Due to the group law invariance we have

$$L_{a^{-1}}B_H(a,r) = B(0,r),$$

and hence it suffices to study only the balls centered at the origin.

The first observation is that the unit Euclidean ball is included into the Heisenberg unit ball, $B_{Eu}(0,1) \subset B_H(0,1)$. This can be shown as follows. Let $x \in B_{Eu}(0,1)$, so $x_1^2+x_2^2+x_3^2 \leq 1$. Since $(x_1^2+x_2^2)^2 \leq x_1^2+x_2^2$ it follows that $(x_1^2+x_2^2)^2+x_3^2 \leq 1$, i.e. $x \in B_H(0,1)$. The same argument can be used to show that $B_{Eu}(0,r) \subset B_H(0,r)$ for any $r \leq 1$. This inclusion does not hold for any value of r. In general, we have the following double inclusions:

Lemma 7.1.4 *The inscribed and circumscribed Euclidean balls to the Heisenberg ball $B_H(0,r)$satisfy the following inclusions:*

(*i*) $B_{Eu}(0,r) \subset B_H(0,r) \subset B_{Eu}(0,\sqrt{r})$ *if* $0 < r \leq 1/2$;

(*ii*) $B_{Eu}(0,r) \subset B_H(0,r) \subset B_{Eu}(0,\sqrt{r^2+1/4})$ *if* $1/2 < r \leq 1$;

(*iii*) $B_{Eu}(0,\sqrt{r}) \subset B_H(0,r) \subset B_{Eu}(0,\sqrt{r^2+1/4})$ *if* $1 < r < \infty$.

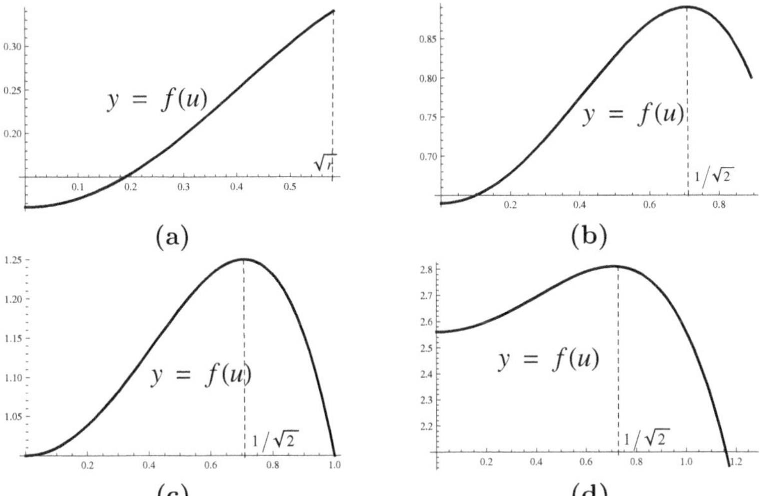

Figure 7.1: *The graph of the function* $f(u) = -u^4 + u^2 + r^2$ *in the following cases:* (a) $0 < r \leq 1/2$; (b) $1/2 < r < 1$; (c) $r = 1$; (d) $1 < r$.

Proof: In order to show the aforementioned inclusions, we need to investigate the global minima and maxima of the function $\|x\|^2 = x_1^2 + x_2^2 + x_3^2$ on the closed surface $S_H(0,r) = \{x \in \mathbb{R}^3; (x_1^2 + x_2^2)^2 + x_3^2 = r^2\}$. The level surfaces given by $\|x\|^2$, i.e. Euclidean balls, which are tangent to $S_H(0,r)$ correspond to a minimum (for the inscribed Euclidean sphere), or to a maximum (for the circumscribed Euclidean sphere). After the elimination of the variable x_3 and the substitution $u = \sqrt{x_1^2 + x_2^2}$, the problem reduces to finding the global extremes of the function $f(u) = -u^4 + u^2 + r^2$, with domain $0 \leq u \leq \sqrt{r}$. The global extremes of f are located either at the critical points $\{0, 1/\sqrt{2}\}$, or at the end point $\sqrt{r}$. Their behavior (i.e being minima or maxima) depends on the profile of the function f, which varies with the values of r as follows:

1. If $0 < r \leq \frac{1}{2}$, then there is a global maximum at $u = \sqrt{r}$ and a global minimum at $u = 0$, see Fig.7.1 a. In this case the inner ball has the radius r and the circumscribed ball has radius $\sqrt{r}$, see Fig. 7.2 a.

2. If $\frac{1}{2} < r < 1$, then there is a global maximum at $u = 1/\sqrt{2}$ and a global minimum at $u = 0$, see Fig.7.1 b. In this case the inner ball has the radius r and the circumscribed ball has radius $\sqrt{r^2 + 1/4}$, see Fig. 7.2 b.

3. If $r = 1$, then there is a global maximum at $u = 1/\sqrt{2}$ and two global minima at $u = 0, \sqrt{r}$, see Fig.7.1 c. The balls look like in the previous case.

4. If $1 < r$, then there is a global maximum at $u = 1/\sqrt{2}$ and a global minimum at $u = 0, \sqrt{r}$, see Fig.7.1 d. In this case the inner ball has the radius $\sqrt{r}$ and the circumscribed ball has radius $\sqrt{r^2 + 1/4}$, see Fig. 7.2 c.

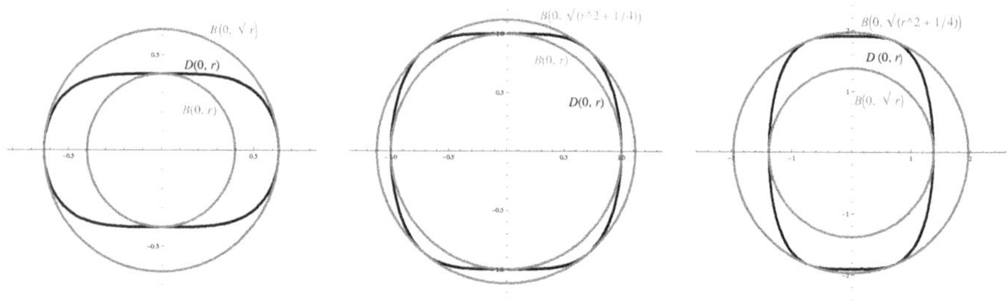

(a) (b) (c)

Figure 7.2: *The contour graphs for* $\{u^4 + z^2 = r^2\}$ *and its inscribed and circumscribed circles in the cases:* (a) $0 < r \leq 1/2$; (b) $1/2 < r \leq 1$; (c) $1 < r$.

The conclusion of the proposition follows from the previous case discussion. ■

We ask now the following question: How long does it take to a Heisenberg diffusion that starts at some given $a \in \mathbb{R}^3$, with $\|a\|_H < r$, to exit from the Heisenberg ball $B_H(0, r)$? Since an explicit formula might be hard to produce, we shall provide next a lower bound.

Proposition 7.1.5 *Let H_t be a Heisenberg diffusion starting at $H_0 = a$, with $\|a\|_H < r$. Consider the first exit time of H_t from the ball $B_H(0, r)$*

$$\tau_r = \inf\{t > 0; H_t \notin B_H(0, r)\}.$$

Then

$$\mathbb{E}^a[\tau_r] \geq \frac{r^2 - \|a\|_H^2}{12r}. \tag{7.1.5}$$

Proof: Standard properties of exit times show that τ_r is a stopping time. First, we show that $\mathbb{E}[\tau_r] < \infty$. For this we consider the ball $B_H(0, r)$ included into a Euclidean ball, $B_{Eu}(0, R)$, with R large enough, depending on r (see Lemma 7.1.4). Let $\widetilde{\tau}$ be the first exit time of the Heisenberg diffusion H_t from the Euclidean ball $B_{Eu}(0, R)$. Obviously, $\tau(\omega) \leq \widetilde{\tau}(\omega)$ for each state ω. Since the first two components of H_t are one-dimensional Brownian motions, then $\mathbb{E}^a[\widetilde{\tau}] \leq \mathbb{E}[\tau_0]$, where $\tau_0 = \inf\{t > 0; \|(H_1(t), H_2(t))\|_{Eu} > R\}$, namely, it is expected to take longer for the (x, y)-projection to leave the disk rather than H_t to leave the ball (this follows from the reasoning: the states ω for which $(H_1(\omega), H_2(\omega))$ leaves the disk, are among the states on which $H_t(\omega)$ leaves the ball, but the converse implication does not necessarily hold). Since $\mathbb{E}[\tau_0] < \infty$, i.e. the exit time from a disk of a two-dimensional Brownian motion is finite, then the above inequalities imply

$$\mathbb{E}[\tau_r] \le \mathbb{E}[\widetilde{\tau}] \le \mathbb{E}[\tau_0] < \infty.$$

Next we shall apply Dynkin's formula, see Theorem 1.1.1, for diffusion H_t, stopping time τ_r and function $f(x) = \|x\|_H^2$

$$\mathbb{E}[f(H_{\tau_r})] = f(a) + \mathbb{E}\left[\int_0^{\tau_r} (\Delta_H)(f)(H_s)\, ds\right].$$

Using Lemma 7.1.3, part (i), the previous identity becomes

$$r^2 = \|a\|_H^2 + 12\mathbb{E}\left[\int_0^{\tau_r} \left(H_1(s)^2 + H_2(s)^2\right) ds\right].$$

Since for any $s < \tau_r$ the Heisenberg diffusion H_t is still inside the ball $B_H(0,r)$, we have

$$\left(H_1(s)^2 + H_2(s)^2\right)^2 \le \left(H_1(s)^2 + H_2(s)^2\right)^2 + H_3(s)^2 = \|H_s\|_H^2 \le r^2.$$

Substituting the square root of this inequality in the previous relation yields

$$r^2 \le \|a\|_H^2 + 12r\mathbb{E}[\tau_r],$$

which implies $\mathbb{E}[\tau_r] \ge \dfrac{r^2 - \|a\|_H^2}{12r}$. ■

Remark 7.1.6 We could have applied Dynkin's formula using the function $f(x) = \|x\|_H$ and then use Lemma 7.1.3, part (ii) to obtain the inequality

$$\begin{aligned} r &= \|a\|_H + 2\mathbb{E}\left[\int_0^{\tau_r} \frac{H_1(s)^2 + H_2(s)^2}{\|H_t\|_H}\, ds\right] \\ &\le \|a\|_H + 2\mathbb{E}[\tau_r], \end{aligned}$$

which implies $\mathbb{E}[\tau_r] \ge \dfrac{r - \|a\|_H}{2}$, which is a weaker bound than (7.1.5).

The exit probability We shall compute first the second moment of the Lévy area S_t based on the probability density provided by Proposition 2.11.2. Considering $k = 1$ in Corollary 2.11.3 yields

$$\int_{\mathbb{R}} y^2 \mathrm{sech}\, y\, dy = \frac{\pi^3}{4}.$$

Using this integral and the substitution $y = \frac{\pi x}{2t}$, we obtain

$$\mathbb{E}[S_t^2] = \int_{\mathbb{R}} \frac{1}{2t} x^2 \operatorname{sech}\Big(\frac{\pi x}{2t}\Big)\, dx$$
$$= \frac{(2t)^2}{\pi^3} \int_{\mathbb{R}} y^2 \operatorname{sech} y\, dy = t^2. \tag{7.1.6}$$

The following result produces an estimation for the exit probability of a Heisenberg diffusion from a ball centered at the origin.

Proposition 7.1.7 *Let H_t be a Heisenberg diffusion starting at the origin, $H_0 = (0,0,0)$, and let $r > 0$. We have*

(i) $P(H_t \notin B_{Eu}(0,r)) \le \dfrac{2t + 4t^2}{r^2}$;

(ii) $P(H_t \notin B_H(0,r)) \le \dfrac{6t^2(t+1)}{r^2}$ *for* $t > 0$.

Proof: *(i)* Let $A_t = \{\omega; \|H_t(\omega)\|_{Eu} > r\} = \{H_t \notin B_{Eu}(0,r)\}$. Then we have

$$P(A_t) = \int_{A_t} dP \le \frac{1}{r^2}\int_{A_t} \|H_t\|_{Eu}^2 dP \le \frac{1}{r^2}\int_{\Omega} \|H_t\|_{Eu}^2 dP$$
$$= \frac{1}{r^2}\mathbb{E}[\|H_t\|_{Eu}^2] = \frac{1}{r^2}\mathbb{E}[W_1(t)^2 + W_2(t)^2 + 4S_t^2]$$
$$= \frac{1}{r^2}(2\mathbb{E}[W_1(t)^2] + 4\mathbb{E}[S_t^2]) = \frac{2t + 4t^2}{r^2}.$$

(ii) Similarly, taking $A_t = \{\omega; \|H_t(\omega)\|_H > r\} = \{H_t \notin B_H(0,r)\}$, we have the estimations

$$P(A_t) = \int_{A_t} dP \le \frac{1}{r^2}\int_{\Omega} \|H_t\|_H^2 dP = \frac{1}{r^2}\mathbb{E}[\|H_t\|_H^2]$$
$$= \frac{1}{r^2}\mathbb{E}[W_1(t)^4 + W_2(t)^4 + 2W_1(t)^2W_2(t)^2 + 4S_t^2]$$
$$= \frac{1}{r^2}(2\mathbb{E}[W_1(t)^4] + 2\mathbb{E}[W_1(t)^2]\mathbb{E}[W_2(t)^2] + 4\mathbb{E}[S_t^2])$$
$$= \frac{1}{r^2}(6t^3 + 2t^2 + 4t^2) = \frac{6t^2(t+1)}{r^2}.$$

■

These inequalities state that the exit probability of H_t is not larger than a quadratic function of time for Euclidean balls and a cubic function of time for Heisenberg balls. This estimation is more accurate for small values of t. Closed form evaluations for these probabilities can be obtained using the joint density of (R_r, S_t), see the Section 7.2, page 392.

The transitivity of the Heisenberg diffusion We consider the annulus

$$A_{a,b} = \{x \in \mathbb{R}^3; a < \|x\|_H < b\},$$

with $0 < a < b$. Let $\tau_{a,b}$ be the first exit time of the Heisenberg diffusion H_t from $A_{a,b}$

$$\tau_{a,b} = \inf\{t > 0; H_t \notin A_{a,b}\},$$

where we assume $H_0 = x^0 \in A_{a,b}$. We apply Dynkin's formula, see Theorem 1.1.1, to get

$$\mathbb{E}^{x^0}[f(H_{\tau_{a,b}})] = f(x^0) + \mathbb{E}\left[\int_0^{\tau_{a,b}} (\Delta_H f)(H_s)\, ds\right]$$

with the function $f(x) = \frac{1}{\|x\|_H}$. Using relation (7.1.3) yields

$$\mathbb{E}^{x_0}\left[\frac{1}{\|H_{\tau_{a,b}}\|_H}\right] = \frac{1}{\|x^0\|_H}. \tag{7.1.7}$$

The Heisenberg diffusion H_t can leave the annulus either towards the interior, with probability

$$p_{a,b} = P\big(\|H_{\tau_{a,b}}\|_H = a\big),$$

or, towards its exterior, with probability

$$q_{a,b} = P\big(\|H_{\tau_{a,b}}\|_H = b\big) = 1 - p_{a,b}.$$

Since the previous expectation can be evaluated as

$$\mathbb{E}^{x^0}\left[\frac{1}{\|H_{\tau_{a,b}}\|_H}\right] = \frac{1}{a}p_{a,b} + \frac{1}{b}q_{a,b},$$

then (7.1.7) becomes

$$\frac{1}{a}p_{a,b} + \frac{1}{b}q_{a,b} = \frac{1}{\|x^0\|_H}.$$

Taking the limit $b \to \infty$, we obtain

$$p_a = \lim_{b\to\infty} p_{a,b} = \frac{a}{\|x^0\|_H}.$$

This is the probability that a Heisenberg diffusion H_t, which starts at $H_0 = x^0$, with $\|x^0\|_H > a$, enters the Heisenberg ball $B_H(0, a)$ in finite time. Using the invariance of the norm $\|\cdot\|_H$ with respect to the Lie group law, we obtain the main result of this section:

Proposition 7.1.8 *(i) Let H_t be a Heisenberg diffusion starting at the origin and let $a > 0$, $\xi \in \mathbb{R}^3$ with $a < \|\xi\|_H$. Then the probability that H_t reaches the Heisenberg ball $B(\xi, a)$ is given by $p_a = \dfrac{a}{\|\xi\|_H}$.*

(ii) Let H_t be a Heisenberg diffusion starting at $x^0 \in \mathbb{R}^3$ and let $a > 0$, with $a < \|x^0 - \xi\|_H$. Then the probability that H_t reaches the Heisenberg ball $B(\xi, a)$ is given by $p_a = \dfrac{a}{\|x^0 - \xi\|_H}$.

This result states the transitivity property of the Heisenberg diffusion in $\mathbb{R}^3$. It says that the probability that the Heisenberg diffusion reaches a Heisenberg ball of radius a is equal to the ratio between the radius a and the distance in the norm $\|\cdot\|_H$ between the center of the ball and the initial point of diffusion.

Using Lemma 7.1.4 we can show that the Heisenberg diffusion visits any Euclidean ball in finite time with probability less than 1. The result of Proposition 7.1.8 is in striking similarity with an analog transitivity property of the Brownian motion in $\mathbb{R}^3$.

We consider now the particular case when the ball is centered at the point $\xi = (0, 0, t)$, which is situated on the "missing direction" ∂_{x_3}. If $0 < a < t$, then the Heisenberg diffusion starting at the origin reaches $B_H(\xi, a)$ with the probability $p = \dfrac{a}{t} < 1$. With the very same probability, a 3-dimensional Brownian motion will reach the Euclidean ball $B_{Eu}(\xi, a)$. Since for $a < 1$ we have the inclusion $B_{Eu}(\xi, a) \subset B_H(\xi, a)$, it follows that a 3-dimensional Brownian motion reaches $B_H(\xi, a)$ with a probability larger than a/t. This can be stated by saying that the 3-dimensional Brownian motion spreads "faster" into the direction ∂_{x_3} rather than the Heisenberg diffusion H_t does. This agrees with the fact that the Heisenberg diffusion spreads under some horizontal constraints, while the 3-dimensional Brownian motions spreads freely in $\mathbb{R}^3$.

7.2 The Joint Distribution of (R_t, S_t)

As an application, we provide the computation of the joint distribution of (R_t, S_t), where $R_t = R_t^{(2)}$ is a two-dimensional Bessel process and S_t is the Lévy area process. To recall definitions, if $W_1(t)$ and $W_2(t)$ are two independent Brownian motions, then

$$R_t = \sqrt{W_1(t)^2 + W_2(t)^2}, \qquad S_t = \int_0^t W_1(s)dW_2(s) - W_2(s)dW_1(s).$$

We shall see that even if S_t is not an Ito diffusion, the pair process (R_t, S_t) is. By Ito's formula we have

$$dR_t = \frac{1}{2r_t}dt + \frac{W_1(t)dW_1(t) + W_2(t)dW_2(t)}{\sqrt{W_1(t)^2 + W_2(t)^2}}$$

$$dS_t = R_t \frac{W_1(t)dW_2(t) - W_2(t)dW_1(t)}{\sqrt{W_1(t)^2 + W_2(t)^2}}.$$

By Proposition 1.1.9 combined with the argument used at page 48, there are two independent one-dimensional Brownian motions, β_t and γ_t, such that

$$dR_t = \frac{1}{2R_t}dt + d\beta_t$$
$$dS_t = R_t d\gamma_t.$$

Therefore, (R_t, S_t) is an Ito diffusion with drift $b(R_t, S_t) = \Big(\frac{1}{2R_t}, 0\Big)^T$ and dispersion $\sigma = \begin{pmatrix} 1 & 0 \\ 0 & R_t \end{pmatrix}$. It follows that the associated generator is given by

$$\mathcal{A}_{r,s} = \frac{1}{2}\partial_r^2 + \frac{1}{2}r^2\partial_s^2 + \frac{1}{2r}\partial_r. \tag{7.2.8}$$

The notation $\mathcal{A}_{r,s}$ means that the operator acts in the variables r and s. In order to find the joint density of (R_t, S_t) it suffices to find the heat kernel of the operator $\mathcal{A}_{r,s}$. More explicitly, if $K(r_0, s_0; r, s; t)$ is the heat kernel, namely

$$\partial_t K(r_0, s_0; r, s; t) = \mathcal{A}_{r,s} K(r_0, s_0; r, s; t)$$
$$\lim_{t \searrow 0} K(r_0, s_0; r, s; t) = \delta(r - r_0) \otimes \delta(s - s_0),$$

then $p_t(r_0, s_0; r, s) = K(r_0, s_0; r, s; t)$ is the joint transition density of (R_t, S_t).

We start by noting that the partial Fourier transform with respect to s of the heat kernel, which is

$$u(r_0, s_0; r, \lambda; t) = (\mathcal{F}_s K)(r_0, s_0; r, \lambda; t) = \int_{\mathbb{R}} e^{-is\lambda} K(r_0, s_0; r, s; t)\, ds,$$

satisfies the initial value problem

$$\partial_t u(r_0, s_0; r, \lambda; t) = L\, u(r_0, s_0; r, \lambda; t)$$
$$\lim_{t \searrow 0} u(r_0, s_0; r, \lambda; t) = e^{-is_0\lambda}\, \delta(r - r_0),$$

where

$$L = \frac{1}{2}\partial_r^2 + \frac{1}{2r}\partial_r - \frac{\lambda^2}{2}r^2, \qquad r > 0$$

is the two-dimensional Bessel operator with quadratic potential, which was studied at page 214. We also used the Fourier transform of the Dirac delta function, which is

$$\mathcal{F}_s\delta(s-s_0) = \int_{\mathbb{R}} e^{-is\lambda}\delta(s-s_0)\,ds = e^{-is_0\lambda}.$$

Then the function $e^{is_0\lambda}u(r_0, s_0; r, \lambda; t)$ is the heat kernel for the operator L. Substituting $n=2$, $\nu=0$, $x=r_0$, $y=r$, and $\gamma=\lambda$ in formula (4.14.106) yields

$$e^{is_0\lambda}u(r_0, s_0; r, \lambda; t) = \frac{\lambda r}{\sinh(t\lambda)}e^{-\frac{\lambda}{2}(r_0^2+r^2)\coth(t\lambda)}I_0\Big(\frac{\lambda r_0 r}{\sinh(t\lambda)}\Big), \quad t>0.$$

Solving for $u(r_0, s_0; r, \lambda; t)$ and applying the inverse Fourier transform in s yields

$$\begin{aligned} K(r_0, s_0; r, s; t) &= \frac{1}{2\pi}\int_{\mathbb{R}} e^{is\lambda}u(r_0, s_0; r, \lambda; t)\,d\lambda \\ &= \frac{1}{2\pi}\int_{\mathbb{R}} e^{i(s-s_0)\lambda-\frac{\lambda}{2}(r_0^2+r^2)\coth(\lambda t)}\frac{\lambda r}{\sinh(t\lambda)}I_0\Big(\frac{\lambda r_0 r}{\sinh(t\lambda)}\Big)\,d\lambda. \end{aligned}$$

Therefore, the transition density of (R_t, S_t) is given by the following integral formula

$$p_t(r_0, s_0; r, s) = \frac{1}{2\pi}\int_{\mathbb{R}} e^{i(s-s_0)\lambda-\frac{\lambda}{2}(r_0^2+r^2)\coth(\lambda t)}\frac{\lambda r}{\sinh(t\lambda)}I_0\Big(\frac{\lambda r_0 r}{\sinh(t\lambda)}\Big)\,d\lambda, \tag{7.2.9}$$

with $r_0, r>0$ and $t>0$.

The joint distribution of (R_t, S_t) is useful in the study of the Heisenberg diffusion H_t. In this case, due to the group law invariance it suffices to assume that the diffusion starts at the origin, which means that $r_0=0$ and $s_0=0$. Under this assumption the density (7.2.9) has the more simple form

$$p_t(r, s) = \frac{1}{2\pi}\int_{\mathbb{R}} e^{is\lambda-\frac{\lambda}{2}r^2\coth(\lambda t)}\frac{\lambda r}{\sinh(t\lambda)}\,d\lambda, \qquad r>0, \tag{7.2.10}$$

where we used that $I_0(0)=1$. The probability that a Heisenberg diffusion starting at the origin is inside the Heisenberg ball of radius ρ at time t can be evaluated as

$$P\big(H_t\in B_H(0,\rho)\big) = P(\|H_t\|_H<\rho) = P(R_t^4+4S_t^2<\rho^2).$$

We shall deal with the computation of this probability in the Section 7.3.

Interpretation of (R_t, S_t) We shall show that (R_t, S_t) is a diffusion associated with the radial part of the Heisenberg Laplacian. To this end, we consider the vector fields on $\mathbb{R}^3$

$$X = \partial_x - y\partial_s, \qquad Y = \partial_y + x\partial_s$$

and consider the sum of squares operator

$$\begin{aligned}\Delta_H &= \frac{1}{2}(X^2 + Y^2) \\ &= \frac{1}{2}(\partial_x^2 + \partial_y^2) + \frac{1}{2}(x^2 + y^2)\partial_s^2 + (x\partial_y - y\partial_x)\partial_s,\end{aligned}$$

which is rotational invariant with respect to variables (x, y). If let $r^2 = x^2+y^2$, then the radial part of the previous operator becomes

$$\Delta_H^{rad} = \frac{1}{2}\partial_r^2 + \frac{1}{2r}\partial_r + \frac{1}{2}r^2\partial_s^2.$$

Therefore, using (7.2.8) yields $\mathcal{A}_{r,s} = \Delta_H^{rad}$, and hence, the radial Heisenberg Laplacian becomes the infinitesimal generator of diffusion (R_t, S_t).

7.3 Koranyi Process

The next concept is the analog of the standard Bessel process, i.e., the Euclidean distance measured from the origin to a Brownian motion in space, for the case of the Heisenberg group.

Definition 7.3.1 *A Koranyi process is the process given by the Koranyi distance from the origin of a Heisenberg diffusion* $H_t = (H_1(t), H_2(t), H_3(t))$ *starting at* $H_0 = (0, 0, 0)$, *i.e.*

$$C_t = \sqrt{\left(H_1(t)^2 + H_2(t)^2\right)^2 + H_3(t)^2}, \qquad t \geq 0. \tag{7.3.11}$$

The process can be described equivalently as

$$C_t = \|H_t\|_H = \sqrt{\left(W_1(t)^2 + W_2(t)^2\right)^2 + 4S_t^2},$$

where $W_1(t)$ and $W_2(t)$ are two independent Brownian motions and S_t stands for the Lévy area process. This also writes as

$$C_t = \sqrt{R_t^4 + 4S_t^2},$$

where $R_t = \sqrt{W_1(t)^2 + W_2(t)^2}$ denotes the two-dimensional Bessel process starting at the origin.

The process law

To understand this process we shall find its law. The cumulative distribution function of C_t can be computed using the joint distribution of (R_t, S_t) given by (7.2.10). Therefore, for any $\rho \geq 0$, using Fubini's theorem and the substitution $u = r^2/2$, we have

$$\begin{aligned}
F_{C_t}(\rho) &= P(C_t \leq \rho) = P(R_t^4 + 4S_t^2 \leq \rho^2) = \iint_{\{r^4+4s^2\leq\rho^2,\, r\geq 0\}} p_t(r,s)\,drds \\
&= \int_{-\rho/2}^{\rho/2}\int_0^{(\rho^2-4s^2)^{1/4}} p_t(r,s)\,drds \\
&= \frac{1}{2\pi}\int_{\mathbb{R}} \frac{\lambda}{\sinh(t\lambda)}\int_{-\rho/2}^{\rho/2} e^{is\lambda}\left(\int_0^{(\rho^2-4s^2)^{1/4}} e^{-\frac{\lambda}{2}r^2\coth(\lambda t)} rdr\right) dsd\lambda \\
&= \frac{1}{2\pi}\int_{\mathbb{R}} \frac{\lambda}{\sinh(t\lambda)}\int_{-\rho/2}^{\rho/2} e^{is\lambda}\left(\int_0^{\frac{1}{2}\sqrt{\rho^2-4s^2}} e^{-\lambda u\coth(\lambda t)} du\right) dsd\lambda \\
&= \frac{1}{2\pi}\int_{\mathbb{R}} \frac{\lambda}{\sinh(t\lambda)}\int_{-\rho/2}^{\rho/2} e^{is\lambda}\left(\frac{1-e^{-\frac{\lambda}{2}\sqrt{\rho^2-4s^2}\coth(\lambda t)}}{\lambda\coth(\lambda t)}\right) dsd\lambda \\
&= \frac{1}{2\pi}\int_{\mathbb{R}} \frac{1}{\cosh(t\lambda)}\int_{-\rho/2}^{\rho/2} e^{is\lambda}\left(1-e^{-\frac{\lambda}{2}\sqrt{\rho^2-4s^2}\coth(\lambda t)}\right) dsd\lambda \\
&= \frac{1}{2\pi}\int_{\mathbb{R}} \frac{1}{\cosh(t\lambda)}\int_{-\rho/2}^{\rho/2} \cos(s\lambda)\left(1-e^{-\frac{\lambda}{2}\sqrt{\rho^2-4s^2}\coth(\lambda t)}\right) dsd\lambda \\
&= \frac{1}{\pi}\int_{\mathbb{R}} \frac{1}{\cosh(t\lambda)}\int_0^{\rho/2} \cos(s\lambda)\left(1-e^{-\frac{\lambda}{2}\sqrt{\rho^2-4s^2}\coth(\lambda t)}\right) dsd\lambda.
\end{aligned} \tag{7.3.12}$$

Using Leibniz' formula of differentiation

$$\frac{d}{d\rho}\int_0^{b(\rho)} f(\rho,t)dt = f(\rho,b(\rho))b'(\rho) + \int_0^{b(\rho)} \frac{\partial}{\partial\rho}f(\rho,t)dt,$$

we obtain the probability density of C_t

$$\begin{aligned}
p_{C_t}(\rho) &= \frac{d}{d\rho}F_{C_t}(\rho) \\
&= \frac{1}{\pi}\int_{\mathbb{R}}\frac{1}{\cosh(\lambda t)}\int_0^{\rho/2}\cos(s\lambda)\frac{\lambda}{2}\coth(\lambda t)\frac{\rho}{\sqrt{\rho^2-4s^2}}e^{-\frac{\lambda}{2}\sqrt{\rho^2-4s^2}\coth(\lambda t)}dsd\lambda \\
&= \frac{\rho}{4\pi t}\int_{\mathbb{R}}\frac{\lambda t}{\sinh(\lambda t)}\int_0^{\rho}\frac{\cos(\tau\lambda/2)}{\sqrt{\rho^2-\tau^2}}e^{-\frac{\lambda}{2}\sqrt{\rho^2-\tau^2}\coth(\lambda t)}d\tau d\lambda,
\end{aligned}$$

where we used the substitution $\tau = 2s$. Substituting $\tau = \rho\cos\theta$, with $\theta \in [0, \frac{\pi}{2}]$, the inner integral becomes

$$\int_0^\rho \frac{\cos(\tau\lambda/2)}{\sqrt{\rho^2-\tau^2}} e^{-\frac{\lambda}{2}\sqrt{\rho^2-\tau^2}\coth(\lambda t)}\,d\tau = \int_0^{\pi/2} \cos\Big(\frac{\lambda\rho}{2}\cos\theta\Big) e^{-\frac{\lambda\rho}{2}\sin\theta\coth(\lambda t)}\,d\theta,$$

and hence we arrive at the following law

$$\begin{aligned} p_{C_t}(\rho) &= \frac{\rho}{4\pi t}\int_{\mathbb{R}} \frac{\lambda t}{\sinh(\lambda t)} \int_0^{\pi/2} \cos\Big(\frac{\lambda\rho}{2}\cos\theta\Big) e^{-\frac{\lambda\rho}{2}\sin\theta\coth(\lambda t)}\,d\theta d\lambda \\ &= \frac{\rho}{2\pi t}\int_{\mathbb{R}} \frac{2\xi t}{\sinh(2\xi t)} \int_0^{\pi/2} \cos\big(\xi\rho\cos\theta\big) e^{-\xi\rho\sin\theta\coth(2\xi t)}\,d\theta d\xi, \quad (7.3.13) \end{aligned}$$

where we used $\xi = \lambda/2$.

The probability of H_t to be in a ball

Now, we can answer the question asked at the end of the previous section. Using formula (7.3.12), we obtain a closed form expression for the probability that a Heisenberg diffusion H_t starting at the origin belongs to the Heisenberg ball of radius ρ at time t as

$$\begin{aligned} P\big(H_t{\in}B_H(0,\rho)\big) &= P(C_t \le \rho) \\ &= \frac{1}{\pi}\int_{\mathbb{R}} \frac{1}{\cosh(t\lambda)} \int_0^{\rho/2} \cos(s\lambda)\Big(1 - e^{-\frac{\lambda}{2}\sqrt{\rho^2-4s^2}\coth(\lambda t)}\Big)\,ds d\lambda \\ &= \frac{2}{\pi}\int_0^\infty \frac{1}{\cosh(t\lambda)} \int_0^{\rho/2} \cos(s\lambda)\Big(1 - e^{-\frac{\lambda}{2}\sqrt{\rho^2-4s^2}\coth(\lambda t)}\Big)\,ds d\lambda, \end{aligned}$$

where we used that the integrand is an even function in λ. Since this integral can't be computed explicitly using elementary functions, we shall compute an upper bound. Using the inequality $1 - e^{-x} \le x$ and $\cos x \le 1$, we have

$$\begin{aligned} P\big(H_t{\in}B_H(0,\rho)\big) &\le \frac{2}{\pi}\int_0^\infty \frac{1}{\cosh(t\lambda)} \int_0^{\rho/2} \frac{\lambda}{2}\sqrt{\rho^2-4s^2}\coth(\lambda t)\,ds d\lambda \\ &= \frac{1}{\pi}\int_0^\infty \frac{\lambda}{\sinh(t\lambda)} \int_0^{\rho/2} \sqrt{\rho^2-4s^2}\,ds d\lambda \\ &= \frac{1}{\pi}\int_0^\infty \frac{\lambda}{\sinh(t\lambda)}\frac{\rho^2}{2} \int_0^{\pi/2} \sin^2\theta d\theta\, d\lambda \\ &= \frac{\rho^2}{8}\int_0^\infty \frac{\lambda}{\sinh(t\lambda)}d\lambda \\ &= \frac{\rho^2\pi^2}{32t^2}, \end{aligned}$$

where we employed the substitution $2s = \rho\cos\theta$ and used in the last row the value of the improper integral given by Remark 2.10.2, page 88.

The diffusion (R_t, S_t, C_t)

The process C_t is not an Ito diffusion, but the triplet (R_t, S_t, C_t) is an Ito diffusion, as we shall show next. Since

$$dR_t = \frac{1}{2R_t}dt + d\beta_t, \qquad dS_t = R_t d\gamma_t,$$

with β_t and γ_t independent Brownian motions, then

$$(dR_t)^2 = dt, \quad (dS_t)^2 = R_t^2 dt, \quad dR_t dS_t = 0.$$

Let $f_t = f(R_t, S_t)$ with f differentiable function. Then Ito's formula yields

$$\begin{aligned} df_t &= \partial_r f\, dR_t + \partial_s f\, dS_t + \frac{1}{2}\partial_r^2 f(dR_t)^2 + +\frac{1}{2}\partial_s^2 f(dS_t)^2 + \frac{1}{2}\partial_r\partial_s f dR_t dS_t \\ &= \Big(\frac{1}{2R_t}\partial_r f + \frac{1}{2}\partial_r^2 f + \frac{1}{2}R_t^2\partial_s^2 f\Big)dt + \partial_r f\, d\beta_t + \partial_s f R_t\, d\gamma_t. \qquad (7.3.14) \end{aligned}$$

We choose the function $f(r,s) = \sqrt{r^4 + 4s^2}$ and compute its derivatives

$$\begin{aligned} \partial_r f &= \frac{2r^3}{\sqrt{r^4+4s^2}} & \partial_r^2 f &= \frac{2r^2(r^4+12s^2)}{(r^4+4s^2)^{3/2}} \\ \partial_s f &= \frac{4s}{\sqrt{r^4+4s^2}} & \partial_s^2 f &= \frac{4r^4}{(r^4+4s^2)^{3/2}}. \end{aligned}$$

Since $C_t = \sqrt{R_t^4 + 4S_t^2} = f(R_t, S_t)$, using formula (7.3.14) we obtain

$$dC_t = R_t^2\Big(\frac{1}{C_t} + 3\sqrt{C_t}\Big)dt + \frac{2R_t^3}{C_t}d\beta_t + \frac{4R_tS_t}{C_t}d\gamma_t.$$

Therefore, the triplet (R_t, S_t, C_t) satisfies the following system of differential stochastic equations

$$\begin{aligned} dR_t &= \frac{1}{2R_t}dt + d\beta_t \\ dS_t &= R_t d\gamma_t \\ dC_t &= R_t^2\Big(\frac{1}{C_t} + 3\sqrt{C_t}\Big)dt + \frac{2R_t^3}{C_t}d\beta_t + \frac{4R_tS_t}{C_t}d\gamma_t. \end{aligned}$$

Since this can be written in the equivalent matrix form

$$d\begin{pmatrix} R_t \\ S_t \\ C_t \end{pmatrix} = \begin{pmatrix} \frac{1}{2R_t} \\ 0 \\ R_t^2\Big(\frac{1}{C_t} + 3\sqrt{C_t}\Big) \end{pmatrix} dt + \begin{pmatrix} 1 & 0 \\ 0 & R_t \\ \frac{2R_t^3}{C_t} & \frac{4R_tS_t}{C_t} \end{pmatrix} \begin{pmatrix} d\beta_t \\ d\gamma_t \end{pmatrix},$$

then the generator of the diffusion (R_t, S_t, C_t) takes the form

$$\begin{aligned} L =& \frac{1}{2}\partial_r^2 + \frac{1}{2}r^2\partial_s^2 + 2r^2\partial_c^2 + \frac{2r^3}{c}\partial_r\partial_c + \frac{4r^2 s}{c}\partial_s\partial_c \\ &+ \frac{1}{2r}\partial_r + r^2\big(\frac{1}{c} + 3\sqrt{c}\big)\partial_c. \end{aligned}$$

7.4 Grushin Diffusion

We shall study in the following the diffusion process associated with the Grushin operator.

7.4.1 The Two-Dimensional Case

We recall from Section 4.3 that the vector fields

$$X_1 = \partial_{x_1}, \qquad X_2 = x_1\partial_{x_2}$$

generate the Grushin distribution on $\mathbb{R}^2$. The span of X_1 and X_2 drops rank along the x_2-axis. Since their Lie bracket-generates this missing direction, $[X_1, X_2] = \partial_{x_2}$, then the vector fields $\{X_1, X_2\}$ satisfy the bracket-generating condition on $\mathbb{R}^2$.

Definition 7.4.1 *The Grushin diffusion $G_t = (X(t), Y(t))$ on $\mathbb{R}^2$ is the diffusion whose generator is the hypoeliptic operator*

$$\Delta_G = \frac{1}{2}(X^2 + Y^2) = \frac{1}{2}(\partial_x^2 + x^2\partial_y^2).$$

This diffusion satisfies the following system

$$\begin{aligned} dX(t) &= dW_1(t) \\ dY(t) &= X(t)dW_2(t), \end{aligned}$$

where $W_1(t)$ and $W_2(t)$ are two independent one-dimensional Brownian motions. By integration, we obtain

$$\begin{aligned} X(t) &= x_0 + W_1(t) \\ Y(t) &= y_0 + x_0W_2(t) + \int_0^t W_1(t)dW_2(t), \quad \forall t \geq 0, \end{aligned}$$

where $G_0 = (x_0, y_0)$. The first component is just a one-dimensional Brownian motion, while the second component is the sum between a one-dimensional scaled Brownian motion and the asymmetric area process

$$\tilde{S}_t = \int_0^t W_1(s)dW_2(s),$$

investigated in Section 1.13. We also note that $\mathbb{E}[G_t] = (x_0, y_0)$.

The infinitesimal increments satisfy

$$\begin{aligned} dG_t &= \big(dX(t), dY(t)\big) = \big(dW_1(t), X(t)dW_2(t)\big) \\ &= (1,0)dW_1(t) + (0, X(t))dW_2(t) \\ &= \partial_{x_1|G_t} dW_1(t) + x_1 \partial_{x_2|G_t} dW_2(t) \\ &= X_{1|G_t} dW_1(t) + X_{2|G_t} dW_2(t) \in \mathcal{D}_{G_t}, \end{aligned}$$

where $\mathcal{D} = span\{X_1, X_2\}$ is the horizontal distribution defined by the vector fields X_1 and X_2. Therefore, dG_t belongs to the horizontal distribution, and the trajectories $t \to G_t(\omega)$ are horizontal curves, for each state ω. At each infinitesimal time step dt the Grushin diffusion moves independent and normally distributed amounts $dW_1(t)$ and $dW_2(t)$ into the $X_1$and X_2 directions, respectively. As it will be noticed later, this remarkable property does not hold in all cases of sub-elliptic diffusions.

The transience property We recall the Grushin norm from Section 5.7

$$\|\mathbf{x}\|_G = \|(x_1, x_2)\|_G = \Big(x_2^2 + \frac{x_1^4}{4}\Big)^{1/4},$$

and consider the annulus

$$A_{a,b} = \{(x_1, x_2); a < \|(x_1, x_2)\|_G < b\}$$

for $0 < a < b$. Consider the function $f(x) = \dfrac{1}{\|\mathbf{x}\|_G}$. By Corollary 5.7.2 we have $\Delta_G f = 0$, for any $\mathbf{x} \neq 0$.

Let $G_t = (G_1^1, G_t^2)$ be the Grushin diffusion that starts at $(x_1^0, x_2^0) \in A_{a,b}$ and denote by $T_{a,b}$ the exit time of the diffusion G_t from the annulus $A_{a,b}$. We denote by $A_{a,b}^1$ the projection of $A_{a,b}$ on the x_1-axis and consider $K > 0$ such that $A_{a,b}^1 \subset [-K, K]$. If τ_K represents the exit time of $G_t^1 = x_1^0 + W_1(t)$ from the compact $[-K, K]$, we have $\mathbb{E}[T_{a,b}] \leq \mathbb{E}[\tau_K] < \infty$, by using a property of Brownian motions. Then applying Dynkin's formula, see Theprem 1.1.1, we have

$$\begin{aligned} &\mathbb{E}^{\mathbf{x}^0}[f(G_{T_{a,b}})] = f(G_0) + \mathbb{E}^{\mathbf{x}^0}\Big[\int_0^{T_{a,b}} (\Delta_G f)(G_s)\, ds\Big] \Longleftrightarrow \\ &\frac{1}{a}p_a + \frac{1}{b}p_b = \frac{1}{\|\mathbf{x}^0\|_G}, \end{aligned}$$

where p_a and p_b are the probabilities that the Grushin diffusion G_t reaches the annulus boundaries $\{\mathbf{x}; \|\mathbf{x}\|_G = a\}$ and $\{\mathbf{x}; \|\mathbf{x}\|_G = b\}$, respectively. Taking the limit $b \to \infty$ and using that $\frac{1}{b}p_b \to 0$, we obtain the probability that a Grushin diffusion reaches a Grushin disk centered at the origin:

Proposition 7.4.2 *Let $D_G(0,a) = \{\mathbf{x}; \|\mathbf{x}\|_G < a\}$. Then the probability that a Grushin diffusion starting at $G_0 = \mathbf{x}^0 \notin D_G(0,a)$ reaches the disk $D_G(0,a)$ is*

$$p_a = \frac{a}{\|\mathbf{x}^0\|_G}.$$

Since $a < \|x^0\|_G$, it follows that $p_a < 1$, which expresses the transitivity property of the Grushin diffusions starting outside the origin with respect to neighborhoods of the origin.

Using the invariance property of the Grushin distribution with respect to x_2-translations, a similar argument provides:

Corollary 7.4.3 *Let $D_G((0,c),a) = \{\mathbf{x}; \|(x_1, x_2 - c)\|_G < a\}$. Then the probability that a Grushin diffusion starting at $G_0 = (x_1^0, x_2^0) \notin D_G((0,c),a)$ reaches the disk $D_G((0,c),a)$ is*

$$p_a = \frac{a}{(c - x_2^0)^2 + \frac{(x_1^0)^4}{4}}.$$

The exit probability We shall compute first the second moment of the asymmetric area process, $\tilde{S}_t = \int_0^t W_1(s)dW_2(s)$. We start by computing the expectation of the product process $U_t = W_1(t)W_2(t)S_t$, where S_t denotes the Lévy area associated with the one-dimmensional independent Brownian motions $W_1(t)$ and $W_2(t)$. Using the product rule for stochastic processes together with Ito's lemma, we have

$$\begin{aligned}
dU_t &= S_t d(W_1(t)W_2(t)) + W_1(t)W_2(t)dS_t + dS_t d(W_1(t)W_2(t)) \\
&= S_t W_1(t)dW_2(t) + S_t W_2(t)dW_1(t) \\
&\quad + W_1(t)W_2(t)\big(W_1(t)dW_2(t) - W_2(t)dW_1(t)\big) \\
&\quad + \big(W_1(t)dW_2(t) - W_2(t)dW_1(t)\big)\big(W_1(t)dW_2(t) + W_2(t)dW_1(t)\big) \\
&= \big(W_1(t)^2 - W_2(t)^2\big)dt \\
&\quad + \big(W_2(t)S_t(t) - W_1(t)W_2(t)^2\big)dW_1(t) \\
&\quad + \big(W_1(t)S_t + W_1(t)^2 W_2(t)\big)dW_2(t).
\end{aligned}$$

Integrating and using $U_0 = 0$ yields

$$\begin{aligned}
U_t = &\int_0^t (W_1(u)^2 - W_2(u)^2)du + \int_0^t (W_2(u)S_t(u) - W_1(u)W_2(u)^2)dW_1(u) \\
&+ \int_0^t \big(W_1(u)S_t + W_1(u)^2 W_2(u)\big)dW_2(u),
\end{aligned}$$

and hence

$$\mathbb{E}[U_t] = \int_0^t \big(\mathbb{E}[W_1(u)^2] - \mathbb{E}[W_2(u)^2]\big)du = 0. \tag{7.4.15}$$

Using the computation done in the proof of Proposition 1.13.1 at page 49, we can represent the asymmetric area in terms of Lévy area as

$$\tilde{S}_t = \frac{1}{2}S_t + \frac{1}{2}W_1(t)W_2(t).$$

Then the second moment of $\tilde{S}_t$ can be computed as follows

$$\begin{aligned}
\mathbb{E}[\tilde{S}_t^2] &= \frac{1}{4}\mathbb{E}[S_t^2 + W_1(t)^2W_2(t)^2 + 2W_1(t)W_2(t)S_t] \\
&= \frac{1}{4}\Big(\mathbb{E}[S_t^2] + \mathbb{E}[W_1(t)^2]\mathbb{E}[W_2(t)^2] + 2\mathbb{E}[W_1(t)W_2(t)S_t]\Big) \\
&= \frac{1}{4}(t^2 + t^2 + 0) = \frac{t^2}{2},
\end{aligned}$$

where we used relations (7.1.6) and (7.4.15).

The next result provides an estimation of the exit probability of the Grushin distribution from an Euclidean ball centered at the origin.

Proposition 7.4.4 *Let G_t be a Grushin diffusion starting at the origin and let $r > 0$. Then*

$$P\big(G_t \notin B_{Eu}(0,r)\big) \leq \frac{1}{r^2}\Big(t + \frac{t^2}{2}\Big).$$

Proof: Consider the event $A_t = \{\omega; \|G_t(\omega)\|_{Eu} > r\}$. Its probability satisfies the inequality

$$\begin{aligned}
P(A_t) &= \int_{A_t} dP(\omega) < \frac{1}{r^2}\int_{A_t} \|G_t(\omega)\|_{Eu}^2\, dP(\omega) < \frac{1}{r^2}\int_{\Omega} \|G_t(\omega)\|_{Eu}^2\, dP(\omega) \\
&= \frac{1}{r^2}\mathbb{E}[\|G_t\|_{Eu}^2] = \frac{1}{r^2}\mathbb{E}\Big[W_1(t)^2 + \Big(\int_0^t W_1(u)dW_2(u)\Big)^2\Big] \\
&= \frac{1}{r^2}\Big(t + \mathbb{E}[\tilde{S}_t^2]\Big) = \frac{1}{r^2}\Big(t + \frac{t^2}{2}\Big).
\end{aligned}$$

■

7.4.2 The n-Dimensional Case

We recall the $(n+1)$-dimensional Grushin operator with one missing direction

$$\Delta_G^{(n,1)} = \frac{1}{2}\sum_{k=1}^{n} \partial_{x_k}^2 + 2\|x\|^2\partial_y^2,$$

where $\|x\|^2 = x_1^2 + \cdots + x_n^2$, from Section 5.7, page 235. This operator is a sum of squares of vector fields

$$\Delta_G^{(n,1)} = \frac{1}{2}(X_1^2 + \cdots + X_n^2 + Y^2),$$

with $X_j = \partial_{x_j}$ and $Y = 2\|x\|\partial_y$. However, for $n \geq 2$, the vector fields $\{X_1, \cdots, X_n, Y\}$ do not satisfy the bracket-generating condition along the line $\{x = 0\}$. For this reason the associated diffusion does not qualify as sub-elliptic. But it is elliptic away from the y-axis and we shall still investigate it in the following. The associated Grushin diffusion G_t satisfies the system

$$\begin{aligned} dG_1(t) &= dW_1(t) \\ &\cdots\cdots\cdots\cdots \\ dG_n(t) &= dW_n(t) \\ dG_{n+1}(t) &= 2\sqrt{G_1(t)^2 + \cdots + G_n(t)^2}\, dW_{n+1}(t), \end{aligned}$$

where $W_j(t)$ are independent one-dimensional Brownian motions.

We shall investigate the transience property of G_t. We define the Grushin norm

$$\|(x, y)\|_{G_{n,1}} = (\|x\|^4 + y^2)^{n/4}$$

and consider the function $f(x, y) = 1/\|(x, y)\|_{G_{n,1}}$. By Proposition 5.7.1, page 235, we have

$$\Delta_G^{(n,1)} f(x, y) = 0, \qquad \forall x, y \neq 0.$$

Let $0 < a < b$ and denote $A_{a,b} = \{(x, y) \in \mathbb{R}^{n+1}; a < \|(x, y)\|_{G_{n,1}} < b\}$. We consider a Grushin diffusion starting at $G_0 = (x^0, y^0)$, with $(x^0, y^0) \in A_{a,b}$ and away from the y-axis. We denote by $T_{a,b}$ the first exit time of G_t from the annulus $A_{a,b}$. Applying Dynkin's formula

$$\mathbb{E}^{(x^0,y^0)}[f(G_{T_{a,b}})] = f(x^0, y^0) + \mathbb{E}^{(x^0,y^0)}\Big[\int_0^{T_{a,b}} (\Delta_G^{(n,1)} f)(G_s)\, ds\Big]$$

we obtain

$$\frac{1}{a} p_a + \frac{1}{b}(1 - p_a) = \frac{1}{\|(x^0, y^0)\|_{G_{n,1}}}.$$

Talking $b \to \infty$ yields

$$p_a = \frac{a}{(\|x^0\|^4 + (y^0)^2)^{n/4}} < 1.$$

This represents the probability that the Grushin distribution G_t, which starts outside the Grushin ball $\{(x, y); \|(x, y)\|_{G_{n,1}} < a\}$ enters the ball with probability $p_a < 1$. This shows that G_t is transitive with respect to balls centered at the origin.

7.5 Martinet Diffusion

Consider the following two vector fields on $\mathbb{R}^3$

$$X = \partial_x + \frac{1}{2}y^2\partial_z, \qquad Y = \partial_y.$$

The distribution spanned by $\{X, Y\}$ is called the *Martinet distribution.* As we have the following commutation relations

$$[X, Y] = -y\partial_z, \qquad [[X, Y], Y] = \partial_z,$$

the distribution is step 2 along the plane $\{y = 0\}$ and step 3 outside of it. Since the bracket-generating condition holds everywhere on $\mathbb{R}^3$, the *Martinet Laplacian*, which is defined by

$$\begin{aligned}\Delta_M &= \frac{1}{2}(X^2 + Y^2)\\ &= \frac{1}{2}\Big(\partial_x^2 + \partial_y^2 + \frac{1}{4}y^4\partial_z^2 + z^2\partial_x\partial_z\Big),\end{aligned}$$

is hypoelliptic. The associated subelliptic diffusion to the operator Δ_M, denoted by $M_t = (M_1(t), M_2(t), M_3(t))$, is called the *Martinet diffusion.* This is given by

$$\begin{aligned}dM_1(t) &= dW_1(t)\\ dM_2(t) &= dW_2(t)\\ dM_3(t) &= \frac{1}{2}M_2(t)^2 dW_1(t),\end{aligned}$$

with $W_j(t)$ independent one-dimensional Brownian motions.

We note that the Martinet diffusion spreads along the Martinet distribution as it follows from the next computation:

$$\begin{aligned}dM_t &= \big(dM_1(t), 0, 0\big) + \big(0, dM_2(t), 0\big) + \big(0, 0, dM_3(t)\big)\\ &= (1, 0, 0)dW_1(t) + (0, 1, 0)dW_2(t) + (0, 0, 1)\frac{1}{2}M_2(t)^2 dW_1(t)\\ &= (1, 0, \frac{1}{2}M_2(t)^2)dW_1(t) + (0, 1, 0)dW_2(t)\\ &= X_{|M_t}dW_1(t) + Y_{|M_t}dW_2(t) \in \mathcal{D}_{M_t}.\end{aligned}$$

Finding the transition probability for M_t is equivalent to finding the heat kernel for the operator Δ_M. However, the last one is still an unsolved problem. In the following it is worth investigating where the difficulty lies.

We consider the Martinet diffusion starting at the origin

$$\begin{aligned} M_1(t) &= W_1(t) \\ M_2(t) &= W_2(t) \\ M_3(t) &= \frac{1}{2}\int_0^t W_2(s)^2 \, dW_1(s). \end{aligned}$$

The component $M_3(t)$ is a continuous martingale with the second moment

$$\mathbb{E}[M_3(t)^2] = \frac{1}{4}\mathbb{E}\Big[\int_0^t W_2(s)^4 \, ds\Big] = \frac{t^3}{2} < \infty$$

and quadratic variation

$$\langle M_3 \rangle_t = \frac{1}{4}\int_0^t W_2(s)^4 \, ds.$$

By Theorem 1.1.11 there is a DDS Brownian motion, B_t, independent of $W_2(s)$ such that $M_3(t) = B_{\frac{1}{4}\int_0^t W_2(s)^4 \, ds}$. The characteristic function of $M_3(t)$ can be reduced to the Laplace transform of the Brownian motion functional $\int_0^t W_2(s)^4 \, ds$ as follows

$$\begin{aligned} \mathbb{E}[e^{i\lambda M_3(t)}] &= \mathbb{E}[e^{i\lambda B_{\langle M_3 \rangle_t}}] \\ &= \mathbb{E}[\mathbb{E}[e^{i\lambda B_{\langle M_3 \rangle_t}}] | W_2[0,t]] \\ &= \mathbb{E}[e^{-\frac{1}{2}(\frac{\lambda}{2})^2 \int_0^t W_2(s)^4 \, ds}]. \end{aligned}$$

By Feynman-Kac's formula, the last expectation is the solution of a one-dimensional heat equation with quartic potential, see Calin et al. [29], page 197. More precisely, the function

$$u(t,x) = \mathbb{E}[e^{-\frac{1}{2}(\frac{\lambda}{2})^2 \int_0^t W_2(s)^4 \, ds} | W_2(0) = x]$$

satisfies the initial value problem

$$\frac{\partial u(t,x)}{\partial t} = \frac{1}{2}\frac{\partial^2 u(t,x)}{\partial x^2} - \frac{1}{2}\Big(\frac{\lambda}{2}\Big)^2 x^4 u(t,x) \tag{7.5.16}$$

$$u(0,x) = 1. \tag{7.5.17}$$

Therefore, solving the Cauchy problem (7.5.16) and then setting $x = 0$, we obtain

$$\mathbb{E}[e^{-\frac{1}{2}(\frac{\lambda}{2})^2 \int_0^t W_2(s)^4 \, ds}] = u(t,0).$$

The difficulty of the problem consists in finding a closed form solution using elementary functions. In the case when the quartic coefficient x^4 is replaced by the quadratic coefficient x^2, then there is a closed form solution, which leads to the celebrated Cameron-Martin's formula, see Section 1.7, page 26. Unfortunately, no Cameron-Martin type formulas are known for the quartic potential!

It is worth mentioning that this problem has a Quantum mechanics significance, that consists in the fact that Schrödinger's equation with a special type of quartic potential models the tunneling effect. The approach of the problem so far involved the WKB approximation method, see Voros [134] and Schulman [125], page 96.

Martinet introduced this distribution in 1970. Since then a great deal of work has been done on the geometry of this sub-Riemannian model. This involves the computation of sub-Riemannian geodesics done by Agrachev et al. [2], proof of the global connectivity by normal geodesics by Greiner and Calin [64] as well as Gaveau and Greiner [62],[63]. These advances aim to represent the heat kernel as an integral of all modified complex action contributions along all normal geodesics, see Calin et al. [38]. However, as far as the author knows, the heat kernel for the operator Δ_M is still not represented in a closed form.

7.6 A Step 3 Grushin Diffusion

We consider the following two vector fields on $\mathbb{R}^3$

$$X = \partial_x, \qquad Y = \frac{1}{2}x^2\partial_y. \tag{7.6.18}$$

Since $[X, Y] = x\partial_y$ and $[X, [X, Y]] = \partial_y$, the bracket-generating condition is satisfied, the distribution being of step 3 on the y-axis and step 1 in rest. Consequently, the associated Laplacian

$$L = \frac{1}{2}\Big(\partial_x^2 + \frac{x^4}{4}\partial_y^2\Big)$$

is hypoelliptic and has a smooth heat kernel. The associated subelliptic diffusion G_t satisfies

$$\begin{aligned} dG_1(t) &= dW_1(t) \\ dG_2(t) &= \frac{1}{2}G_1(t)^2 dW_2(t), \end{aligned}$$

with $W_1(t)$, $W_2(t)$ independent Brownian motions. Assuming $G_0 = (x, y)$, we obtain

$$\begin{aligned} G_1(t) &= x + W_1(t) \\ G_2(t) &= y + \frac{1}{2}\int_0^t (x + W_1(s))^2\, dW_2(s). \end{aligned}$$

Similarly with the Martinet case, there is a Brownian motion B_t such that

$$G_2(t) = y + B_{\frac{1}{4}\int_0^t (x+W_1(s))^4 ds}$$

and

$$\mathbb{E}[e^{i\lambda G_2(t)}] = \mathbb{E}[e^{-\frac{1}{2}(\frac{\lambda}{2})^2 \int_0^t (x+W_1(s))^4\, ds}] = u(t,x),$$

which satisfies the initial value problem

$$\begin{aligned}\frac{\partial u(t,x)}{\partial t} &= \frac{1}{2}\frac{\partial^2 u(t,x)}{\partial x^2} - \frac{1}{2}\Big(\frac{\lambda}{2}\Big)^2 x^4 u(t,x)\\ u(0,x) &= 1.\end{aligned}$$

A closed form solution using elementary functions is missing, so an approach using the WKB approximation method may be applied. Small time asymptotics for the heat kernel had been already computed by Chang and Li [44], while the geometry of distribution generated by (7.6.18) involving a characterization of the sub-Riemannian geodesics has been investigated by Calin and Chang [24], [26].

7.7 The Engel Diffusion

We shall consider the vector fields

$$X = \partial_x, \qquad Y = \partial_y + x\partial_z + \frac{1}{2}x^2\partial_u \tag{7.7.19}$$

whose span defines the *Engel distribution* on $\mathbb{R}^4$. This distribution is step 3 everywhere, since the vector fields

$$X, \quad Y, \quad [X,Y] = \partial_z + x\partial_u, \quad [X,[X,Y]] = \partial_u$$

generate the tangent space of $\mathbb{R}^4$ at each point. This distribution has two missing directions, since the codimension is equal to 2. The associate sub-elliptic operator

$$\begin{aligned}\Delta_E &= \frac{1}{2}(X^2+Y^2)\\ &= \frac{1}{2}\Big(\partial_x^2 + \partial_y^2 + x^2\partial_z^2 + \frac{1}{4}x^4\partial_u^2 + 2x\partial_y\partial_z + x^2\partial_y\partial_u + x^3\partial_z\partial_u\Big)\end{aligned}$$

is called the *Engel subLaplacian*, which is hypoelliptic. It is known that the vector fields (7.7.19) are left invariant with respect to a noncommutative Lie group law on $\mathbb{R}^4$ given by

$$(x, y, z, u)*(x', y', z', u') = \Big(x + x', y + y', z + z' + \frac{1}{2}(xy' - x'y),$$
$$u + u' + \frac{1}{2}(xz' - zx') + \frac{1}{12}\big(x^2y' - xx'(y + y') + y(x')^2\big)\Big),$$

see for instance Calin and Chang [26], page 314. The pair $(\mathbb{R}^4, *)$ is called the *Engel group* and is a noncommutative Lie group.

The *Engel diffusion*, $\mathcal{E}_t$, is the diffusion associated with the operator Δ_L

$$\begin{aligned}
d\mathcal{E}_1(t) &= dW_1(t)\\
d\mathcal{E}_2(t) &= dW_2(t)\\
d\mathcal{E}_3(t) &= \mathcal{E}_1(t)dW_2(t)\\
d\mathcal{E}_4(t) &= \frac{1}{2}\mathcal{E}_1(t)^2 dW_2(t),
\end{aligned}$$

with $W_1(t)$ and $W_2(t)$ independent Brownian motions. We also note that the infinitesimal increment occurs along the Engel distribution

$$d\mathcal{E}_t = X_{|\mathcal{E}_t} dW_1(t) + Y_{|\mathcal{E}_t} dW_2(t).$$

Due to the group law invariance it suffices to study the diffusion starting just from the origin. The Engel diffusion starting at the origin is given by

$$\begin{aligned}
\mathcal{E}_1(t) &= W_1(t)\\
\mathcal{E}_2(t) &= W_2(t)\\
\mathcal{E}_3(t) &= \int_0^t W_1(s)dW_2(s)\\
\mathcal{E}_4(t) &= \frac{1}{2}\int_0^t W_1(s)^2 dW_2(s).
\end{aligned}$$

The probability law of $\mathcal{E}_t$ provides the heat kernel for the Engel subLaplacian Δ_L. Unfortunately, the same curse present in the Martinet's case haunts this case too. This consists mainly on involving the law of the Brownian functional $\int_0^t W(s)^4\, ds$, which is not expressible in an elementary form.

It is worth noting that these type of Brownian functionals appear in the study of other subLaplacians of step larger than or equal to 3. The only case where an elementary close form expression for the law of $\int_0^t W(s)^{2k}\, ds$ is available is the case $k = 1$, which corresponds to a step 2 subLaplacian.

After becoming familiar with several examples of sub-elliptic diffusions, in the next section we shall try to construct a general theory.

7.8 General Sub-elliptic Diffusions

Let $X_1, \cdots, X_k$ be k linearly independent smooth vector fields on $\mathbb{R}^n$, with $k < n$, and let $\mathcal{D} = span\{X_1, \cdots, X_k\}$ be the distribution generated by them. This distribution has rank k, namely at each point $x \in \mathbb{R}^n$ the set $\mathcal{D}_p$ is a subspace of $T_x\mathbb{R}^n$ of dimension k. This means that if the vector fields write in local coordinates as

$$X_j = \sum_{i=1}^{n} \sigma_{ij}(x)\partial_{x_i}, \qquad j = 1, \dots, k, \tag{7.8.20}$$

then $rank(\sigma_{ij}(x)) = k$ for each $x \in \mathbb{R}^n$.

We also assume that the bracket-generating conditions is satisfied, i.e., the vector fields X_i together with a finite number of their iterated Lie brackets generate the tangent space $T_x\mathbb{R}^n$ at each point $x \in \mathbb{R}^n$. Under these conditions we define the sub-elliptic diffusion Z_t to be the diffusion on $\mathbb{R}^n$, whose infinitesimal generator is the subLaplacian

$$\Delta_X = \frac{1}{2}(X_1^2 + \cdots + X_k^2). \tag{7.8.21}$$

This means that the sub-elliptic diffusion Z_t starting at $Z_0 = x$ satisfies the relation

$$\lim_{t \to 0} \frac{\mathbb{E}^x[f(Z_t)] - f(x)}{t} = (\Delta_X f)(x),$$

for any smooth function f, where $\mathbb{E}^x$ denotes the expectation operator, given that Z_t starts at x.

We have seen that all examples of sub-elliptic diffusions investigated so far (Heisenberg, Grushin, Martinet and Engel diffusions) have the property that they spread along the horizontal distribution $\mathcal{D}$. But not necessarily all sub-elliptic diffusions spread this way. We shall use the notation

$$X_{j|Z_t} = \sum_{i=1}^{n} \sigma_{ij}(Z_t)e_i,$$

with $e_i = (0, \dots, 0, 1, 0, \dots, 0)$ having 1 on the ith slot, to denote the vector field X_j along the diffusion Z_t.

Definition 7.8.1 *Let $Z_t = (Z_t^1, \dots, Z_t^n)$ be the sub-elliptic diffusion associated with the vector fields (7.8.20). Then Z_t is called horizontal if*

$$dZ_t = X_{1|Z_t}dW_1(t) + \cdots + X_{k|Z_t}dW_k(t), \tag{7.8.22}$$

where $W_j(t)$ are k independent Brownian motions.

The following result shows that a horizontal diffusion is driftless and its dispersion matrix is obtained from the coefficients of the vector fields (7.8.20).

Proposition 7.8.2 *Let Z_t be a sub-elliptic diffusion. Then Z_t is horizontal if and only if*

$$dZ_t^r = \sum_{j=1}^{k} \sigma_{rj}(Z_t)\, dW_j(t), \quad r = 1, \dots, n. \tag{7.8.23}$$

Proof: Writing relation (7.8.22) on components, we have

$$\begin{aligned} dZ_t^r &= X_{1|Z_t}^r dW_1(t) + \cdots + X_{k|Z_t}^r dW_k(t) \\ &= \sigma_{r1}(Z_t) dW_1(t) + \cdots + \sigma_{rk}(Z_t) dW_k(t), \end{aligned}$$

which is equation (7.8.23). The computation can be easily inversed. ■

We note that a horizontal diffusion Z_t can be written more compactly as

$$dZ_t = \sigma(Z_t) dW_t.$$

Its infinitesimal generator is the differential operator

$$L = \frac{1}{2} \sum_{i,j=1}^{n} (\sigma\sigma^T)_{ij} \partial_{x_i} \partial_{x_j}, \tag{7.8.24}$$

where

$$g^{ij} = (\sigma\sigma^T)_{ij} = \sum_{r=1}^{k} \sigma_{ir}\sigma_{rj}^T = \sum_{r=1}^{k} \sigma_{ir}\sigma_{jr}$$

is the contravariant diffusion metric. This metric can be used to measure distances along curves tangent to the distribution $\mathcal{D}$. The next result shows that this metric is naturally induced from $\mathbb{R}^n$ onto the distribution $\mathcal{D}$.

Proposition 7.8.3 *Let $\langle\, , \rangle$ denote the Euclidean inner product on $\mathbb{R}^n$. Then*

$$\langle X_i, X_j \rangle = g^{ij}.$$

Proof: Writing $X_i = \sum_{r=1}^{n} \sigma_{ir} e_r$, we have

$$\begin{aligned} \langle X_i, X_j \rangle &= \sum_{r,p} \langle \sigma_{ir} e_r, \sigma_{jp} e_p \rangle = \sum_{r,p} \sigma_{ir}\sigma_{jp}\delta_{rp} \\ &= \sum_{r} \sigma_{ir}\sigma_{jr} = (\sigma\sigma^T)_{ij} = g^{ij}. \end{aligned}$$

■

The matrix $g = g^{ij}$ is degenerated, fact that results from the following rank evaluation

$$rank(g) = rank(\sigma\sigma^T) = rank(\sigma) = k < n.$$

The fact that g^{ij} does not have an inverse marks the difference between the sub-Riemannian geometry of non-integrable distributions and Riemannian geometry.

Remark 7.8.4 The metric g^{ij} measures the rate of change in the covariance matrix of the displacement vector $Z_t - x$ as follows

$$\mathbb{E}^x[(Z_t^i - x_i)(Z_t^j - x_j)] = tg^{ij}(x) + o(t), \qquad t \to 0.$$

Let ∇^0 denote the Levi-Civita connection on the Euclidean space $\mathbb{R}^n$, which is given by

$$\nabla^0_U V = \sum_{r=1}^n U(V^r)e_r,$$

with $U = \sum_i U^i e_i$ and $V = \sum_j V^j e_j$ vector fields on $\mathbb{R}^n$ and $\{e_j\}$ the usual orthonormal system of coordinates on $\mathbb{R}^n$.

In particular, we have

$$\nabla^0_{X_j} X_i = \sum_{r=1}^n X_j(X_i^r)e_r = \sum_{r=1}^n X_j(\sigma_{ri})e_r.$$

Proposition 7.8.5 *The subLaplacian (7.8.21) is given by*

$$\Delta_X = L + \frac{1}{2}\sum_{j=1}^k \nabla^0_{X_j} X_j,$$

where L is the operator (7.8.24).

Proof: We start with the evaluation of X_j^2

$$\begin{aligned} X_j(X_j) &= \Big(\sum_{i=1}^n \sigma_{ij}\partial_{x_i}\Big)\Big(\sum_{r=1}^n \sigma_{rj}\partial_{x_r}\Big) \\ &= \sum_{i,r} \sigma_{ij}\sigma_{rj}\partial_{x_i}\partial_{x_r} + \sum_{i,r} \sigma_{ij}(\partial_{x_i}\sigma_{rj})\partial_{x_r}. \end{aligned}$$

Summing over j and switching the sums we obtain

$$\begin{aligned}
\Delta_X &= \frac{1}{2}\sum_{j=1}^{k} X_j^2 \\
&= \frac{1}{2}\sum_{i,r}\sum_{j}\sigma_{ij}\sigma_{rj}\partial_{x_i}\partial_{x_r} + \frac{1}{2}\sum_{r}\Big[\sum_{j}\sum_{i}\sigma_{ij}(\partial_{x_i}\sigma_{rj})\Big]\partial_{x_r} \\
&= \frac{1}{2}\sum_{i,r}(\sigma\sigma^T)_{ir}\partial_{x_i}\partial_{x_r} + \frac{1}{2}\sum_{r}\Big[\sum_{j} X_j(\sigma_{rj})\Big]\partial_{x_r} \\
&= L + \frac{1}{2}\sum_{j}\Big[\sum_{r} X_j(X_j^r)\partial_{x_r}\Big] \\
&= L + \frac{1}{2}\sum_{j}\nabla^0_{X_j}X_j.
\end{aligned}$$

■

Proposition 7.8.6 *Let Z_t be the sub-elliptic diffusion associated with the vector fields (7.8.20). Then Z_t is horizontal if and only if*

$$\sum_{j=1}^{k}\nabla^0_{X_j}X_j = 0. \tag{7.8.25}$$

Proof: If (7.8.25) holds, then by Proposition 7.8.5 we obtain $\Delta_X = L$, namely the sub-elliptic diffusion Z_t has the infinitesimal generator given by L. Then we can write $dZ_t = \sigma(Z_t)dW_t$, which by Proposition 7.8.2 implies that Z_t is horizontal. The reasoning can be easily reversed. ■

Relation (7.8.25) states that the linear term of the subLaplacian Δ_X vanishes, which is equivalent with the fact that the sub-elliptic diffusion Z_t is driftless. This can be equivalently stated as

$$\mathbb{E}^x[Z_t] = x + o(t).$$

We note that all examples of sub-elliptic diffusion investigated in the previous sections (Heisenberg, Grushin, Martinet and Engel diffusions) satisfy the driftless property (7.8.25), and therefore, they live on the horizontal distribution $\mathcal{D}$.

Proposition 7.8.7 *Let Z_t be the sub-elliptic diffusion associated with the vector fields (7.8.20). Then Z_t satisfies*

$$dZ_t = X_{1|Z_t}dW_1(t) + \cdots + X_{k|Z_t}dW_k(t) + \frac{1}{2}\sum_{j=1}^{k}\nabla^0_{X_j}X_{j\,|Z_t}\,dt. \tag{7.8.26}$$

Proof: This is a consequence of Proposition 7.8.5. The term $\frac{1}{2}\sum_{j=1}^{k}\nabla^0_{X_j}X_j$ represents the linear part of Δ_X and produces drift for the diffusion Z_t. ■

Remark 7.8.8 Due to the resemblance between relation (7.8.26) and the kinematic decomposition given by Corollary 2.18.4, page 115,

$$dY_t = T_t\,dW_t + \frac{1}{2}\kappa(W_t)N_t dt,$$

we may regard the expression $\sum_{j=1}^{k}\nabla^0_{X_j}X_j$ as some sort of normal curvature of the distribution $\mathcal{D}$.

Then Proposition 3.6.3 implies:

Proposition 7.8.9 *Let Z_t be the sub-elliptic diffusion associated with the vector fields (7.8.20). Then the probability density $p_t(x_0, x)$ of Z_t is a smooth function for any $t > 0$.*

Transience property In the following we shall deal with the transience property of sub-elliptic diffusions. We have seen that this property has been proved for the Heisenberg and Grushin cases by constructing a norm and considering balls and annulae with respect to that norm. Then an application of Dynkin's formula leads to the desired result. However, in the general case we do not have access to a closed form expression for the heat kernel or fundamental solution, so constructing the associated norm requires a few extra assumptions.

We start by assuming the integral $F(x_0, x) = \int_0^\infty p_t(x_0, x)\,dt$ convergent. Then we define the norm

$$\|x - x_0\| = \frac{1}{F(x_0, x)}. \tag{7.8.27}$$

The ball centered at x_0 is defined by $B(x_0, r) = \{x; \|x - x_0\| \le r\}$. In order to have balls nested ascendentally, namely

$$B(x_0, a) \subset B(x_0, b) \ \text{ for } \ a < b,$$

the gradient of the contour function has to point outside the ball. This is satisfied if

$$\langle x - x_0, \,\text{grad}(\|x - x_0\|)\rangle \ge 0,$$

where the gradient of a function is defined by $\text{grad}f(x) = \sum_j \partial_{x_j} f(x)\, e_j$. Since

$$\begin{aligned}\text{grad}(\|x - x_0\|) &= \text{grad}\Big(\frac{1}{F(x_0, x)}\Big) = \frac{-1}{F^2(x_0, x)}\text{grad}F(x_0, x)\\ &= \frac{-1}{F^2(x_0, x)}\int_0^\infty \text{grad}\, p_t(x_0, x)\,dt,\end{aligned}$$

then

$$\langle x - x_0, \operatorname{grad}(\|x - x_0\|)\rangle = \frac{-1}{F^2(x_0, x)} \int_0^\infty \langle x - x_0, \operatorname{grad} p_t(x_0, x)\rangle \, dt.$$

Definition 7.8.10 *A diffusion with transition probability $p_t(x_0, x)$ is called expansible at x_0 if*

$$\langle x - x_0, \operatorname{grad} p_t(x_0, x)\rangle \leq 0, \qquad \forall x.$$

This condition can be also written as $D_{x-x_0} p_t(x_0, x) \leq 0$, namely, the rate of change of the density in the direction $x - x_0$ is negative. Roughly speaking, it means that the density of a diffusion initiated at x_0 decreases when moving away from x_0.

Example 7.8.1 The n-dimensional Brownian motion W_t has the transition density $p_t(x_0, x) = \frac{1}{(2\pi t)^{n/2}} e^{-\frac{\|x-x_0\|^2}{2t}}$. Since we have

$$\langle x - x_0, \operatorname{grad} p_t(x_0, x)\rangle = -\|x - x_0\|^2 p_t(x_0, x) \leq 0,$$

then W_t is expansible at any point x_0.

Let $0 < a < b$ and consider the annulus

$$A_{a,b} = \{x \in \mathbb{R}^n; a < \|x\| < b\},$$

taken with respect to the norm (7.8.27). We assume the diffusion Z_t expansible and starting at $Z_0 = x_0 \in A_{a,b}$. We denote by $T_{a,b}$ the first exit time of Z_t from the annulus $A_{a,b}$. We apply Dynkin's formula

$$\mathbb{E}^{x_0}[f(Z_{T_{a,b}})] = f(x_0) + \mathbb{E}^{x_0}\Big[\int_0^{T_{a,b}} (\Delta_X f)(Z_s)\, ds\Big]$$

with $f(x) = \frac{1}{\|x\|} = F(0, x)$. By the computation done in Section 5.1, page 225, we have $\Delta_X f(x) = 0$. Therefore, the aforementioned formula becomes

$$\frac{1}{a} p_{a,b} + \frac{1}{b} q_{a,b} = \frac{1}{\|x_0\|},$$

where $p_{a,b}$ is the probability that Z_t enters the ball $B(0, a)$ and $q_{a,b}$ is the probability that Z_t leaves the ball $B(0, b)$. Taking $b \to \infty$ yields

$$p_a = \lim_{b \to \infty} p_{a,b} = \frac{a}{\|x_0\|} < 1.$$

We conclude with the following result:

Proposition 7.8.11 *Let Z_t be a sub-elliptic diffusion which is expansible at the origin, which satisfies $\int_0^\infty p_t(x_0, x)\, dt < \infty$, with $Z_0 = x_0 \neq 0$. Then the probability that Z_t enters any ball centered at the origin of radius $r < \|x_0\|$ is strictly less than 1.*

Notes and remarks 1. The state of a mechanical system with constraints is described by a point on a sub-Riemannian distribution. The movements of the system are described by curves tangent to the distribution $\mathcal{D}$. For a few examples the reader is referred to Calin et al. [40]. Chow's theorem of connectivity states that if the bracket-generating condition is satisfied, then the mechanical system can be steered continuously between any two given states A and B. In the case when the movement of the system is random, then this is described by a horizontal sub-elliptic diffusion. The transitivity property of the diffusion means that a system left under random evolution visits any neighborhood with a probability between 0 and 1.

2. The transience of a sub-elliptic diffusion is influenced by the codimension of the horizontal distribution as well as the step of the distribution. The larger these parameters, the less transitive the diffusion seems to be.

7.9 Nonholonomic Diffusions

Sub-elliptic diffusions of nonholonomic origin, i.e. diffusions associated with non-integrable distributions modeling nonholonomic mechanical systems, will be called *nonholonomic diffusions.* In this chapter we shall encounter a few examples of these type of diffusions. From the physical point of view, they represent trajectories in the phase-space of mechanical systems with constraints, which are left free under uncertain evolution conditions, such as stochastic controls.

7.10 A Sub-elliptic Diffusion

This section serves as an introduction to the subsequent section, focusing on a sub-elliptic diffusion on the Lie group $SE(2)$. The example presented here illustrates a "slice" of the $SE(2)$ group, analogous to the case of the Grushin distribution, which represents a "slice" of the Heisenberg group.

We consider the following two vector fields

$$X = \sin\theta\, \partial_y, \qquad Y = \partial_\theta \tag{7.10.28}$$

on $\mathbb{R} \times \mathbb{S}^1$. Then Y represents an angular rotation vector, while X is the projection of the direction vector on the y-axis. With notation $Z = \cos\theta\, \partial_y$, we have

$$[X, Y] = -Z, \qquad [X, Z] = 0, \qquad [Y, Z] = X.$$

Then the distribution $\mathcal{D} = \text{span}\{X, Y\}$ is of rank 2 everywhere but at the points $\{\theta = n\pi; n \in \mathbb{Z}\}$, where the rank drops to 1. Since $\{X, Y, Z\}$ are linearly independent at every point, then the bracket-generating condition holds and the operator

$$L = \frac{1}{2}(X^2 + Y^2) = \frac{1}{2}\big((\sin\theta)^2\, \partial_y^2 + \partial_\theta^2\big)$$

is hypoelliptic on $\mathbb{R} \times \mathbb{S}^1$. The sub-Riemannian geometry of the distribution $\mathcal{D}$ has been studied in Calin and Chang [26], pages 288-296. It is shown there that the aforementioned distribution is not geodesically complete, in the sense that there are points, such as $(0, 0)$ and $(n\pi, y)$, $n \geq 1$, which cannot be connected by a normal geodesic (i.e. a projection of the solution of the Hamilton system on the coordinate space, $\mathbb{R} \times \mathbb{S}^1$).

The goal of studying the Hamiltonian system associated to a bracket-generating distribution is to obtain closed form expressions for the normal geodesics. It is accepted that locally the heat flows mainly along normal geodesics, so the heat kernel expression of the operator L depends on the associated Carnot-Carathéodory distance, which is the length of the normal geodesic, see Leandre [97, 96]. This behavior holds globally for the Heisenberg group, but in the present case fails, due to geodesical incompleteness. This makes the problem of finding the heat kernel more challenging. We shall show that a closed form expression for the heat kernel can be found in terms of Mathieu's functions of first kind. We will shift to the equivalent problem of finding the transition probability for the associated diffusion.

The associated nonholonomic diffusion The diffusion (X_t, T_t) on $\mathbb{R} \times \mathbb{S}^1$ associated to the hypoelliptic operator L is given by

$$\begin{aligned} dX_t &= \sin T_t \, dW_t \\ dT_t &= dB_t, \end{aligned}$$

with W_t standard Brownian motion on $\mathbb{R}$ and B_t Brownian motion on the circle $\mathbb{S}^1$, independent of W_t. Integrating yields

$$\begin{aligned} X_t &= x_0 + \int_0^t \sin T_u \, dW_u \\ T_t &= \theta_0 + B_t, \qquad t \geq 0. \end{aligned}$$

By Proposition 3.6.3, its transition probability

$$p_t(x_0, \theta_0; x, \theta) d\theta dx = P(X_t \in dx, dT_t \in d\theta | X_0 = x_0, T_0 = \theta_0)$$

is C^∞-smooth. Since T_t is a Brownian motion on the circle $\mathbb{S}^1$, then the following periodicity condition is also required

$$p_t(x_0, \theta_0; x, \theta) = p_t(x_0, \theta_0; x, \theta + 2\pi).$$

We shall denote by $B[0,t] = \{B_s; 0 \le s \le t\}$ the history of the Brownian motion B_s until time t.

The following conditional expectation will be useful shortly. Since the quadratic variation of $\int_0^t \sin T_u\, dW_u$ is

$$Q_t = \left\langle \int_0^t \sin T_u\, dW_u \right\rangle_t = \int_0^t \sin^2 T_u\, du,$$

by Theorem 1.1.11, there is a DDS Brownian motion, γ_t, such that we have $\int_0^t \sin T_u\, dW_u = \beta_{Q_t}$. Then

$$\mathbb{E}[e^{i\lambda\gamma_{Q_t}} | B[0,t]] = e^{-\frac{1}{2}\lambda^2 Q_t}. \tag{7.10.29}$$

We shall compute next the characteristic function of diffusion (X_t, T_t)

$$\begin{aligned}
\Phi_{X_t,T_t}(\lambda,\eta) &= \mathbb{E}[e^{i\lambda X_t} e^{i\eta T_t}] = \mathbb{E}[\mathbb{E}[e^{i\lambda X_t} e^{i\eta T_t} | B[0,t]]] \\
&= \mathbb{E}[e^{i\eta T_t}\mathbb{E}[e^{i\lambda X_t}|B[0,t]]] = e^{i\lambda x_0}\mathbb{E}[e^{i\eta T_t}\mathbb{E}[e^{i\lambda \int_0^t \sin T_u\, dW_u}|B[0,t]]] \\
&= e^{i\lambda x_0}\mathbb{E}[e^{i\eta T_t}\mathbb{E}[e^{i\lambda\gamma_{Q_t}}|B[0,t]]] = e^{i\lambda x_0}\mathbb{E}[e^{i\eta T_t} e^{-\frac{1}{2}\lambda^2\int_0^t \sin^2 T_u\, du}] \\
&= e^{i\lambda x_0}\Big(\mathbb{E}[\cos(\eta T_t) e^{-\frac{1}{2}\lambda^2\int_0^t \sin^2 T_u\, du}] + i\mathbb{E}[\sin(\eta T_t) e^{-\frac{1}{2}\lambda^2\int_0^t \sin^2 T_u\, du}]\Big),
\end{aligned}$$

where we used the properties of the conditional expectations, formula (7.10.29) and Euler's formula.

Now, we note that for $\lambda, \eta \in \mathbb{R}$ fixed, the previous expression can be written as

$$\Phi_{X_t,T_t}(\lambda,\eta) = e^{i\lambda x_0}(u(t,\theta_0) + i\, v(t,\theta_0)), \tag{7.10.30}$$

where

$$\begin{aligned}
u(t,\theta) &= \mathbb{E}^\theta[\cos(\eta T_t) e^{-\frac{1}{2}\lambda^2\int_0^t \sin^2(T_u)\, du}] = \mathbb{E}[\cos(\eta(\theta + B_u)) e^{-\frac{1}{2}\lambda^2\int_0^t \sin^2(\theta + B_u)\, du}] \\
v(t,\theta) &= \mathbb{E}^\theta[\sin(\eta T_t) e^{-\frac{1}{2}\lambda^2\int_0^t \sin^2(T_u)\, du}] = \mathbb{E}[\sin(\eta(\theta + B_u)) e^{-\frac{1}{2}\lambda^2\int_0^t \sin^2(\theta + B_u)\, du}].
\end{aligned}$$

By Feynman-Kac's formula, $u(t,\theta)$ and $v(t,\theta)$ solve the following initial condition heat equations, subject to periodicity conditions

$$\frac{\partial u}{\partial t} = \frac{1}{2}\frac{\partial^2 u}{\partial \theta^2} - U(\theta) u \tag{7.10.31}$$

$$u(0,\theta) = \cos(\eta\theta), \tag{7.10.32}$$

$$u(t,\theta + 2\pi) = u(t,\theta), \tag{7.10.33}$$

and

$$\frac{\partial v}{\partial t} = \frac{1}{2}\frac{\partial^2 v}{\partial \theta^2} - U(\theta)v \tag{7.10.34}$$

$$v(0,\theta) = \sin(\eta\theta), \tag{7.10.35}$$

$$v(t,\theta+2\pi) = u(t,\theta), \tag{7.10.36}$$

with the potential function

$$U(\theta) = \frac{1}{2}\lambda^2 \sin^2\theta = \frac{\lambda^2}{4}(1-\cos 2\theta). \tag{7.10.37}$$

Both previous equations can be solved similarly. We shall start with the first one.

The equation (7.10.31) will be solved using the separation of variables method subject to the periodicity condition (7.10.33) as follows. We assume $u(t,\theta) = f(t)g(\theta)$, with $f(t)$ and $g(\theta)$ differentiable functions. Substituting in (7.10.31) and dividing by $f(t)g(\theta)$, we obtain

$$\frac{f'(t)}{f(t)} = \frac{1}{2}\frac{g''(\theta)}{g(\theta)} - \frac{\lambda^2}{4}(1-\cos 2\theta) = \frac{E}{2},$$

where E is a separation constant, subject to be determined as an eigenvalue from the condition (7.10.33). The previous expression leads to two differential equations. The first one, $\dfrac{f'(t)}{f(t)} = \dfrac{E}{2}$, has the exponential solution $f(t) = f(0)e^{\frac{Et}{2}}$. The second equation can be written as

$$g''(\theta) + (a - 2q\cos 2\theta)g(\theta) = 0, \tag{7.10.38}$$

where

$$a = -\frac{\lambda^2}{2} - E, \qquad q = -\frac{\lambda^2}{4}.$$

The relation (7.10.38) is known under the name of *Mathieu's equation*, which originally emerged while studying vibrations on an elliptical drumhead [105]. The amplitude coefficient q depends on the input variable λ, while the characteristic coefficient, a, depends on the energy constant E, which is quantified as an eigenvalue depending on the periodicity condition $g(\theta+2\pi) = g(\theta)$.

If q is considered fixed and a is treated as the eigenvalue, then the Mathieu equation is of Sturm-Liouville form. Thus, there are two sequences of eigenvalues, $a_{2n+1}(q)$ and $b_{2n+1}(q)$ with the corresponding eigenfunctions $\mathrm{ce}_{2n+1}(\theta,q)$ and $\mathrm{se}_{2n+1}(\theta,q)$, which are called *Mathieu functions of the first kind* of odd order.[1] It can be shown that since $q < 0$, these eigenvalues are ordered increasingly

$$a_1(q) < b_1(q) < a_3(q) < b_3(q) < \dots$$

The relation with the separation constant E is given by

$$E_{2n+1} = -\frac{\lambda^2}{2} - a_{2n+1}(q), \quad E'_{2n+1} = -\frac{\lambda^2}{2} - b_{2n+1}(q).$$

Since only the odd order Mathieu functions are 2π-periodic, then the general solution $u(t,\theta)$ of (7.10.31) can be expressed as a superposition of products $f_{2n+1}(t)g_{2n+1}(\theta)$ over the two sequences of eigenvalues as[2]

$$\begin{aligned} u(t,\theta) &= e^{-\frac{\lambda^2 t}{4}}\Big[\sum_{n\geq 0} \alpha_{2n+1} e^{-\frac{t}{2}a_{2n+1}(q)} \mathrm{ce}_{2n+1}(\theta,q) \\ &\quad + \sum_{n\geq 0} \beta_{2n+1} e^{-\frac{t}{2}b_{2n+1}(q)} \mathrm{se}_{2n+1}(\theta,q)\Big], \end{aligned} \tag{7.10.39}$$

with $\alpha_{2n+1}, \beta_{2n+1} \in \mathbb{R}$. The involved Mathieu functions can be expanded as a series

$$\mathrm{ce}_{2n+1}(\theta,q) = \sum_{r\geq 0} A_{2r+1}^{(2n+1)}(q)\cos[(2r+1)\theta]$$
$$\mathrm{se}_{2n+1}(\theta,q) = \sum_{r\geq 0} B_{2r+1}^{(2n+1)}(q)\sin[(2r+1)\theta],$$

where the coefficients $A_{2r+1}^{(2n+1)}$ and $B_{2r+1}^{(2n+1)}$ obey a three-term recurrence relations in the lower index as follows:

$$(a-1-q)A_1 = qA_3$$
$$[a-(2r+1)^2]A_{2r+1} = q(A_{2r-1} + A_{2r+3}),$$

where $r = 1,2,3,\dots$, $a = a_{2n+1}(q)$, $A_{2r+1} = A_{2r+1}^{(2n+1)}$, and

$$(a-1-q)B_1 = qB_3$$
$$[a-(2r+1)^2]B_{2r+1} = q(B_{2r-1} + B_{2r+3}),$$

where $r = 1,2,3,\dots$, $a = b_{2n+1}(q)$, $B_{2r+1} = B_{2r+1}^{(2n+1)}$.

[1]ce and se stand for cosine-elliptic and sine-elliptic functions.

[2]A great deal of information about Mathieu's functions can be found at https://dlmf.nist.gov/28.2

To accommodate for the initial condition (7.10.32), we let $t = 0$ in the expression (7.10.39) and obtain

$$\sum_{n\geq 0} \alpha_{2n+1}\mathrm{ce}_{2n+1}(\theta, q) + \sum_{n\geq 0} \beta_{2n+1}\mathrm{se}_{2n+1}(\theta, q) = \cos(\eta\theta).$$

Using the orthogonality conditions of Mathieu functions

$$\int_0^{2\pi} \mathrm{ce}_n(\theta, q)\mathrm{ce}_m(\theta, q)d\theta = \pi\delta_{nm}, \qquad \int_0^{2\pi} \mathrm{se}_n(\theta, q)\mathrm{se}_m(\theta, q)d\theta = \pi\delta_{nm}$$

$$\int_0^{2\pi} \mathrm{ce}_n(\theta, q)\mathrm{se}_m(\theta, q)d\theta = 0,$$

we can compute the coefficients as follows

$$\begin{aligned}
\alpha_{2n+1} &= \frac{1}{\pi}\int_0^{2\pi} \cos(\eta\theta)\mathrm{ce}_{2n+1}(\theta, q)\, d\theta \\
&= \frac{1}{\pi}\sum_{r\geq 0} A_{2r+1}^{(2n+1)}(q)\int_0^{2\pi} \cos[(2r+1)\theta]\cos(\eta\theta)\, d\theta \\
&= \frac{1}{2\pi}\sum_{r\geq 0} A_{2r+1}^{(2n+1)}(q)\int_0^{2\pi} \{\cos[(2r+1+\eta)\theta] + \cos[(2r+1-\eta)\theta]\}\, d\theta \\
&= \frac{1}{2\pi}\sum_{r\geq 0} A_{2r+1}^{(2n+1)}(q)\Big[\frac{\sin[(2\pi(2r+1+\eta))]}{2r+1+\eta} + \frac{\sin[2\pi(2r+1-\eta)]}{2r+1-\eta}\Big],
\end{aligned} \tag{7.10.40}$$

$$\begin{aligned}
\beta_{2n+1} &= \frac{1}{\pi}\int_0^{2\pi} \cos(\eta\theta)\mathrm{se}_{2n+1}(\theta, q)\, d\theta \\
&= \frac{1}{\pi}\sum_{r\geq 0} B_{2r+1}^{(2n+1)}(q)\int_0^{2\pi} \sin[(2r+1)\theta]\cos(\eta\theta)\, d\theta \\
&= \frac{1}{2\pi}\sum_{r\geq 0} B_{2r+1}^{(2n+1)}(q)\int_0^{2\pi} \{\sin[(2r+1+\eta)\theta] + \sin[(2r+1-\eta)\theta]\}\, d\theta \\
&= \frac{1}{2\pi}\sum_{r\geq 0} B_{2r+1}^{(2n+1)}(q)\Big[\frac{1-\cos[(2\pi(2r+1+\eta))]}{2r+1+\eta} + \frac{1-\cos[2\pi(2r+1-\eta)]}{2r+1-\eta}\Big].
\end{aligned} \tag{7.10.41}$$

It is worth noting that if $\eta \in \mathbb{Z}$, then $\alpha_{2n+1} = \beta_{2n+1} = 0$.

To conclude the previous computation, the expression of $u(t, \theta)$ given by the series (7.10.39), where the coefficients are given by (7.10.40)-(7.10.41), solves the system (7.10.31)-(7.10.33).

To find the function $v(t,\theta)$ that satisfies the problem (7.10.34)-(7.10.36) we proceed similarly and obtain a relation analog to (7.10.39)

$$
\begin{aligned}
v(t,\theta) &= e^{-\frac{\lambda^2}{4}t}\Big[\sum_{n\geq 0}\tilde{\alpha}_{2n+1}e^{-\frac{t}{2}a_{2n+1}(q)}\mathrm{ce}_{2n+1}(\theta,q) \\
&\quad +\sum_{n\geq 0}\tilde{\beta}_{2n+1}e^{-\frac{t}{2}b_{2n+1}(q)}\mathrm{se}_{2n+1}(\theta,q)\Big], \qquad (7.10.42)
\end{aligned}
$$

The coefficients can be identified from the boundary condition (7.10.35), which writes as

$$
\sum_{n\geq 0}\tilde{\alpha}_{2n+1}\mathrm{ce}_{2n+1}(\theta,q)+\sum_{n\geq 0}\tilde{\beta}_{2n+1}\mathrm{se}_{2n+1}(\theta,q)=\sin(\eta\theta).
$$

Using the orthonormality, we obtain

$$
\tilde{\alpha}_{2n+1}=\frac{1}{2\pi}\sum_{r\geq 0}A_{2r+1}^{(2n+1)}(q)\Big[\frac{1-\cos[(2\pi(2r+1+\eta))]}{2r+1+\eta}-\frac{1-\cos[2\pi(2r+1-\eta)]}{2r+1-\eta}\Big], \qquad (7.10.43)
$$

$$
\tilde{\beta}_{2n+1}=\frac{1}{2\pi}\sum_{r\geq 0}B_{2r+1}^{(2n+1)}(q)\Big[\frac{\sin[(2\pi(2r+1-\eta))]}{2r+1-\eta}-\frac{\sin[2\pi(2r+1+\eta)]}{2r+1+\eta}\Big]. \qquad (7.10.44)
$$

It is worth noting that the eigenvalues $a_{2n+1}(q)$ and $b_{2n+1}(q)$ depend only on λ, since $q=-\lambda^2/4$. The coefficients α_{2n+1}, β_{2n+1}, $\tilde{\alpha}_{2n+1}$, $\tilde{\beta}_{2n+1}$ depend on both λ and η. These dependencies are quite intricate and we shall not attempt to write an explicit formula for them.

Consider now the complex numbers $\alpha^c_{2n+1}=\alpha_{2n+1}+i\tilde{\alpha}_{2n+1}$ and $\beta^c_{2n+1}=\beta_{2n+1}+i\tilde{\beta}_{2n+1}$. Then, substituting in (7.10.30), the characteristic function of (X_t,T_t) becomes

$$
\begin{aligned}
\Phi_{X_t,T_t}(\lambda,\eta) &= e^{i\lambda x_0}(u(t,\theta_0)+iv(t,\theta_0)) \\
&= e^{i\lambda x_0}e^{-\frac{\lambda^2}{4}t}\Big(\sum_{n\geq 0}\alpha^c_{2n+1}(\lambda,\eta)e^{-\frac{t}{2}a_{2n+1}(\lambda)}\mathrm{ce}_{2n+1}(\theta_0,\lambda) \\
&\quad +\sum_{n\geq 0}\beta^c_{2n+1}(\lambda,\eta)e^{-\frac{t}{2}b_{2n+1}(\lambda)}\mathrm{se}_{2n+1}(\theta_0,\lambda)\Big), \qquad (7.10.45)
\end{aligned}
$$

where we indicated in parentheses the dependencies of λ and η.

Inverting the double integral

$$
\Phi_{X_t,T_t}(\lambda,\eta)=\mathbb{E}[e^{i\lambda X_t}e^{i\eta T_t}]=\iint_{\mathbb{R}^2}e^{i\lambda x}e^{i\eta\theta}p_t(x_0,\theta_0;x,\theta)\,dxd\theta
$$

and using (7.10.45), we obtain the transition density of (X_t, T_t)

$$
\begin{aligned}
p_t(x_0,\theta_0;x,\theta) &= \frac{1}{(2\pi)^2}\iint_{\mathbb{R}^2} e^{-i\lambda x}e^{-i\eta\theta}\Phi_{X_t,T_t}(\lambda,\eta)\,d\lambda d\eta \\
&= \frac{1}{(2\pi)^2}\int_{\mathbb{R}} e^{-i\lambda(x-x_0)}\sum_{n\geq 0} e^{-t(\frac{\lambda^2}{4}+\frac{1}{2}a_{2n+1}(\lambda))}\mathrm{ce}_{2n+1}(\theta_0,\lambda)\phi_{2n+1}(\theta,\lambda)d\lambda \\
&\quad+\frac{1}{(2\pi)^2}\int_{\mathbb{R}} e^{-i\lambda(x-x_0)}\sum_{n\geq 0} e^{-t(\frac{\lambda^2}{4}+\frac{1}{2}b_{2n+1}(\lambda))}\mathrm{se}_{2n+1}(\theta_0,\lambda)\psi_{2n+1}(\theta,\lambda)d\lambda,
\end{aligned}
$$

with

$$
\begin{aligned}
\phi_{2n+1}(\theta,\lambda) &= \int_{\mathbb{R}} e^{-i\eta\theta}\alpha^c_{2n+1}(\lambda,\eta)\,d\eta \\
\psi_{2n+1}(\theta,\lambda) &= \int_{\mathbb{R}} e^{-i\eta\theta}\beta^c_{2n+1}(\lambda,\eta)\,d\eta.
\end{aligned}
$$

The next section is built on the experience gained from the study of the vector fields (7.10.28).

7.11 The Knife Edge Diffusion

The Lie Group $SE(2)$ Let $M=\mathbb{R}^2\times\mathbb{S}^1$ and consider on M the following composition law

$$
(x,e^{i\theta})\circ(x',e^{i\theta'}) = (x+\mathcal{R}(\theta)x', e^{i(\theta+\theta')}),\qquad x,x'\in\mathbb{R}^2,\theta,\theta'\in\mathbb{R}
$$

where

$$
\mathcal{R}(\theta)=\begin{pmatrix}\cos\theta & -\sin\theta\\ \sin\theta & \cos\theta\end{pmatrix}
$$

denotes the rotation of angle θ. This is a group law on M with the inverse given by

$$
(x,e^{i\theta})^{-1} = \big(-\mathcal{R}^{-1}(\theta)x, e^{-i\theta}\big),
$$

and the neutral element $e=(0_{\mathbb{R}^2},1)$. Therefore, the pair $SE(2)=(M,\circ)$ forms a non-commutative Lie group of dimension 3, called the *2D Euclidean motion group*. Each element of this group represents a position $x\in\mathbb{R}^2$ and an orientation of angle θ, and thus can be used to parameterize nonholonomic systems provided by the motions of a knife edge, a skate, or a boat. Its application to image processing, such as contour enhancement or contour completion can be found in Mumford [112], Citti et al. [47] and Duits and van Almsick [52].

The associated Lie algebra is generated by 3 left invariant vector fields, $\mathfrak{se}(2) = \text{span}\{X_1, X_2, X_3\}$, where

$$X_1 = \cos\theta\,\partial_{x_1} + \sin\theta\,\partial_{x_2}, \quad X_2 = \partial_\theta, \quad X_3 = \sin\theta\,\partial_{x_1} - \cos\theta\,\partial_{x_2}. \quad (7.11.46)$$

The commutation relations are given by the following table:

$[\ ,\ ]$	X_1	X_2	X_3
X_1	0	X_3	0
X_2	$-X_3$	0	X_1
X_3	0	$-X_1$	0

We are interested in the distribution $\mathcal{D} = \text{span}\{\mathrm{X}_1, \mathrm{X}_2\}$. This is not involutive, and hence not integrable. Since $\{X_1, X_2, X_3\}$ forms a basis in the tangent space T_gM for any $g = (x, e^{i\theta}) \in M$, it follows that the distribution $\mathcal{D}$ is bracket-generating. However, the distribution $\mathcal{D}$ is not nilpotent, since

$$[\dots[[[X_1, X_2], X_2], X_2], \dots, X_2] \neq 0.$$

This is the same distribution as the one introduced to study the nonholonomy of the knife edge, see page 153.

The associated nonholonomic diffusion The sum of squares operator

$$\begin{aligned} L &= \frac{1}{2}(X_1^2 + X_2^2) = \frac{1}{2}(\cos\theta\,\partial_{x_1} + \sin\theta\,\partial_{x_2})^2 + \frac{1}{2}\partial_\theta^2 \qquad (7.11.47)\\ &= \frac{1}{2}\Big((\cos\theta)^2\partial_{x_1}^2 + (\sin\theta)^2\partial_{x_2}^2 + 2\sin\theta\cos\theta\partial_{x_1}\partial_{x_2} + \partial_\theta^2\Big) \end{aligned}$$

is hypoelliptic by Hörmander's theorem. The sub-Riemannian metric (in contravariant form) can be collected from the coefficients of the principal symbol as

$$g^{ij} = (\sigma\sigma^T)_{ij} = \begin{pmatrix} \cos^2\theta & \sin\theta\cos\theta & 0 \\ \sin\theta\cos\theta & \sin^2\theta & 0 \\ 0 & 0 & 1 \end{pmatrix},$$

and has the significance of diffusion metric. The associated dispersion matrix takes the standard form

$$\sigma = \begin{pmatrix} \cos\theta & 0 & 0 \\ \sin\theta & 0 & 0 \\ 0 & 0 & 1 \end{pmatrix}.$$

Therefore, the associated nonholonomic diffusion (X_t, Y_t, T_t) satisfies

$$\begin{aligned} dX_t &= \cos T_t\,dW_t \\ dY_t &= \sin T_t\,dW_t \\ dT_t &= dB_t, \end{aligned}$$

with W_t standard Brownian motion on $\mathbb{R}$ and B_t Brownian motion on $\mathbb{S}^1$, independent of W_t. Due to the Lie group law, it suffices to assume that the diffusion starts at $X_0 = Y_0 = T_0 = 0$. Integrating yields

$$\begin{aligned} X_t &= \int_0^t \cos B_u \, dW_u \\ Y_t &= \int_0^t \sin B_u \, dW_u \\ T_t &= B_t, \qquad t \geq 0. \end{aligned}$$

Since

$$\begin{aligned} (dX_t, dY_t, dT_t) &= (\cos T_t, \sin T_t, 0) dW_t + (0, 0, 1) dB_t \\ &= X_1 dW_t + X_2 dB_t, \end{aligned}$$

it follows that (X_t, Y_t, T_t) is a horizontal diffusion. Therefore, this diffusion models the knife edge when the controls are independent Brownian motions. Its transition probability is C^∞-smooth, see Proposition 3.6.3. The closed form expression of the transition density is provided in Agrachev et al. [3] as an integral involving series containing Mathieu cosines and sines. In the following we shall present our approach towards finding the transition density of diffusion (X_t, Y_t, T_t) using a similar method with the one employed in section 7.10.

We consider $\|\lambda\| = \sqrt{\lambda_1^2 + \lambda_2^2}$ and $B[0,t] = \{B_s; 0 \leq s \leq t\}$. Then the characteristic function of the diffusion is given by

$$\begin{aligned} \Phi_{X_t,Y_t,T_t}(\lambda_1, \lambda_2, \eta) &= \mathbb{E}[e^{i\lambda_1 X_t} e^{i\lambda_2 Y_t} e^{i\eta T_t}] = \mathbb{E}[\mathbb{E}[e^{i\lambda_1 X_t} e^{i\lambda_2 Y_t} e^{i\eta T_t} | B[0,t]]] \\ &= \mathbb{E}[e^{i\eta T_t} \mathbb{E}[e^{i\lambda_1 X_t} e^{i\lambda_2 Y_t} | B[0,t]]] \\ &= \mathbb{E}[e^{i\eta T_t} \mathbb{E}[e^{i\int_0^t (\lambda_1 \cos T_s + \lambda_2 \sin T_s) dW_s} | B[0,t]]] \\ &= \mathbb{E}[e^{i\eta T_t} \mathbb{E}[e^{i\|\lambda\| \int_0^t \sin(T_s + \varphi) dW_s} | B[0,t]]], \end{aligned}$$

where $\varphi = \tan^{-1}(\lambda_1/\lambda_2)$. By Theorem 1.1.11 there is a DDS Brownian motion, γ_t, such that

$$\int_0^t \sin(T_s + \varphi) dW_s = \gamma_{\int_0^t \sin^2(B_u + \varphi) du},$$

fact that simplifies the computation of the characteristic function as

$$\begin{aligned} \Phi_{X_t,Y_t,T_t}(\lambda_1, \lambda_2, \eta) &= \mathbb{E}[e^{i\eta T_t} \mathbb{E}[e^{i\|\lambda\| \gamma_{\int_0^t \sin^2(B_u+\varphi) du}} | B[0,t]]] \\ &= \mathbb{E}[e^{i\eta B_t} e^{-\frac{1}{2}\|\lambda\|^2 \int_0^t \sin^2(B_u + \varphi) du}] \\ &= u(t,0) + iv(t,0), \end{aligned}$$

where

$$u(t,\theta) = \mathbb{E}[\cos(\eta(B_t+\theta))e^{-\frac{1}{2}\|\lambda\|^2\int_0^t \sin^2(B_u+\theta+\varphi)du}]$$
$$v(t,\theta) = \mathbb{E}[\sin(\eta(B_t+\theta))e^{-\frac{1}{2}\|\lambda\|^2\int_0^t \sin^2(B_u+\theta+\varphi)du}].$$

By the Feynman-Kac formula the aforementioned two expectations satisfy the following initial value heat equations

$$\frac{\partial u}{\partial t} = \frac{1}{2}\frac{\partial^2 u}{\partial \theta^2} - U(\theta)u \tag{7.11.48}$$
$$u(0,\theta) = \cos(\eta\theta) \tag{7.11.49}$$

and

$$\frac{\partial v}{\partial t} = \frac{1}{2}\frac{\partial^2 v}{\partial \theta^2} - U(\theta)v \tag{7.11.50}$$
$$v(0,\theta) = \sin(\eta\theta), \tag{7.11.51}$$

where $U(\theta) = \frac{1}{2}\|\lambda\|^2\sin^2(\theta+\varphi)$. Both problems are subject to the periodicity constraints $u(t,\theta+2\pi) = u(t,\theta)$ and $v(t,\theta+2\pi) = v(t,\theta)$.

Since in these problems λ_1, λ_2 and η are kept fixed, then φ is regarded as a constant. With the translation $\theta \to \theta+\varphi$ the potential $U(\theta)$ becomes similar with the one given by (7.11.52), i.e.

$$U(\theta) = \frac{1}{2}\|\lambda\|^2\sin^2\theta = \frac{\|\lambda\|^2}{4}(1-\cos 2\theta). \tag{7.11.52}$$

Then we can solve the problems (7.11.48)-(7.11.51) as in section 7.10, obtaining solutions $\tilde{u}(t,\theta)$ and $\tilde{v}(t,\theta)$ given by (7.10.39) and (7.10.42), where λ is replaced by $\|\lambda\|$

$$\begin{aligned}\tilde{u}(t,\theta) = e^{-\frac{\|\lambda\|^2 t}{4}}\Big[&\sum_{n\geq 0}\alpha_{2n+1}e^{-\frac{t}{2}a_{2n+1}(q)}\mathrm{ce}_{2n+1}(\theta,q)\\ &+\sum_{n\geq 0}\beta_{2n+1}e^{-\frac{t}{2}b_{2n+1}(q)}\mathrm{se}_{2n+1}(\theta,q)\Big],\end{aligned} \tag{7.11.53}$$

and

$$\begin{aligned}\tilde{v}(t,\theta) = e^{-\frac{\|\lambda\|^2}{4}t}\Big[&\sum_{n\geq 0}\tilde{\alpha}_{2n+1}e^{-\frac{t}{2}a_{2n+1}(q)}\mathrm{ce}_{2n+1}(\theta,q)\\ &+\sum_{n\geq 0}\tilde{\beta}_{2n+1}e^{-\frac{t}{2}b_{2n+1}(q)}\mathrm{se}_{2n+1}(\theta,q)\Big],\end{aligned} \tag{7.11.54}$$

The solutions of problems (7.11.48)-(7.11.51) are obtained from $\tilde{u}$ and $\tilde{v}$ after applying the inverse translation

$$u(t,\theta) = \tilde{u}(t,\theta-\varphi), \quad v(t,\theta) = \tilde{v}(t,\theta-\varphi).$$

Substituting back in the formula of the characteristic function obtained previously

$$\begin{aligned}\Phi_{X_t,Y_t,T_t}(\lambda_1,\lambda_2,\eta) &= u(t,0) + iv(t,0)\\ &= \tilde{u}(t,-\varphi) + i\,\tilde{v}(t,-\varphi),\end{aligned}$$

whose expression can be obtained by substituting θ by $-\varphi$ in the relations (7.11.53) and (7.11.54).

Taking into consideration that $x_0 = 0$ and $\theta_0 = 0$, and denoting by $p_t(x,y,\theta)$ the probability density of (X_t,Y_t,T_t), we invert the triple integral

$$\Phi_{X_t,Y_t,T_t}(\lambda_1,\lambda_2,\eta) = \iiint_{\mathbb{R}^3} e^{i\lambda_1 x} e^{i\lambda_2 y} e^{i\eta\theta} p_t(x,y,\theta)\,dxdyd\theta$$

to obtain

$$\begin{aligned}p_t(x,y,\theta) &= \frac{1}{(2\pi)^3}\iiint_{\mathbb{R}^2} e^{-i\lambda_1 x} e^{-i\lambda_2 y} e^{-i\eta\theta}\Phi_{X_t,Y_t,T_t}(\lambda,\eta)\,d\lambda_1 d\lambda_2 d\eta\\ &= \frac{1}{(2\pi)^3}\iint_{\mathbb{R}^2} e^{-i(\lambda_1 x+\lambda_2 y)}\sum_{n\ge 0} e^{-t(\frac{\|\lambda\|^2}{4}+\frac{1}{2}a_{2n+1}(\lambda))}\mathrm{ce}_{2n+1}(-\varphi,\lambda)\phi_{2n+1}(\theta,\lambda)d\lambda_1 d\lambda_2\\ &+ \frac{1}{(2\pi)^3}\iint_{\mathbb{R}^2} e^{-i(\lambda_1 x+\lambda_2 y)}\sum_{n\ge 0} e^{-t(\frac{\|\lambda\|^2}{4}+\frac{1}{2}b_{2n+1}(\lambda))}\mathrm{se}_{2n+1}(-\varphi,\lambda)\psi_{2n+1}(\theta,\lambda)d\lambda_1 d\lambda_2,\end{aligned} \tag{7.11.55}$$

with

$$\begin{aligned}\phi_{2n+1}(\theta,\lambda) &= \int_{\mathbb{R}} e^{-i\eta\theta}\alpha^c_{2n+1}(\lambda,\eta)\,d\eta\\ \psi_{2n+1}(\theta,\lambda) &= \int_{\mathbb{R}} e^{-i\eta\theta}\beta^c_{2n+1}(\lambda,\eta)\,d\eta.\end{aligned}$$

This represents also the heat kernel for the sum of squares operator (7.11.47) at $(x_0,y_0,\theta_0) = (0,0,0)$.

Remark 7.11.1 The Brownian motion T_t on the unit circle $\mathbb{S}^1$, starting at $T_0 = 0$, has the probability density function given by a series of Gaussians that sum up the contributions of all integer rotations about the circle

$$p_t(\theta) = \frac{1}{\sqrt{2\pi t}}\sum_{n\in\mathbb{Z}} e^{-\frac{(\theta+2n\pi)^2}{2t}}, \qquad t\ge 0, \quad \theta\in\mathbb{R},$$

see, for instance, Calin et al. [29], page 182. Then the characteristic function of T_t can be computed as

$$\mathbb{E}[e^{i\lambda T_t}] = \int_{\mathbb{R}} e^{i\lambda\theta} p_t(\theta)\,d\theta = \frac{1}{\sqrt{2\pi t}} \sum_{n\in\mathbb{Z}} \int_{\mathbb{R}} e^{i\lambda\theta} e^{-\frac{(\theta+2n\pi)^2}{2t}}\,d\theta.$$

Employing the substitution $\psi = (\theta + 2n\pi)/\sqrt{2t}$, the characteristic function can be further simplified into the form

$$\begin{aligned}
\mathbb{E}[e^{i\lambda T_t}] &= \frac{\sqrt{2t}}{\sqrt{2\pi t}} \sum_{n\in\mathbb{Z}} \int_{\mathbb{R}} e^{i\lambda(\sqrt{2t}\psi - 2n\pi) - \psi^2}\,d\psi \\
&= \frac{1}{\sqrt{\pi}} \sum_{n\in\mathbb{Z}} e^{-2n\pi\lambda i} \int_{\mathbb{R}} e^{-\psi^2 + i\lambda\sqrt{2t}\psi}\,d\psi \\
&= e^{-\frac{\lambda^2 t}{2}} \sum_{n\in\mathbb{Z}} e^{-2n\pi\lambda i} = e^{-\frac{\lambda^2 t}{2}}\Big(1 + 2\sum_{n\geq 1} \cos(2n\pi\lambda)\Big),
\end{aligned}$$

where we used the improper integral

$$\int_{\mathbb{R}} e^{-a\psi^2 + i\xi\psi}\,d\psi = \sqrt{\frac{\pi}{a}}\,e^{-\frac{\xi^2}{4a}}, \qquad a > 0$$

and the fact that sine is an odd function.

7.12 Nonholonomic Hyperbolic Diffusion

In this section we shall introduce a hyperbolic variant of the knife edge nonholonomic system and we shall compute the transition density of the associated diffusion. We introduce first the underling Lie group law.

Hyperbolic rotations A *hyperbolic plane rotation* of angle $\varphi \in \mathbb{R}$ is given by the matrix

$$h(\varphi) = \begin{pmatrix} \cosh\varphi & \sinh\varphi \\ \sinh\varphi & \cosh\varphi \end{pmatrix}.$$

Since $\det h(\varphi) = \cos^2\varphi - \sinh^2\varphi = 1$, then $h(\varphi)$ is invertible, with $h(\varphi)^{-1} = h(-\varphi)$. Since $h(\varphi_1)h(\varphi_2) = h(\varphi_1 + \varphi_2)$, the set of hyperbolic plane rotation forms a group with respect to the multiplication of matrices, having the unit element $\mathbb{I}_2$. It is worth noting that a hyperbolic rotation slides a hyperbola $\{x = \cosh\theta, y = \sin\theta; \theta \in \mathbb{R}\}$ into itself, via formula

$$\begin{pmatrix} \cosh\varphi & \sinh\varphi \\ \sinh\varphi & \cosh\varphi \end{pmatrix}\begin{pmatrix} \cosh\theta \\ \sinh\theta \end{pmatrix} = \begin{pmatrix} \cosh(\varphi+\theta) \\ \sinh(\varphi+\theta) \end{pmatrix},$$

which is a property similar to the one of an usual rotation that rotates a circle into itself.

The Lie group $SH(2)$ We define the following noncommutative group law on $\mathbb{R}^3 = \mathbb{R}^2_x \times \mathbb{R}_\theta$ by

$$(x, \theta) \circ (x', \theta') = (x + h(\theta)x', \theta + \theta'), \quad \forall x, x' \in \mathbb{R}^2, \theta, \theta' \in \mathbb{R} \tag{7.12.56}$$

with the inverse $(x, \theta)^{-1} = (-h(-\theta)x, -\theta)$ and neutral element $(0_{\mathbb{R}^2}, 0)$. This is a Lie group that represents the group of hyperbolic motions in the plane.

In order to find the left invariant vector fields and the associated Lie algebra, we consider the left translation $L_a : SH(2) \to SH(2)$, $L_a g = ag$, for all $g = (x_1, x_2, \theta) \in SH(2)$. If $a = (a_1, a_2, \alpha)$, then

$$L_a g = (a_1 + x_1 \cosh \alpha + x_2 \sinh \alpha,\ a_2 + x_1 \sinh \alpha + x_2 \cosh \alpha,\ \alpha + \theta).$$

The left invariant vector fields are constructed by translating a vector at the origin as $X_a = (L_a)_* X_e$. Using

$$(L_a)_* = \begin{pmatrix} \cosh \alpha & \sinh \alpha & 0 \\ \sinh \alpha & \cosh \alpha & 0 \\ 0 & 0 & 1 \end{pmatrix}$$

and considering $X_e = \xi_1 \partial_{a_1} + \xi_2 \partial_{a_2} + \xi_3 \partial_\alpha$, we obtain

$$\begin{aligned} X_a &= (L_a)_* X_e \\ &= \xi_1(\cosh \alpha\, \partial_{a_1} + \sinh \alpha\, \partial_{a_2}) + \xi_2(\sinh \alpha\, \partial_{a_1} + \cosh \alpha\, \partial_{a_2}) + \xi_3 \partial_\alpha \\ &= \xi_1 X_{1|a} + \xi_2 X_{2|a} + \xi_3 X_{3|a}, \end{aligned}$$

so the left invariant vector field X is a linear combination of the left invariant vector fields X_1, X_2 and X_3. Hence, we arrived at the following result.

Proposition 7.12.1 *The vector fields*

$$X = \cosh \theta\, \partial_{x_1} + \sinh \theta\, \partial_{x_2}, \quad Y = \partial_\theta, \quad Z = \sinh \theta\, \partial_{x_1} + \cosh \theta\, \partial_{x_2} \tag{7.12.57}$$

are left invariant on the Lie group $SH(2) = (\mathbb{R}^2_x \times \mathbb{R}_\theta, \circ)$.

The horizontal distribution We consider the rank 2 distribution given by $\mathcal{D} = \text{span}\{X, Y\}$, which is not integrable, since

$$[X, Y] = -Z, \qquad [Y, Z] = X, \qquad [X, Z] = 0.$$

Since $\{X, Y, Z\}$ are linearly independent at every point, the distribution $\mathcal{D}$ is bracket-generating. Then, by Chow-Rashevskii's theorem, the global connectivity by piece-wise horizontal curves holds. Since any smooth curve $\gamma(s) = \big(x_1(s), x_2(s), \theta(s)\big)$ has the velocity given by

$$\dot\gamma(s) = (\dot x \cosh \theta - \dot y \sinh \theta)X + \dot\theta Y + (\dot y \cosh \theta - \dot x \sinh \theta)Z,$$

it follows that the curve $\gamma(s)$ is horizontal if and only if $\dot{y}\cosh\theta = \dot{x}\sinh\theta$, which implies $\theta = \tanh^{-1}\dfrac{\dot{y}}{\dot{x}} = \ln\sqrt{\dfrac{\dot{x}+\dot{y}}{\dot{x}-\dot{y}}}$.

The distribution $\mathcal{D}$ being bracket-generating, the associated subLaplacian

$$L = \frac{1}{2}(\cosh\theta\,\partial_{x_1} + \sinh\theta\partial_{x_2})^2 + \frac{1}{2}\partial_\theta^2$$

is hypoelliptic by Hörmander's theorem, and hence its heat kernel is $\mathcal{C}^\infty$-smooth. Since we have

$$\sigma\sigma^T = \begin{pmatrix} \cosh^2\theta & \sinh\theta\cosh\theta & 0 \\ \sinh\theta\cosh\theta & \sinh^2\theta & 0 \\ 0 & 0 & 1 \end{pmatrix}, \qquad \sigma = \begin{pmatrix} \cosh\theta & 0 & 0 \\ \sinh\theta & 0 & 0 \\ 0 & 0 & 1 \end{pmatrix},$$

then the associated nonholonomic diffusion (X_t, Y_t, T_t) satisfies

$$\begin{aligned} dX_t &= \cosh T_t\, dW_t \\ dY_t &= \sinh T_t\, dW_t \\ dT_t &= dB_t, \end{aligned}$$

with W_t and B_t independent standard Brownian motions on $\mathbb{R}$. Due to the Lie group law invariance, it suffices to assume $X_0 = Y_0 = T_0 = 0$, i.e. the diffusion starts at the origin. Integrating yields

$$\begin{aligned} X_t &= \int_0^t \cosh B_u\, dW_u \\ Y_t &= \int_0^t \sinh B_u\, dW_u \\ T_t &= B_t, \qquad t \geq 0. \end{aligned}$$

Since

$$\begin{aligned} (dX_t, dY_t, dT_t) &= (\cosh T_t, \sinh T_t, 0)dW_t + (0,0,1)dB_t \\ &= XdW_t + YdB_t \in \mathcal{D}_{(X_t,Y_t,T_t)}, \end{aligned}$$

then (X_t, Y_t, T_t) is a horizontal diffusion, which spreads along the horizontal distribution having independent normal components along the vector fields X and Y.

The diffusion law The procedure follows the first part of Section 7.11, without the periodicity requirements. We shall include it in the following for the sake of completeness. With notation $\|\lambda\| = \sqrt{\lambda_1^2 + \lambda_2^2}$ and $B[0,t] = \{B_s; 0 \leq$

$s \leq t\}$, the characteristic function of the diffusion can be processed via tower property of conditional expectations

$$\begin{aligned}\Phi_{X_t,Y_t,T_t}(\lambda_1,\lambda_2,\eta) &= \mathbb{E}[e^{i\lambda_1 X_t}e^{i\lambda_2 Y_t}e^{i\eta T_t}] = \mathbb{E}[\mathbb{E}[e^{i\lambda_1 X_t}e^{i\lambda_2 Y_t}e^{i\eta T_t}|B[0,t]]] \\ &= \mathbb{E}[e^{i\eta T_t}\mathbb{E}[e^{i\lambda_1 X_t}e^{i\lambda_2 Y_t}|B[0,t]]] \\ &= \mathbb{E}[e^{i\eta T_t}\mathbb{E}[e^{i\int_0^t(\lambda_1\cosh T_s+\lambda_2\sinh T_s)dW_s}|B[0,t]]] \\ &= \mathbb{E}[e^{i\eta T_t}\mathbb{E}[e^{i\|\lambda\|\int_0^t\sinh(T_s+\varphi)dW_s}|B[0,t]]],\end{aligned}$$

where $\varphi = \tanh^{-1}(\lambda_1/\lambda_2) = \ln\sqrt{\frac{\lambda_1+\lambda_2}{\lambda_2-\lambda_1}}$. By Theorem 1.1.11 there is a DDS Brownian motion, γ_t, such that

$$\int_0^t \sinh(T_s+\varphi)dW_s = \gamma_{\int_0^t \sinh^2(B_u+\varphi)du},$$

fact that simplifies the computation of the characteristic function as

$$\begin{aligned}\Phi_{X_t,Y_t,T_t}(\lambda_1,\lambda_2,\eta) &= \mathbb{E}[e^{i\eta T_t}\mathbb{E}[e^{i\|\lambda\|\gamma_{\int_0^t\sinh^2(B_u+\varphi)du}}|B[0,t]]] \\ &= \mathbb{E}[e^{i\eta B_t}e^{-\frac{1}{2}\|\lambda\|^2\int_0^t\sinh^2(B_u+\varphi)du}] \\ &= u(t,0)+iv(t,0),\end{aligned} \tag{7.12.58}$$

where

$$\begin{aligned}u(t,\theta) &= \mathbb{E}[\cosh(\eta(B_t+\theta))e^{-\frac{1}{2}\|\lambda\|^2\int_0^t\sinh^2(B_u+\theta+\varphi)du}] \\ v(t,\theta) &= \mathbb{E}[\sinh(\eta(B_t+\theta))e^{-\frac{1}{2}\|\lambda\|^2\int_0^t\sinh^2(B_u+\theta+\varphi)du}].\end{aligned}$$

Via Feynman-Kac's formula, we have

$$\frac{\partial u}{\partial t} = \frac{1}{2}\frac{\partial^2 u}{\partial\theta^2} - U(\theta)u \tag{7.12.59}$$

$$u(0,\theta) = \cosh(\eta\theta) \tag{7.12.60}$$

and

$$\frac{\partial v}{\partial t} = \frac{1}{2}\frac{\partial^2 v}{\partial\theta^2} - U(\theta)v \tag{7.12.61}$$

$$v(0,\theta) = \sinh(\eta\theta), \tag{7.12.62}$$

where $U(\theta) = \frac{1}{2}\|\lambda\|^2\sinh^2(\theta+\varphi)$. There are no periodicity constraints for these initial value problems. The problems will be solved by the method of separation of variables. Setting $u(t,\theta) = f(t)g(\theta)$ and substituting into (7.12.59) we obtain

$$\frac{f'(t)}{f(t)} = \frac{1}{2}\frac{g''(\theta)}{b(\theta)} - U(\theta) = \frac{1}{2}E$$

where E is a separation constant. Then $f(t) = f(0)e^{\frac{1}{2}Et}$, while $g(\theta)$ satisfies the equation

$$g''(\theta) - (\|\lambda\|^2 \sinh^2(\theta + \varphi) + E)g(\theta) = 0.$$

Using $\sinh^2 x = \dfrac{\cosh(2x) - 1}{2}$, the previous equations becomes

$$g''(x) - (a - 2q\cosh(2x))g(x) = 0, \tag{7.12.63}$$

where $a = E - \frac{\|\lambda\|^2}{2}$ and $q = -\frac{\|\lambda\|^2}{4}$ and $x = \theta + \varphi$. Equation (7.12.63) is called the *modified Mathieu's equation.* The solutions of this equation are called *radial Mathieu functions* and are provided in Abramowitz and Stegun [1], page 732, as follows

$$\mathrm{Ce}_{2r+p}(x, q) = \mathrm{ce}_{2r+p}(ix, q) = \sum_{k\geq 0} A_{2k+p}^{2r+p}(q)\cosh(2k + p)x$$

associated with the eigenvalue a_r and $p \in \{0, 1\}$ and

$$\mathrm{Se}_{2r+p}(x, q) = -i\mathrm{se}_{2r+p}(ix, q) = \sum_{k\geq 0} B_{2k+p}^{2r+p}(q)\sinh(2k + p)x$$

associated with the eigenvalue b_r and $p \in \{0, 1\}$. The functions $\mathrm{Ce}_{2r+p}(x, q)$ are even solutions, while $\mathrm{Se}_{2r+p}(x, q)$ are odd solutions.

Thus, we obtained two families of solutions for the equation (7.12.59)

$$u_r(t, \theta) = f_r(t)g_r(\theta+\varphi) = e^{(\frac{1}{2}a_r + \frac{\|\lambda\|^2}{4})t}\Big(\alpha_{2r}\mathrm{Ce}_{2r}(\theta+\varphi, q) + \alpha_{2r+1}\mathrm{Ce}_{2r+1}(\theta+\varphi, q)\Big)$$

$$u_r(t, \theta) = f_r(t)g_r(\theta+\varphi) = e^{(\frac{1}{2}b_r + \frac{\|\lambda\|^2}{4})t}\Big(\beta_{2r}\mathrm{Se}_{2r}(\theta+\varphi, q) + \beta_{2r+1}\mathrm{Se}_{2r+1}(\theta+\varphi, q)\Big).$$

By superposition we obtain the general solution of equation (7.12.59)

$$\begin{aligned} u(t, \theta) = &\sum_{r\geq 0}\Big\{e^{(\frac{1}{2}a_r + \frac{\|\lambda\|^2}{4})t}\Big(\alpha_{2r}\mathrm{Ce}_{2r}(\theta + \varphi, q) + \alpha_{2r+1}\mathrm{Ce}_{2r+1}(\theta + \varphi, q)\Big)\Big\} \\ &+ \sum_{r\geq 0}\Big\{e^{(\frac{1}{2}b_r + \frac{\|\lambda\|^2}{4})t}\Big(\beta_{2r}\mathrm{Se}_{2r}(\theta + \varphi, q) + \beta_{2r+1}\mathrm{Se}_{2r+1}(\theta + \varphi, q)\Big)\Big)\Big\}. \end{aligned}$$

The constants α_j and β_j are subject to be determined from the initial relation (7.12.60), which in this case becomes

$$\sum_{j\geq 0}\Big(\alpha_j\mathrm{Ce}_j(\theta + \varphi, q) + \beta_j\mathrm{Se}_j(\theta + \varphi, q)\Big) = \cosh(\eta\theta).$$

This identification can be done, at least theoretically, by expanding both terms as a power series over the variable θ.

Solving the initial value problem (7.12.61)-(7.12.62) is similar, obtaining the general solution

$$\begin{aligned} v(t,\theta) =& \sum_{r\geq 0}\Big\{e^{(\frac{1}{2}a_r+\frac{\|\lambda\|^2}{4})t}\Big(\tilde{\alpha}_{2r}\mathrm{Ce}_{2r}(\theta+\varphi,q)+\tilde{\alpha}_{2r+1}\mathrm{Ce}_{2r+1}(\theta+\varphi,q)\Big)\Big\} \\ &+\sum_{r\geq 0}\Big\{e^{(\frac{1}{2}b_r+\frac{\|\lambda\|^2}{4})t}\Big(\tilde{\beta}_{2r}\mathrm{Se}_{2r}(\theta+\varphi,q)+\tilde{\beta}_{2r+1}\mathrm{Se}_{2r+1}(\theta+\varphi,q)\Big)\Big)\Big\}, \end{aligned}$$

subject to the coefficients identification

$$\sum_{j\geq 0}\Big(\tilde{\alpha}_j\mathrm{Ce}_j(\theta+\varphi,q)+\tilde{\beta}_j\mathrm{Se}_j(\theta+\varphi,q)\Big)=\sinh(\eta\theta).$$

Substituting now in relation (7.12.58) we obtain the characteristic function of diffusion (X_t,Y_t,T_t) as

$$\begin{aligned} \Phi_{X_t,Y_t,T_t}(\lambda_1,\lambda_2,\eta) &= u(t,0)+iv(t,0) \\ &= \sum_{r\geq 0}\Big\{e^{(\frac{1}{2}a_r+\frac{\|\lambda\|^2}{4})t}\Big(\alpha^c_{2r}\mathrm{Ce}_{2r}(\varphi,q)+\alpha^c_{2r+1}\mathrm{Ce}_{2r+1}(\varphi,q)\Big)\Big\} \\ &\quad+\sum_{r\geq 0}\Big\{e^{(\frac{1}{2}b_r+\frac{\|\lambda\|^2}{4})t}\Big(\beta^c_{2r}\mathrm{Se}_{2r}(\varphi,q)+\beta^c_{2r+1}\mathrm{Se}_{2r+1}(\varphi,q)\Big)\Big)\Big\}, \end{aligned}$$

where $\alpha_j^c=\alpha_j+i\alpha_j$, and $\beta_j^c=\beta_j+i\beta_j$. The probability density is obtained inverting the integral defining the characteristic function as

$$p_t(x,y,\theta)=\frac{1}{(2\pi)^3}\iiint_{\mathbb{R}^3}e^{-i\lambda_1 x}e^{-i\lambda_2 y}e^{-i\eta\theta}\Phi_{X_t,Y_t,T_t}(\lambda_1,\lambda_2,\eta)\,d\lambda_1 d\lambda_2 d\eta.$$

7.13 The Rolling Coin Diffusion

This section deals with the subelliptic diffusion induced by a rolling disk of radius R, which rolls without slipping on a horizontal plane and is constrained to be vertical at all times. The position of the disk is parameterized by coordinates $(x,y,\phi,\psi)\in\mathbb{R}^2\times\mathbb{S}^1\times\mathbb{S}^1$, as described at page 151. The associated distribution is generated by the vector fields

$$X_1=\partial_\psi,\qquad X_2=R(\cos\psi\,\partial_x+\sin\psi\,\partial_y)+\partial_\phi.$$

Since the vector fields $\{X_1, X_2\}$ are bracket-generating, the operator

$$
\begin{aligned}
L &= \frac{1}{2}\Big(R(\cos\psi\,\partial_x + \sin\psi\,\partial_y) + \partial_\phi\Big)^2 + \frac{1}{2}\partial_\psi^2 \\
&= \left\langle \begin{pmatrix} R^2\cos^2\psi & R^2\cos\psi\sin\psi & R\cos\psi & 0 \\ R^2\cos\psi\sin\psi & R^2\sin^2\psi & R\sin\psi & 0 \\ R\cos\psi & R\sin\psi & 1 & 0 \\ 0 & 0 & 0 & 1 \end{pmatrix} \begin{pmatrix} \partial_x \\ \partial_y \\ \partial_\phi \\ \partial_\psi \end{pmatrix}, \begin{pmatrix} \partial_x \\ \partial_y \\ \partial_\phi \\ \partial_\psi \end{pmatrix} \right\rangle
\end{aligned}
$$

is hypoelliptic. Since the dispersion matrix is given by

$$
\sigma = \begin{pmatrix} R\cos\psi & 0 & 0 & 0 \\ R\sin\psi & 0 & 0 & 0 \\ 1 & 0 & 0 & 0 \\ 0 & 0 & 0 & 1 \end{pmatrix},
$$

then the associated diffusion $(X_t, Y_t, \Phi_t, \Psi_t)$ satisfies

$$
\begin{aligned}
dX_t &= R\cos\Psi_t\,dW_t \\
dY_t &= R\sin\Psi_t\,dW_t \\
d\Phi_t &= dW_t \\
d\Psi_t &= dB_t,
\end{aligned}
$$

where W_t and B_t are two independent Brownian motions on the unit circle $\mathbb{S}^1$. We note that the diffusion is horizontal, since its increments belong to the horizontal distribution

$$
\begin{aligned}
(dX_t, dY_t, d\Phi_t, d\Psi_t) &= (dX_t, dY_t, d\Phi_t, 0) + (0, 0, 0, d\Psi_t) \\
&= (R\cos\Psi_t, R\sin\Psi_t, 1)dW_t + (0, 0, 0, 1)dB_t \\
&= X_{2|(X_t,Y_t,\Phi_t,\Psi_t)}dW_t + X_{1|(X_t,Y_t,\Phi_t,\Psi_t)}dB_t.
\end{aligned}
$$

This diffusion models the dynamics of the rolling coin with stochastic independent controls, dW_t and dB_t.

Some simplifying assumptions regarding the initial data can always be done. Thus, applying a rotation and a translation in the plane $\mathbb{R}^2$, we can assume that the disk starts rolling from the origin towards the x-direction, i.e., $X_0 = 0$, $Y_0 = 0$, $\Psi_0 = 0$. We can also choose the distinguished radius from which the angle ϕ is measured to be vertical, so $\Phi_0 = 0$. Therefore, integrating the aforementioned equations we obtain the following strong solution

$$
\begin{aligned}
X_t &= R\int_0^t \cos B_s\, dW_s \\
Y_t &= R\int_0^t \sin B_s\, dW_s \\
\Phi_t &= W_t \\
\Psi_t &= B_t, \qquad t \geq 0.
\end{aligned}
$$

It is worth noting that for $R = 1$ the diffusion (X_t, Y_t, Ψ_t) is just the knife-edge diffusion studied in Section 7.11 at page 424. This observation is useful in the computation of the probability density of the diffusion associated to the rolling coin, which will be reduced to the probability density of the knife-edge. To this end, let $\|\lambda\| = \sqrt{\lambda_1^2 + \lambda_2^2}$ and consider the sigma-algebras

$$
\begin{aligned}
B[0,t] &= \{B_s; 0 \leq s \leq t\} \\
\{W[0,t], B[0,t]\} &= \{W_s, B_s; 0 \leq s \leq t\},
\end{aligned}
$$

which represent the history of B_s as well as the joint history of both Brownian motions, W_s, B_s, until time t. The computation of the characteristic function will done in a couple of steps.

Step 1: Using a computation similar with the one done at page 424, we shall compute the conditional expectation as follows

$$
\begin{aligned}
\mathbb{E}[e^{i\lambda_1 X_t} e^{i\lambda_2 Y_t} | B[0,t]] &= \mathbb{E}[e^{iR\int_0^t (\lambda_1 \cos T_s + \lambda_2 \sin T_s) dW_s} | B[0,t]] \\
&= \mathbb{E}[e^{i\|\lambda\| R \int_0^t \sin(T_s + \varphi) dW_s} | B[0,t]] \\
&= \mathbb{E}[e^{i\|\lambda\| R \gamma_{\int_0^t \sin^2(B_u+\varphi)du}} | B[0,t]] \\
&= e^{-\frac{1}{2}R^2 \|\lambda\|^2 \int_0^t \sin^2(B_u+\varphi)du},
\end{aligned}
$$

where γ_t is a Brownian motion and $\tan\varphi = \lambda_1/\lambda_2$.

Step 2: We shall compute the conditional expectation

$$
\mathbb{E}[e^{i\lambda_1 X_t} e^{i\lambda_2 Y_t} | W[0,t], B[0,t]]
$$

using tower property and Step 1 as follows

$$
\begin{aligned}
\mathbb{E}[e^{i\lambda_1 X_t} e^{i\lambda_2 Y_t} | W[0,t], B[0,t]] &= \mathbb{E}[\,\mathbb{E}[e^{i\lambda_1 X_t} e^{i\lambda_2 Y_t} | B[0,t]]\, | W[0,t], B[0,t]] \\
&= \mathbb{E}[\, e^{-\frac{1}{2}R^2 \|\lambda\|^2 \int_0^t \sin^2(B_u+\varphi)du} \, | W[0,t], B[0,t]] \\
&= \mathbb{E}[\, e^{-\frac{1}{2}R^2 \|\lambda\|^2 \int_0^t \sin^2(B_u+\varphi)du} \, | B[0,t]] \\
&= e^{-\frac{1}{2}R^2 \|\lambda\|^2 \int_0^t \sin^2(B_u+\varphi)du},
\end{aligned}
$$

where we dropped the independent condition $W[0,t]$ and then used that the exponential is $B[0,t]$-measurable.

Step 3: The characteristic function of the diffusion $(X_t, Y_t, \Phi_t, \Psi_t)$ can be computed using tower property, factoring out the measurable part, and then using Step 2 and the independence property as follows

$$\begin{aligned}
\Phi_{X_t,Y_t,\Phi_t,\Psi_t}(\lambda_1,\lambda_2,\eta_1,\eta_2) &= \mathbb{E}[e^{i\lambda_1 X_t}e^{i\lambda_2 Y_t}e^{i\eta_1 B_t}e^{i\eta_2 W_t}] \\
&= \mathbb{E}[\,\mathbb{E}[e^{i\lambda_1 X_t}e^{i\lambda_2 Y_t}e^{i\eta_1 B_t}e^{i\eta_2 W_t}|W[0,t],B[0,t]]] \\
&= \mathbb{E}[e^{i\eta_1 B_t}e^{i\eta_2 W_t}\mathbb{E}[e^{i\lambda_1 X_t}e^{i\lambda_2 Y_t}|W[0,t],B[0,t]]] \\
&= \mathbb{E}[e^{i\eta_1 B_t}e^{i\eta_2 W_t}e^{-\frac{1}{2}R^2\|\lambda\|^2\int_0^t \sin^2(B_u+\varphi)du}] \\
&= \mathbb{E}[e^{i\eta_2 W_t}]\mathbb{E}[e^{i\eta_1 B_t}e^{-\frac{1}{2}R^2\|\lambda\|^2\int_0^t \sin^2(B_u+\varphi)du}] \\
&= e^{-\frac{1}{2}\eta_2^2 t}\Phi_{X_t,Y_t,T_t}(\lambda_1/R,\lambda_2/R,\eta_1),
\end{aligned}$$

where Φ_{X_t,Y_t,T_t} is the characteristic function of the knife-edge diffusion (X_t, Y_t, T_t). Hence, the relation between the characteristic functions of the rolling coin and the knife-edge diffusions is given by

$$\Phi_{X_t,Y_t,\Phi_t,\Psi_t}(\lambda_1,\lambda_2,\eta_1,\eta_2) = e^{-\frac{1}{2}\eta_2^2 t}\Phi_{X_t,Y_t,T_t}(\lambda_1/R,\lambda_2/R,\eta_1). \tag{7.13.64}$$

Step 4: To retrieve the probability density $p_t(x,y,\phi,\psi)$ of diffusion $(X_t, Y_t, \Phi_t, \Psi_t)$ we shall invert the following integral

$$\Phi_{X_t,Y_t,\Phi_t,\Psi_t}(\lambda_1,\lambda_2,\eta_1,\eta_2) = \int_{\mathbb{R}^4} e^{i\lambda_1 x}e^{i\lambda_2 y}e^{i\eta_1\psi}e^{i\eta_2\phi}p_t(x,y,\phi,\psi)\,dx dy d\phi d\psi$$

and use formula (7.13.64)

$$\begin{aligned}
&p_t(x,y,\phi,\psi) \\
&= \frac{1}{(2\pi)^4}\int_{\mathbb{R}^4} e^{-i\lambda_1 x}e^{-i\lambda_2 y}e^{-i\eta_1\psi}e^{-i\eta_2\phi}\Phi_{X_t,Y_t,\Phi_t,\Psi_t}(\lambda_1,\lambda_2,\eta_1,\eta_2)d\lambda_1 d\lambda_2 d\eta_1 d\eta_2 \\
&= \frac{1}{(2\pi)^4}\int_{\mathbb{R}^4} e^{-i\lambda_1 x}e^{-i\lambda_2 y}e^{-i\eta_1\psi}e^{-i\eta_2\phi}e^{-\frac{1}{2}\eta_2^2 t}\Phi_{X_t,Y_t,T_t}\Big(\frac{\lambda_1}{R},\frac{\lambda_2}{R},\eta_1\Big)d\lambda_1 d\lambda_2 d\eta_1 d\eta_2 \\
&= \frac{R^2}{(2\pi)^4}\int_{\mathbb{R}^4} e^{-iR\tilde\lambda_1 x}e^{-iR\tilde\lambda_2 y}e^{-i\eta_1\psi}e^{-\frac{1}{2}\eta_2^2 t-i\eta_2\phi}\Phi_{X_t,Y_t,T_t}(\tilde\lambda_1,\tilde\lambda_2,\eta_1)d\tilde\lambda_1 d\tilde\lambda_2 d\eta_1 d\eta_2 \\
&= \frac{R^2}{(2\pi)^4}\int_{\mathbb{R}} e^{-\frac{1}{2}\eta_2^2 t-i\eta_2\phi}d\eta_2\int_{\mathbb{R}^3} e^{-iR\tilde\lambda_1 x}e^{-iR\tilde\lambda_2 y}e^{-i\eta_1\psi}\Phi_{X_t,Y_t,T_t}(\tilde\lambda_1,\tilde\lambda_2,\eta_1)d\tilde\lambda_1 d\tilde\lambda_2 d\eta_1 \\
&= \frac{R^2}{(2\pi)^4}\sqrt{\frac{2\pi}{t}}e^{-\frac{\phi^2}{2t}}(2\pi)^3 p_t(Rx,Ry,\psi) = \frac{R^2}{\sqrt{2\pi t}}e^{-\frac{\phi^2}{2t}}p_t(Rx,Ry,\psi),
\end{aligned}$$

where $p_t(x,y,\psi)$ is the density of the knife-edge diffusion which is computed at page 426 and is has an explicit expression involving Mathieu functions as given by formula[3] (7.11.55). Since the vector fields $\{X_1, X_2\}$ are bracket-generating, by Proposition 3.6.3 the probability density $\rho_t(x, y, \phi, \psi)$ is $\mathcal{C}^\infty$-smooth.

7.14 Summary

This chapter explores the topic of sub-elliptic diffusions, which are diffusions governed by sub-elliptic operators. Specifically, we focus on operators that can be expressed as the sum of squares of vector fields and satisfy the bracket-generating condition. This condition ensures the smoothness of the diffusion's transition probability.

We begin by exploring the prototype example of sub-elliptic diffusions, namely the Heisenberg diffusion. We also introduce and analyze the Koranyi process. Additionally, we delve into another example known as the Grushin distribution. While similar to the Heisenberg diffusion, the Grushin distribution lacks the underlying invariance of a group law. Furthermore, we explore additional examples of sub-elliptic diffusions such as the Martinet and Engel diffusions. These examples possess the unique property of evolving or spreading along the horizontal distribution on a sub-Riemannian manifold. Not all diffusions satisfy this condition, and thus a general case of sub-elliptic diffusions is thoroughly examined.

A notable subcategory within sub-elliptic diffusions is nonholonomic diffusions, which are associated with nonholonomic mechanical systems like the knife edge or the rolling coin. In these cases, we derive explicit expressions for their transition densities, which involve Mathew functions. Remarkably, all the sub-elliptic diffusions discussed in this chapter share a common property—their transitivity. This means that the diffusion spreads throughout the space without returning.

7.15 Exercises

Exercise 7.15.1 Consider the vector fields $\{X, Y\}$ on $\mathbb{S}^3$ given by Example 3.8.5 from page 150. Write the stochastic differential equations of the diffusion associated with the operator $L = \frac{1}{2}(X^2 + Y^2)$.

Exercise 7.15.2 Write the stochastic differential equations of the diffusion describing the two-wheel cart nonholonomic system described at page 152. This is the diffusion associated with the subelliptic operator $\frac{1}{2}(X_1^2 + X_2^2)$.

[3]Formula (7.11.55) used θ instead of ψ.

Exercise 7.15.3 Let $(G, \circ)$ be a Lie group of dimension n and denote by $X_1, \dots, X_n$ the left invariant vector fields on G. Consider the sub-elliptic operator $L = \frac{1}{2}\sum_{j=1}^{k} X_j^2$, $1 \le k \le n$, and denote by $p_t(x, x')$ the transition density of the associated sub-elliptic diffusion. Show that this density is left invariant, namely, $p_t(g \circ x, g \circ x') = p_t(x, x')$, $\forall g \in G$.

Exercise 7.15.4 Let $L = \dfrac{1}{2}x_1^2\partial_{x_2}^2 + \lambda\partial_{x_1}$ with $\lambda \in \mathbb{R}$ parameter.

(a) Show that the diffusion $X(t) = \big(X_1(t), X_2(t)\big)$, which is associated to the operator L and starting at $x^0 = (x_1^0, x_2^0)$, is given by

$$\begin{aligned} X_1(t) &= x_1^0 + \lambda t \\ X_2(t) &= x_2^0 + x_1^0 W_t + \lambda t W_t - \lambda \int_0^t W_s\, ds, \qquad t \ge 0. \end{aligned}$$

(b) Using Proposition 4.1.1, page 160, find the transition probability of diffusion $X(t)$.

Exercise 7.15.5 Let $L = \dfrac{1}{2}\displaystyle\sum_{j=1}^{n} x_j^2\partial_{x_{j+n}}^2 + \sum_{j=1}^{n}\lambda_j\partial_{x_j}$ with $\lambda_j \in \mathbb{R}$ parameters.

(a) Find the diffusion $X(t)$ on $\mathbb{R}^{2n}$ associated with the operator L.

(b) Using Exercise 7.15.4 find the transition probability of diffusion $X(t)$.

Exercise 7.15.6 Consider the degenerate diffusion $X_t = (W_t, \sinh W_t)$, starting at $X_0 = (0, 0)$, where W_t is a one-dimensional Brownian motion.

(a) Find the infinitesimal generator of X_t.

(b) Using Bougerol's identity (page 54) and Hartman-Watson density (page 187), find the characteristic function of X_t.

Exercise 7.15.7 Consider the Heisenberg vector fields on $\mathbb{R}^{2n+1}$

$$\begin{aligned} X_{2j-1} &= \partial_{x_{2j-1}} + 2x_{2j}\partial_t \\ X_{2j} &= \partial_{x_{2j}} - 2x_{2j-1}\partial_t, \qquad 1 \le j \le n \end{aligned}$$

and the Heisenberg Laplacian $\Delta_H = \frac{1}{2}\sum_{k=1}^{2n} X_k^2$. The Heisenberg norm is defined by $\|(x,t)\|_H = (|x|^4 + t^2)^{-n/2}$, where $|x|^2 = \sum_{j=1}^{2n} x_j^2$. Let $B_H(\xi, r) = \{y; \|y - \xi\|_H \le r\}$.

(a) Show that $\Delta_H \dfrac{1}{\|(x,t)\|_H} = 0$, for all $(x,t) \neq 0_{\mathbb{R}^{2n+1}}$.

(b) Let $H_t = (H_1(t), \dots, H_{n+1}(t))$ be the associated diffusion to the operator Δ_H. Find the probability that H_t reaches the neighborhood $B_H(\xi, r)$, given that $H_0 = 0$.

Chapter 8

Systems of Sub-Elliptic Differential Equations

The chapter commences with a comprehensive overview of elliptic differential systems of equations and subsequently expands to encompass sub-elliptic differential systems of equations. The former means that the number of equations equals the dimension of the space, while the latter refers to systems with fewer equations than the dimension of the space. A sub-Riemannian adaptation of the Poincaré lemma is introduced, which establishes equivalent conditions for closeness and exactness in sub-elliptic differential systems, including detailed examinations of the Heisenberg, Grushin, and Engel distributions.

8.1 Poincaré's Lemma

All sub-elliptic systems of equations that will be approached in this chapter will be completed to elliptic systems where one of the variants of the Poincaré's lemma will be applied. Therefore, the present section is dedicated to some notions regarding the classical Poincaré lemma. We shall present it in two variants, involving either vector fields or one-forms.

Poincaré's lemma in vector fields variant Let $\Omega \subseteq \mathbb{R}^n$ be an open, connected and simply connected set and $a_j : \Omega \to \mathbb{R}$ be $\mathcal{C}^1$-functions, where $1 \leq j \leq n$. Consider the vector field $U = \sum_{k=1}^n a_k(x)\partial_{x_k}$ on Ω. Then there is a $\mathcal{C}^2$-smooth function u on Ω such that

$$\nabla_{Eu} u = U \text{ in } \Omega \tag{8.1.1}$$

if and only if

$$\partial_{x_i} a_j = \partial_{x_j} a_i, \text{ for every } 1 \leq i, j \leq n. \tag{8.1.2}$$

The vector field U given as the gradient of a function u is called a *potential vector field*, while relations (8.1.2) are the integrability conditions for the system (8.1.1). The system (8.1.1) can be written equivalently on components as $\partial_{x_j} u = a_j(x)$, with $1 \le j \le n$. We recall that $\nabla_{Eu} u$ denotes the Euclidean gradient, i.e. the vector field $\sum_{i=1}^n \frac{\partial u}{\partial x_i} \partial_{x_i}$.

Remark 8.1.1 It is worth reminding that a contractible set Ω is both connected and simply connected. Sometimes we shall use Poincaré's lemma under this hypothesis.

It is remarkable that the the solution u can be constructed explicitly and has the physical significance of work done by the force U along a curve, as it will be shown next for the cases $n = 2$ and $n = 3$.

The case $n = 2$ The system (8.1.1) takes the form

$$\partial_x u = a, \qquad \partial_y u = b$$

where a and b are $\mathcal{C}^1$-smooth functions. Let $r(s) = (x(s), y(s)) = (sx, sy)$, $0 \le s \le 1$ be a curve that joins the origin with the given point (x, y). Let u be the work done by the force field U along the curve $r(s)$

$$u(x, y) = \int_0^1 \langle U(r(s)), \dot{r}(s) \rangle \, ds = \int_0^1 (ax + by) \, ds,$$

where $\langle \, , \, \rangle$ denotes the Euclidean inner product. A straightforward computation yields

$$\partial_x u = a(x, y) + \int_0^1 ty \Big(\frac{\partial b}{\partial x} - \frac{\partial a}{\partial y} \Big) \, ds$$
$$\partial_y u = b(x, y) + \int_0^1 tx \Big(\frac{\partial a}{\partial y} - \frac{\partial b}{\partial x} \Big) \, ds,$$

so, these equations can be written in the following vector form

$$\nabla_{Eu} u = (a, b) + \int_0^1 \Big(\frac{\partial b}{\partial x} - \frac{\partial a}{\partial y} \Big) (sy, \, -sx) \, ds.$$

Therefore, if $\frac{\partial a}{\partial y} = \frac{\partial b}{\partial x}$ then $\partial_x u = a$ and $\partial_y u = b$. The converse statement also holds true as a direct consequence of the symmetry of the mixed derivative.

The case $n = 3$ Let $r = (x_1, x_2, x_3)$ be a given point in $\mathbb{R}^3$. A computation provides the formula

$$U(r) = \nabla_{Eu} \Big(\int_0^1 U(sr) \cdot r \, ds \Big) + \int_0^1 sr \times \operatorname{curl} U(sr) \, ds.$$

Considering the work done by the force U along the curve $\{sr; 0 \le s \le 1\}$

$$u(r) = \int_0^1 \langle U(sr), r\rangle\, ds,$$

the previous formula is equivalent to

$$\nabla_{Eu} u(r) = U(r) + \int_0^1 sr \times \operatorname{curl} U(sr)\, ds. \tag{8.1.3}$$

Then, if $\operatorname{curl} U = 0$, i.e. if

$$\frac{\partial a_2}{\partial x_1} - \frac{\partial a_1}{\partial x_2} = 0, \quad \frac{\partial a_3}{\partial x_1} - \frac{\partial a_1}{\partial x_3} = 0, \quad \frac{\partial a_2}{\partial x_3} - \frac{\partial a_3}{\partial x_2} = 0,$$

then the work $u(r)$ is a solution of the system (8.1.1)

$$\frac{\partial u}{\partial x_1} = a_1, \quad \frac{\partial u}{\partial x_2} = a_2, \quad \frac{\partial u}{\partial x_3} = a_3.$$

It is worth noting that the aforementioned integral does not depend on the chosen path, provided (8.1.1) holds. To this end, let $c : [0,1] \to \mathbb{R}^n$ be an arbitrary fixed smooth curve with $c(0) = 0$ and $c(1) = x$. Then

$$\begin{aligned}\int_0^1 \langle U_{|c(s)}, \dot c(s)\rangle\, ds &= \int_0^1 \sum_j \partial_{x_j} u_{|c(s)}\, \dot c^j(s)\, ds = \int_0^1 \frac{d}{ds} u(c(s))\, ds \\ &= u(x) - u(0),\end{aligned}$$

namely, the integral along any other path provides the same solution.

Poincaré's lemma on Riemannian manifolds Let (M, g) be a Riemannian manifold and X be a smooth vector field on it. The curl of X is a two-covariant, antisymmetric tensor, $A = \operatorname{curl} \mathrm{X}$, defined by

$$A(U, V) = g(\nabla_V X, U) - g(\nabla_U X, V), \tag{8.1.4}$$

for any vector fields U and V on M. Here, ∇ stands for the Levi-Civita connection. This means the curl of X is the antisymmetric difference of the projections of the covariant derivative of X with respect to a pair of vector fields. The relation with the Lie bracket is given by the following result, see Calin and Chang [36], page 138.

Lemma 8.1.2 *If $A = \operatorname{curl} X$, then*

$$A(U, V) = V g(X, U) - U g(X, V) + g(X, [U, V])$$

for any two smooth vector fields U and V on M.

Proof: Since the Levi-Civita connection is metrical, we have

$$Vg(X,U) = g(\nabla_V X, U) + g(X, \nabla_V U)$$
$$Ug(X,V) = g(\nabla_U X, V) + g(X, \nabla_U V).$$

Subtracting and using the torsion-free property $\nabla_V U - \nabla_U V = [V,U]$, we obtain

$$Vg(X,U) - Ug(X,V) = A(U,V) + g(X,[U,V]),$$

which is leads to the desired formula. ■

Remark 8.1.3 As a consequence, we obtain the expression of curl in a local system of coordinates

$$A_{ij} = A(\partial_{x_i}, \partial_{x_j}) = \partial_{x_j} X^i - \partial_{x_i} X^j.$$

We recall that the gradient of a differentiable function f on the Riemannian manifold (M,g) is given by the vector field $\nabla_g f = \sum_{k=1}^n (\nabla_g f)^k \partial_{x_k}$, where $(\nabla_g f)^k = \sum_{i=1}^n g^{ki}\partial_{x_i} f$. This is equivalent to the fact that

$$g(\nabla_g f, U) = U(f), \tag{8.1.5}$$

for any vector field U on (M,g).

A vector field X on (M,g) is call *potential* vector field if it is a gradient vector field, i.e., there is a smooth function f on M such that $\nabla_g f = X$. The following result represents a variant of Poincaré's lemma on Riemannian manifolds.

Theorem 8.1.4 *Let (M,g) be a connected and simply connected Riemannian manifold. Then the vector field X is potential if and only if* curl X = 0.

Proof: " $\Rightarrow$ " Let $X = \nabla_g f$, with $f : M \to \mathbb{R}$ be C^2-differentiable. Then using Lemma 8.1.2 yields

$$\begin{aligned} A(U,V) &= Vg(X,U) - Ug(X,V) + g(X,[U,V]) \\ &= Vg(\nabla_g f, U) - Ug(\nabla_g f, V) + g(\nabla_g f, [U,V]) \\ &= VU(f) - UV(f) + [U,V](f) = ([V,U] + [U,V])(f) = 0, \end{aligned}$$

where we also used relation (8.1.5) and the skew-symmetry of the Lie bracket. Therefore, curl X = 0.

" $\Leftarrow$ " We assume now curl X = 0. Then by Lemma 8.1.2

$$Vg(X,U) - Ug(X,V) + g(X,[U,V]) = 0$$

for any two vector fields U and V. In particular, considering the coordinates vector fields $U = \partial_{x_i}$, $V = \partial_{x_j}$, and using $X = \sum_k X^k \partial_{x_k}$, the previous relation becomes

$$\partial_{x_j} g\Big(\sum_k X^k \partial_{x_k}, \partial_{x_i}\Big) - \partial_{x_i} g\Big(\sum_k X^k \partial_{x_k}, \partial_{x_j}\Big) = 0 \Leftrightarrow$$

$$\partial_{x_j}\big(\sum_k X^k g_{ki}\big) - \partial_{x_i}\big(\sum_k X^k g_{kj}\big) = 0.$$

Let $a_j = \sum_k X^k g_{kj}$. Then the previous relation becomes $\partial_{x_j} a_i = \partial_{x_i} a_j$, which are the integrability conditions for the system $\partial_{x_j} f = a_j$, $1 \leq j \leq n$. Therefore, there is a smooth function f such that $\partial_{x_j} f = \sum_k X^k g_{kj}$. Inverting, this yields $X^k = \sum_i g^{ki} \partial_{x_i} f = (\nabla_g f)^k$, and hence $X = \nabla_g f$, i.e. X is a potential vector field.

■

Elliptic systems A system of linear first order differential equations is called *elliptic* if the number of equations equals the dimension of the space and the equations are independent, i.e., the vector fields defining these equations are linearly independent at each point.

The next result shows that the integrability conditions for an elliptic system are Schwartz-type conditions, provided the vector fields commute. We first prove the following lemma:

Lemma 8.1.5 *Let $\{X_1, \ldots, X_n\}$ be orthonormal vector fields on the Riemannian manifold (M, g) of dimension n. Then for any smooth function f on M we have*

$$\nabla_g f = \sum_k X_k(f) X_k.$$

Proof: For any vector field $Z = \sum Z^k X_k$ we have

$$\begin{aligned} g(\nabla_g f, Z) = Z(f) &= \sum_j Z^j X_j(f) = \sum_{i,j} Z^i X_j(f) g(X_i, X_j) \\ &= g\big(\sum_i X_i(f) X_i, \sum_j Z^j X_j\big) = g\big(\sum_i X_i(f) X_i, Z\big). \end{aligned}$$

Since g is non-degenerate, dropping the argument Z, we obtain the required conclusion. ■

Theorem 8.1.6 *Let (M, g) be a connected and simply connected Riemannian manifold and $\{X_1, \ldots, X_n\}$ be orthonormal vector fields such that $[X_i, X_j] = 0$,*

$i, j \in \{1, \dots, n\}$. Consider $\mathcal{C}^1$-differentiable functions a_j on M, $1 \le j \le n$. Then there is a $\mathcal{C}^2$-differentiable function f on M such that

$$X_1 f = a_1, \quad X_2 f = a_2, \ \dots, \ X_n f = a_n \tag{8.1.6}$$

if and only if $X_i a_j = X_j a_i$, for all $i, j \in \{1, \dots, n\}$.

Proof: " $\Rightarrow$ " Let f be a solution of the system (8.1.6). Using $[X_i, X_j] = 0$, we obtain

$$0 = [X_i, X_j](f) = X_i X_j f - X_j X_i f = X_i a_j - X_j a_i,$$

which implies the desired integrability conditions.

" $\Leftarrow$ " We assume that $X_i a_j = X_j a_i$, for all $i, j \in \{1, \dots, n\}$ and construct the vector field $V = \sum_k a_k X_k$. We shall prove that curl V = 0. To show this it suffices to show it on a basis. Since

$$\begin{aligned}\text{curl } V(X_i, X_j) &= X_j g(V, X_i) - X_i g(V, X_j) + g(V, [X_i, X_j]) \\ &= X_j a_i - X_i a_j = 0.\end{aligned}$$

By Theorem 8.1.4 the vector field V is potential, namely there is a smooth function f on M such that $\nabla_g f = V$. By Lemma 8.1.5, this is equivalent to

$$\sum_k X_k(f) X_k = \sum_k a_k X_k,$$

which implies $X_k(f) = a_k$, for $1 \le k \le n$, which is the system (8.1.8). ■

Remark 8.1.7 The integral curves of the vector fields X_j can be considered as an orthonormal coordinate system on the manifold (M, g).

Poincaré's lemma in one-forms variant Let $\Omega \subseteq \mathbb{R}^n$ be an open, connected and simply connected set and $a_k : \Omega \to \mathbb{R}$ be $\mathcal{C}^1$-functions, $1 \le k \le n$. Consider the one-form $\theta = \sum_{k=1}^n a_k(x) dx_k$. We recall the following two notions:

- The form θ is called *exact* if there is a $\mathcal{C}^2$-differentiable function f on Ω such that $\theta = df$, where the differential of f is defined by the one-form $df = \sum_{k=1}^n \frac{\partial f}{\partial x_k} dx_k$.
- The form θ is called *closed* if $d\theta = 0$, where the exterior derivative is defined by

$$d\theta(u, v) = u\theta(v) - v\theta(u) = \theta([u, v]), \qquad \forall u, v \in T_p\Omega.$$

The closeness condition can be written on the basis $\{\partial_{x_1}, \dots, \partial_{x_n}\}$ as $d\theta(\partial_i, \partial_j) = 0$, for all $1 \le i, j \le n$. Since $[\partial_{x_i}, \partial_{x_j}] = 0$, this condition is equivalent to

$\partial_{x_i} a_j = \partial_{x_j} a_i$. Then Poincaré's lemma states that θ is closed on Ω if and only if θ is exact. Equivalently, we have

$$\partial_{x_i} a_j = \partial_{x_j} a_i,\ 1 \le i, j \le n \Longleftrightarrow \begin{cases} \exists \text{ a } \mathcal{C}^2\text{–smooth function } f \\ \text{on } \Omega, \text{ such that} \\ \partial_{x_k} f = a_k \text{ for } 1 \le k \le n. \end{cases}$$

We consider next n linear independent vector fields $\{X_1, \dots, X_n\}$ on $\mathbb{R}^n$ and we consider the dual one-forms, $\{\omega_1, \dots, \omega_n\}$, given by $\omega(X_k) = \delta_{jk}$, for $1 \le j, k \le n$, where δ_{jk} denotes the Kroneker's symbol. Then given the one-form θ, we can expand it as a linear combination $\theta = \sum_{j=1}^n \theta^j \omega_j$. The jth component is given by

$$\theta(X_j) = \sum_{i=1}^n \theta^i \omega_i(X_j) = \theta^j,$$

so we can write the one-form θ in terms of the vector fields $\{X_k\}$ and one-forms $\{\omega_k\}$ as $\theta = \sum_{k=1}^n \theta(X_k)\omega_k$. Now, for a given smooth function f, the one-form df can be written as $df = \sum_{j=1}^n f^j \omega_j$, with the component f^k given by

$$X_k(f) = df(X_k) = \sum_{j=1}^n f^j \omega_j(X_k) = f^k,$$

so $df = \sum_{j=1}^n X_j(f)\omega_j$. Therefore, the exactness condition on θ, which is $\theta = df$, can be written as $\sum_{k=1}^n \theta^k \omega_k = \sum_{k=1}^n X_k(f)\omega_k$, which is equivalent to $\theta^k = X_k(f)$.

On the other side, the closeness condition $d\theta = 0$ can be written on the basis $\{X_k\}$ as

$$0 = d\theta(X_i, X_j) = X_i\theta(X_j) - X_j(X_i) - \theta([X_i, X_j]).$$

Let c_{ij}^k be functions defined by $[X_i, X_j] = \sum_{k=1}^n c_{ij}^k X_k$. Then the closeness condition becomes

$$X_i(\theta^j) - X_j(\theta^i) = \sum_{k=1}^n c_{ij}^k \theta^k, \qquad 1 \le i, j \le n. \tag{8.1.7}$$

Substituting $a_j = \theta^j$, Poincaré's lemma in this case takes the following form:

Theorem 8.1.8 *Let M be a connected and simply connected manifold and $\{X_1, \dots, X_n\}$ be linear independent vector fields on M. Consider $\mathcal{C}^1$-differentiable* *functions a_j on M, $1 \le j \le n$. Then there is a $\mathcal{C}^2$-differentiable function f on M such that*

$$X_1 f = a_1, \quad X_2 f = a_2, \, \dots, \, X_n f = a_n \tag{8.1.8}$$

if and only if

$$X_i(a_j) - X_j(a_i) = \sum_{k=1}^{n} c_{ij}^k \theta^k, \qquad 1 \le i, j \le n.$$

We note that in the case of commuting vector fields, $c_{ij}^k = 0$, we retrieve Theorem 8.1.6. The Riemannian metric g with respect to which the vector fields form an orthonormal system is the diffusion metric introduced at page 130.

8.2 Definition of Sub-elliptic Systems

The main topic of this chapter deals with systems of differential equations of first order. The attribute "sub-elliptic" refers to the fact that the number of differential equations is less than the dimension of the coordinates space.

To set the ideas, we shall start by considering the following two first order equations on $\mathbb{R}^3$, written as

$$\alpha_{11}\partial_x u + \alpha_{12}\partial_y u + \alpha_{13}\partial_z u = a(x, y, z)$$
$$\alpha_{21}\partial_x u + \alpha_{22}\partial_y u + \alpha_{23}\partial_z u = b(x, y, z),$$

with given functions $a, b \in \mathcal{C}^\infty(\mathbb{R}^3)$ and smooth coefficients $\alpha_{ij} \in \mathcal{C}^\infty(\mathbb{R}^3)$. The aforementioned system can be also written in the more simple form

$$Xu = a$$
$$Yu = b,$$

where we considered the vector fields

$$X = \alpha_{11}\partial_x + \alpha_{12}\partial_y + \alpha_{13}\partial_z, \qquad Y = \alpha_{21}\partial_x + \alpha_{22}\partial_y + \alpha_{23}\partial_z$$

on $\mathbb{R}^3$. We shall study whether the aforementioned problem is well-possed. By this we mean the following three items:

1. *Existence of solution:* We need to find necessary and sufficient conditions satisfied by the terms a and b such that the system has at least one solution. These are called *integrability conditions* and they involve algebraic relations containing combinations of the vector fields X, Y and their bracket, $[X, Y]$, applied onto the functions a and b.

2. *Uniqueness of solution:* This means that if a solution exists, then it is unique. This is equivalent to showing that the associated homogeneous system, $Xu = 0, Yu = 0$, has only the trivial solution, $u = 0$.

3. *Smoothness of solution:* The solutions (if exist) are smooth with respect to the underlying variables. In other words, if $a, b \in \mathcal{C}^\infty(\mathbb{R}^3)$ and X and Y are $\mathcal{C}^\infty$-vector fields, then the solution is also smooth, $u \in \mathcal{C}^\infty(\mathbb{R}^3)$.

Most cases treated explicitly in this chapter will cover examples of sub-elliptic systems containing two equations and three coordinates.

8.3 Nonholonomic systems of equations

The general case of sub-elliptic systems of differential equations on $\mathbb{R}^n$ containing k equations, with $k < n$, are given by

$$\alpha_{11}(x)\partial_{x_1}u + \alpha_{12}(x)\partial_{x_2}u + \cdots + \alpha_{1n}(x)\partial_{x_n}u = a_1(x)$$
$$\cdots\cdots\cdots\cdots\cdots\cdots\cdots\cdots\cdots\cdots\cdots\cdots\cdots\cdots\cdots$$
$$\alpha_{k1}(x)\partial_{x_1}u + \alpha_{k2}(x)\partial_{x_2}u + \cdots + \alpha_{kn}(x)\partial_{x_n}u = a_k(x)$$

with given $a_j, \alpha_{ij} \in \mathcal{C}^\infty(\mathbb{R}^n)$ and $\text{rank}(\alpha_{ij}(x)) = k$ for any $x \in \mathbb{R}^n$. The maximal rank condition was added to secure the independence of the equations.

Equivalently, if $X_j = \sum_{i=1}^n \alpha_{ji}(x)\partial_{x_i}$, with $1 \le j \le k$, are independent vector fields on $\mathbb{R}^n$ and $a_j \in \mathcal{C}^\infty(\mathbb{R}^n)$ are k given functions, then the sub-elliptic system of equations can be written as

$$X_1 u = a_1, \quad X_2 u = a_2, \ldots, X_k u = a_k. \tag{8.3.9}$$

The sub-elliptic system (8.3.9) is called *nonholonomic* if the vector fields $\{X_1, \ldots, X_k\}$ satisfy the bracket-generating condition at each point $x \in \mathbb{R}^n$.

The problem of finding integrability conditions (involving only relations between the vector fields X_j and the functions a_j's) for the nonholonomic system (8.3.9) was posed for the first time in the book [26] at page 55.

The horizontal gradient Let $\mathcal{D}$ be the distribution generated by the vector fields $\{X_1, \ldots, X_k\}$.

In order to characterize the rates of change of a smooth function f with respect to the vector fields X_j, we introduce the *horizontal gradient* of f by the formula

$$\nabla_h f = \sum_{j=1}^k X_j(f)\, X_j. \tag{8.3.10}$$

This vector field is introduced by analogy with the Euclidean gradient of f on $\mathbb{R}^n$, given by $\nabla_{Eu} f = \sum_{p=1}^n \partial_{x_p}(f)\, \partial_{x_p}$. Given a smooth function f on $\mathbb{R}^n$, there is an intrinsic relation between the vector field $\nabla_h f$ and the diffusion metric g^{ij}. We recall that g^{ij} is a degenerate matrix containing the quadratic form coefficients that defines the principal symbol of the operator $\frac{1}{2}\sum_{j=1}^{k} X_j^2$, see Chapter 3, page 127.

Proposition 8.3.1 *The horizontal gradient of the function f is the gradient with respect to the diffusion metric g^{ij}, i.e.*

$$\nabla_h f = \sum_{p=1}^{n} V^p \partial_{x_p}, \qquad (8.3.11)$$

with $V^p = \sum_{\ell=1}^{n} g^{\ell p}\partial_{x_\ell} f$.

Proof: The proof follows from a straightforward computation using $\sigma_{\ell j} = \alpha_{j\ell}$ and $g^{ij} = (\sigma\sigma^T)_{ij}$ as follows

$$\begin{aligned}
\nabla_h f &= \sum_{j=1}^{k} X_j(f) X_j = \sum_{j=1}^{k} X_j(f) \sum_{p=1}^{n} \alpha_{jp}(x)\partial_{x_p} \\
&= \sum_j \sum_\ell \alpha_{j\ell}(x)\partial_{x_\ell} f \sum_p \alpha_{jp}(x)\partial_{x_p} = \sum_p \Big[\sum_\ell \sum_j \alpha_{j\ell}(x)\alpha_{jp}(x)\partial_{x_\ell} f\Big]\partial_{x_p} \\
&= \sum_p \Big[\sum_\ell \sum_j \sigma_{\ell j}(x)\sigma_{pj}(x)\partial_{x_\ell} f\Big]\partial_{x_p} = \sum_p \Big[\sum_\ell (\sigma\sigma^T)_{\ell p}\partial_{x_\ell} f\Big]\partial_{x_p} \\
&= \sum_p \Big[\sum_\ell g^{\ell p}\partial_{x_\ell} f\Big]\partial_{x_p} = \sum_{p=1}^{n} V^p \partial_{x_p}.
\end{aligned}$$

■

Problem reformulation Given the smooth functions a_j, we consider the vector field $U = \sum_{k=1}^{k} a_j X_j$, which is horizontal, i.e. it is tangent to the distribution $\mathcal{D}$. Then the sub-elliptic system (8.3.9) can be written equivalently using (8.3.10) as $\nabla_h u = U$.

If we represent $U = \sum_p U^p \partial_{x_p}$ in Cartesian coordinates on $\mathbb{R}^n$, then using Proposition 8.3.1 we obtain

$$\sum_{\ell=1}^{n} g^{\ell p}(x)\partial_{x_\ell} u = U^p(x), \quad 1 \le p \le n.$$

Since $g^{\ell p}(x)$ is degenerated, we cannot invert it to obtain the partial derivatives $\partial_{x_\ell} u$. Working around this issue can be done for a series of examples.

8.4 Uniqueness and Smoothness

If $\nabla_h f = 0$, then $\nabla_{Eu} f = (\partial_{x_1} f, \dots, \partial_{x_n} f)^T$ satisfies the linear system

$$\sum_{\ell=1}^{n} g^{\ell p}(x)\partial_{x_\ell} f = 0.$$

Since for any $x \in \mathbb{R}^n$ we have

$$\operatorname{rank} g^{\ell p}(x) = \operatorname{rank}(\sigma\sigma^T)(x) = \operatorname{rank}\sigma(x) = \operatorname{rank}\alpha_{ij}(x) = k < n,$$

then the space of solutions is $n-k$ dimensional. Assuming that the first k equations are independent, it follows that

$$\partial_{x_1} f = \cdots = \partial_{x_k} f = 0.$$

Then there is a function $\varphi \in \mathcal{C}^\infty(\mathbb{R}^n)$ such that $f(x) = \varphi(x_{k+1}, \dots, x_n)$. Therefore, the functions f satisfying $\nabla_h f = 0$ depend on $n-k$ coordinates. We shall see in the next result that if the bracket-generating condition is satisfied, then the solutions f are just constants.

Proposition 8.4.1 (Uniqueness) *The nonholonomic sub-elliptic system (8.3.9) has a unique solution up to a constant.*

Proof: By contradiction, we assume that the sub-elliptic system (8.3.9) has two distinct solutions, u and v. Then the difference, $\phi = u - v$, satisfies the associated homogeneous sub-elliptic system

$$X_1\phi = 0, \;\; X_2\phi = 0, \;\; \dots, \;\; X_k\phi = 0.$$

Since $\langle X_j, \nabla_{Eu}\phi\rangle = X_j\phi$, it follows that $\nabla_{Eu}\phi$ is perpendicular to the vector field X_j with respect to the Euclidean scalar product in $\mathbb{R}^n$. Using $[X_i, X_j]\phi = X_i(X_j\phi) - X_j(X_i\phi) = 0$, if follows that $\nabla_{Eu}\phi$ is perpendicular to the bracket vector field $[X_i, X_j]$ in $\mathbb{R}^n$. Iterating the brackets we obtain

$$[X_{j_1}, [X_{j_2}, [X_{j_3}, \dots]]]\phi = 0.$$

Therefore, $\nabla_{Eu}\phi$ is perpendicular to the vector fields

$$X_{j_1}, [X_{j_1}, X_{j_2}], [X_{j_1}, [X_{j_2}, X_{j_3}]], \cdots, [X_{j_1}, [X_{j_2}, [X_{j_3}, \dots]]], \cdots \quad , 0 \le j_k \le r.$$

Since these vector fields span the tangent space $T_x\mathbb{R}^n$ at each $x \in \mathbb{R}^n$, it follows that $\nabla_{Eu}\phi = 0$ on $\mathbb{R}^n$ and hence $\phi = c$, constant. ■

Remark 8.4.2 The uniqueness of solution can be also addressed by Chow-Rashevskii's theorem, Theorem 3.7.2. Since the horizontal distribution $\mathcal{D} = \text{span}\{X_1, X_2, \dots X_k\}$ is bracket-generating, then any two points can be connected by a piece-wise horizontal path. Therefore, for any point $x \in \mathbb{R}^n$, there is a piece-wise horizontal path γ such that

$$\gamma(0) = 0_{\mathbb{R}^n}, \qquad \gamma(t) = x$$

and γ is obtained by the concatenation of integral curves of the vector fields X_i as

$$\gamma(t) = \phi_{X_{i_1}}(s_1) \circ \phi_{X_{i_2}}(s_2) \circ \cdots \circ \phi_{X_{i_p}}(s_p),$$

where $\dot{\phi}_{X_j}(s) = X_{j|\phi_{X_j}(s)}$, $i_r \in \{1, \dots, k\}$ and $s_1 + \cdots + s_p = t$. Assuming that

$$X_1 u = 0, \quad X_2 u = 0, \; \dots \;, X_k u = 0,$$

we obtain that u is preserved along the integral curves $\phi_{X_j}(s)$. This follows from the following application of the chain rule

$$\begin{aligned} \frac{d}{ds} u(\phi_{X_j}(s)) &= \sum_i \frac{\partial u}{\partial x_i} \frac{\partial \phi^i_{X_j}(s)}{ds} = \langle \nabla_E u, \dot{\phi}_{X_j}(s) \rangle \\ &= \langle \nabla_E u, X_{j|\phi_{X_j}(s)} \rangle = X_j(u) = 0. \end{aligned}$$

By the continuity of γ, the transitivity implies

$$\begin{aligned} u(0) &= u(\gamma(0)) = u(\phi_{X_{i_1}}(0)) = u(\phi_{X_{i_1}}(s_1)) = u(\phi_{X_{i_2}}(0)) = u(\phi_{X_{i_2}}(s_2)) = \cdots \\ &= u(\phi_{X_{i_p}}(0)) = u(\phi_{X_{i_p}}(s_p)) = u(\gamma(t)) = u(x), \qquad \forall x \in \mathbb{R}^n. \end{aligned}$$

From here, it follows that u is a constant function.

The next result deals with the smoothness of solutions.

Proposition 8.4.3 (Smoothness) *Let $a_j \in \mathcal{C}^\infty(\mathbb{R}^n)$ and consider the smooth vector fields X_j on $\mathbb{R}^n$ satisfying the bracket-generating condition. If the non-holonomic sub-elliptic system (8.3.9) has a solution u, then $u \in \mathcal{C}^\infty(\mathbb{R}^n)$.*

Proof: Consider the sub-elliptic operator $\Delta_X = \frac{1}{2}\sum_{j=1}^k X_j^2$. By Hörmander's theorem, Theorem 3.4.4, the operator Δ_X is hypoelliptic. Since

$$\Delta_X u = \frac{1}{2}\sum_{j=1}^k X_j(a_j) \in \mathcal{C}^\infty(\mathbb{R}^n)$$

then $u \in \mathcal{C}^\infty(\mathbb{R}^n)$. ∎

8.5 Solutions Existence

The solution of the nonholonomic sub-elliptic system (8.3.9) exists provided some integrability conditions hold. These conditions depend on the properties of the distribution $\mathcal{D}$, such as rank, step, etc., and their structure is specific to each particular case of distribution. We shall investigate in the following several examples.

8.5.1 The Heisenberg Distribution

We recall that the Heisenberg distribution on $\mathbb{R}^3$ is generated by the vector fields

$$X_1 = \partial_x + 2y\partial_z, \qquad X_2 = \partial_y - 2x\partial_z.$$

The associated sub-elliptic system of equations becomes

$$\partial_x u + 2y\partial_z u = a(x, y, z) \tag{8.5.12}$$

$$\partial_y u - 2x\partial_z u = b(x, y, z) \tag{8.5.13}$$

for two given functions $a, b \in \mathcal{C}^\infty(\mathbb{R}^3)$. Since $[X_1, X_2] = -4\partial_z$, then the system of vector fields $\{X_1, X_2, [X_1, X_2]\}$ are linearly independent on $\mathbb{R}^3$, and hence the distribution is bracket-generating. Consequently, by Proposition 8.4.1 the system has at most one solution and by Proposition 8.4.3 the solution is smooth. We still need to investigate the existence of solutions. This shall be done by the next result, which follows Calin and Chang [26].

Proposition 8.5.1 *The nonholonomic sub-elliptic system (8.5.12)-(8.5.13) has a solution $u \in \mathcal{C}^\infty(\mathbb{R}^3)$ if and only if a and b satisfy the following integrability conditions*

$$X_1^2 b = (X_1X_2 + [X_1, X_2])a \tag{8.5.14}$$

$$X_2^2 a = (X_2X_1 + [X_2, X_1])b. \tag{8.5.15}$$

Proof: Let u be a solution of the system (8.5.12)-(8.5.13) and denote $\phi = X_2 a - X_1 b$. Then $4\partial_z u = [X_2, X_1]u = X_2X_1u - X_1X_2u = X_2a - X_1b = \phi$. We have

$$X_1 u = \partial_x u + 2y\partial_z u = a \;\Leftrightarrow\; \partial_x u = a - 2y\partial_z u = a - \frac{1}{2}y\phi$$

$$X_2 u = \partial_y u - 2x\partial_z u = b \;\Leftrightarrow\; \partial_y u = b + 2x\partial_z u = b + \frac{1}{2}x\phi.$$

Therefore the sub-elliptic system (8.5.12)-(8.5.13) can be written as an equivalent elliptic system as

$$\partial_x u = \tilde{a}, \qquad \partial_y u = \tilde{b}, \qquad \partial_z u = \tilde{\phi}, \tag{8.5.16}$$

where

$$\tilde{a} = a - \frac{1}{2}y\phi, \quad \tilde{b} = b + \frac{1}{2}x\phi, \qquad \tilde{\phi} = \frac{1}{4}\phi. \tag{8.5.17}$$

It is known that the differential system (8.5.16) has solutions if and only if the following integrability conditions are satisfied

$$\tilde{a}_y = \tilde{b}_x, \qquad \tilde{a}_z = \tilde{\phi}_x, \qquad \tilde{b}_z = \tilde{\phi}_y. \tag{8.5.18}$$

We shall write these conditions in terms of the functions a, b and vector fields X_1 and X_2 as follows

$$\tilde{a}_y = \tilde{b}_x \Leftrightarrow a_y - b_x = \phi + \frac{1}{2}(x\phi_x + y\phi_y) \tag{8.5.19}$$

$$\begin{aligned}
\tilde{a}_z = \tilde{\phi}_x &\Leftrightarrow a_z - \frac{1}{2}y\phi_z = \frac{1}{4}\phi_x \Leftrightarrow 4a_z = \phi_x + 2y\phi_z \\
&\Leftrightarrow [X_2, X_1]a = X_1\phi \Leftrightarrow [X_2, X_1]a = X_1X_2a - X_1^2 b \\
&\Leftrightarrow X_1^2 b = (X_1X_2 + [X_1, X_2])a
\end{aligned} \tag{8.5.20}$$

$$\begin{aligned}
\tilde{b}_z = \tilde{\phi}_y &\Leftrightarrow b_z + \frac{1}{2}x\phi_z = \frac{1}{4}\phi_y \Leftrightarrow 4b_z = \phi_y - 2x\phi_z \\
&\Leftrightarrow [X_2, X_1]b = X_2\phi \Leftrightarrow [X_2, X_1]b = X_2^2 a - X_2X_1b \\
&\Leftrightarrow X_2^2 a = (X_2X_1 + [X_2, X_1])b.
\end{aligned} \tag{8.5.21}$$

In the following we show that the integrability relation (8.5.19) is a consequence of relations (8.5.20)-(8.5.21) and hence, there are only two independent integrability relations. To this end, we evaluate

$$\begin{aligned}
b_x - a_y &= (a_y - 2xa_z) + 2xa_z - (b_x + 2yb_z) + 2yb_z \\
&= X_2a - X_1b + 2xa_z + 2yb_z \\
&= \phi + \frac{1}{2}x[X_2, X_1]a + \frac{1}{2}y[X_2, X_1]b \qquad (\text{use } (8.5.20) - (8.5.21)) \\
&= \phi + \frac{1}{2}(xX_1\phi + yX_2\phi) \\
&= \phi + \frac{1}{2}(x\phi_x + 2xy\phi_z + y\phi_y - 2xy\phi_z) \\
&= \phi + \frac{1}{2}(x\phi_x + y\phi_y),
\end{aligned}$$

which recovers relation (8.5.19). We note that the integrability relations (8.5.20)-(8.5.21) have been used in their equivalent forms

$$[X_2, X_1]a = X_1\phi, \qquad [X_2, X_1]b = X_2\phi.$$

■

Explicit construction of solutions We shall provide a closed form expression for the solution u of the sub-elliptic system (8.5.12)-(8.5.13). The computation follows Calin et al. [31]. We consider the vector field $U = \tilde{a}\partial_x + \tilde{b}\partial_y + \tilde{\phi}\partial_z$ with the coefficients given by (8.5.17). Let

$$A = \begin{pmatrix} 1 & 0 & 2y \\ 0 & 1 & -2x \end{pmatrix}$$

be the coefficients matrix of the vector fields X_1 and X_2. Then

$$AU = \begin{pmatrix} \tilde{a} + 2y\tilde{\phi} \\ \tilde{b} - 2x\tilde{\phi} \end{pmatrix} = \begin{pmatrix} a \\ b \end{pmatrix}.$$

Set $r = (x, y, z)$ and consider the work of the field U along the curve $\{sr; 0 \leq s \leq 1\}$

$$u(r) = \int_0^1 \langle U(sr), r \rangle ds. \tag{8.5.22}$$

Multiplying by the matrix A to the left of relation (8.1.3)

$$\nabla_{Eu} u(r) = U(r) + \int_0^1 sr \times \text{curl}\, U(sr)\, ds \tag{8.5.23}$$

for the aforementioned vector field U, we obtain

$$\begin{aligned} A\nabla_{Eu} u(r) &= AU(r) + \int_0^1 A(sr \times \text{curl}\, U(sr))\, ds \Leftrightarrow \\ \begin{pmatrix} X_1 u \\ X_2 u \end{pmatrix}(r) &= \begin{pmatrix} a \\ b \end{pmatrix} + \int_0^1 \begin{pmatrix} \langle (1,0,2y), sr \times \text{curl}\, U(sr) \rangle \\ \langle (0,1,-2x), sr \times \text{curl}\, U(sr) \rangle \end{pmatrix} ds \Leftrightarrow \\ \begin{pmatrix} X_1 u \\ X_2 u \end{pmatrix}(r) &= \begin{pmatrix} a \\ b \end{pmatrix} + \int_0^1 \begin{pmatrix} \langle \text{curl}\, U(sr), (1,0,2y) \times sr \rangle \\ \langle \text{curl}\, U(sr), (0,1,-2x) \times sr \rangle \end{pmatrix} ds \end{aligned} \tag{8.5.24}$$

where we used the invariance property of the mixed product $\langle u, v \times w \rangle = \langle w, u \times v \rangle$.

We shall show by a direct computation that the integrand matrix can be written in terms of the integrability conditions. The first two components of the vector field

$$\text{curl U} = \big((\text{curl U})^1, (\text{curl U})^2, (\text{curl U})^3 \big)$$

are computed as

$$\begin{aligned} (\text{curl U})^1 &= \frac{\partial \tilde{\phi}}{\partial y} - \frac{\partial \tilde{b}}{\partial z} = \frac{1}{4}\partial_y \phi - \partial_z\big(b + \frac{1}{2}x\phi\big) = \frac{1}{4}(\partial_y \phi - 2x\partial_z \phi) - \partial_z b \\ &= \frac{1}{4}X_2\phi - \partial_z b = \frac{1}{4}X_2(X_2 a - X_1 b) + \frac{1}{4}[X_1, X_2]b \\ &= \frac{1}{4}\Big\{X_2^2 a - \big(X_2 X_1 + [X_2, X_1]\big)b\Big\}. \end{aligned}$$

Similarly,

$$
\begin{aligned}
(\text{curl U})^2 &= \frac{\partial \tilde{a}}{\partial z} - \frac{\partial \tilde{\phi}}{\partial x} = \partial_z\Big(a - \frac{1}{2}y\phi\Big) - \frac{1}{4}\partial_x\phi \\
&= -\frac{1}{4}\Big(- 4\partial_z a + \partial_x\phi + 2y\partial_z\phi\Big) = -\frac{1}{4}\Big\{[X_1, X_2]a + X_1\phi\Big\} \\
&= -\frac{1}{4}\Big\{[X_1, X_2]a + X_1(X_2 a - X_1 b)\Big\} \\
&= \frac{1}{4}\Big\{X_1^2 b - \Big(X_1X_2 + [X_1, X_2]\Big)a\Big\}.
\end{aligned}
$$

The third component of the curl is a linear combination of the first two components. To show this, we first solve for $\partial_x\phi$ and $\partial_y\phi$ from the previous expressions as

$$
\begin{aligned}
\partial_x\phi &= -4(\text{curl U})^2 + 4\partial_z a - 2y\partial_z\phi \\
\partial_y\phi &= 4(\text{curl U})^1 + 2x\partial_z\phi + 4\partial_z b
\end{aligned}
$$

and combine them to get

$$
x\partial_x\phi + y\partial_y\phi = 4y(\text{curl U})^1 - 4x(\text{curl U})^2 + 4y\partial_z b + 4x\partial_z a.
$$

Then the third component of the curl vector becomes

$$
\begin{aligned}
(\text{curl U})^3 &= \frac{\partial \tilde{b}}{\partial x} - \frac{\partial \tilde{a}}{\partial y} = \partial_x\Big(b + \frac{1}{2}x\phi\Big) - \partial_y\Big(a - \frac{1}{2}y\phi\Big) \\
&= (\partial_x b - \partial_y a) + \phi + \frac{1}{2}(x\partial_x\phi + y\partial_y\phi) \\
&= (\partial_x b - \partial_y a) + \phi + 2y(\text{curl U})^1 - 2x(\text{curl U})^2 + 2y\partial_z b + 2x\partial_z a \\
&= X_1 b - X_2 a + \phi + 2y(\text{curl U})^1 - 2x(\text{curl U})^2 \\
&= 2y(\text{curl U})^1 - 2x(\text{curl U})^2.
\end{aligned}
$$

Evaluating the mixed product as a determinant and expanding over the first row we obtain

$$
\langle \text{curl}\, U(r), (1, 0, 2y) \times r\rangle = \begin{vmatrix} (\text{curl}\, U)^1 & (\text{curl}\, U)^2 & (\text{curl}\, U)^3 \\ 1 & 0 & 2y \\ x & y & z \end{vmatrix}
$$

$$
\begin{aligned}
&= (\text{curl}\, U)^1(-2y^2) - (\text{curl}\, U)^2(z - 2xy) + (\text{curl}\, U)^3 y \\
&= -z(\text{curl}\, U)^2 \\
&= -\frac{z}{4}\Big\{X_1^2 b - (X_1X_2 + [X_1, X_2])a\Big\}.
\end{aligned}
$$

A similar computation provides

$$\begin{aligned}\langle \text{curl}\, U(r), (0,1,-2x)\times r\rangle &= (\text{curl}\, U)^1(z+2xy) - 2x^2(\text{curl}\, U)^2 - x(\text{curl}\, U)^3 \\ &= z(\text{curl}\, U)^1 \\ &= \frac{z}{4}\Big\{X_2^2 a - \big(X_2X_1 + [X_2, X_1]\big)b\Big\}.\end{aligned}$$

Substituting into (8.5.24) we obtain

$$(X_1u)(r) = a(r) - \int_0^1 \frac{sz}{4}\Big\{X_1^2 b - \big(X_1X_2 + [X_1, X_2]\big)a\Big\}(sr)\, ds \qquad (8.5.25)$$

$$(X_2u)(r) = b(r) + \int_0^1 \frac{sz}{4}\Big\{X_2^2 a - \big(X_2X_1 + [X_2, X_1]\big)b\Big\}(sr)\, ds. \qquad (8.5.26)$$

Therefore, if the integrability conditions (8.5.14)-(8.5.15) are satisfied, then $u(r)$ given by (8.5.22) becomes a solution of the system

$$X_1 u = a, \qquad X_2 u = b.$$

The solution as a Cesáro-Voltera path integral Furthermore, if $\gamma(s)$ is a horizontal curve such that $\gamma(0) = 0$ and $\gamma(1) = (x, y, z)$ and g denotes the subRiemannin metric in which $\{X_1, X_2\}$ are orthonormal, then the solution can be also expressed as

$$u(x, y, z) = \int_0^1 g(V_{|\gamma(s)}, \dot\gamma(s))\, ds, \qquad (8.5.27)$$

where $V = aX_1 + bX_2$ is the horizontal field associated with the system. The solution represents the work done by the vector field V along any horizontal curve starting at the origin and ending at (x, y, z).

To show this, we note first that if the integrability conditions (8.5.14)-(8.5.15) are satisfied, then the vector field U is potential, and then by the Fundamental Theorem of Calculus we obtain the dependence of the integral only on the end points of γ as

$$\int_0^1 \langle U(\gamma(s)), \dot\gamma(s)\rangle ds = \int_0^1 \langle \nabla_E u_{|\gamma(s)}, \dot\gamma(s)\rangle ds = \int_0^1 \frac{d}{ds} u(\gamma(s)) = u(x,y,z) - u(0),$$

On the other hand, using the horizontal constraint we obtain

$$
\begin{aligned}
\int_0^1 \langle U(\gamma(s)), \dot\gamma(s)\rangle ds &= \int_0^1 [\tilde a\dot x(s) + \tilde b\dot y(s) + \tilde\phi\dot z(s)]\, ds \\
&= \int_0^1 \Big[\big(a - \frac{1}{2}y\phi\big)\dot x(s) + \big(b + \frac{1}{2}x\phi\big)\dot y(s) + \frac{1}{4}\phi\dot z(s)\Big]\, ds \\
&= \int_0^1 \big(a\dot x(s) + b\dot y(s)\big) + \frac{1}{4}\phi \underbrace{\big(\dot z(s) + 2x(s)\dot y(s) - 2y(s)\dot x(s)\big)}_{=0}\, ds \\
&= \int_0^1 g\big(aX_1 + bX_2, \dot x(s)X_1 + \dot y(s)X_2\big)\, ds \\
&= \int_0^1 g(V_{|\gamma(s)}, \dot\gamma(s))\, ds.
\end{aligned}
$$

Comparing the last two relations yields

$$
u(x, y, z) = u(0) + \int_0^1 g(V_{|\gamma(s)}, \dot\gamma(s))\, ds.
$$

The constant $u(0)$ can be neglected, as the uniqueness of solution holds up to an additive constant.

The next section will show that the integrability conditions (8.5.14)-(8.5.15) are specific to nilpotent distributions of class 2.

8.5.2 Nilpotent Distributions

A distribution $\mathcal{D} = \mathrm{span}\{X_1, \dots, X_k\}$ in $\mathbb{R}^n$ is called *nilpotent* at the point $x \in \mathbb{R}^n$ if there is an integer $p(x) \geq 1$ such that all the Lie brackets of vector fields $\{X_i\}$ iterated $p(x)$ times vanish. The smallest integer $p(x)$ with this property is called the nilpotence class of $\mathcal{D}$ at x. If the class $p = p(x)$ is independent of the point x, then the distributions $\mathcal{D}$ is called nilpotent of class p on $\mathbb{R}^n$.

It is worth noting that the nilpotence class describes the functional nature of the distribution, dealing with the degree of non-commutativity of the vector fields, containing information about the coefficients of the vector fields X_i, such as the degree of the polynomial coefficients. This should not be misunderstood with the *step* of the distribution, which describes the degree of non-integrability of the distribution and is applied to bracket-generating distributions, the step at x being given by 1 plus the number of brackets needed to span $T_x\mathbb{R}^n$.

The fact that the nilpotence class and the step of a distribution are distinct entities can be seen from the following examples.

Example 8.5.1 The nilpotence class and the step can be equal, as in the case of the Heisenberg distribution that is generated by $X_1 = \partial_x + 2y\partial_z$ and $X_2 = \partial_y - 2x\partial_z$. In this case both the step and the nilpotence class are equal to 2 on $\mathbb{R}^3$.

Example 8.5.2 The distribution which is generated by $X_1 = \partial_x + 2e^y\partial_z$ and $X_2 = \partial_y$ has the step be equal to 2 everywhere, while the nilpotence class is infinite.

Example 8.5.3 The distribution generated by $X_1 = \partial_y + zx\partial_z$ and $X_2 = \partial_x$ is nilpotent with the nilpotence class 2, but the distribution is not bracket-generating along the plane $\{z = 0\}$, i.e. it does not have a finite step there.

Distributions of rank 2 Consider the distribution $\mathcal{D} = \text{span}\{X_1, X_2\}$ on $\mathbb{R}^n$, with $n \geq 3$. We assume that $\mathcal{D}$ has rank 2, i.e. $\{X_1, X_2\}$ are linearly independent. The first few nilpotence class distributions are characterized as follows:

The nilpotence class $p = 1$ The distribution $\mathcal{D}$ has nilpotence class $p = 1$ if and only if $[X_1, X_2] = 0$, i.e. the vector fields commute.

The nilpotence class $p = 2$ The distribution $\mathcal{D}$ has nilpotence class $p = 2$ if and only if

$$[X_1, [X_1, X_2]] = 0, \qquad [X_2, [X_2, X_1]] = 0, \qquad [X_1, X_2] \neq 0.$$

Using the definition of the Lie bracket, these relations are equivalent to

$$X_1^2 X_2 + X_2 X_1^2 = 2X_1 X_2 X_1 \tag{8.5.28}$$
$$X_2^2 X_1 + X_1 X_2^2 = 2X_2 X_1 X_2 \tag{8.5.29}$$
$$X_1 X_2 \neq X_2 X_1. \tag{8.5.30}$$

The nilpotence class $p = 3$ The distribution $\mathcal{D}$ has nilpotence class $p = 3$ if and only if

$$\begin{aligned}
[X_1, [X_1, [X_1, X_2]]] &= 0, \qquad & [X_2, [X_1, [X_1, X_2]]] &= 0 \\
[X_1, [X_2, [X_1, X_2]]] &= 0, \qquad & [X_2, [X_2, [X_1, X_2]]] &= 0 \\
[X_1, [X_1, X_2]] &\neq 0, \qquad & [X_2, [X_2, X_1]] &\neq 0 \\
[X_1, X_2] &\neq 0. & &
\end{aligned}$$

Expanding the Lie brackets, these become

$$
\begin{aligned}
X_1^3X_2 + 3X_1X_2X_1^2 &= X_2X_1^3 + 3X_1^2X_2X_1\\
X_2^2X_1^2 + 2(X_1X_2)^2 &= X_1^2X_2^2 + 2(X_2X_1)^2\\
X_1^2X_2^2 + 2(X_2X_1)^2 &= X_2^2X_1^2 + 2(X_1X_2)^2\\
X_2^3X_1 + 3X_2X_1X_2^2 &= X_1X_2^3 + 3X_2^2X_1X_2\\
X_1^2X_2 + X_2X_1^2 &\neq 2X_1X_2X_1\\
X_2^2X_1 + X_1X_2^2 &\neq 2X_2X_1X_2\\
X_1X_2 &\neq X_2X_1.
\end{aligned}
$$

These relations involve some symmetries. Thus, the second and the third equations are identical, while the fourth is obtained from the first equation by swapping X_1 and X_2. Similar swapping symmetry applies to the fifth and sixth equations.

The following results deal with necessary conditions for the sub-elliptic system to have a solution for the cases $p \in \{1, 2\}$.

Proposition 8.5.2 *Let $\mathcal{D} = \text{span}\{X_1, X_2\}$ be a nilpotent rank 2 distribution with nilpotence class $p = 1$ on $\mathbb{R}^n$, $n \geq 3$. If the system*

$$X_1u = a, \qquad X_2u = b \tag{8.5.31}$$

has $\mathcal{C}^2$-solutions if and only if the functions a and b satisfy the integrability condition $X_1b = X_2a$.

Proof: Since the nilpotence class is $p = 1$, then X_1 and X_2 commute. Then an application of Proposition 8.1.6 leads to the desired result. ∎

Remark 8.5.3 In this case the system (8.5.31) is not non-holonomic and the solution uniqueness (up to an additive constant) does not hold anymore. Since $[X_1, X_2] = 0 \in \mathcal{D}$, the distribution $\mathcal{D}$ is involutive and hence integrable, so for any $q \in \mathbb{R}^n$ there is a two-dimensional manifold S_q passing through q such that X_1 and X_2 are tangent to S_q. If S_q is represented locally by

$$x_1 = \varphi_1(x), \quad x_2 = \varphi_2(x), \quad x_3 = c_3, \quad \dots, \quad x_n = c_n$$

with c_i constants, then any smooth function $f(x_3, \dots, x_n)$ is a solution of the homogeneous system $X_1f = 0$, $X_2f = 0$.

Proposition 8.5.4 *Let $\mathcal{D} = \text{span}\{X_1, X_2\}$ be a nilpotent rank 2 distribution with nilpotence class $p = 2$. If the system*

$$X_1u = a, \qquad X_2u = b$$

has $\mathcal{C}^3$-solutions, then the functions a and b satisfy the following integrability conditions

$$X_1^2 b = (X_1 X_2 + [X_1, X_2])a \tag{8.5.32}$$
$$X_2^2 a = (X_2 X_1 + [X_2, X_1])b. \tag{8.5.33}$$

Proof: Let u be a solution of the system. Applying relation (8.5.28) on u we obtain

$$\begin{aligned} X_1^2(X_2 u) + X_2 X_1(X_1 u) = 2X_1 X_2(X_1 u) \Leftrightarrow \\ X_1^2 b + X_2 X_1 a = 2X_1 X_2 a \Leftrightarrow \\ X_1^2 b = ([X_1, X_2] + X_1 X_2)a, \end{aligned}$$

which recovers (8.5.32). Similarly, applying equation (8.5.29) on u leads to relation (8.5.33). ■

A partial converse of the previous result is given next. We note first that the vector fields

$$X_1 = \partial_x - (\alpha y + \beta)\partial_z, \qquad X_2 = \partial_y + (\gamma x + \delta)\partial_z, \tag{8.5.34}$$

with $\alpha, \beta, \gamma, \delta \in \mathbb{R}$ form a nilpotent distribution of class 2 on $\mathbb{R}^3$.

Proposition 8.5.5 *Let*

$$X_1 = \partial_x - A_1(y)\partial_z, \quad X_2 = \partial_y + A_2(x)\partial_z$$

and $a, b \in \mathcal{C}^\infty(\mathbb{R})$ such that

$$X_1^2 b = (X_1 X_2 + [X_1, X_2])a \tag{8.5.35}$$
$$X_2^2 a = (X_2 X_1 + [X_2, X_1])b \tag{8.5.36}$$
$$X_1 b \neq X_2 a. \tag{8.5.37}$$

Then if the system

$$X_1 f = a, \quad X_2 f = b$$

has solutions, then the distribution $\mathcal{D} = \mathrm{span}\{X_1, X_2\}$ is nilpotent of class 2.

Proof: Let f be a solution of the system. Then $[X_1, X_2]f = X_1 b - X_2 a \neq 0$ by (8.5.37). On the other side, a computation provides $[X_1, X_2]f = \lambda \partial_z f$, with $\lambda = \frac{\partial A_2}{\partial x} + \frac{\partial A_1}{\partial y}$. It follows that $\lambda \partial_z f \neq 0$, which implies that $\lambda \neq 0$ and that the solution f depends on the variable z. Denoting $\phi = X_1 b - X_2 a$, we have

$$\partial_z f = \frac{1}{\lambda}\phi = \tilde{\phi}.$$

The equation $X_1 f = a$ can be written equivalently as

$$\partial_x f = a + A_1 \partial_z f = \tilde{a},$$

where $\tilde{a} = a + \frac{1}{\lambda} A_1 \phi = a + A_1 \tilde{\phi}$. Similarly, the equation $X_2 f = b$ writes as

$$\partial_y f = b - A_2 \partial_z f = \tilde{b},$$

with $\tilde{b} = b - A_2 \tilde{\phi}$. Since f is a solution of the system

$$\partial_x f = \tilde{a}, \qquad \partial_y f = \tilde{b}, \qquad \partial_z f = \tilde{\phi}$$

then the following integrability conditions must hold

$$\tilde{a}_z = \tilde{\phi}_x, \qquad \tilde{b}_z = \tilde{\phi}_y, \qquad \tilde{a}_y = \tilde{b}_x. \tag{8.5.38}$$

We shall show next that the system (8.5.38) implies λ = constant. To this end, we transform the first equation as

$$\tilde{a}_z = \tilde{\phi}_x \Leftrightarrow a_z = X_1 \tilde{\phi} \Leftrightarrow \frac{1}{\lambda}[X_1, X_2]a = \frac{1}{\lambda} X_1 \phi + \phi X_1 \Big(\frac{1}{\lambda}\Big) \Leftrightarrow$$
$$[X_1, X_2]a = X_1(X_1 b - X_2 a) + \lambda \phi X_1 \Big(\frac{1}{\lambda}\Big) \Longleftrightarrow$$
$$([X_1, X_2] + X_1 X_2)a = X^2 b + \lambda \phi X_1 \Big(\frac{1}{\lambda}\Big) \Longleftrightarrow \lambda \phi X_1 \Big(\frac{1}{\lambda}\Big) = 0 \Longleftrightarrow$$
$$X_1 \Big(\frac{1}{\lambda}\Big) = 0 \Longleftrightarrow \partial_x \Big(\frac{1}{\lambda}\Big) = 0,$$

where we used relation (8.5.35) and the fact that $\lambda \phi \neq 0$. Similarly, the second equation of the system (8.5.38) is equivalent to

$$\tilde{b}_z = \tilde{\phi}_y \Longleftrightarrow b_z = X_2 \tilde{\phi} \Longleftrightarrow$$
$$X_2^2 a = (X_2 X_1 + [X_2, X_1])b + \lambda \phi X_1 \Big(\frac{1}{\lambda}\Big) \Longleftrightarrow \partial_y \Big(\frac{1}{\lambda}\Big) = 0,$$

by relation (8.5.36). Since λ is independent of z, the last two identities imply that λ is constant. Then separating variable in the equation

$$\lambda = \frac{\partial A_2(x)}{\partial x} + \frac{\partial A_1(y)}{\partial y}$$

there is a constant C such that

$$\lambda - \frac{\partial A_2(x)}{\partial x} = C, \qquad \frac{\partial A_1(y)}{\partial y} = C,$$

which after integration yields $A_1(y) = Cy + C_0$ and $A_2(x) = (\lambda - C)x + C_1$, with C_0, C_1 constants. Therefore, the vector fields X_1 and X_2 are like in relation (8.5.34), and hence the distribution generated by X_1 and X_2 is nilpotent of class 2.

■

The next section deals with a result which is a reciprocal of Proposition 8.5.4 and a generalization of Proposition 8.5.1.

8.5.3 Heisenberg-type Distributions

A rank 2 distribution, $\mathcal{D} = \text{span}\{X,Y\}$, on $\mathbb{R}^3$ is of *Heisenberg-type* if

$$[X,Y] \neq 0, \qquad [X,[X,Y]] = [Y,[X,Y]] = 0$$

and $\{X, Y, [X,Y]\}$ are linearly independent at each point $x \in M$.

Equivalently, a *Heisenberg-type* distribution is a rank 2 distribution on $\mathbb{R}^3$, which is nilpotent of class 2 and has step 2 at each point.

Using the local character of the definition, the concept can be easily extended to differentiable manifolds of dimension 3.

Proposition 8.5.6 *Let $\mathcal{D} = \text{span}\{X,Y\}$ be a Heisenberg-type distribution on $\mathbb{R}^3$. Then for a pair of smooth functions a and b defined on $\mathbb{R}^3$, we have*

$$\left\{\begin{array}{ll} X^2b & = (XY + [X,Y])a \\ Y^2a & = (YX + [Y,X])b \end{array}\right. \Longleftrightarrow \left\{\begin{array}{l} \exists \text{ a smooth function } f \text{ on } \mathbb{R}^3 \\ \text{such that } Xf = a \text{ and } Yf = b. \end{array}\right.$$

Proof: " $\Longleftarrow$ " It follows from Proposition 8.5.4.

" $\Longrightarrow$ " We shall complete the sub-elliptic system to an equivalent elliptic system as follows. Let $T = [X,Y]$ and consider the Riemannian metric g with respect to which $\{X, Y, T\}$ is an orthonormal basis at each point (The existence of this type of metric has been discussed at page 131). Then by Lemma 8.1.5 the gradient of f can be written as $\nabla_g f = X(f)X + Y(f)Y + T(f)T$. We let $c = Xb - Ya$, then we complete to an elliptic system and apply Theorem 8.1.4 to obtain the sequence of equivalences

$$\left\{\begin{array}{l} Xf = a \\ Yf = b \end{array}\right. \Longleftrightarrow \left\{\begin{array}{l} Xf = a \\ Yf = b \\ Tf = c \end{array}\right. \Longleftrightarrow \nabla_g f = U \Longleftrightarrow \text{curl U} = 0,$$

where $U = aX + bY + cT$. Denote $A = \text{curl U}$. Since a tensor is zero if and only if vanishes on a basis, then

$$\text{curl U} = 0 \Longleftrightarrow \left\{\begin{array}{l} A(X,Y) = 0 \\ A(T,X) = 0 \\ A(T,Y) = 0. \end{array}\right.$$

Then it suffices to show the previous vanishing identities. We shall proceed by applying Lemma 8.1.2. Thus we have

$$\begin{aligned}A(X,Y) &= Yg(U,X) - Xg(U,Y) + g(U,[X,Y])\\&= Ya - Xb + c = Ya - Xb + Xb - Ya\\&= 0,\end{aligned}$$

$$\begin{aligned}A(T,X) &= Xg(U,T) - Tg(U,X) + g(U,[T,X])\\&= Xc - Ta = X^2b - XYa - Ta\\&= X^2b - (XY + [X,Y])a\\&= 0,\end{aligned}$$

where we used the nilpotency condition $[T,X] = [[X,Y],X] = 0$ and the integrability condition. Similarly, we can show that $A(T,Y) = 0$, by reducing it to the second integrability condition. Hence, curl U $= 0$, which ends the proof. ∎

8.6 The Engel Distribution

The Engel distribution on $\mathbb{R}^4$ is the distribution $\mathcal{E} = \text{span}\,\{Y_1, Y_2\}$, where the generating vector fields are given by

$$Y_1 = \partial_{x_2} + x_1\partial_{x_3} + \frac{1}{2}x_1^2\partial_{x_4} \tag{8.6.39}$$

$$Y_2 = \partial_{x_1}. \tag{8.6.40}$$

This is a rank 2 distribution with nilpotence class $p = 3$ and step $k = 3$ everywhere on $\mathbb{R}^4$. We have encountered this distribution at page 407, where we studied its associated sub-elliptic diffusion.

The reduced Engel vector fields The vector field Y_1 has a quadratic coefficient, $x_1^2/2$. To alleviate the difficulty of the problem, we shall perform a change of coordinates under which the new vector fields will have only linear coefficients.

Lemma 8.6.1 *Let $\phi : \mathbb{R}^4 \to \mathbb{R}^4$ be given by $y = \phi(x)$, with*

$$\begin{aligned}y_1 &= x_2\\y_2 &= x_1\\y_3 &= x_1x_2 - x_3\\y_4 &= \frac{1}{2}x_1^2x_2 - x_1x_3 + x_4.\end{aligned}$$

Then ϕ is a diffeomorphism, satisfying

$$\phi_*(Y_1) = \partial_{y_1}, \qquad \phi_*(Y_2) = \partial_{y_2} + y_1\partial_{y_3} + y_3\partial_{y_4},$$

where $\phi_* : T\mathbb{R}^4 \to T\mathbb{R}^4$ is the differential mapping of ϕ.

Proof: The jacobian matrix of ϕ is

$$[\phi_*] = \begin{pmatrix} 0 & 1 & 0 & 0 \\ 1 & 0 & 0 & 0 \\ x_2 & x_1 & -1 & 0 \\ x_1x_2 - x_3 & \frac{1}{2}x_1^2 & -x_1 & 1 \end{pmatrix}.$$

By swapping the first two rows, $[\phi_*]$ becomes a diagonal matrix with $\det[\phi_*] = 1$, and hence ϕ is a diffeomorphism. To show the desired relations it suffices to apply the jacobian matrix of ϕ to the component vectors of Y_1 and Y_2 as follows

$$[\phi_*][Y_1] = \begin{pmatrix} 0 & 1 & 0 & 0 \\ 1 & 0 & 0 & 0 \\ x_2 & x_1 & -1 & 0 \\ x_1x_2 - x_3 & \frac{1}{2}x_1^2 & -x_1 & 1 \end{pmatrix}\begin{pmatrix} 0 \\ 1 \\ x_1 \\ \frac{x_1^2}{2} \end{pmatrix} = \begin{pmatrix} 1 \\ 0 \\ x_1 - x_1 \\ \frac{1}{2}x_1^2 - x_1^2 + \frac{1}{2}x_1^2 \end{pmatrix} = \begin{pmatrix} 1 \\ 0 \\ 0 \\ 0 \end{pmatrix}$$

$$[\phi_*][Y_2] = \begin{pmatrix} 0 & 1 & 0 & 0 \\ 1 & 0 & 0 & 0 \\ x_2 & x_1 & -1 & 0 \\ x_1x_2 - x_3 & \frac{1}{2}x_1^2 & -x_1 & 1 \end{pmatrix}\begin{pmatrix} 1 \\ 0 \\ 0 \\ 0 \end{pmatrix} = \begin{pmatrix} 0 \\ 1 \\ x_2 \\ x_1x_2 - x_3 \end{pmatrix} = \begin{pmatrix} 0 \\ 1 \\ y_1 \\ y_3 \end{pmatrix}.$$

■

In the following we shall denote

$$Z_1 = \partial_{y_1} \tag{8.6.41}$$

$$Z_2 = \partial_{y_2} + y_1\partial_{y_3} + y_3\partial_{y_4} \tag{8.6.42}$$

and called them the *reduced Engel vector fields.*

Remark 8.6.2 We note that the coefficients of Y_1 depend on one variable, x_1, at the second power, while the coefficients of Z_2 depend on two variables, y_1 and y_3, at the first power. Thus, using the transformation ϕ we traded the degree for the number of variables.

Since for any two vector fields X and Y we have $\phi_*[X, Y] = [\phi_*X, \phi_*Y]$, it follows that the rank, nilpotence class and distribution step are preserved by the diffeomorphism ϕ. Therefore, the reduced Engel distribution $\tilde{\mathcal{E}} = \text{span}\,\{Z_1, Z_2\}$ shares the same characteristics as the Engel distribution $\mathcal{E} =$

span $\{Y_1, Y_2\}$, i.e., it has rank 2, nilpotence class $p = 2$ and step $k = 3$ everywhere on $\mathbb{R}^4$. More precisely, the commutation relations are

$$[Z_1, Z_2] = Z_3 = \partial_{y_3}, \qquad [Z_1, Z_3] = 0, \qquad [Z_2, Z_3] = -\partial_{y_4} = Z_4.$$

The sum of squares operator associated with the reduced Engel vector fields is given by

$$\begin{aligned}
\Delta_Z &= \frac{1}{2}(Z_1^2 + Z_2^2) = \frac{1}{2}\partial_{y_1}^2 + \frac{1}{2}(\partial_{y_2} + y_1\partial_{y_3} + y_3\partial_{y_4})^2 \\
&= \frac{1}{2}\Big(\partial_{y_1}^2 + \partial_{y_2}^2 + y_1^2\partial_{y_3}^2 + y_3^2\partial_{y_4}^2 + 2y_1\partial_{y_2}\partial_{y_3} + 2y_3\partial_{y_2}\partial_{y_4} + 2y_1y_3\partial_{y_3}\partial_{y_4} + y_1\partial_{y_4}\Big) \\
&= \frac{1}{2}\langle\sigma\sigma^T\partial_y, \partial_y\rangle + \frac{1}{2}y_1\partial_{y_4},
\end{aligned}$$

where

$$\sigma\sigma^T = \begin{pmatrix} 1 & 0 & 0 & 0 \\ 0 & 1 & y_1 & y_3 \\ 0 & y_1 & y_1^2 & y_1y_3 \\ 0 & y_4 & y_1y_3 & y_3^2 \end{pmatrix} \quad \text{and} \quad \sigma = \begin{pmatrix} 1 & 0 & 0 & 0 \\ 0 & 1 & 0 & 0 \\ 0 & y_1 & 0 & 0 \\ 0 & y_3 & 0 & 0 \end{pmatrix}.$$

Therefore, the associated diffusion with the reduced Engel vector fields is given by

$$\begin{aligned}
dY_1(t) &= dB_1(t) \\
dY_2(t) &= dB_2(t) \\
dY_3(t) &= Y_1(t)\,dB_2(t) \\
dY_4(t) &= Y_3(t)\,dB_2(t) + \frac{1}{2}Y_1(t)\,dt,
\end{aligned}$$

where $B_1(t)$ and $B_2(t)$ are two independent Brownian motions. The diffusion starting at the origin is given by

$$\begin{aligned}
Y_1(t) &= B_1(t) \\
Y_2(t) &= B_2(t) \\
Y_3(t) &= \int_0^t B_1(s)\,dB_2(s) \\
Y_4(t) &= \int_0^t Y_3(s)\,dB_2(s) + \frac{1}{2}\int_0^t B_1(s)\,ds.
\end{aligned}$$

We note that

$$\begin{aligned}
dY(t) &= (1,0,0,0)dB_1(t) + (0,1,y_1,y_3)dB_2(t) + \frac{1}{2}Y_1(t)dt \\
&= Z_{1|Y(t)}dB_1(t) + Z_{2|Y(t)}dB_2(t) + \frac{1}{2}B_1(t)dt \notin \mathcal{D}_{Y(t)},
\end{aligned}$$

which shows that the diffusion $Y(t)$ does not spread horizontally, due to the last drift term. Consequently, a diffeomorphism of the ambient space $\mathbb{R}^4$ might not preserve the diffusion horizontality (since ϕ changes the Engel diffusion $\mathcal{E}(t)$, which is horizontal, into the reduced Engel diffusion $Y(t)$, which is not horizontal).

This concept can be generalized as follows.

Engel pair of vector fields Two vector fields Z_1 and Z_2 on $\mathbb{R}^4$ form an *Engel pair* if they satisfy the following commutation relations

$$[Z_1, Z_2] = Z_3, \qquad [Z_1, Z_3] = 0, \qquad [Z_2, Z_3] = Z_4, \tag{8.6.43}$$

and all other iterated brackets vanish.

An *Engel-type distribution* is a distribution on $\mathbb{R}^4$ generated by an Engel pair $\{Z_1, Z_2\}$ that satisfies the bracket-generating condition.

For instance, the vector fields (8.6.39)-(8.6.40) and (8.6.41)-(8.6.42) form Engel pairs.

We shall study in the following the integrability conditions of the non-holonomic system

$$Z_1 u = a_1, \qquad Z_2 u = a_2$$

for an Engel pair of vector fields $\{Z_1, Z_2\}$ following Calin et al. [32].

Theorem 8.6.3 *Let $\mathcal{D} = \mathrm{span}\,\{Z_1, Z_2\}$ be an Engel-type distribution on $\mathbb{R}^4$. For a pair of smooth functions a_1 and a_2 on $\mathbb{R}^4$, we have*

$$\begin{cases} Z_1^2 a_2 = (Z_1 Z_2 + [Z_1, Z_2]) a_1 \\ Z_2^3 a_1 = (Z_2^2 Z_1 - Z_2 [Z_1, Z_2] - [Z_2, [Z_1, Z_2]]) a_2 \end{cases}$$

$$\Longleftrightarrow \begin{cases} \exists \text{ a smooth function } u \\ \text{such that } Z_1 u = a_1 \text{ and } Z_1 u = a_1 \end{cases}$$

Proof: " $\Longleftarrow$ " We assume there is a smooth function u on $\mathbb{R}^4$ such that $Z_1 u = a_1$ and $Z_1 u = a_1$. We shall show the integrability conditions by direct computation as follows

$$\begin{aligned} Z_1^2 a_2 &= (Z_1 Z_2 + [Z_1, Z_2]) a_1 \Longleftrightarrow \\ Z_1^2 Z_2 u &= (Z_1 Z_2 + [Z_1, Z_2]) Z_1 u \Longleftrightarrow \\ (Z_1^2 Z_2 - Z_1 Z_2 Z_1) u &= [Z_1, Z_2] Z_1 u \Longleftrightarrow \\ Z_1 (Z_1 Z_2 - Z_2 Z_1) u &= [Z_1, Z_2] Z_1 u \Longleftrightarrow \\ Z_1 [Z_1, Z_2] u &= [Z_1, Z_2] Z_1 u \Longleftrightarrow [Z_1, [Z_1, Z_2]] u = 0 \Longleftrightarrow \\ [Z_1, Z_3] u &= 0, \end{aligned}$$

which holds true since Z_1 and Z_3 commute. For the second integrability condition, we have

$$\begin{aligned} Z_2^3 a_1 &= (Z_2^2 Z_1 - Z_2[Z_1, Z_2] - [Z_2, [Z_1, Z_2]])a_2 \iff \\ Z_2^3 a_1 &= (Z_2^2 Z_1 - Z_2 Z_3 - [Z_2, Z_3])a_2 \iff \\ Z_2^3 a_1 - Z_2^2 Z_1 a_2 &= -Z_2 Z_3 a_2 - Z_2 Z_3 a_2 + Z_3 Z_2 a_2 \iff \\ Z_2^2(Z_2 a_1 - Z_1 a_2) &= -Z_2 Z_3 a_2 - Z_4 a_2 \iff \\ Z_2^2(Z_2 Z_1 - Z_1 Z_2)u &= -Z_2 Z_3 Z_2 u - Z_4 Z_2 u \iff \\ -Z_2^2 Z_3 u &= -Z_2 Z_3 Z_2 u - Z_4 Z_2 u \iff \\ (Z_2 Z_3 Z_2 - Z_2 Z_2 Z_3)u &= -Z_4 Z_2 u \iff Z_2[Z_3, Z_2] = -Z_4 Z_2 u \iff \\ -Z_2 Z_4 u &= -Z_4 Z_2 u \iff [Z_4, Z_2]u = 0, \end{aligned}$$

since Z_2 and Z_4 commute.

" $\Longrightarrow$ " We assume

$$\begin{aligned} Z_1^2 a_2 &= (Z_1 Z_2 + [Z_1, Z_2])a_1 \\ Z_2^3 a_1 &= (Z_2^2 Z_1 - Z_2[Z_1, Z_2] - [Z_2, [Z_1, Z_2]])a_2 \end{aligned}$$

and consider the vector fields Z_3 and Z_4 given by (8.6.43). Since $\{Z_1, Z_2\}$ is bracket-generating, the set $\{Z_1, Z_2, Z_3, Z_4\}$ forms a basis in $\mathbb{R}^4$. We endow $\mathbb{R}^4$ with the Riemannian metric g under which $\{Z_j\}$ is an orthonormal system and consider the gradient, see Lemma 8.1.5

$$\nabla_g u = \sum_{j=1}^{4} Z_j(u) Z_j.$$

Consider the vector field $U = \sum_{j=1}^4 a_j Z_j$, where

$$\begin{aligned} a_3 &= Z_1 a_2 - Z_2 a_1 \\ a_4 &= Z_2 a_3 - Z_3 a_2. \end{aligned}$$

Then the nonholonomic system is equivalent to the following elliptic system

$$\begin{cases} Z_1 u = a_1 \\ Z_2 u = a_2 \end{cases} \iff \begin{cases} Z_1 u = a_1 \\ Z_2 u = a_2 \\ Z_3 u = a_3 \\ Z_4 u = a_4 \end{cases} \iff \nabla_g u = U.$$

By Theorem 8.1.4 the last condition is equivalent to the condition curl U $= 0$. This fact will be proved on the basis $\{Z_1, Z_2, Z_3, Z_4\}$. If let $A =$ curl U, then it suffices to show

$$A(Z_1, Z_2) = 0, \qquad A(Z_1, Z_3) = 0, \qquad A(Z_1, Z_4) = 0$$

$$A(Z_2, Z_3) = 0, \qquad A(Z_2, Z_4) = 0, \qquad A(Z_3, Z_4) = 0.$$

We shall prove each of these identities one by one using Lemma 8.1.2.

(i) Using that $[Z_1, Z_2] = Z_3$, we have

$$\begin{aligned} A(Z_1, Z_2) &= Z_2 g(U, Z_1) - Z_1 g(U, Z_2) + g(U, [Z_1, Z_2]) \\ &= Z_2 a_1 - Z_1 a_2 + a_3 = 0. \end{aligned}$$

(ii) Using that $[Z_2, Z_3] = Z_4$, we have

$$\begin{aligned} A(Z_2, Z_3) &= Z_3 g(U, Z_2) - Z_2 g(U, Z_3) + g(U, [Z_2, Z_3]) \\ &= Z_3 a_2 - Z_2 a_3 + a_4 = 0. \end{aligned}$$

(iii) Using the first integrability condition and $[Z_1, Z_3] = 0$, we have

$$\begin{aligned} A(Z_1, Z_3) &= Z_3 g(U, Z_1) - Z_1 g(U, Z_3) + g(U, [Z_1, Z_3]) \\ &= Z_3 a_1 - Z_1 a_3 = [Z_1, Z_2] a_1 - Z_1 (Z_1 a_2 - Z_2 a_1) \\ &= ([Z_1, Z_2] + Z_1 Z_2) a_1 - Z_1^2 a_2 = 0. \end{aligned}$$

(iv) Using the second integrability condition and $[Z_2, Z_4] = 0$, we have

$$\begin{aligned} A(Z_2, Z_4) &= Z_4 g(U, Z_2) - Z_2 g(U, Z_4) + g(U, [Z_2, Z_4]) \\ &= Z_4 a_2 - Z_2 a_4 = [Z_2, Z_3] a_2 - Z_2 (Z_2 a_3 - Z_3 a_2) \\ &= [Z_2, Z_3] a_2 - Z_2^2 (Z_1 a_2 - Z_2 a_1) + Z_2 Z_3 a_2 \\ &= Z_2^3 a_1 - (Z_2^2 Z_1 - Z_2 Z_3 + [Z_3, Z_2]) a_2 \\ &= Z_2^3 a_1 - (Z_2^2 Z_1 - Z_2 [Z_1, Z_2] - [Z_2, [Z_1, Z_2]]) a_2 = 0. \end{aligned}$$

(v) Using $[Z_1, Z_4] = 0$, we have

$$\begin{aligned} A(Z_1, Z_4) &= Z_4 g(U, Z_1) - Z_1 g(U, Z_4) + g(U, [Z_1, Z_4]) \\ &= Z_4 a_1 - Z_1 a_4. \end{aligned}$$

Therefore, it suffices to show $Z_4 a_1 = Z_1 a_4$. We have shown in cases (iii) and (iv) that

$$Z_1 a_3 = Z_3 a_1, \qquad Z_2 a_4 = Z_4 a_3. \tag{8.6.44}$$

Since Z_1 and Z_3 commute, we evaluate

$$Z_1 a_4 = Z_1 (Z_2 a_3 - Z_3 a_2) = Z_1 Z_2 a_3 - Z_1 Z_3 a_2 = Z_1 Z_2 a_3 - Z_3 Z_1 a_2 \tag{8.6.45}$$

$$Z_4 a_1 = [Z_2, Z_3] a_1 = Z_2 Z_3 a_1 - Z_3 Z_2 a_1 = Z_2 Z_1 a_3 - Z_3 Z_2 a_1, \tag{8.6.46}$$

where we used the first equation of (8.6.44). Using (8.6.45) and (8.6.46) we have the following sequence of equivalencies

$$\begin{aligned}
Z_1 a_4 &= Z_4 a_1 \Longleftrightarrow \\
Z_1 Z_2 a_3 - Z_3 Z_1 a_2 &= Z_2 Z_1 a_3 - Z_3 Z_2 a_1 \Longleftrightarrow \\
(Z_1 Z_2 - Z_2 Z_1) a_3 &= Z_3 Z_1 a_2 - Z_3 Z_2 a_1 \Longleftrightarrow \\
[Z_1, Z_2] a_3 &= Z_3 (Z_1 a_2 - Z_2 a_1) \Longleftrightarrow \\
Z_3 a_3 &= Z_3 a_3,
\end{aligned}$$

which holds true. Hence, $A(Z_1, Z_4) = 0$.

(vi) Using $[Z_3, Z_4] = 0$, we have

$$\begin{aligned}
A(Z_3, Z_4) &= Z_4 g(U, Z_3) - Z_3 g(U, Z_4) + g(U, [Z_3, Z_4]) \\
&= Z_4 a_3 - Z_3 a_4.
\end{aligned}$$

In order to show that $Z_4 a_3 = Z_3 a_4$, we recall that $[Z_1, Z_4] = [Z_2, Z_4] = 0$, and from part (v) we have $Z_4 a_1 = Z_1 a_4$. Then a computation provides

$$\begin{aligned}
Z_4 a_3 &= Z_4 (Z_1 a_2 - Z_2 a_1) = Z_4 Z_1 a_2 - Z_4 Z_2 a_1 \\
&= Z_1 Z_4 a_2 - Z_2 Z_4 a_1 = Z_1 Z_2 a_4 - Z_2 Z_1 a_4 \\
&= (Z_1 Z_2 - Z_2 Z_1) a_4 = [Z_1, Z_2] a_4 = Z_3 a_4.
\end{aligned}$$

Hence, $A(Z_3, Z_4) = 0$, which ends the proof. ■

All examples of distributions encountered so far (such as Heisenberg an Engel) had a constant rank and step everywhere. This made possible the extension to an elliptic system where a simple application of the Poincaré's lemma provides integrability conditions. However, in the case when the distribution drops rank along a certain subset, this application is not straightforward. We shall deal with this phenomenon in the case of Grushin and Martinet distributions. The method will be similar with the one used in the previous sections, the exposition following Calin et. al. [28]. Another equivalent variant of approach that uses Rumin and Engel complexes of differential operators has been developed independently by Eastwood [27].

8.7 Grushin-Type Distributions

We shall start with some preliminary results on nilpotent distributions of class 2 on a two-dimensional Riemannian manifold. The next result provides a relation between the curl of a vector field and the integrability conditions.

Theorem 8.7.1 *Let X and Y be two orthonormal vector fields on the Riemannian manifold (M, g) satisfying*

$$[X, Y] \neq 0, \qquad [X, [X, Y]] = [Y, [X, Y]] = 0.$$

Consider the vector field $U = aX + bY$ with a and b smooth functions on M and denote $A = \operatorname{curl} U$. Then

$$\begin{aligned} -X^2 b + (XY + [X, Y])a &= A([Y, X], X) - XA(Y, X) && (8.7.47) \\ Y^2 a - (YX + [Y, X])b &= A([Y, X], Y) - YA(Y, X). && (8.7.48) \end{aligned}$$

Proof: Applying the curl formula given by Lemma 8.1.2 we have

$$\begin{aligned} A(X, Y) &= Yg(U, X) - Xg(U, Y) + g(U, [X, Y]) \\ &= Ya - Xb - g(U, [Y, X]), \end{aligned}$$

which implies

$$g(U, [Y, X]) = A(Y, X) + Ya - Xb, \tag{8.7.49}$$

where we used the skew-symmetry of the curl and the Lie bracket. By the curl definition formula (8.1.4) we have

$$g(\nabla_X U, [Y, X]) = g(\nabla_{[Y,X]} U, X) + A([Y, X], X), \tag{8.7.50}$$

where ∇ denotes the Levi-Civita connection on (M, g). Applying X to the left side of relation (8.7.49) and using the properties of ∇ and formula (8.7.50) we obtain

$$\begin{aligned} Xg(U, [Y, X]) &= g(\nabla_X U, [Y, X]) + g(U, \nabla_X [Y, X]) \\ &= g(\nabla_{[Y,X]} U, X) + A([Y, X], X) + g(U, \nabla_{[Y,X]} X + \underbrace{[X, [Y, X]]}_{=0}) \\ &= A([Y, X], X) + g(\nabla_{[Y,X]} U, X) + g(U, \nabla_{[Y,X]} X) \\ &= A([Y, X], X) + [Y, X]g(U, X) \\ &= A([Y, X], X) + [Y, X]a. \end{aligned} \tag{8.7.51}$$

Applying X to the right side of relation (8.7.49) yields

$$XA(Y, X) + XYa - X^2 b. \tag{8.7.52}$$

Equating relations (8.7.51) and (8.7.52) we obtain relation (8.7.47).

By applying Y to both sides of relation relation (8.7.49), a similar computation leads to relation (8.7.48). ■

Remark 8.7.2 The previous theorem holds on Riemannian manifolds on any dimension $n \geq 2$. However, in the next applications we shall use this result only for the case of two-dimensional manifolds.

Remark 8.7.3 If curl $U = 0$ then

$$X^2 b = (XY + [X,Y])a \tag{8.7.53}$$
$$Y^2 a = (YX + [Y,X])b. \tag{8.7.54}$$

This follows from Theorem 8.7.1 by taking $A = 0$.

The converse of the result stated by Remark 8.7.3 is not necessarily true, unless extra conditions are required. We shall deal with this issue in the following.

We consider $\dim M = 2$. Then there are two functions α and β on M such that $[X,Y] = \alpha X + \beta Y$. By the skew-symmetry of tensor A we have

$$\begin{aligned} A([Y,X],X) &= -\beta A(Y,X) \\ A([Y,X],Y) &= \alpha A(Y,X). \end{aligned}$$

Substituting into relations (8.7.47)-(8.7.48) yields

$$-X^2 b + (XY + [X,Y])a = -\beta A(Y,X) - XA(Y,X) \tag{8.7.55}$$
$$Y^2 a - (YX + [Y,X])b = \alpha A(Y,X) - YA(Y,X). \tag{8.7.56}$$

Therefore, conditions (8.7.53)-(8.7.54) are equivalent to

$$XA(Y,X) = -\beta A(Y,X) \tag{8.7.57}$$
$$YA(Y,X) = \alpha A(Y,X). \tag{8.7.58}$$

In order to prove that $A = 0$ it suffices to show it on the basis $\{X, Y\}$. Since the skew-symmetry provides $A(X,X) = 0$, $A(Y,Y) = 0$, and $A(X,Y) = -A(Y,X)$, then the only component we need to study is $\rho = A(X,Y)$. We need a condition which implies that the solution of the system

$$X\rho = -\beta\rho \tag{8.7.59}$$
$$Y\rho = \alpha\rho \tag{8.7.60}$$

is $\rho = 0$. To this end, we recall the divergence with respect to the Riemannian metric g of a vector field $U = aX + bY$ is $\operatorname{div} U = Xa + Yb$. In particular, the divergence of the Lie bracket of X and Y is $\operatorname{div}[X,Y] = X\alpha + Y\beta$.

Theorem 8.7.4 *Let (M, g) be a two-dimensional connected and simply connected Riemannian manifold. Consider X and Y be two orthonormal vector fields with respect to g satisfying*

$$[X, Y] \neq 0, \qquad [X, [X, Y]] = [Y, [X, Y]] = 0$$

and $\operatorname{div}[X, Y] \neq 0$. *Let a and b be smooth functions on M such that*

$$X^2 b = (XY + [X, Y])a \tag{8.7.61}$$

$$Y^2 a = (YX + [Y, X])b. \tag{8.7.62}$$

Then we have:

(*i*) $\rho = 0$;

(*ii*) *There is a smooth function f on M such that $Xf = a$ and $Yf = b$.*

Proof: (*i*) We multiply relations (8.7.59)-(8.7.60) by Y and X respectively to get

$$YX\rho = -(Y\beta)\rho - \beta(Y\rho) \tag{8.7.63}$$

$$XY\rho = (X\alpha)\rho + \alpha(X\rho). \tag{8.7.64}$$

Subtracting and using $[X, Y] = \alpha X + \beta Y$ yields

$$\begin{aligned}[X, Y]\rho &= (X\alpha + Y\beta)\rho + \alpha(X\rho) + \beta(Y\rho) \\ &= \operatorname{div}[X, Y]\,\rho + (\alpha X + \beta Y)\rho \\ &= \operatorname{div}[X, Y]\,\rho + [X, Y]\rho,\end{aligned}$$

which after reduction implies $\operatorname{div}[X, Y]\,\rho = 0$. Since $\operatorname{div}[X, Y] \neq 0$, it follows that $\rho = 0$ on M.

(*ii*) The identity $\rho = 0$ implies $A = 0$. Then Theorem 8.1.4 implies that the vector field $U = aX + bY$ is potential, namely there is a smooth function f such that $U = \nabla_g f$, i.e. $Xf = a$ and $Yf = b$. ■

The condition $\operatorname{div}[X, Y] \neq 0$ from the hypothesis of Theorem 8.7.4 is not satisfied by all distributions on $\mathbb{R}^2$ (a counterexample being the Grushin distribution). The next section develops a method which is meant to fix this situation.

Completing to a zero-curl vector field Given the vector field $U = aX + bY$, with $A(X, Y) = \rho \neq 0$, we shall construct another vector field, $\bar{U} = \bar{a}X + \bar{b}Y$, such that $\operatorname{curl}(U + \bar{U}) = 0$. This means $\bar{A}(X, Y) = -\rho$, where $\bar{A} = \operatorname{curl} \bar{U}$. We note that if ρ is a solution of the system (8.7.59)-(8.7.60), then $-\rho$ is also a solution. This is equivalent to the fact that the system (8.7.57)-(8.7.58) holds if A is replaced by $\bar{A}$. This is implied by the condition that $\bar{a}$ and $\bar{b}$ satisfy

$$X^2 \bar{b} = (XY + [X, Y])\bar{a} \tag{8.7.65}$$

$$Y^2 \bar{a} = (YX + [Y, X])\bar{b}. \tag{8.7.66}$$

Since $\mathrm{curl}(U+\bar{U})=0$, by Theorem 8.1.4 the vector field $U+\bar{U}$ is a potential vector field, i.e., there is a smooth function f on M (which is assumed connected and simply connected) such that

$$Xf = a+\bar{a} \tag{8.7.67}$$
$$Yf = b+\bar{b}. \tag{8.7.68}$$

The summary of the method is as in the following:
If we find two functions $\bar{a}$ and $\bar{b}$ that satisfy (8.7.65)-(8.7.66) and such that $\mathrm{curl}\,\bar{U}(X,Y)=-\rho$, where $\bar{U}=\bar{a}X+\bar{b}Y$, then the system (8.7.67)-(8.7.68) has a solution f if and only if

$$X^2b = (XY+[X,Y])a \tag{8.7.69}$$
$$Y^2a = (YX+[Y,X])b. \tag{8.7.70}$$

We note that the system (8.7.67)-(8.7.68) for which we found the integrability conditions (8.7.69)-(8.7.70) is not the same as the initial system $Xf=a$, $Yf=b$. We shall see that at the best behavior one of the either $\bar{a}$ or $\bar{b}$ is zero and the other is a constant.

Grushin vector fields Let $X=\partial_x$ and $Y=x\partial_y$ be two vector fields on a contractible open set $\Omega\subset\mathbb{R}^2\backslash\{x=0\}$. The metric g on Ω with respect to which $\{X,Y\}$ are orthonormal is given by

$$g_{ij}=\begin{pmatrix}1 & 0\\ 0 & x^{-2}\end{pmatrix}.$$

It is easy to check that $[X,Y]\neq 0$ and $[X,[X,Y]]=[Y,[X,Y]]=0$. The bracket satisfies

$$[X,Y]=\partial_y=\alpha X+\beta Y=\alpha\partial_x+\beta x\partial_y,$$

with $\alpha=0$ and $\beta=1/x$. Since $\mathrm{div}[X,Y]=X\alpha+Y\beta=0$, then Theorem 8.7.4 cannot be applied here.

We shall continue with the method of completion to a zero-curl vector field as described before. First we find ρ by solving the system (8.7.59)-(8.7.60).

$$\begin{cases} X\rho=-\beta\rho\\ Y\rho=\ \alpha\rho\end{cases} \Longleftrightarrow \begin{cases}\partial_x\rho=-\frac{1}{x}\rho\\ x\partial_y\rho=0\end{cases}\quad \text{on } \Omega.$$

The second equation implies $\rho=\rho(x)$. Substituting in the first equation we obtain the separable equation $\rho'(x)=-\dfrac{\rho(x)}{x}$ with solution $\rho(x)=\dfrac{C}{x}$, where C is a real constant. We shall write next a condition for the coefficient functions

$\bar{a}$ and $\bar{b}$ such that $\operatorname{curl} \bar{U}(X,Y) = -\rho$, where $\bar{U} = \bar{a}X + \bar{b}Y$. Applying Lemma 8.1.2 we have

$$\operatorname{curl} \bar{U}(X,Y) = Y\bar{a} - X\bar{b} - g(\bar{U}, [Y,X]) = Y\bar{a} - X\bar{b} + \bar{a}\alpha + \bar{b}\beta.$$

Therefore, the condition $\operatorname{curl}(U+\bar{U}) = 0$, which is equivalent to $\operatorname{curl} \bar{U}(X,Y) = -\rho$, becomes

$$Y\bar{a} - X\bar{b} + \bar{a}\alpha + \bar{b}\beta = -\frac{C}{x} \quad \text{on } \Omega.$$

We need to find some functions $\bar{a}$ and $\bar{b}$ such that the following three conditions are satisfied

$$\begin{cases} Y\bar{a} - X\bar{b} + \bar{a}\alpha + \bar{b}\beta = -C/x \\ X^2\bar{b} = (XY + [X,Y])\bar{a} \\ Y^2\bar{a} = (YX + [Y,X])\bar{b}. \end{cases} \iff \begin{cases} x\partial_y\bar{a} - \partial_x\bar{b} - \bar{b}/x = -C/x \\ \partial_x^2\bar{b} - (\partial_x(x\partial_y) + \partial_y)\bar{a} = 0 \\ x^2\partial_x^2\bar{a} - (x\partial_y\partial_x - \partial_y)\bar{b} = 0. \end{cases}$$

It is a straightforward verification that $\bar{a} = 0$ and $\bar{b} = C$ satisfy the previous system. Therefore, $\bar{U} = CY = Cx\partial_y$, and we have $\operatorname{curl}(U + \bar{U}) = 0$. Writing this condition equivalently by stating that $U + \bar{U}$ is a potential vector field, using (8.7.67)-(8.7.68), we obtain the following result.

Theorem 8.7.5 *Let $\Omega \subset \mathbb{R}^2 \backslash \{x = 0\}$ be a contractible set. Consider the vector fields $X = \partial_x$ and $Y = x\partial_y$ on Ω. Then for a pair of smooth functions a, b on Ω*

$$\begin{cases} X^2 b = (XY + [X,Y])a \\ Y^2 a = (YX - [X,Y])b \end{cases} \iff \begin{cases} \exists \text{ a smooth function } f \text{ on } \Omega \\ \text{and a constant } C \text{ such that} \\ Xf = a \text{ and } Yf = b + C. \end{cases}$$

We shall continue with a result analog to Theorem 8.7.1 for the case of a nilpotent distribution of class 3.

Theorem 8.7.6 *Let X and Y be two orthonormal vector fields on the Riemannian manifold (M,g) satisfying*

$$[X,Y] \neq 0, \qquad [X,[X,Y]] \neq 0,$$

$$[Y,[X,Y]] = 0, \qquad [X,[X,[X,Y]]] = 0, \quad [Y,[X,[X,Y]]] = 0.$$

Consider the vector field $U = aX + bY$ with a and b smooth functions on M and denote $A = \operatorname{curl} U$. Then

$$\begin{aligned} -X^3 b + (X^2 Y + X[X,Y] + [X,[X,Y]])a =& - X^2 A(Y,X) + XA([Y,X],X) \\ &+ A([X,[Y,X]],X) \qquad (8.7.71) \\ Y^2 a - (YX + [Y,X])b =& A([Y,X],Y) - YA(Y,X). \end{aligned}$$

(8.7.72)

Proof: Similarly with the proof of Theorem 8.7.1 we obtain

$$g(U, [Y, X]) = A(Y, X) + Ya - Xb, \tag{8.7.73}$$

Applying X on the left of (8.7.73) and using the curl definition formula (8.1.4)

$$g(\nabla_X U, [Y, X]) = g(\nabla_{[Y,X]} U, X) + A([Y, X], X), \tag{8.7.74}$$

where ∇ denotes the Levi-Civita connection on (M, g), we have

$$\begin{aligned} Xg(U, [Y, X]) &= g(\nabla_X U, [Y, X]) + g(U, \nabla_X [Y, X]) \\ &= A([Y, X], X) + g(\nabla_{[Y,X]} U, X) + g(U, \nabla_{[Y,X]} X + [X, [Y, X]]) \\ &= A([Y, X], X) + g(\nabla_{[Y,X]} U, X) + g(U, \nabla_{[Y,X]} X) + g(U, [X, [Y, X]]) \\ &= A([Y, X], X) + [Y, X] g(U, X) + g(U, [X, [Y, X]]) \\ &= A([Y, X], X) + [Y, X] a + g(U, [X, [Y, X]]). \end{aligned} \tag{8.7.75}$$

Applying X to (8.7.75) again yields

$$X^2 g(U, [Y, X]) = XA([Y, X], X) + X[Y, X]a + Xg(U, [X, [Y, X]]). \tag{8.7.76}$$

The last term can be computed using the properties of the Levi-Civita connection and the nilpotency property of X and Y as follows

$$\begin{aligned} Xg(U, [X, [Y, X]]) &= g(\nabla_X U, [X, [Y, X]]) + g(U, \nabla_X [X, [Y, X]]) \\ &= g(\nabla_X U, [X, [Y, X]]) + g(U, \nabla_{[X,[Y,X]]} X + \underbrace{[X, [X, [Y, X]]]}_{=0}) \\ &= g(\nabla_X U, [X, [Y, X]]) + g(U, \nabla_{[X,[Y,X]]} X) \\ &= g(\nabla_X U, [X, [Y, X]]) + [X, [Y, X]] g(U, X) - g(\nabla_{[X,[Y,X]]} U, X) \\ &= A([X, [Y, X]], X) + [X, [Y, X]] g(U, X) \\ &= A([X, [Y, X]], X) + [X, [Y, X]] a. \end{aligned}$$

We substitute in (8.7.75) to obtain

$$X^2 g(U, [Y, X]) = XA([Y, X], X) + A([X, [Y, X]], X) + (X[Y, X] + [X, [Y, X]])a. \tag{8.7.77}$$

On the other hand, applying X^2 on the right side of (8.7.73) yields

$$X^2 g(U, [Y, X]) = X^2 A(Y, X) + X^2 Ya - X^3 b, \tag{8.7.78}$$

Equating the right sides on (8.7.77) and (8.7.78) we obtain (8.7.71).

To prove relation (8.7.72) we apply Y on both sides of relation (8.7.73)

$$Yg(U, [Y, X]) = YA(Y, X) + Y^2 a - YXb. \tag{8.7.79}$$

Using $[Y, [X, Y]] = 0$ the left side of (8.7.79) becomes

$$\begin{aligned} Yg(U, [Y, X]) &= g(\nabla_Y U, [Y, X]) + g(U, \nabla_Y [Y, X]) \\ &= A([Y, X], Y) + g(\nabla_{[Y,X]} U, Y) + g(U, \nabla_{[Y,X]} Y + \underbrace{[Y, [Y, X]]}_{=0}) \\ &= A([Y, X], Y) + [Y, X] g(U, Y) = A([Y, X], Y) + [Y, X] b. \end{aligned}$$

Substituting back in (8.7.79) yields relation (8.7.72). ∎

Remark 8.7.7 In the hypotheses of Theorem 8.7.6, the condition curl $U = 0$ implies

$$X^3 b = (X^2 Y + X[X, Y] + [X, [X, Y]])a \tag{8.7.80}$$
$$Y^2 a = (YX + [Y, X])b. \tag{8.7.81}$$

By Theorem 8.7.6 the system (8.7.80)-(8.7.81) is equivalent to

$$X^2 A(Y, X) = XA([Y, X], X) + A([X, [Y, X]], X) \tag{8.7.82}$$
$$YA(Y, X) = A([Y, X], Y). \tag{8.7.83}$$

In order to apply Poincaré's lemma we would need $A = 0$. To this end we assume $\dim M = 2$, and using the skew-symmetry of A it suffices to show that $A(X, Y) = 0$. Since this would not hold true for any vector fields X and Y we shall use the zero-curl method completion.
For this we denote $\rho = A(X, Y)$ and let α and β be two functions on M such that $[X, Y] = \alpha X + \beta Y$. Then the following nested Lie bracket can be written in terms of α and β as

$$\begin{aligned} [X, [X, Y]] &= X[X, Y] - [X, Y]X \\ &= X(\alpha X + \beta Y) - (\alpha X + \beta Y)X \\ &= (X\alpha)X + \alpha X^2 + (X\beta)Y + \beta XY - \alpha X^2 - \beta YX \\ &= (X\alpha)X + (X\beta)Y + \beta(XY - YX) \\ &= (X\alpha)X + (X\beta)Y + \beta[X, Y] \\ &= (X\alpha)X + (X\beta)Y + \beta(\alpha X + \beta Y) \\ &= (X\alpha + \alpha\beta)X + (X\beta + \beta^2)Y. \end{aligned} \tag{8.7.84}$$

Using relation (8.7.84) and the skew-symmetry of the tensor A we have

$$\begin{aligned} A([X, [Y, X]], X) &= (X\beta + \beta^2)\rho \\ A([Y, X], X) &= \beta\rho \\ A([Y, X], Y) &= -\alpha\rho. \end{aligned}$$

Then substituting the previous relations in the system (8.7.82)-(8.7.83) yields

$$X^2\rho + \beta X\rho + (2X\beta + \beta^2)\rho = 0 \tag{8.7.85}$$
$$Y\rho = \alpha\rho. \tag{8.7.86}$$

We look for two functions $\bar{a}$ and $\bar{b}$ such that that $\text{curl}(U + \bar{U}) = 0$, where $\bar{U} = \bar{a}X + \bar{b}Y$. The zero-curl condition is equivalent to $\bar{A}(X,Y) = -\rho$, where $\bar{A} = \text{curl}\,\bar{U}$. We shall perform the computation for the following particular case of vector fields.

The step 3 Grushin distribution We consider the vector fields $X = \partial_x$ and $Y = x^2\partial_y$ on $\mathbb{R}^2$. A computation shows

$$[X,Y] = 2x\partial_y, \quad [X,[X,Y]] = 2\partial_y, \quad [Y,[X,Y]] = 0$$

$$[X,[X,[X,Y]]] = 0, \qquad [Y,[X,[X,Y]]] = 0.$$

Therefore, the hypothesis of Theorem 8.7.6 holds on any open set Ω included in $\mathbb{R}^2\backslash\{x = 0\}$.

Since in this case $\alpha = 0$ and $\beta = \frac{2}{x}$, the system (8.7.85)-(8.7.86) becomes

$$\partial_x^2\rho + \frac{2}{x}\partial_x\rho + \Big(2\partial_x\Big(\frac{2}{x}\Big) + \frac{4}{x^2}\Big)\rho = 0$$
$$x^2\partial_x\rho = 0 \ \text{ on } \Omega.$$

From the second equation it follows that $\rho = \rho(x)$. Then the first equation becomes an Euler equation after multiplication by x^2

$$x^2\rho''(x) + 2x\rho'(x) = 0.$$

The solution is given by $\rho(x) = \frac{C_1}{x} + C_2$, with C_1, C_2 real constants. Therefore

$$\text{curl U} = \frac{C_1}{x} + C_2.$$

We look for a vector field $\bar{U} = \bar{a}X + \bar{b}Y$ such that curl $(U + \bar{U}) = 0$, fact equivalent to $\bar{A}(X,Y) = -\Big(\frac{C_1}{x} + C_2\Big)$. Since

$$\bar{A}(X,Y) = Y\bar{a} - X\bar{b} - g(\bar{U},[Y,X]) = Y\bar{a} - X\bar{b} + \bar{a}\alpha + \bar{b}\beta,$$

this becomes

$$Y\bar{a} - X\bar{b} + \bar{a}\alpha + \bar{b}\beta = -\Big(\frac{C_1}{x} + C_2\Big).$$

Therefore, we look for a pair of functions $\bar{a}$ and $\bar{b}$ such that the following three equations are satisfied

$$\begin{cases} Y\bar{a} - X\bar{b} + \bar{a}\alpha + \bar{b}\beta = -(\frac{C_1}{x} + C_2) \\ X^3\bar{b} = (X^2Y + X[X,Y] + [X,[X,Y]])\bar{a} \\ Y^2\bar{a} = (YX + [Y,X])\bar{b} \end{cases} \Leftrightarrow$$

$$\begin{cases} x^2\partial_y\bar{a} - \partial_x\bar{b} + \frac{2}{x}\bar{b} = -(\frac{C_1}{x} + C_2) \\ \partial_x^3\bar{b} = (\partial_x^2(x^2\partial_y) + \partial_x(2x\partial_y) + 2\partial_y)\bar{a} \\ x^4\partial_y^2\bar{a} = (x^2\partial_y\partial_x - 2x\partial_y)\bar{b}. \end{cases}$$

We can check by a direct computation that $\bar{a} = 0$ and $\bar{b} = -C_2x - \frac{C_1}{2}$ satisfies the aforementioned system. Therefore $\bar{U} = \bar{a}X + \bar{b}Y = 0 \cdot X + (-C_2x - \frac{C_1}{2})Y$ with C_1 and C_2 real constants. Then curl $(U + \bar{U}) = 0$ and if Ω is assumed contractible, Poincaré's lemma provides the existence of a smooth function f on Ω such that

$$Xf = a + \bar{a}, \qquad Yf = b + \bar{b}.$$

Renaming the constants we obtain the following result:

Proposition 8.7.8 *Let $X = \partial_x$ and $Y = x^2\partial_y$. For a pair of smooth functions a and b defined on the contractible set $\Omega \subset \mathbb{R}^2\backslash\{x = 0\}$ we have the equivalence*

$$\begin{cases} X^3 = (X^2Y + X[X,Y] + [X,[X,Y]])a \\ Y^2a = (YX + [Y,X])b \end{cases} \Leftrightarrow \begin{cases} \exists \text{ a smooth function } f \text{ on } \Omega \\ \text{and constants } C \text{ and } D \\ \text{such that} \\ Xf = a \text{ and } Yf = b + Cx + D. \end{cases}$$

Remark 8.7.9 There might be more pair solutions $(\bar{a}, \bar{b})$ for the previous system of equations. It is worth noting that looking for a solution $\bar{a} = 0$ is equivalent to keepying the first exactness condition in the simple form $Xf = a$.

8.8 Summary

The chapter focuses on the introduction and analysis of elliptic and sub-elliptic systems of equations. Elliptic systems involve a set of n independent linear equations in an n-dimensional space. On the other hand, sub-elliptic systems consist of k independent linear equations in an n-dimensional space, where $k < n$. Among the sub-elliptic systems, particular attention is given to non-holonomic systems, which satisfy the bracket-generating condition for their vector fields.

In this chapter, we extensively examine the uniqueness and regularity of solutions for non-holonomic systems. Integrability conditions are studied for various common distributions, including Heisenberg, Grushin, Engel, Martinet, and others. The case of constant step distributions, or regular distributions, is relatively straightforward as it can be treated by extending the system to an elliptic one over the entire space and employing a generalized version of Poincaré's lemma.

However, for distributions where the step drops along certain sets, a series of innovative techniques are employed to tackle the challenges. These techniques are devised to handle the intricacies associated with such distributions and provide insights into the regularity and properties of solutions.

Overall, the chapter offers a comprehensive exploration of both elliptic and sub-elliptic systems, with a focus on non-holonomic systems and their integrability conditions. It provides a detailed analysis of various distributions and presents approaches tailored to different scenarios, enhancing our understanding of these systems and their unique characteristics.

8.9 Exercises

Exercise 8.9.1 Consider two vector fields on $\mathbb{R}^3$

$$X = z\partial_x + \partial_y, \qquad Y = \partial_z.$$

Find the integrability conditions for the system $Xf = a, Yf = b$.

Exercise 8.9.2 (a) Consider the composition law on $\mathbb{R}^3$

$$(a, b, c) * (a', b', c') = (a + a', b + b', c + c' + b'e^a + a'e^b).$$

Show that $(\mathbb{R}^3, *)$ is a noncommutative Lie group.

(b) Verify whether the vector fields $X = \partial_{x_1} + e^{x_2}\partial_{x_3}$ and $Y = \partial_{x_2} - e^{x_1}\partial_{x_3}$ are left invariant on $(\mathbb{R}^3, *)$.

(c) Find the integrability conditions for the system

$$Xf = a, \qquad Yf = b.$$

Exercise 8.9.3 Let $Z_i = \sum_{j=1}^n \zeta_i^j \partial_{x_j}$, $1 \leq j \leq n$, be n linearly independent vector fields on an contractible domain U in $\mathbb{R}^n$. We denote by $\omega_k = \sum_{j=1}^n \omega_k^j dx_j$ the dual one-forms, i.e. $\omega_k(Z_j) = \delta_{ij}$. Let $\omega = (\omega_k^j)$ and $\zeta = (\zeta_i^j)$ be the $n \times n$ coefficients matrices.

(a) Show that $\omega\zeta^T = \zeta\omega^T = \mathbb{I}_n$.

(b) Let $a_j \in \mathcal{C}^\infty(U)$, $1 \leq j \leq n$. Show that the elliptic system

$$Z_1(f) = a_1, \cdots, Z_n(f) = a_n \tag{8.9.87}$$

can be written in local coordinates as $\nabla_{Eu} f = \omega^T a^T$, where $a = (a_1, \cdots, a_n)$.
(c) Prove that the integrability conditions for the elliptic system (8.9.87) can be written as

$$\sum_{j=1}^{n}(\partial_{x_k}\omega_j^i - \partial_{x_i}\omega_j^k)a_j = \sum_{j=1}^{n}\big(\omega_j^k\partial_{x_i}a_j - \omega_j^i\partial_{x_k}a_j\big).$$

Exercise 8.9.4 (Kristaly [91]) Let (M, g) be an n-dimensional Riemannian manifold and $U \subseteq M$ be a simply connected open set.
(a) Given a $\mathcal{C}^1$-vector field V on U, show that the elliptic system $\nabla_g f = V$ in Ω has a solution $f \in \mathcal{C}^2(U)$ if and only if

$$\partial_{x_i}\tilde{V}_j = \partial_{x_j}\tilde{V}_i \text{ in } U, \ \ 1 \le i, j \le n$$

where $\tilde{V}_j = g_{jk}V_k$.
(b) Let $x_0 \in U$ and and assume the integrability conditions given by (a) hold. Then the solution $f : U \to \mathbb{R}$ is given by

$$f(x) = f(x_0) + \int_0^1 g(V(\gamma(t)), \dot{\gamma}(t))\, dt,$$

where $\gamma : [0, 1] \to U$ is a curve joining x_0 and x.

Exercise 8.9.5 (sub-elliptic system on $(\mathbb{S}^3)$, Calin et. al [33]) Consider the following vector fields on $\mathbb{R}^4$

$$\begin{aligned}
X &= x_2\partial_{x_1} - x_1\partial_{x_2} - x_4\partial_{x_3} + x_3\partial_{x_4}\\
Y &= x_4\partial_{x_1} - x_3\partial_{x_2} + x_2\partial_{x_3} - x_1\partial_{x_4}\\
T &= x_3\partial_{x_1} + x_4\partial_{x_2} - x_1\partial_{x_3} - x_2\partial_{x_4}.
\end{aligned}$$

(a) Show that $\{X, Y, T\}$ is an orthonormal system of vector fields on the unit sphere $\mathbb{S}^3$ with respect to the Euclidean scalar product.

(b) Show that the distribution $\mathcal{D} = \mathrm{span}\{X, Y\}$ is non-integrable of step 2 at each point of $\mathbb{S}^3$.

(c) Let $a, b \in \mathcal{C}^\infty(\mathbb{S}^3)$. Show that the system $Xf = a$, $Yf = b$ has at most one solution.

(d) Show that if the solution in part (c) exists, then $f \in \mathcal{C}^\infty(\mathbb{S}^3)$.

(e) Show that the system $Xf = a, Yf = b$ is equivalent to $\nabla_g f = U$, where $U = aX + bY + cT$, with $c = (Xb - Ya)/2$ and $\nabla_g f = X(f)X + Y(f)Y + T(f)T$ is the gradient of f in the induced metric from $\mathbb{R}^4$.

(f) Using the equation curl $U = 0$, prove that the system $Xf = a, Yf = b$ has solutions if and only if the following integrability conditions hold

$$X^2 b + 4b = (XY + [X,Y])a$$
$$Y^2 a + 4a = (YX - [X,Y])b.$$

(g) Show that in spherical coordinates (α, β, θ) on $\mathbb{S}^3$ we have

$$X = -\partial_\alpha + \partial_\beta$$
$$Y = -\cos(\alpha - \beta)\Big(\tan\theta\,\partial_\alpha + \cot\theta\,\partial_\beta\Big) + \sin(\alpha-\beta)\partial_\theta$$
$$T = -\sin(\alpha - \beta)\Big(\tan\theta\,\partial_\alpha + \cot\theta\,\partial_\beta\Big) - \cos(\alpha-\beta)\partial_\theta.$$

(e) Consider the work $f(\mathbf{r}) = \int_0^1 V(t\mathbf{r})\cdot\mathbf{r}\,dt$, where $\mathbf{r} = (\alpha, \beta, \theta)$. Show that

$$(Xf)(\mathbf{r}) = a(\mathbf{r}) + \int_0^1 \phi(\mathbf{r})\Big(X^2 b + 4b - (XY + [X,Y])a\Big)(t\mathbf{r})\,dt$$
$$(Yf)(\mathbf{r}) = b(\mathbf{r}) + \int_0^1 \psi(\mathbf{r})\Big(Y^2 a + 4a - (YX - [X,Y])b\Big)(t\mathbf{r})\,dt,$$

where

$$\phi(\mathbf{r}) = \frac{1}{2}\big(-\theta\cos(\alpha-\beta) + (\alpha+\beta)\sin\theta\cos\theta\sin(\alpha-\beta)\big)$$
$$\psi(\mathbf{r}) = \frac{1}{2}\big(\theta\cos(\alpha+\beta) + (\alpha+\beta)\sin\theta\cos\theta\sin(\alpha-\beta)\big).$$

Exercise 8.9.6 (The group $SE(2)$, Isagurova [81]) Consider two vector fields on $\mathbb{R}^3$ given by

$$X_1 = \cos\theta\partial_x + \sin\theta\partial_y, \qquad X_2 = \partial_\theta.$$

Show that the integrability conditions for the system $X_1 f = a_1, X_2 f = a_2$ are given by

$$X_2^2 a_1 + a_1 = [X_2, X_1]a_2 + X_2 X_1 a_2$$
$$X_1^2 a_2 = [X_1, X_2]a_1 + X_1 X_2 a_1.$$

Exercise 8.9.7 (Isagurova [81]) Let $z > 0$ and consider two vector fields on $\mathbb{R}^2 \times \mathbb{R}_+$

$$X_1 = z\partial_x + \partial_y, \qquad X_2 = -z\partial_z.$$

Prove that the integrability conditions for the system $X_1 f = a_1, X_2 f = a_2$ are given by

$$X_2^2 a_1 - 2X_2 X_1 a_2 + X_1 X_2 a_2 = X_1 a_2 - X_2 a_1$$
$$X_1^2 a_2 - 2X_1 X_2 a_1 + X_2 X_1 a_1 = 0.$$

Exercise 8.9.8 (The free Carnot group, Isagurova [81]) Consider the distribution on $\mathbb{R}^5$ generated by

$$X_1 = \partial_{x_1} - \frac{1}{2}x_2\partial_{x_3} - \frac{1}{2}(x_1^2 + x_2^2)\partial_{x_5}$$
$$X_2 = \partial_{x_2} + \frac{1}{2}x_1\partial_{x_3} + \frac{1}{2}(x_1^2 + x_2^2)\partial_{x_4}.$$

(a) Show the following noncommutativity relations

$$[X_1, X_2] = X_3, \quad [X_1, X_3] = X_4, \quad [X_2, X_3] = X_5.$$

(b) Prove that the integrability conditions for the system $X_1 f = a_1$, $X_2 f = a_2$ are given by

$$X_1 a_4 = X_4 a_1, \quad X_1 a_5 = X_5 a_1, \quad X_2 a_5 = X_5 a_2,$$

where

$$a_3 = X_1 a_2 - X_2 a_1, \quad a_4 = X_1 a_3 - X_3 a_1, \quad a_5 = X_2 a_3 - X_3 a_2.$$

Exercise 8.9.9 (Goursat-Darboux distribution, Isagurova [81]) Let

$$X_1 = \partial_{x_1}, \qquad X_2 = \sum_{j=2}^{k+2} \frac{x_1^{j-2}}{(j-2)!}\partial_{x_j}$$

be two vectors on $\mathbb{R}^{k+2}$. Let $X_i = [X_1, X_{i-1}]$, $i = 3, \dots, k+2$. For two given functions, a_1 and a_2, define inductively $a_3, \dots, a_{k+2}$ by

$$a_{i+1} = X_1 a_i - X_i a_1, \qquad i = 2, \dots, k+1.$$

Show that the sub-elliptic system $X_1 f = a_1$, $X_2 f = a_2$ has the integrability conditions given by

$$X_1 a_{k+2} = X_{k+2} a_1, \qquad X_2 a_i = X_i a_2, \qquad i = 3, \dots, k+2.$$

Chapter 9

Applications to LC-Circuits

This chapter explores the analysis of inductor-capacitor electrical circuits from multiple perspectives: the Lagrangian and Hamiltonian variational perspective, as well as the stochastic perspective. We delve into the associated Lagrangian, establish a classical action, and solve the Euler-Lagrange equations to obtain the system's solutions. We further utilize these solutions to determine the heat kernel of the associated Hermite operator and examine its stochastic interpretation.

9.1 The LC-circuit

The Lagrangian and Hamiltonian formalisms are powerful alternatives of describing the dynamics of electrical circuits. Equivalently, the same dynamics can be obtained using a combination of Kirchhoff's circuit laws regarding the current conservation and the voltage law.

We consider an electric circuit made up of an inductor of inductance L and a capacitor of capacitance C, which are connected in series, see Fig. 9.1. The inductor is a coiled wire, which resists to the change of the passing current, while the capacitor is made up of two conducting plates isolated by a a dielectric medium. The current through the inductor and the charge on the capacitor at time t are denoted by $I(t)$ and $q(t)$, respectively. The resistance in the wire is considered negligible.

We start the experiment considering the capacitor fully charged with the maximum charge q_{max}. This means that one of the capacitor plates has an excess of electrons, while the other plate has a lack of electrons. The electrons in excess will move through the electrical circuit to the opposite plate of the capacitor. This movement produces an electrical current, $I(t) = \dot{q}(t)$, which can be seen as the velocity of the electrons through the wire. When this current passes through the inductor's coil, a magnetic field is produced. At some

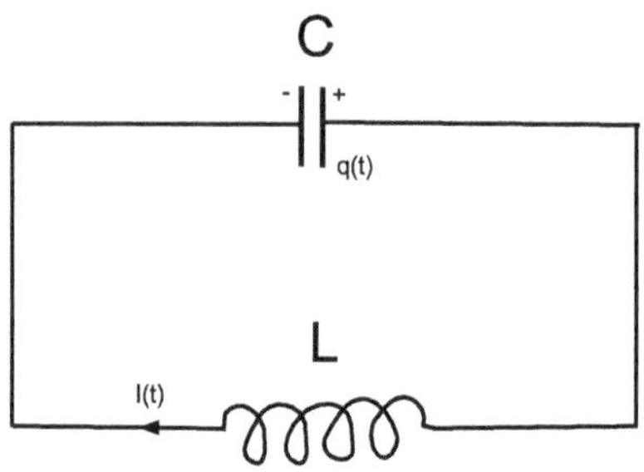

Figure 9.1: *An LC-circuit can be described at any time $t > 0$ by the pair* $\big(q(t), I(t)\big)$.

moment in time the capacitor is fully discharged, the entire system energy being now stocked into magnetic energy. As a reaction to the variation of the magnetic field, the inductor induces its own current, which flows in the reversed order, charging the capacitor again. The process repeats cyclically, and if the resistance through the wire is negligible, it lasts forever. This is the reason why this system is considered as an example of a harmonic oscillator.

9.2 The Lagrangian Formalism

We shall approach the previous electric oscillating system from the Lagrangian point of view. The state of the circuit at time t is given by the pair $(q(t), \dot{q}(t))$, i.e, by the amount of charge at the capacitor and current $I(t) = \dot{q}(t)$ through the inductor.

The Lagrangian We shall denote the inductance and capacitance by L and C respectively, where L and C are positive constants. Then the inductor stores the energy $E_M = \frac{1}{2}LI^2$ in the magnetic field and the capacitor stores the potential energy $E_P = \frac{1}{2}CV^2$ in the electric field, where V denotes the voltage at the capacitor's arms. Since $I = \dot{q}$ and $V = q/C$, the natural Lagrangian associated with the aforementioned LC-circuit is given by the difference between the aforementioned two energies

$$\mathcal{L}(q, \dot{q}) = E_M - E_P = \frac{1}{2}L\dot{q}^2 - \frac{1}{2C}q^2. \tag{9.2.1}$$

This relation was written by analogy with the natural Lagrangian from classical mechanics, which is the difference between the kinetic and potential energy of a mechanical system.

The Euler-Lagrange equation The dynamics of the charge and current through the LC-circuit are given by the associated Euler-Lagrange equations as follows. Let $T > 0$ be fixed. The critical paths $q : [0, T] \to \mathbb{R}$ of the action functional

$$q \to \int_0^T \mathcal{L}\big(q(t), \dot{q}(t)\big)\, dt$$

satisfies the Euler-Lagrange equation

$$\frac{d}{dt}\Big(\frac{\partial \mathcal{L}}{\partial \dot{q}}\Big) = \frac{\partial \mathcal{L}}{\partial q},$$

which takes the form of the following second order constant coefficients differential equation

$$\ddot{q} + \frac{1}{LC} q = 0. \tag{9.2.2}$$

For immediate practical purposes, this equation is solved as

$$q(t) = q_{\max} \cos\Big(\frac{t}{\sqrt{LC}} - \phi_0\Big),$$

where $q_{\max}$ is the maximum charge on the capacitor and ϕ_0 is a phase shift. This solution form is unfortunately not useful for the next treatment of this problem involving variational calculus. The desired solution will be written in terms of the following two boundary conditions on the initial and final capacitor charges

$$q(0) = q_0, \qquad q(T) = q_T. \tag{9.2.3}$$

For simplicity reasons, we denote $\omega = \dfrac{1}{\sqrt{LC}}$ and call it *pulsation*. The solution of equation (9.2.2) depends on two real parameters A and B, and pulsation ω, as

$$q(t) = A\cos(\omega t) + B\sin(\omega t). \tag{9.2.4}$$

Parameters A and B will be determined in terms of the boundary conditions (9.2.3) as follows. Making $t = 0$, we easily get $A = q_0$. Then, making $t = T$, we obtain

$$B = \frac{q_T - q_0\cos(\omega T)}{\sin(\omega T)}.$$

Substituting in (9.2.4) we have

$$\begin{aligned} q(t) &= q_0\cos(\omega t) + \frac{q_T - q_0\cos(\omega T)}{\sin(\omega T)}\sin(\omega t) \\ &= q_T\frac{\sin(\omega t)}{\sin(\omega T)} + q_0\frac{\cos(\omega t)\sin(\omega T) - \cos(\omega T)\sin(\omega t)}{\sin(\omega T)}, \end{aligned}$$

and applying a trigonometric formula, we obtain

$$q(t) = \frac{1}{\sin(\omega T)} \big[q_T \sin(\omega t) + q_0 \sin\big(\omega(T-t)\big)\big], \qquad 0 \le t \le T. \tag{9.2.5}$$

This provides a closed form solution for the charge $q(t)$ on the capacitor at time t in terms of the capacitor charge at two time instances, $t = 0$ and $t = T$. The interpretation is the following: the entire charge evolution at the capacitor during the time interval $[0, T]$ is determined by the initial and final charge measurements, q_0 and q_T.

Parity conditions If the charge q_T measured at time $t = T$ is equal to the charge q_0 measured at time $t = 0$, or equal to its negative, $-q_0$, then we say the *parity condition* holds on the capacitor. Mathematically, this condition can be expressed by

$$q_T = (-1)^k q_0, \qquad k = 0, 1, 2, \cdots \tag{9.2.6}$$

We shall see that parity condition plays a determinant role in the behavior of the LC-circuit under resonances.

Resonances There are specific pulsation values, $\omega_k = \dfrac{k\pi}{T}$, $k = 1, 2, \ldots,$ for which the solution $q(t)$ becomes unbounded almost everywhere. This is due to the fact that for these values of the pulsation the denominator in the formula (9.2.5) vanishes. We shall study this effect in the following.

We are interested in studying the limit $\lim\limits_{\omega T \to k\pi} q(t)$, for $k \ge 1$ integer. First, we write the solution $q(t)$, which is given by formula (9.2.5), as a linear combination of $\sin(\omega t)$ and $\cos(\omega t)$ as

$$q(t) = \frac{q_T - q_0 \cos(\omega T)}{\sin(\omega T)} \sin(\omega t) + q_0 \cos(\omega t).$$

Then the limit applies only on the first term as follows

$$\lim_{\omega T \to k\pi} q(t) = \lim_{\omega T \to k\pi} \frac{q_T - q_0 \cos(\omega T)}{\sin(\omega T)} \sin(\omega t) + q_0 \cos(\omega t).$$

This limit is finite if the parity condition (9.2.6) is satisfied. Under this hypothesis, using L'Hospital's rule, we have

$$\lim_{\omega T \to k\pi} \frac{q_T - q_0 \cos(\omega T)}{\sin(\omega T)} = \lim_{u \to k\pi} q_0 \frac{(-1)^k - \cos(u)}{\sin(u)} = \lim_{u \to k\pi} q_0 \frac{\sin(u)}{\cos(u)} = 0.$$

If (9.2.6) is not satisfied, then the previous limit is either ∞ or $-\infty$, fact that makes the value $q(t)$ to be infinite for almost all values of t.

The case $k = 0$ is not covered by the previous discussion. This corresponds to the case $\omega = 0$, which is obtained by taking the limit in relation (9.2.5)

$$\begin{aligned}\lim_{\omega \to 0} q(t) &= q_T \lim_{\omega \to 0} \frac{\sin(\omega t)}{\sin(\omega T)} + q_0 \lim_{\omega \to 0} \frac{\sin\big(\omega(T-t)\big)}{\sin(\omega T)} \\ &= q_T \frac{t}{T} + q_0 \frac{T-t}{T} \\ &= \frac{t}{T}(q_T - q_0) + q_0,\end{aligned}$$

which is a line segment joining q_0 and q_T, parameterized by t. Using the parity condition, we further obtain

$$\begin{aligned}\lim_{\omega \to 0} q(t) &= \frac{t}{T}(q_T - q_0) + q_0 \\ &= \frac{t}{T}\big((-1)^k - 1\big)q_0 + q_0 = q_0,\end{aligned}$$

since $k = 0$.

The conclusions of the previous analysis are given in the next result.

Proposition 9.2.1 *(i) Assume the parity condition (9.2.6) is satisfied. Then*

$$\lim_{\omega T \to k\pi} q(t) = q_0 \cos\left(\frac{k\pi t}{T}\right), \quad k = 1, 2, 3, \cdots$$

and

$$\lim_{\omega \to 0} q(t) = q_0, \quad \textit{for } k = 0.$$

(ii) If parity condition (9.2.6) does not hold, then $\lim_{\omega T \to k\pi} q(t)$ *is infinite for almost all values of* t*, if* $k = 1, 2, \dots$*. For* $k = 0$*, the limit is finite*

$$\lim_{\omega \to 0} q(t) = \frac{t}{T}(q_T - q_0) + q_0, \quad 0 \le t \le T.$$

Remark 9.2.2 Part *(ii)* corresponds to the resonance case. Therefore, for the resonance to occur, the following resonance conditions have to hold

$$q_T \neq (-1)^k q_0, \tag{9.2.7}$$

$$\omega = \frac{k\pi}{T}, \qquad k = 1, 2, \cdots \tag{9.2.8}$$

The current The associated current flowing through the inductor is obtained by differentiating with respect to t in relation (9.2.5)

$$I(t) = \dot{q}(t) = \frac{\omega}{\sin(\omega T)} \big[q_T \cos(\omega t) - q_0 \cos\big(\omega(T-t)\big)\big], \quad 0 \le t \le T. \quad (9.2.9)$$

This is also determined by the measurements q_0 and q_T and might become unbounded if ω is an integer multiple of π/T. A similar computation as before, or just a mere differentiation with respect to t in formulas of Proposition 9.2.1 provides the following result.

Proposition 9.2.3 (*i*) *Assume the parity condition (9.2.6) is satisfied. Then*

$$\lim_{\omega T \to k\pi} I(t) = -q_0 \frac{k\pi}{T} \sin\Big(\frac{k\pi t}{T}\Big), \quad k = 1, 2, 3, \cdots$$

and

$$\lim_{\omega \to 0} I(t) = 0, \quad \textit{for } k = 0.$$

(*ii*) *Assume the parity condition (9.2.6) does not hold. Then $I(t)$ is infinite for almost all values of t if $k = 1, 2, 3, \cdots$. If $k = 0$ we have*

$$\lim_{\omega \to 0} I(t) = \frac{q_T - q_0}{T}.$$

Remark 9.2.4 (*a*) In the case of a large product between the inductance, L, and capacitance, C, the pulsation $\omega = 1/\sqrt{LC}$ becomes small. Consequently, if $q_T \neq \pm q_0$, the charge $q(t)$ tends to an affine function of t, while the current $I(t)$ approaches a constant, which is proportional to the difference of measurements, $q_T - q_0$.
(*b*) When the pulsation ω reaches one of the resonance values, $\omega_k = \frac{k\pi}{T}$, and the parity conditions are satisfied, the magnitude of the current grows unbounded as the order of the harmonic, k, increases.

The formula for the charge $q(t)$ in the case when the parity condition holds is more simple and it is given by the next result.

Proposition 9.2.5 *If the parity condition (9.2.6) is satisfied, then the charge is given by*

$$q(t) = q_0 \frac{F\big(\omega(\frac{T}{2} - t)\big)}{F(\frac{\omega T}{2})},$$

where

$$F(x) = \begin{cases} \cos x, & \textit{if } k \textit{ is even} \\ \sin x, & \textit{if } k \textit{ is odd.} \end{cases}$$

Proof: Substituting (9.2.6) into (9.2.5) and applying the sum-to-product trigonometric identities, we have

$$\begin{aligned} q(t) &= \frac{1}{\sin(\omega T)}\big[q_T \sin(\omega t) + q_0 \sin\big(\omega(T-t)\big)\big] \\ &= \frac{q_0}{\sin(\omega T)}\big[\sin\big(\omega(T-t)\big) + (-1)^k \sin(\omega t)\big] \\ &= \frac{2q_0}{\sin(\omega T)} \sin\Big(\frac{\omega T + ((-1)^k - 1)\omega t}{2}\Big) \cos\Big(\frac{\omega T - ((-1)^k + 1)\omega t}{2}\Big). \end{aligned}$$

Using the parity of k and the sine double angle formula, yields

$$\begin{aligned} q(t) &= \begin{cases} \frac{2q_0}{\sin(\omega T)} \sin\frac{\omega T}{2} \cos\frac{\omega T - 2\omega t}{2}, & \text{if } k \text{ even} \\ \\ \frac{2q_0}{\sin(\omega T)} \sin\frac{\omega T - 2\omega t}{2} \cos\frac{\omega T}{2}, & \text{if } k \text{ odd} \end{cases} \\ &= \begin{cases} q_0 \frac{\cos\omega(\frac{T}{2}-t)}{\cos\frac{\omega T}{2}}, & \text{if } k \text{ even} \\ \\ q_0 \frac{\sin\omega(\frac{T}{2}-t)}{\sin\frac{\omega T}{2}}, & \text{if } k \text{ odd.} \end{cases} \end{aligned}$$

■

It is worth noting that the statements of Proposition 9.2.1 follow also from Proposition 9.2.5, as the reader can easily verify.

Conjugate points In the following we consider the pulsation ω fixed, i.e., the inductor and capacitor are not variable. The initial reading q_0 at the capacitor is also considered fixed. The only quantity allowed to be variable is the value of the initial current, $I_0 = I(0) = \dot{q}(0)$, through the inductor. For each initial current value, I_0, the charge on the capacitor at time t has the value $q(t; I_0)$ depending on I_0.
The question now is:

For which time values T is the charge $q(T; I_0)$ independent on the initial current value I_0?

In this case, the points $q(T)$ are conjugate with respect to $q(0)$ along the solution $q(t)$. These points will be found in the following.

To this end, we consider a smooth variation of the solution $q(t)$ depending on a parameter ϵ, denoted by $q_\epsilon(t)$. Since the initial point is fixed, then $q_\epsilon(0) = q(0)$. The variation vector field along the solution $q(t)$ is given by the Jacobi vector field

$$Y(t) = \frac{dq_\epsilon(t)}{d\epsilon}_{|\epsilon=0}.$$

We obviously have $Y(0) = 0$. We are looking for those values $T > 0$ such that $Y(T) = 0$, since these are the points that remain unchanged when I_0 is variable. A differential equation can be set for the vector $Y(t)$ by differentiating the equation (9.2.2)

$$\ddot{q}_\epsilon(t) + \omega^2 q_\epsilon(t) = 0. \tag{9.2.10}$$

with respect to ϵ. Then taking $\epsilon = 0$, we obtain

$$\ddot{Y}(t) + \omega^2 Y(t) = 0. \tag{9.2.11}$$

Imposing the vanishing boundary conditions

$$Y(0) = 0, \qquad Y(T) = 0,$$

we obtain the solution $Y(t) = A\sin(\omega t)$, with $A \in \mathbb{R}$. Then the condition $Y(T) = 0$ implies

$$T \in \Big\{\frac{k\pi}{\omega}; k = 1, 2, \cdots.\Big\}. \tag{9.2.12}$$

Therefore, the points $q\big(\frac{k\pi}{\omega}\big)$ are conjugate with respect to $q(0)$ along the solution $q(t)$.

The physical significance of this fact is formulated in the following. As long as q_0 is fixed, the capacitor's charge at time instances $T_k = \frac{k\pi}{\omega}$ is not affected by the values of the initial current, I_0. However, all the other charge values, $q(t)$, are affected by variations of I_0 for

$$t \notin \Big\{\frac{k\pi}{\omega}; k = 0, 1, 2, \cdots.\Big\}.$$

The action We shall now assume that $T < \pi/\omega$ and $q : [0, T] \to \mathbb{R}$, so $q(t)$ does not have any conjugate points with respect to q_0 for $0 < t < T$. Furthermore, if q_0 and q_T are given, there is only one solution $q(t)$ of the differential equation (9.2.2) such that $q(0) = q_0$ and $q(T) = q_T$. We shall compute the action

$$S(q_0, q_T; 0, T) = \int_0^T \mathcal{L}\big(q(t), \dot{q}(t)\big)\, dt$$

along this solution. This action minimizes the integral of the Lagrangian over the space of continuous paths

$$\mathcal{P} = \{x : [0, T] \to \mathbb{R}, \, x(0) = q_0, \, x(T) = q_T\},$$

namely, we have the inequality

$$S(q_0, q_T; 0, T) \leq \int_0^T \mathcal{L}\big(x(t), \dot{x}(t)\big)\, dt, \qquad \forall x \in \mathcal{P}.$$

Proposition 9.2.6 *The action along $q(t)$ from $q_0 = q(0)$ to $q_T = q(T)$ is given by*

$$S(q_0, q_T; 0, T) = \frac{L}{2}\frac{\omega}{\sin(\omega T)}\Big\{(q_T^2 + q_0^2)\cos(\omega T) - 2q_0 q_T\Big\}. \tag{9.2.13}$$

Proof: First we prepare the Lagrangian

$$\begin{aligned}\mathcal{L}\big(q(t), \dot{q}(t)\big) &= \frac{1}{2}L\dot{q}(t)^2 - \frac{1}{2C}q(t)^2 = \frac{L}{2}\{\dot{q}(t)^2 - \omega^2 q(t)^2\}\\ &= \frac{L}{2}\Big\{\frac{\omega^2}{(\sin\omega T)^2}\big[q_T\cos\omega t - q_0\cos\big(\omega(T-t)\big)\big]^2\\ &\quad - \frac{\omega^2}{(\sin\omega T)^2}\big[q_T\sin\omega t + q_0\sin\big(\omega(T-t)\big)\big]^2\Big\}\\ &= \frac{L}{2}\frac{\omega^2}{(\sin\omega T)^2}\Big\{q_T^2\cos(2\omega t) + q_0^2\cos(2\omega(T-t))\\ &\qquad - 2q_0q_T\cos\big(\omega(T-2t)\big)\Big\}.\end{aligned}$$

Then we compute the action by integrating with respect to t

$$\begin{aligned}S(q_0, q_T; 0, T) &= \int_0^T \mathcal{L}\big(q(t), \dot{q}(t)\big)\,dt\\ &= \frac{L}{2}\frac{\omega^2}{(\sin\omega T)^2}\int_0^T \Big\{q_T^2\cos(2\omega t) + q_0^2\cos(2\omega(T-t))\\ &\qquad - 2q_0q_T\cos\big(\omega(T-2t)\big)\Big\}\,dt\\ &= \frac{L}{2}\frac{\omega^2}{(\sin\omega T)^2}\Big\{q_T^2\frac{\sin(2\omega T)}{2\omega} + q_0^2\frac{\sin(2\omega T)}{2\omega} - 2q_0q_T\frac{\sin(\omega T)}{\omega}\Big\}.\end{aligned}$$

Using the sine of double angle formula, after simplification, we arrive at

$$S(q_0, q_T; 0, T) = \frac{L}{2}\frac{\omega}{\sin(\omega T)}\Big\{(q_T^2 + q_0^2)\cos(\omega T) - 2q_0 q_T\Big\}.$$

This formula is well defined for $0 < \omega T < \pi$.

∎

Remark 9.2.7 Using $\omega = 1/\sqrt{LC}$, the action can be written in terms of inductance and capacitance as

$$S(q_0, q_T; 0, T) = \frac{1}{2}\sqrt{\frac{L}{C}}\frac{1}{\sin\left(\frac{T}{\sqrt{LC}}\right)}\left\{(q_T^2 + q_0^2)\cos\left(\frac{T}{\sqrt{LC}}\right) - 2q_0q_T\right\}.$$

The relation between the previous action and the classical action

$$S_{cls}(q_0, q_T; 0, T) = \frac{1}{2T}(q_T - q_0)^2$$

is presented in the next result.

Proposition 9.2.8 (Action for large capacitance) *If the capacitance increases unbounded, then the action becomes proportional with the classical action*

$$\lim_{C \to \infty} S(q_0, q_T; 0, T) = \frac{L}{2T}(q_T - q_0)^2,$$

where L is the inductance. This corresponds to the case when the charge path is the line segment

$$q(t) = q_0 + (q_T - q_0)\frac{t}{T}, \quad 0 \le t \le T.$$

Proof: As $C \to \infty$, then $\omega \to 0$. Applying the limit $\lim_{\omega \to 0} \frac{\omega}{\sin(\omega T)} = \frac{1}{T}$ in formula (9.2.13) yields the desired result. Taking the limit $\omega \to 0$ in formula (9.2.5) and using that $\lim_{\omega \to 0} \dfrac{\sin(\omega t)}{\sin(\omega T)} = \dfrac{t}{T}$, provides the second statement. ■

Another particular expression for the action occurs when parity condition holds.

Proposition 9.2.9 (Action with parity condition) *We assume the parity condition (9.2.6) holds. Then the action is given by*

$$S(q_0, q_T; 0, T) = \begin{cases} L\omega q_0^2 \cot \dfrac{\omega T}{2}, & \text{if } k \text{ odd} \\ \\ -L\omega q_0^2 \tan \dfrac{\omega T}{2}, & \text{if } k \text{ even.} \end{cases}$$

Proof: Substituting $q_T = (-1)^k q_0$ in formula (9.2.13) yields

$$S(q_0, q_T; 0, T) = \frac{\omega L}{\sin(\omega T)}\Big\{ \cos(\omega T) - (-1)^k \Big\} q_0^2.$$

Assume k is odd. Then using formula $1 + \cos(\omega T) = 2\cos^2 \dfrac{\omega T}{2}$, we have

$$\begin{aligned} S(q_0, q_T; 0, T) &= \frac{\omega L}{\sin(\omega T)}\Big\{ \cos(\omega T) + 1 \Big\} q_0^2 \\ &= \frac{\omega L}{2\sin(\frac{\omega T}{2})\cos(\frac{\omega T}{2})} 2\cos^2 \frac{\omega T}{2} q_0^2 = L\omega q_0^2 \cot \frac{\omega T}{2}. \end{aligned}$$

The computation for the case when k is odd is very similar, using formula $1 - \cos(\omega T) = 2\sin^2 \frac{\omega T}{2}$.

■

The van Vleck determinant The mixed derivative of the action with respect to the end points, q_0 and q_T, will play a role in the study of heat kernels. In general, this mixed derivative is captured by the *van Vleck determinant*

$$V(T) = \sqrt{\det\Big(-\frac{1}{2\pi}\frac{\partial^2 S(q_0, q_T; 0, T)}{\partial q_0 \partial q_T}\Big)}. \tag{9.2.14}$$

In our one-dimensional case we obtain

$$V(T) = \sqrt{-\frac{1}{2\pi}\frac{\partial^2 S(q_0, q_T; 0, T)}{\partial q_0 \partial q_T}}.$$

A straightforward computation using the definition of the van Vleck determinant and action formula (9.2.13) yields the following result:

Proposition 9.2.10 *For any $0 < T < \pi/\omega$, the van Vleck determinant is given by*

$$V(T) = \sqrt{\frac{L\omega}{2\pi \sin(\omega T)}}.$$

The interpretations of the van Vleck determinant are given as follows.

(i) First we notice that the derivative of the action with respect to q_T provides the current $I(T)$. To show this, we differentiate in the action formula (9.2.13) to obtain

$$\begin{aligned}\frac{\partial S(T)}{\partial q_T} &= \frac{\partial S(q_0, q_T; 0, T)}{\partial q_T} = \frac{L\omega}{\sin(\omega T)}\big(q_T \cos(\omega T) - q_0\big)\\ &= L\dot{q}(T) = L\, I(T),\end{aligned}$$

where we substituted $t = T$ in the current formula (9.2.9). Therefore, after differentiating again with respect to q_0, we obtain

$$\frac{\partial^2 S(T)}{\partial q_0 \partial q_T} = L\frac{\partial I(T)}{\partial q_0} = L\frac{\partial \dot{q}(T)}{\partial q_0}.$$

Thus, the electrical interpretation states that the van Vleck determinant for this problem is related to the sensitivity of the current with respect to changes in the initial charge by the formula

$$\frac{\partial I(T)}{\partial q_0} = -\frac{2\pi}{L}V(T)^2.$$

(*ii*) We differentiate in the action formula (9.2.13) with respect to the initial position q_0 to obtain

$$\begin{aligned}\frac{\partial S(T)}{\partial q_0} = \frac{\partial S(q_0, q_T; 0, T)}{\partial q_0} &= \frac{L\omega}{\sin(\omega T)}(q_0 \cos(\omega T) - q_T) \\ &= -L\dot{q}(0) = -L\, I(0),\end{aligned}$$

where we substituted $t = 0$ in the current formula (9.2.9). Then differentiating one more time and inverting, yields

$$\frac{\partial^2 S(T)}{\partial q_0 \partial q_T} = -L\frac{\partial I(0)}{\partial q_T} = -L\Big(\frac{\partial q_T}{\partial I(0)}\Big)^{-1} = -L\Big(\frac{\partial q_T}{\partial \dot{q}_0}\Big)^{-1}.$$

Now, the derivative $\dfrac{\partial q_T}{\partial \dot{q}_0}$ has the interpretation of a Jacobi vector along the solution $q(t)$ at $t = T$. To this end we consider the family of solutions $q_\epsilon(t)$ defined by the initial conditions

$$q_\epsilon(0) = q_0, \quad \dot{q}_\epsilon(0) = \dot{q}(0) + \epsilon, \quad \epsilon > 0.$$

Then we write

$$\frac{\partial q_\epsilon(t)}{\partial \epsilon} = \frac{\partial q_\epsilon(t)}{\partial \dot{q}_\epsilon(0)}$$

and then take the limit $\epsilon \to 0$ to obtain a Jacobi field along $q(t)$

$$Y(t) = \frac{\partial q_\epsilon(t)}{\partial \epsilon}_{|\epsilon=0} = \frac{\partial q_0(t)}{\partial \dot{q}_0(0)} = \frac{\partial q(t)}{\partial \dot{q}(0)}, \quad 0 \le t \le T.$$

Therefore,

$$\frac{\partial^2 S(T)}{\partial q_0 \partial q_T} = -\frac{L}{Y(T)}.$$

This yields the following interpretation of the van Vleck determinant as an inverse of a Jacobi vector field

$$\frac{1}{Y(T)} = \frac{2\pi}{L} V(T)^2.$$

Since the Jacobi vector field vanishes at the conjugate points to $q(0)$ along $q(t)$, then the singularities of the van Vleck determinant correspond to the conjugate points to $q(0)$. Thus, the lateral limits $V(0+)$ and $V\big(\dfrac{\pi}{\omega} -\big)$ are equal to $+\infty$, which indicates the fact that $q(0)$ and $q(\frac{\pi}{\omega})$ are conjugate points along $q(t)$.

9.3 The Hamiltonian Formalism

The Hamiltonian function

The Hamiltonian function, $\mathcal{H}$, is defined on the phase plane $\{q, p\}$. In the mechanical context, q stands for the position coordinate, while p represents the momentum. In the present LC-circuit setup, q describes the charge at the capacitor and p represents the magnetic flux. The function $\mathcal{H}$ is obtained from the Lagrangian (9.2.1) by applying the Legendre transform

$$\mathcal{H}(p, q) = p\dot{q} - \mathcal{L}(q, \dot{q}), \tag{9.3.15}$$

where $\dot{q}$ is expressed in terms of p by solving the equation $p = \dfrac{\partial \mathcal{L}}{\partial \dot{q}}$. In the case of the Lagrangian (9.2.1)

$$\mathcal{L}(q, \dot{q}) = \frac{1}{2}L\dot{q}^2 - \frac{1}{2C}q^2$$

this equation becomes $p = L\dot{q}$, so $\dot{q} = p/L$. Substituting in (9.3.15) we obtain the Hamiltonian function

$$\begin{aligned} \mathcal{H}(p, q) &= p\dot{q} - \mathcal{L}(q, \dot{q}, t) = \frac{p^2}{L} - \Big(\frac{1}{2}\frac{p^2}{L} - \frac{1}{2C}q^2\Big) \\ &= \frac{1}{2L}p^2 + \frac{1}{2C}q^2. \end{aligned}$$

This can be written equivalently as

$$\mathcal{H}(p, q) = \frac{1}{2}LI^2 + \frac{1}{2C}q^2 = E_M + E_P,$$

which is the sum between the the magnetic and potential electric energies, i.e., it is the total energy of the electrical system.

It is worth to note that the magnetic flux, which is given by $p = L\dot{q} = L\,I$, is consistent with the definition of the inductance as the quotient between flux and current.

The Hamiltonian system

We are interested to find the solution $\big(q(t), p(t)\big)$ of the Hamiltonian system, which is a system of 2 first order ordinary differential equations given by

$$\dot{q} = \frac{\partial H}{\partial p}, \quad \dot{p} = -\frac{\partial H}{\partial q}, \tag{9.3.16}$$

with boundary conditions $q(0) = q_0$ and $q(T) = q_T$. This system will be solved in two different ways: (i) by using the general theory of ODE's systems and (ii) by using the first integral of energy method.

(i) The ODE system solution Using the Hamiltonian function expression

$$\mathcal{H}(p,q) = \frac{1}{2L}p^2 + \frac{1}{2C}q^2 \tag{9.3.17}$$

the system (9.3.16) becomes

$$\dot{q} = \frac{1}{L}p, \quad \dot{p} = -\frac{1}{C}q. \tag{9.3.18}$$

It is worth noting that this system can be transformed into a second order equation for q by differentiating in the first relation and substituting in the second, fact that recovers equation (9.2.2) with solution (9.2.5). This shows the equivalence between the Lagrangian and Hamiltonian systems in this setup.

In the following we shall solve (9.3.18) as a system. This can be written in the matrix form $\dot{Y} = AY$ with $Y^T = (q,p)$ and coefficients matrix

$$A = \begin{pmatrix} 0 & \frac{1}{L} \\ -\frac{1}{C} & 0 \end{pmatrix}.$$

The solution can be expressed in terms of either the initial state, $Y_0 = Y(0)$, as $Y(t) = e^{At}Y(0)$, or in terms of the final state, $Y_T = Y(T)$, as $Y(t) = e^{-A(T-t)}Y_T$.

The aforementioned exponentials can be computed using the next result:

Lemma 9.3.1 *Consider the matrix*

$$A = \begin{pmatrix} 0 & a \\ -b & 0 \end{pmatrix},$$

with $a, b > 0$. Then

$$e^{At} = \begin{pmatrix} \cos(\sqrt{ab}\,t) & \sqrt{\frac{a}{b}}\sin(\sqrt{ab}\,t) \\ -\sqrt{\frac{b}{a}}\sin(\sqrt{ab}\,t) & \cos(\sqrt{ab}\,t) \end{pmatrix}, \qquad t > 0.$$

Thus, we obtain that

$$Y(t) = \begin{pmatrix} \cos(\omega t) & \sqrt{\frac{C}{L}}\sin(\omega t) \\ -\sqrt{\frac{L}{C}}\sin(\omega t) & \cos(\omega t) \end{pmatrix} Y_0,$$

or equivalently

$$\begin{aligned} q(t) &= \cos(\omega t)q_0 + \sqrt{\frac{C}{L}}\sin(\omega t)\,p(0) \\ p(t) &= -\sqrt{\frac{L}{C}}\sin(\omega t)\,q_0 + \cos(\omega t)\,p(0). \end{aligned}$$

These represent the explicit expressions of the charge and flux at any time t in terms of the initial state (q_0, p_0).

(ii) The first integral of motion Since the Hamiltonian function does not depend explicitly on time t, using chain rule and Hamiltonian equations (9.3.16), we have

$$\frac{dH}{dt} = \frac{\partial H}{\partial q}\dot{q} + \frac{\partial H}{\partial p}\dot{p} = -\dot{p}\dot{q} + \dot{q}\dot{p} = 0.$$

Therefore, the Hamiltonian is constant along the solution $(q(t), p(t))$. This corresponds to the *total energy*, E, as a first integral of motion. Therefore, given $T > 0$ and q_0, q_T, there is a constant $E = E(q_0, q_T, T) > 0$, such that

$$\frac{1}{2L}p(t)^2 + \frac{1}{2C}q(t)^2 = E. \tag{9.3.19}$$

A first qualitative description of solutions is given next.

Proposition 9.3.2 *The solution $(q(t), p(t))$ describes in the phase space an ellipse of area $\mathcal{A} = \dfrac{2\pi E}{\omega}$.*

Proof: Equation (9.3.19) can be written equivalently as

$$\frac{q^2}{a^2} + \frac{p^2}{b^2} = 1,$$

with $a = \sqrt{2CE}$ and $b = \sqrt{2LE}$, which is the equation of an ellipse with semiaxes a and b. Its area is given by $\mathcal{A} = \pi ab = 2\pi E\sqrt{LC} = 2\pi E/\omega$. ■

The next result provides the solution in terms of boundary conditions and energy.

Proposition 9.3.3 *The solution of the Hamiltonian system (9.3.16) is given by*

$$\begin{aligned} q(t) &= q_{\max}\cos(\theta_0 - \omega t) \\ I(t) &= I_{\max}\sin(\theta_0 - \omega t), \end{aligned}$$

where $q_{\max} = \sqrt{2CE}$, $I_{\max} = \sqrt{2E/L}$ and

$$\theta_0 = \tan^{-1}\Big(\frac{1}{\sin\omega T}\frac{q_T}{q_0} - \cot(\omega T)\Big).$$

Proof: From Proposition 9.3.2 the solution belongs to an ellipse, so there is a smooth function $\theta(t)$ such that

$$\begin{aligned} q(t) &= a\cos\theta(t) = \sqrt{2CE}\cos\theta(t) \qquad (9.3.20)\\ p(t) &= b\sin\theta(t) = \sqrt{2LE}\sin\theta(t). \qquad (9.3.21) \end{aligned}$$

Differentiating equation (9.3.20) with respect to t and multiplying by L, we obtain

$$p(t) = L\dot{q}(t) = -L\sqrt{2CE}\sin\theta(t)\dot{\theta}(t).$$

Comparing the last equation against equation (9.3.21), and using $\omega = 1/\sqrt{LC}$, we get $\dot{\theta}(t) = -\omega$, as long as $\sin\theta(t) \neq 0$. Therefore, $\theta(t) = -\omega t + \theta_0$, with the initial parameter θ_0 subject to be determined from the boundary conditions. To accomplish this, we write (9.3.20) as

$$q(t) = \sqrt{2CE}\cos(\omega t - \theta_0)$$

and make $t = 0$ and then $t = T$ to obtain

$$\begin{aligned} q_0 &= \sqrt{2CE}\cos(\theta_0)\\ q_T &= \sqrt{2CE}\cos(\omega T - \theta_0). \end{aligned}$$

We assume $q_0 \neq 0$. Dividing the previous two equations yields

$$\frac{q_T}{q_0} = \frac{\cos(\omega T - \theta_0)}{\cos(\theta_0)} = \cos(\omega T) + \sin(\omega T)\tan\theta_0$$

from where

$$\tan\theta_0 = \frac{1}{\sin\omega T}\frac{q_T}{q_0} - \cot(\omega T)$$

which determines θ_0 univocally in terms of boundary conditions q_0 and q_T. If $q_0 = 0$ we take by convention $\theta_0 = \pi/2$.

■

In the following we shall express the energy E in terms of boundary conditions. This shows that if two measurements of charge are made, q_0 and q_T, these completely determine the energy of the system.

Proposition 9.3.4 *Let $0 < T < \pi/\omega$. Then the total energy of the electrical system in terms of boundary conditions is given by*

$$E(q_0, q_T) = \frac{L}{2}\frac{\omega^2}{\sin(\omega T)^2}\Big(q_0^2 + q_T^2 - 2q_0 q_T\cos(\omega T)\Big). \qquad (9.3.22)$$

Proof: Since E is independent of t, then we can express the energy by substituting $t = 0$ in relation (9.3.19)

$$E = \frac{1}{2L}p(0)^2 + \frac{1}{2C}q(0)^2 = \frac{L}{2}I(0)^2 + \frac{1}{2C}q_0^2. \qquad (9.3.23)$$

It suffices to express the initial current $I(0)$ in terms of q_0 and q_T. To obtain this we make $t = 0$ in relation (9.2.9) and obtain

$$I(0) = \frac{\omega}{\sin(\omega T)}\Big(q_T - q_0\cos(\omega T)\Big).$$

Substituting back into (9.4.28), we have

$$\begin{aligned} E &= \frac{L}{2}I(0)^2 + \frac{1}{2C}q_0^2 \\ &= \frac{L}{2}\frac{\omega^2}{\sin(\omega T)^2}\Big(q_T - q_0\cos(\omega T)\Big)^2 + \frac{1}{2C}q_0^2 \\ &= \frac{q_0^2 - 2q_0q_T\cos(\omega T) + q_T^2}{2C\sin(\omega T)^2} \\ &= \frac{L}{2}\frac{\omega^2}{\sin(\omega T)^2}\Big(q_0^2 + q_T^2 - 2q_0q_T\cos(\omega T)\Big). \end{aligned}$$

■

Corollary 9.3.5 *Let $0 < T < \pi/\omega$. Then the maximum charge and the maximum current are given in terms of boundary conditions by*

$$\begin{aligned} q_{\max} &= \frac{1}{\sin(\omega T)}\sqrt{q_0^2 + q_T^2 - 2q_0q_T\cos(\omega T)} \\ I_{\max} &= \omega\, q_{\max}. \end{aligned}$$

Proof: Using Propositions 9.3.3 and 9.3.4, we have

$$\begin{aligned} q_{\max} &= \sqrt{2CE} = \sqrt{\frac{CL\omega^2}{\sin(\omega T)^2}\big(q_0^2 + q_T^2 - 2q_0q_T\cos(\omega T)\big)} \\ &= \frac{1}{\sin(\omega T)}\sqrt{q_0^2 + q_T^2 - 2q_0q_T\cos(\omega T)}. \\ I_{\max} &= \sqrt{\frac{2E}{L}} = \sqrt{\frac{2CE}{LC}} = \sqrt{2CE\omega^2} = \omega\, q_{\max}. \end{aligned}$$

■

Remark 9.3.6 If the capacitance decreases to zero, $C \to 0$, then the pulsation $\omega \to 0$. Taking the limit in formula (9.3.22) we obtain

$$\lim_{C \to 0} E(q_0, q_T) = \frac{L}{2T^2}(q_T - q_0)^2.$$

Therefore, the energy of the system at low capacitance tends to be proportional to the square of charge difference of two measurements and inverse proportional to the square of the time lag of the measurements.

The next result shows that the flux is provided as a gradient of the action function with respect to charge. This result will be useful in the next section.

Proposition 9.3.7 *The partial derivative of the action $S(q_0, q_T; 0, T)$ with respect to the end point q_T is the magnetic flux value at time $t = T$*

$$\frac{\partial S(q_0, q_T; 0, T)}{\partial q_T} = p(T),$$

where $p(t) = L\,I(t)$.

Proof: It follows from a computation differentiating in (9.2.13) and applying relation (9.2.9)

$$\frac{\partial S(q_0, q_T; 0, T)}{\partial q_T} = \frac{L\omega}{\sin(\omega T)^2}\Big\{q_T \cos(\omega T) - q_0\Big\} = L\,I(T).$$

■

9.4 The Hamilton-Jacobi Equation

Denote by $S = S(q_0, q; 0, \tau)$ the action that starts from q_0 at $t = 0$ and ends at q at time $t = \tau$, with $0 < \tau < \pi/\omega$. It will be shown that S is a solution of the following initial condition *Hamilton-Jacobi equation*

$$\frac{\partial S}{\partial \tau} + \mathcal{H}\Big(q, \frac{\partial S}{\partial q}\Big) = 0 \tag{9.4.24}$$

$$S(q_0, q_0; 0, 0) = 0, \tag{9.4.25}$$

where $\mathcal{H}$ denotes the Hamiltonian function, which is given by (9.3.17)

$$\mathcal{H}(p, q) = \frac{1}{2L}p^2 + \frac{1}{2C}q^2. \tag{9.4.26}$$

Then the equation (9.4.24) writes explicitly as

$$\frac{\partial S}{\partial \tau} + \frac{1}{2L}\Big(\frac{\partial S}{\partial q}\Big)^2 + \frac{1}{2C}q^2 = 0. \qquad (9.4.27)$$

Solving Hamilton-Jacobi We shall solve equation (9.4.27) and retrieve the previous expression of the action. First, we introduce the variable $p = \dfrac{\partial S}{\partial q}$ and consider the Hamiltonian (9.4.26). If $\big(q(t), p(t)\big)$ is a solution of the associated Hamiltonian system, then the total energy is preserved along this solution as in relation (9.3.19)

$$\frac{1}{2L}p(t)^2 + \frac{1}{2C}q(t)^2 = E(q_0, q; \tau). \qquad (9.4.28)$$

Therefore,

$$H\Big(q, \frac{\partial S}{\partial q}\Big) = \frac{1}{2L}\Big(\frac{\partial S}{\partial q}\Big)^2 + \frac{1}{2C}q^2 = E(q_0, q; \tau), \qquad (9.4.29)$$

and the Hamilton-Jacobi equation (9.4.27) becomes

$$\frac{\partial S}{\partial \tau} = -E(q_0, q; \tau).$$

Using the expression of energy given by Proposition 9.3.4, this writes equivalently as

$$\frac{\partial S}{\partial \tau} = -\frac{L}{2}\frac{\omega^2}{\sin(\omega\tau)^2}\Big(q_0^2 + q^2 - 2q_0 q\cos(\omega\tau)\Big).$$

Integrating partially with respect to τ, we obtain

$$\begin{aligned} S &= -\frac{L}{2}\left\{(q_0^2 + q^2)\int \frac{\omega^2\,d\tau}{\sin(\omega\tau)^2} - 2q_0 q\int \frac{\omega^2}{\sin(\omega\tau)^2}\cos(\omega\tau)\,d\tau\right\} + C(q_0, q) \\ &= \frac{L\omega}{2}(q_0^2 + q^2)\cot(\omega\tau) - L\omega q_0 q\frac{1}{\sin(\omega\tau)} + C(q_0, q) \\ &= \frac{L}{2}\frac{\omega}{\sin(\omega\tau)}\Big((q_0^2 + q^2)\cos(\omega\tau) - 2q_0 q\Big) + C(q_0, q), \end{aligned}$$

where $C(q_0, q)$ is a function which does not depend explicitly of τ. In order to retrieve formula (9.2.13), it suffices to show that $C(q_0, q) = 0$.
To accomplish this, we denote

$$\tilde{S}(q_0, q; 0, \tau) = \frac{L}{2}\frac{\omega}{\sin(\omega\tau)}\Big((q_0^2 + q^2)\cos(\omega\tau) - 2q_0 q\Big)$$

and have

$$S(q_0, q; 0, \tau) = \tilde{S}(q_0, q; 0, \tau) + C(q_0, q). \qquad (9.4.30)$$

Taking the partial derivative with respect to q in (9.4.30) we have

$$\frac{\partial}{\partial q}S(q_0, q; 0, \tau) = \frac{\partial}{\partial q}\tilde{S}(q_0, q; 0, \tau) + \frac{\partial}{\partial q}C(q_0, q).$$

Since $\frac{\partial}{\partial q}S(q_0, q; 0, \tau) = p$ by construction and by Proposition (9.3.7) we have $\frac{\partial \tilde{S}}{\partial q} = p(\tau)$, it follows that $\frac{\partial}{\partial q}C(q_0, q) = 0$. Therefore, $C(q_0, q) = C(q_0, q_0)$. In the following we shall show that $C(q_0, q_0) = 0$. For this we take the limit $\tau \searrow 0$ in relation (9.4.30)

$$\lim_{\tau\searrow 0} S(q_0, q; 0, \tau) = \lim_{\tau\searrow 0} \tilde{S}(q_0, q; 0, \tau) + \lim_{\tau\searrow 0} C(q_0, q).$$

The initial condition (9.4.25) implies $\lim_{\tau\searrow 0} S(q_0, q; 0, \tau) = 0$. The second limit can be evaluated as

$$\begin{aligned}
\lim_{\tau\searrow 0} \tilde{S}(q_0, q; 0, \tau) &= \lim_{\tau\searrow 0} \frac{L}{2}\frac{\omega}{\sin(\omega\tau)}\Big((q_0^2 + q^2)\cos(\omega\tau) - 2q_0 q\Big) \\
&= \lim_{\tau\searrow 0} \frac{L}{2\tau}\Big((q_0^2 + q^2)\cos(\omega\tau) - 2q_0 q\Big) \\
&= \frac{L}{2}\lim_{\tau\searrow 0}\frac{q - q_0}{\tau}\lim_{\tau\searrow 0}(q - q_0) \\
&= \frac{L\, I_0}{2}\lim_{\tau\searrow 0}(q - q_0) = 0.
\end{aligned}$$

Therefore, $C(q_0, q_0) = \lim_{\tau\searrow 0} C(q_0, q) = 0$. Hence, the integration function $C(q_0, q) = 0$ and then relation (9.4.30) implies $S = \tilde{S}$. We just shown that

$$S(q_0, q; 0, \tau) = \frac{L}{2}\frac{\omega}{\sin(\omega\tau)}\Big((q_0^2 + q^2)\cos(\omega\tau) - 2q_0 q\Big).$$

This formula is valid for $0 < \tau < \pi/\omega$.

9.5 Complex Lagrangian Mechanics

Many of the formulas encountered in the previous sections have a simpler form in the context of the complex Lagrangian mechanics. In this new framework, instead of working with the real-valued differentiable function $q : [0, T] \to \mathbb{R}$ with fixed end points, $q(0) = q_0$ and $q(T) = q_T$, we employ the complexified version of $q(t)$, which is the complex valued function $z(t) = q(t)e^{i\omega t}$, where ω stands for the pulsation. Its endpoints are given by $z(0) = q_0$ and $z(T) = q_T e^{i\omega T}$. The derivative of $z(t)$ and its square modulus are given by

$$\dot{z}(t) = e^{i\omega t}\big(\dot{q}(t) + i\omega q(t)\big) \tag{9.5.31}$$

$$|\dot{z}(t)|^2 = |\dot{q}(t) + i\omega q(t)|^2 = \dot{q}(t)^2 + \omega^2 q(t)^2. \tag{9.5.32}$$

The complex Lagrangian We shall write the magnetic and electric energies in terms of the path $z(t)$. Solving for $\dot{q}(t)^2$ from (9.5.32) and using that $|z(t)| = |q(t)|$ we obtain

$$\begin{aligned} E_M &= \frac{1}{2}L\dot{q}^2 = \frac{1}{2}L(|\dot{z}|^2 - \omega^2|z|^2) \\ E_P &= \frac{1}{2C}q^2 = \frac{1}{2C}|z|^2. \end{aligned}$$

Then the Lagrangian (9.2.1) becomes

$$\mathcal{L} = E_M - E_P = \frac{1}{2}L|\dot{z}|^2 - \frac{1}{C}|z|^2.$$

The complex Euler-Lagrange equation Using the derivatives

$$\begin{aligned} \dot{z}(t) &= e^{i\omega t}\big(\dot{q}(t) + i\omega q(t)\big) \\ \ddot{z}(t) &= e^{i\omega t}\big(\ddot{q}(t) + 2i\omega\dot{q}(t) - \omega^2 q(t)\big) \end{aligned}$$

we have

$$\ddot{z}(t) - 2i\omega\dot{z}(t) = e^{i\omega t}\big(\ddot{q}(t) + \omega^2 q(t)\big).$$

Therefore, the Euler-Lagrange equation (9.2.2) is equivalent to

$$\ddot{z}(t) - 2i\omega\dot{z}(t) = 0, \tag{9.5.33}$$

subject to the boundary conditions

$$z_0 = z(0) = q_0, \qquad z_T = z(T) = q_T e^{i\omega T}.$$

Substituting formulas (9.2.5) and (9.2.9) into (9.5.32) we obtain the magnitude of the velocity along the complex solution $z(t)$

$$\begin{aligned} |\dot{z}(t)|^2 &= \dot{q}(t)^2 + \omega^2 q(t)^2 \\ &= \frac{\omega^2}{\sin(\omega T)^2}\big[q_T\cos(\omega t) - q_0\cos\big(\omega(T-t)\big)\big]^2 \\ &+ \frac{\omega^2}{\sin(\omega T)^2}\big[q_T\sin(\omega t) + q_0\sin\big(\omega(T-t)\big)\big]^2 \\ &= \frac{\omega^2}{\sin(\omega T)^2}\big[q_0^2 + q_T^2 - 2q_0q_T\cos\big(\omega(T-2t)\big)\big]. \end{aligned}$$

Next we shall compute the energy along $z(t)$, with $0 \le t \le T$. First we compute

$$\begin{aligned} |z_T - z_0|^2 &= \big(z(T) - z(0)\big)\big(\bar{z}(T) - \bar{z}(0)\big) \\ &= |z(T)|^2 + |z(0)|^2 - 2z(0)Re(z(T)) \\ &= q_T^2 + q_0^2 - 2q_0q_T\cos(\omega T). \end{aligned}$$

Substituting in the relation of total energy (9.3.22) yields

$$E(z_0, z_T) = \frac{L}{2}\frac{\omega^2}{\sin(\omega T)^2}|z_T - z_0|^2,$$

where $0 < T < \pi/\omega$. By Corollary 9.3.5 we also obtain

$$q_{max} = \frac{1}{\sin(\omega T)}|z_T - z_0|, \qquad I_{max} = \frac{\omega}{\sin(\omega T)}|z_T - z_0|.$$

9.6 A Stochastic Approach

The associated operator and its heat kernel

We consider the differential operator obtained by replacing the momentum p by the differential operator ∂_x and the charge q by x in the Hamiltonian function of an LC-circuit

$$H(p, q) = \frac{1}{2L}p^2 + \frac{1}{2C}q^2.$$

The resulting operator takes the form of a Hermite operator

$$\mathcal{A} = \frac{1}{2L}\partial_x^2 + \frac{1}{2C}x^2. \tag{9.6.34}$$

The relationship between the operator $\mathcal{A}$, the action

$$S(x_0, x; t) = \frac{L}{2}\frac{\omega}{\sin(\omega t)}\big((x^2 + x_0^2)\cos(\omega t) - 2xx_0\big) \tag{9.6.35}$$

and the van Vleck determinant

$$V(t) = \sqrt{\frac{L\omega}{2\pi\sin(\omega t)}} \tag{9.6.36}$$

is given by the following result.

Proposition 9.6.1 *The heat kernel of the operator (9.6.34) is given by*

$$K(x_0, x; t) = V(t)e^{-S(x_0,x;t)}, \qquad t > 0, x, x_0 \in \mathbb{R}.$$

Proof: We need to show

$$\partial_t K(x_0, x; t) = \frac{1}{2L}\partial_x^2 K(x_0, x; t) + \frac{1}{2C}x^2 K(x_0, x; t). \tag{9.6.37}$$

We start by taking the logarithm in (9.6.36) and then differentiate

$$\partial_t \ln V(t) = -\frac{1}{2}\partial_t(\ln \sin(\omega t)) = -\frac{\omega}{2}\cot(\omega t).$$

Since from the Hamilton-Jacobi equation

$$\partial_t S(x_0, x; t) = -E(x_0, x; t),$$

then the left side of (9.6.37) is given by

$$\begin{aligned} \partial_t K(x_0, x; t) &= \partial_t\big(V(t)e^{-S(x_0,x;t)}\big) \\ &= \Big(\frac{\partial_t V(t)}{V(t)} - \partial_t S(x_0, x; t)\Big)K(x_0, x; t) \\ &= \Big(-\frac{\omega}{2}\cot(\omega t) + E(x_0, x; t)\Big)K(x_0, x; t). \end{aligned} \tag{9.6.38}$$

Differentiate twice in (9.6.35) yields

$$\partial_x^2 S(x_0, x; t) = L\omega \cot(\omega t).$$

Rewriting the fact that the Hamiltonian function is constant along solutions, relation (9.4.29) becomes

$$(\partial_x S)^2 = 2L\Big(E - \frac{1}{2C}x^2\Big).$$

Using the last two relations we compute next the right side of (9.6.37)

$$\partial_x K(x_0, x; t) = -\partial_x S\, K(x_0, x; t)$$

$$\partial_x^2 K(x_0, x; t) = \big((\partial_x S)^2 - \partial_x^2 S\big)K(x_0, x; t)$$

$$\frac{1}{2L}\partial_x^2 K(x_0, x; t) + \frac{1}{2C}x^2 K(x_0, x; t) = (E - \frac{\omega}{2}\cot(\omega t))K(x_0, x; t) \tag{9.6.39}$$

Relation (9.6.37) follows now from the comparison of relations (9.6.38) and (9.6.39).

To finish the proof we still need to show that $K(x_0, x; t)$ behaves like a delta function as $t \to 0$. To this end we use the asymptotic correspondences for $t \to 0$, which follow from (9.6.35) and (9.6.36)

$$S(s_0, x; t) \sim \frac{L}{2t}(x - s_0)^2$$

$$V(t) \sim \sqrt{\frac{L}{2\pi t}}.$$

Therefore

$$\lim_{t\to 0} K(x_0,x;t) = \lim_{t\to 0} V(t)e^{-S(x_0,x;t)} = \lim_{t\to 0}\sqrt{\frac{L}{2\pi t}}e^{-\frac{L}{2t}(x-s_0)^2}$$
$$= \lim_{\tau\to 0}\frac{1}{\sqrt{2\pi\tau}}e^{-\frac{1}{2\tau}(x-s_0)^2} = \delta(x-x_0),$$

where we substituted $\tau = t/L$ and use that $p_\tau(x_0,x) = \frac{1}{\sqrt{2\pi\tau}}e^{-\frac{1}{2\tau}(x-s_0)^2}$ is a heat kernel for the operator $\frac{1}{2}\partial_x^2$, so it behaves like a delta function as $\tau \to 0$. ∎

The total heat

The heat kernel $K(x_0,x;t)$ does not represent the transition density of any diffusion, since $\int_{\mathbb{R}} K(x_0,x;t)dx > 1$ for $t>0$. This occurs due to the fact that the operator $\mathcal{A}$ has a positive quadratic potential, which acts as a heat source.

Thus, it makes sense to introduce the notion of *incoming heat* initiated at the point x_0 at time t, as the integral

$$Q(x_0;t) = \int_{\mathbb{R}} K(x_0,x;t)dx. \tag{9.6.40}$$

This satisfies:
(*i*) $Q(x_0;0) = 1$,
(*ii*) $Q(x_0,t) > 1$ for $t > 0$,
(*iii*) $Q(x_0;t)$ is increasing over time, with a singularity at $t = \dfrac{\pi}{2\omega}$.

These properties follow from the following more precise statement.

Proposition 9.6.2 *The incoming heat is given by*

$$Q(x_0;t) = \frac{1}{\sqrt{\cos(\omega t)}}\, e^{\frac{L}{2}x_0^2\omega\tan(\omega t)}, \qquad 0 < \omega t < \pi/2.$$

Proof: The result is a direct computation using the Gaussian integral

$$\int_{\mathbb{R}} e^{-ax^2+bx}\,dx = \sqrt{\frac{\pi}{a}}\,e^{\frac{b^2}{4a}}, \qquad a > 0,$$

with $a = \dfrac{L}{\omega}\omega\cot(\omega t)$ and $b = \dfrac{L\omega x_0}{\sin(\omega t)}$. Thus, we have

$$\begin{aligned}
Q(x_0;t) &= \int_{\mathbb{R}} K(x_0,x;t)dx = V(t)\int_{\mathbb{R}} e^{-S(x_0,x;t)}\,dx \\
&= V(t)\int_{\mathbb{R}} e^{-\frac{L}{2}\frac{\omega}{\sin(\omega t)}((x^2+x_0^2)\cos(\omega t)-2xx_0)}\,dx \\
&= V(t)e^{-\frac{L}{2}\frac{\omega}{\sin(\omega t)}\cos(\omega t)x_0^2}\int_{\mathbb{R}} e^{-\frac{L}{2}\omega\cot(\omega t)x^2-\frac{L\omega x_0}{\sin(\omega t)}x}\,dx \\
&= \sqrt{\frac{L\omega}{2\pi\sin(\omega t)}}e^{-\frac{L}{2}\frac{\omega}{\sin(\omega t)}\cos(\omega t)x_0^2}\sqrt{\frac{2\pi}{L\omega\cot(\omega t)}}e^{\frac{L}{2}\omega x_0^2\frac{1}{\sin^2(\omega t)\cot(\omega t)}} \\
&= \frac{1}{\sqrt{\cos(\omega t)}}e^{-\frac{L}{2}\omega x_0^2\left(\cot(\omega t)-\frac{1}{\sin(\omega t)\cos(\omega)}\right)} \\
&= \frac{1}{\sqrt{\cos(\omega t)}}e^{\frac{L}{2}x_0^2\omega\tan(\omega t)}.
\end{aligned}$$

■

The stochastic interpretation

The heat $Q(x_0;t)$ can be interpreted as an expectation involving a functional of a Brownian motion as follows. Setting $\gamma = i\omega$ in the Cameron-Martin formula

$$\mathbb{E}_{x_0}[e^{-\frac{\gamma^2}{2}\int_0^t W_s^2 ds}] = \frac{1}{\sqrt{\cosh(t\gamma)}}e^{-\frac{x_0^2}{2}\gamma\tanh(t\gamma)},$$

we obtain

$$\mathbb{E}_{x_0}[e^{\frac{\omega^2}{2}\int_0^t W_s^2 ds}] = \frac{1}{\sqrt{\cos(t\omega)}}e^{\frac{x_0^2}{2}\omega\tan(t\omega)}.$$

Using Proposition 9.6.2 we obtain the following stochastic interpretation of the heat

$$Q(x_0;t) = \mathbb{E}_{\sqrt{L}x_0}[e^{\frac{\omega^2}{2}\int_0^t W_s^2 ds}]. \tag{9.6.41}$$

To make a physical sense of this expression, we shall consider the inductance $L = 1$. Then we assume that the charge on the capacitor at time $t = 0$ is x_0. We further assume that the electrons move from one arm of the capacitor to the other in an arbitrary way due to the tunnelling effect. This phenomenon occurs when the electrons perform a random penetration of the dielectric, passing from one side of a potential barrier to the other. Under these circumstances, the capacitor charge at time t is described by a Brownian motion, namely $q_t = W_t$. Recalling that the electric energy stored between the arms of a capacitor under charge q is given by $E_P = \frac{1}{2C}q^2$, we have

$$\frac{\omega^2}{2}\int_0^t W_s^2 ds = \int_0^t \frac{1}{2C}q_s^2\, ds = \int_0^t E_P(s)\, ds.$$

This provides the interpretation of $\frac{\omega^2}{2}\int_0^t W_s^2 ds$ as the cumulative electric energy on the capacitor until time t. Thus, the heat $Q(x_0;t)$ given by (9.6.41) becomes the expectation of the exponential of the cumulative electric energy. By property (iii) of page 506 the heat $Q(x_0;t)$ reaches a singularity for $t = \frac{\pi}{2\omega} = \frac{\pi}{2}\sqrt{C}$. Therefore, in the case of a large capacity (i.e. a thicker dielectric layer) the singularity occurs later. And in the case of a small capacitance C, the capacitor gets electrically burnt very soon.

The stochastic interpretation of the heat kernel

Using the action formula and the van Vleck determinant given by relations (9.6.35) and (9.6.36), Proposition 9.6.1 provides the following expression for the heat kernel of the operator (9.6.34)

$$\begin{aligned} K(x_0,x;t) &= \sqrt{\frac{L\omega}{2\pi\sin(\omega t)}}e^{-[\frac{L}{2}\frac{\omega}{\sin(\omega t)}((x^2+x_0^2)\cos(\omega t)-2xx_0)]} \\ &= \frac{\sqrt{L\omega}}{\sqrt{2\pi\sin(\omega t)}}e^{-[\frac{L\omega}{2}((x^2+x_0^2)\cot(\omega t)-\omega\frac{xx_0}{\sin(\omega t)})]}, \quad t>0, x, x_0\in\mathbb{R}. \end{aligned} \tag{9.6.42}$$

Substituting $\gamma = i\omega$ in the following formula[1]

$$\mathbb{E}_{x_0}[e^{-\frac{\gamma^2}{2}\int_0^t W_s^2 ds}; W_t\in dx] = \frac{\sqrt{\gamma}}{\sqrt{2\pi\sinh(t\gamma)}}e^{-[\frac{\gamma}{2}(x_0^2+x^2)\coth(t\gamma)-\gamma\frac{xx_0}{\sinh(t\gamma)}]}dx,$$

and using that $\cosh(it\gamma)=\cos(t\gamma)$ and $\sinh(it\gamma)=i\sin(t\gamma)$, we obtain

$$\mathbb{E}_{x_0}[e^{\frac{\omega^2}{2}\int_0^t W_s^2 ds}; W_t\in dx] = \frac{\sqrt{\omega}}{\sqrt{2\pi\sin(\omega t)}}e^{-[\frac{\omega}{2}((x^2+x_0^2)\cot(\omega t)-\omega\frac{xx_0}{\sin(\omega t)})]}.$$

Comparing to (9.6.42) we obtain the following stochastic interpretation of the heat kernel for the case $L=1$

$$\mathbb{E}_{x_0}[e^{\frac{\omega^2}{2}\int_0^t W_s^2 ds}; W_t\in dx] = K(x_0,x;t)dx. \tag{9.6.43}$$

[1] See formula (1.9.7) from Borodin and Salminen [15], page 174

This provides the interpretation of the heat kernel $K(x_0, x; t)$ as an average of the exponential functional $e^{\frac{\omega^2}{2}\int_0^t W_s^2 ds}$ over all Brownian bridges W_s between x_0 and x within time t.

9.7 Summary

In this chapter, we explore the dynamics of inductor-capacitor electrical circuits from a variational perspective. We begin by utilizing the Lagrangian formalism, which allows us to derive the dynamic equations using the Euler-Lagrange equations and calculate the classical action. Alternatively, we employ the Hamiltonian formalism to uncover the dynamics.

We also delve into the Hamilton-Jacobi equation and determine the energy involved. Extensive analysis is conducted on the solutions, including the examination of Jacobi vector fields and their connection to the van Vleck determinant.

We explore the relationship between the van Vleck determinant and the transition probability of the associated diffusion. The Hermite operator, associated with the underlying Hamiltonian, is introduced along with the computation of its associated heat kernel. Furthermore, we provide a stochastic interpretation of this heat kernel.

9.8 Exercises

Exercise 9.8.1 At a certain moment in time, t_0, a lightning strikes an RLC electrical circuit. The corresponding equation is

$$L(t)q''(t) + R(t)q'(t) + \frac{1}{C(t)}q(t) = \delta(t - t_0),$$

where $\delta(t - t_0)$ denotes Dirac's function centered at t_0 and $q(t)$ is the charge on the capacitor at time t. The inductance, resistance and capacitance are considered positive, $L(t) > 0$, $R(t) > 0$, $C(t) > 0$, differentiable and time dependent.

(a) We assume $\dfrac{R(t)}{\sqrt{L(t)}}$ is not constant. Use Hörmander's theorem to show that the operator

$$P = L\frac{d^2}{dx^2} + R\frac{d}{dx} + \frac{1}{C}$$

is hypoelliptic.

(b) Under the hypothesis of part (a), show that $\operatorname{singsupp} q(t) = \{t_0\}$.

(b) We assume $\frac{R(t)}{\sqrt{L(t)}} = k > 0$, constant. Show that in this case Hörmander's condition does not hold. Find in this case $\operatorname{singsupp} q(t)$ directly, by applying the Laplace transform method.

Exercise 9.8.2 Consider the action given by formula (9.2.13). Show that the van Vleck determinant is given by

$$V(T) = \sqrt{\frac{L\omega}{2\pi \sin(\omega T)}}.$$

Exercise 9.8.3 Prove Lemma 9.3.1, where the exponential is defined by the series $e^{At} = \sum_{k\geq 0} \frac{t^n}{n!} A^n$.

Exercise 9.8.4 Consider an electrical system consisting in a set of inductances $L_1, \dots, L_n$ and capacitances $C_1, \dots, C_n$, whose dynamics is defined by the Lagrangian

$$\mathcal{L}(q, \dot{q}) = \frac{1}{2}\sum_{k=1}^{n} L_k \dot{q}_k^2 - \frac{1}{2}\sum_{k=1}^{n} \frac{1}{C_k} q_k^2.$$

(a) Write the Euler-Lagrange equations with boundary conditions $q(0) = q_0$, $q(T) = q_T \in \mathbb{R}^n$ and solve them.

(b) Find the associated action $S_{cls}(x_0, x; t)$;

(c) Compute the van Vleck determinant $V(t)$;

(d) Find the heat kernel $K(x_0, x; t)$ of the operator

$$\frac{1}{2}\sum_{k=1}^{n} L_k \partial_{x_k}^2 - \frac{1}{2}\sum_{k=1}^{n} \frac{1}{C_k} x_k^2;$$

(e) Compute $Q(x_0, t) = \int_{\mathbb{R}^n} K(x_0, x; t)\, dx$.

Exercise 9.8.5 (The RL-circuit) Consider an electric circuit with Lagrangian given by $\mathcal{L}(\dot{q}, q) = \frac{1}{2}L\dot{q}^2 - Rq\dot{q}$, where L and R stand for the inductance and resistance, respectively.

(a) Show that the Euler-Lagrange equation is $\ddot{q} = -k\dot{q}$, where $k = R/L$;

(b) Verify that the solution of the Euler-Lagrange equation with boundary conditions $q(0) = q_0$ and $q(T) = q_T$ is given by

$$q(t) = \frac{e^{-kT} - e^{-kt}}{e^{-kT} - 1} q_0 + \frac{e^{-kt} - 1}{e^{-kT} - 1} q_T;$$

(c) Show that there are no conjugate points to q_0 long $q(t)$;

(d) Find the associated action $S(q_0, q_T; t)$;

(e) Compute the van Vleck determinant $V(t)$;

(f) Suggest a differential operator associated to the Lagrangian $\mathcal{L}$ such that $K(x_0, x; t) = V(t)e^{S(x_0, x_T; t)}$ is its heat kernel.

Epilogue

As we near the end of this extensive journey, it is time for us to provide some final remarks, address open problems, and suggest future research directions.

One of the primary objectives of this book has been to present an analytical approach to finding heat kernels for elliptic and sub-elliptic operators based solely on the properties of diffusions. It is worth noting that there exist several alternative methods, including the Fourier method, the geometric method, eigenfunction expansion, pseudo-differential operators, Laguerre calculus, and more. Each of these methods has its own advantages and disadvantages. The question we aimed to answer was whether the diffusion method surpasses the others in terms of effectiveness.

We have successfully derived closed-form expressions for the heat kernel of familiar operators that could also be addressed using alternative methods. The core idea behind our approach is to associate a diffusion process and compute its transition density. This is supposed to be a smooth function, since we require the bracket-generating condition to hold. While this approach works well for simple cases, it becomes more challenging and less straightforward for higher step scenarios, making it difficult to compute these probability transitions.

For instance, we were unable to compute the heat kernel for the Grushin operator with a step larger than 3 using the diffusion method. This computation requires closed-form expressions for the joint distribution of $(W_t, \int_0^t W_s^n ds)$ with $n \geq 2$. Additionally, a missing component in our framework is a Cameron-Martin type formula for $\mathbb{E}[e^{-\frac{\lambda^2}{2}\int_0^t W_s^n ds}]$, with $n \geq 3$. Similar restrictive considerations apply also for computing the heat kernel of the Martinet operator.

These limitations highlight potential areas for future research and further exploration. By addressing the challenges posed by more complex scenarios and developing additional tools and formulas, it may be possible to extend the diffusion method to cover a broader range of operators and problems.

On the other hand, it is important to note that the diffusion method proves to be effective for operators with a potential. While other alternative methods have successfully computed the cases of linear and quadratic potentials, we have taken a step further by applying the diffusion method to solve for an exponential potential (known as the Liouville potential). Interestingly, the Hamiltonian method, which is another approach employed in these computations, was unable to handle this specific case. Thus, our use of the diffusion method has provided a valuable contribution by extending the scope of solvable potentials and demonstrating its superiority in tackling the challenges posed by the Liouville potential.

Additionally, the diffusion method has proven to be successful in computing the transition density on Lie groups SE(2) and SH(2), showcasing its versatility in handling complex mathematical structures. Furthermore, its application to the knife edge diffusion is particularly noteworthy due to its significance in the field of human vision. This specific case holds practical implications and remains an active area of research, highlighting the ongoing efforts to understand and enhance our understanding of this phenomenon.

In addition to the aforementioned topics, the book also explores the application of stochasticity to Euclidean geometry. An intriguing area of investigation involved the study of Brownian triangles, where the vertices are represented by Brownian motions in the plane. Notably, the probability distribution and expectation of the area of Brownian triangles were computed. Building upon this, further research directions were identified, including the exploration of Brownian triangles on spheres, upper half planes, or other surfaces. These investigations aim to determine if the curvature of a given surface can be inferred from the area of the Brownian triangle. Another significant application involved the behavior of diffusions in the vicinity of a curve. It was observed that the probability of a Brownian motion diffusing in the exterior of a convex curve is dependent on the curvature at nearby points. Extending this concept to surfaces or submanifolds is an interesting area for future research, with the expectation of uncovering similar patterns and results. These investigations highlight the broad range of applications and potential research directions when applying stochastic processes to geometric settings.

One of the chapters in the book focuses on sub-elliptic systems of differential equations, delving into various aspects such as the existence of solutions (integrability conditions), uniqueness, and smoothness. In this chapter, we have included results that pertain to the most commonly employed prototypes in the realm of sub-Riemannian geometry, with the aim of constructing sub-elliptic versions of Poincaré's lemma. While attempts have been made to establish general results, it is important to note that the field remains an active area of scientific investigation.

Bibliography

[1] M. Abramowitz and I.A. Stegun. *Handbook of Mathematical Functions with Formulas, Graphs and Mathematical Tables.* New York, Dover, 1972.

[2] A. Agrachev, B. Bonnard, M. Chyba, and I. Kupka. *Sub-Riemannian sphere in Martinet flat case.* Control Optim. Calc. Var., Vol. 2, 377-448, 1997.

[3] A. Agrachev, U. Boscain, J.-P. Gauthier, and F. Rossi. *The intrinsic hypoellptic Laplacian and its heat kernel on unimodular Lie groups.* J. Funct. Anal. 256, 2621-2655, 2009.

[4] B. Alshamary, O. Calin, and D. C. Chang. *Global connectivity and optimization on an infinite step distribution.* Advances in Applied Mathematics Volume 50, Issue 4, 634-644, April 2013.

[5] D. G. Aronson. *Bounds for the fundamental solution of a parabolic equation.* Bull. of AMS, 73, 890-896, 1967.

[6] I. G. Avramidi. *Heat Kernel Method and its Applications.* Birkhauser, New York, 2015.

[7] R. Azencott. *Behavior of diffusion semi-groups at infinity.* Bull. Soc. Math. (France), 102, 193-240, 1974.

[8] A. Baddeley, I. Bárány, R. Schneider, and W. Weil. *Stochastic Geometry: Lectures given at the C.I.M.E. Summer School held in Martina Franca, Italy, September 13-18, (Lecture Notes in Mathematics Book 1892).* Springer, 2004.

[9] F. Baudoin. *An Introduction to the Geometry of Stochastic Flows.* Imperial College Press, 2004.

[10] R. Beals, B. Gaveau, and P. Greiner. *Hamilton-Jacobi theory and the heat kernel on Heisenberg groups.* J. Math. Pures Appl. 79, 2000, 633-689.

[11] D. R. Bell and S. E. A. Mohammed. *An extension of Hormander's theorem for infinitely degenerate second-order operators.* Duke Math. J. 78 (3) 453-475, 1995.

[12] A. Bellaïche and J. J. Risler (editors). *Sub-Riemannian Geometry.* Progress in Mathematics, 144, Birhkäuser, Basel, 1996.

[13] V. Benes and J. Rataj. *Stochastic Geometry: Selected Topics.* Kluwer Academic Publishers, 2001.

[14] A. M. Bloch. *Nonholonomic Mecanics and Control.* Vol. 24, Interdisciplinary Applied mathematics, Springer, New York, 2003.

[15] A. N. Borodin and P. Salminen. *Handbook of Brownian Motion - facts and Formulae.* Birhäuser, 2nd ed., Probability and Its Applications, 2015.

[16] U. Boscain, R. A. Chertovskih, J.-P. Gauthier, and A. O. Remizov. *Hypoelliptic diffusion and human vision: A semi-discrete and new twist.* SIAM Journal on Imaging Sciences, Vol. 7, Iss. 2, 2014.

[17] U. Boscain, J. Duplaix, J.-P. Gauthier, and F. Rossi. *Antropomorphic image reconstruction via hypoelliptic diffusion.* SIAM Journal on Control Optimization, Vol. 50, No. 3, 1309-1336, 2012.

[18] P. Bougerol. *Examples de théorèmes locaux sur les groupes résoluble.* Ann. Inst. H. Poincaré, 19, 369-391, 1983.

[19] M. Bramanti and L. Brandolini. *Hörmander Operators.* World Scientific, New Jersey, 2023.

[20] B. Błaszczyszyn, M. Haenggi, P. Keeler, and S. Mukh. *Stochastic Geometry Analysis of Cellular Networks.* Cambridge University Press, 2018.

[21] O. Calin. *Transience of diffusions on Heisenberg and Grushin distributions.* J. Geometry and Physics, 77, 131-142, 2014.

[22] O. Calin. *An Informal Introduction to Stochastic Calculus with Applications.* World Scientific, 2015.

[23] O. Calin. *Deterministic and Stochastic Topics in Computational Finance.* World Scientific, 2016.

[24] O. Calin and D. C. Chang. *The geometry on a step 3 Grushin operator.* Appl. Anal., Vol. 84, No. 2, 111-129, 2005.

[25] O. Calin and D. C. Chang. *Stochastic Trigonometry and Stochastic Invariants.* J. Nonlinear and Convex Optimization, 11.1, 157-174, 2010.

[26] O. Calin and D. C. Chang. *Sub-Riemannian Geometry - General Theory and Examples.* Encyclopedia of Mathematics and Its Applications, 126, Cambridge University Press, 2009.

[27] O. Calin, D. C. Chang, and M. Eastwood. *Integrability conditions for the Grushin and Martinet distributions.* Bull. Inst. Math. Acad. Sin. (N.S.), 8, No. 2, 159–168, 2013.

[28] O. Calin, D. C. Chang, and M. Eastwood. *Integrability conditions for Heisenberg and Grushin-type distributions.* Anal. Math. Phys. 4, No. 1-2, 99–114, 2014.

[29] O. Calin, D. C. Chang, K. Furutani, and C. Iwasaki. *Heat Kernels for Elliptic and Sub-elliptic Operators.* Birkhauser, Applied and Numerical Harmonic Analysis, 2011.

[30] O. Calin, D. C. Chang, and P. Greiner. *Geometric Analysis on the Heisenberg Group and Its generalizations.* AMS, IP, 2007.

[31] O. Calin, D. C. Chang, and J. Hu. *Poincaré's lemma on the Heisenberg group.* Adv. in Appl. Math. 60, 90–102, 2014.

[32] O. Calin, D. C. Chang, and J. Hu. *Integrability conditions on Engel-type manifolds.* Anal. Math. Phys. 5, No. 3, 217–231, 2015.

[33] O. Calin, D. C. Chang, and J. Hu. *Integrability conditions on a sub-Riemannian structure on S^3.* Anal. Math. Phys. 7, No. 1, 9–1, 2017.

[34] O. Calin, D. C. Chang, and Y. Li. *On heat kernels of a class of degenerate elliptic operators.* J. Nonlinear Convex Anal., 12 (8), 309-340, 2011.

[35] O. Calin, D. C. Chang, and J. Tie. *Fundamental Solutions for Hermite and Subelliptic Operators.* Journal d'analyse matématique (Jerusalem), Vol. 100(1), 223-248, 2006.

[36] O. Calin and D.C. Chang. *Geometric Mechanics on Riemannian Manifolds.* Birkhauser, Applied and Numerical Harmonic Analysis, Boston, 2004.

[37] O. Calin, D.C. Chang, and H. Fan. *The Keat Kernel for Kolmogorov Type Operators and its Applications.* J. Fourier Anal. Appl. 15, 2009, 816-838.

[38] O. Calin, D.C. Chang, and P. Greiner. *Geometric Analysis on the Heisenberg Group and Its Generalizations.* AMS/IP, 2007.

[39] O. Calin, D.C. Chang, and Y. Li. *Excursions in harmonic analysis.* Appl. Numer. Harmon. Anal., Birkhauser/Springer, New York, Vol. 2, 197-209, 2012.

[40] O. Calin, D.C. Chang, and S.T. Yau. *Nonholonomic systems and sub-Riemannian geometry.* Commun. Inf. Systems, International Press, Vol. 10, 4, 293-316, 2010.

[41] O. Calin and C. Udriste. *Geometric Modeling in Probability and Statistics.* Springer, 2014.

[42] P. Carr and M. Schröder. *Bessel processes, the integral of geometric Brownian motion, and Asian options.* Teor. Veroyatnost. i Primenen. 48 (3), 503–533, 2003.

[43] D.C. Chang and Y. Li. Heat kernels for a family of grushin operators. *Methods and Applications of Analysis 21(3):291-312*, 01 2014.

[44] D.C. Chang and Y. Li. Heat kernel asymptotic expansions for the heisenberg sub-laplacian and the grushin operator. *Proceedings of the Royal Society A: Mathematical, Physical and Engineering Sciences*, 471:20140943–20140943, 01 2015.

[45] I. Chavel. *Riemannian Geometry: A Modern Introduction, 2nd ed.* Cambridge Studies in Advanced Mathematics, No. 98, 2006.

[46] W. L. Chow. *Über Systeme von linearen partiellen Differentialgleichungen erster Ordnung.* Mathematische Annalen, 117: 98–105, doi:10.1007/bf01450011, 1939.

[47] G. Citti and A. Sarti. *A cortical based model of perceptional completion in the roto-translation space.* Journal of Mathematical Imaging and Vision, 24:307-326, 2006.

[48] M. Derridj. *Sur une classe d'opérateurs différentiels hypoelliptiques á coefficients analytiques.* Séminaire Goulaouic-Schwartz 1970-1971: Equations aux dérivées partielles et analyse fonctionnelle, Exp. 12, Centre de math., Ecole Polytechnique, 1971.

[49] M.P. do Carmo. *Riemannian geometry.* Mathematics: Theory and Applications. Translated from the second Portuguese edition by Francis Flaherty. Boston, MA: Birkhäuser Boston, 1992.

[50] C. Donati-Martin, R. Ghomrasni, and M. Yor. . *On certain Markov processes attached to exponential functionals of Brownian motion; application to Asian options.* Rev. Mat. Iberoamericana 17 (1), 179–193, 2001.

[51] H. Doss. *Lien entre équations différentielles stochastiques et ordinaires.* Ann. Inst. H. Poincaré, Prob. Stat. 13, 99-125, 1977.

[52] R. Duits and M. van Almsick. *The Explict SOlutions of left-inveriant Second Order Stochastic Evolution Equationson the 2D Euclidean Motion Group.* Quarterly of Applied Mathematics, Vol. 66, No. 1 (March 2008), 27-67, March 2008.

[53] R. Durrett. *Stochastic Calculus: A Practical Introduction.* CRC Press, Boca Raton, Florida, 1996.

[54] D. Coupier (editor). *Stochastic Geometry: Modern Research Frontiers.* Springer, Lecture Notes in Mathematics, 2237, 2019.

[55] V. S. Fedii. *On a criterion for hypoellipticity.* Mat. Sb. (N.S.), Volume 85(127), Number 1(5), 18–48, 1971.

[56] R. Feynman. *Space-time Approach to Non-relativistic Quantum Mechanics. Rev. Mod. Phys.* 20, 367-387, 1948.

[57] G. B. Folland. *A fundamental solution for a subelliptic operator.* Bull. Amer. Math.Soc., 79, 373–376, 1973.

[58] G. B. Folland. *Introduction to PDEs.* second ed. Princeton University Press, New Jersey, 1995.

[59] G. Frobenius. *Über das Pfaffsche Problem.* J. Reine Angew. Math. (82): 230–315, 1877.

[60] B. Gaveau. *Principe de moindre action, propagation de la chaleur et estimées sous elliptiques sur certains groupes nilpotents.* Acta Math. 139, 95-153, 1977.

[61] B. Gaveau. *Systémes Dynamiques Associés a certains Opérateurs Hyppoelliptiques.* Bull. Sc. Math. 102, 203-229, 1978.

[62] B. Gaveau and P. Greiner. *On geodesics in subRiemannian geometry.* Bull. Inst. Math. Acad. Sinica (New Series), Vol. 1, 79-209, 2006.

[63] B. Gaveau and P. Greiner. *On geodesics in subRiemannian geometry II.* Anal. Appl., Vol. 5, 413-414, 2007.

[64] P. Greiner and O. Calin. *On subRiemannian geodesics.* Annal. Appl. Vol. 1, No. 3, 289-350, 2003.

[65] P. C. Greiner. *A Hamiltonian Approach to the Heat Kernel of a SUb-laplacian on* S^{2n+1}. Analysis and Applications, Vol. 11, No. 6, 2013.

[66] P. C. Greiner and Y. Li. *Heat Kernels, Old and New.* Bulletin of the Institute of Mathematics Academia Sinica (New Series), Vol. 12, No. 1, 1-37, 2017.

[67] A. Grigor'yan. *On stochastically complete manifolds, (in Russian).* DAN SSSR, 290 (1986) No. 3, 534-537. Engl. transl.: Soviet Math. Dokl., 34 (1987) No. 2, 310-313, 1986.

[68] A. Grigor'yan. *Stochastic Completeness of Markov Processes.* Lecture course at CUHK, Feb.- March, 2011.

[69] A. Gulisashvili. *Analytically Tractable Stochastic Stock Price Models.* Springer-Verlag, Berlin, Heidelberg, 2012.

[70] R. Haberman. *Elementary Applied Partial Differential Equations with Fourier Series and Boundary Value Problems, 3rd ed.* Prentice Hall, New Jersey, 1983.

[71] M. Haenggi. *Stochastic Geometry for Wireless Networks.* Cambridge University Press, 2013.

[72] M. Hairer. *On Malliavin's proof of Hörmander's theorem.* Bull. Sci. Math. 135, 650-666, 2011.

[73] P. Hartman and G. S. Watson. *'Normal' distribution functions on spheres and the modified Bessel functions.* Ann. Prob. **2**, 593-607, 1974.

[74] K. Helmes and A. Schwane. *Lévy Stochastic Area Formule in Higher Dimensions.* J. Funct. Anal. 54, 177-192, 1983.

[75] S.L. Heston. *A Closed-Form Solution for Options with Stochastic Volatility with Applications to Bonds ad Currency Options.* Review of Financial Studies, 6, 92, 327-343, 1993.

[76] J. E. Hilliard and L. Lawson. *Stereology and Stochastic Geometry.* Kluwer Academic Publishers, 2004.

[77] L. Hörmander. *Hypoelliptic second order differential equations.* Acta Math. 119, 147–171, 1967.

[78] R. A. Horn and C. R. Johnson. *Matrix Analysis (2nd ed.)*. Cambridge University Press, 2013.

[79] A. Hulanicki. *The distribution of energy in the Brownian motion in the Gaussian field and analytic-hypoellipticity of certain subelliptic operators on the Heisenberg group*. Studia Mathematica, 56, 165-173, 1976.

[80] N. Ikeda and S. Watanabe. *Stochastic Differential Equations and Difussion Processes*. North-Holland (2nd ed., 1989.

[81] D. V. Isagurova. *A Vector Field Potentiality Criterion in Sub-Riemannian Geometry*. Matematicheskie Zametki, Vol 102, No. 6, 949-954, 2017.

[82] K. Ito and H. P. McKean. *Diffusion Processes and Their Sample Paths*. Springer-Verlag, 1965.

[83] K. Ito and H.P. McKean. *Diffusion Processes and their Sample Paths*. Springer-Verlag, Berlin, Heidelberg, New York, 1974.

[84] D. Jeltsema. *Memory Elements: A Paradigm Shift in Lagrangian Modeling of Electrical Circuits*. IFAC Proceedings Volumes Vol. 45, Issue 2, 445-450, 2012.

[85] I. Karatzas and S. E. Shreve. *Brownian Motion and Stochastic Calculus, 2nd ed.* Springer-Verlag, 1991.

[86] K.C. Khandekar and S.V. Lawande. *Path Integral for a Three Body Problem*. J. Phys. A, 5, 812, 1972.

[87] K.C. Khandekar and S.V. Lawande. *Exact Propagator for a Time-Dependent Harmonic Oscillator with and without a Singular Perturbation*. J. Math. Phys. 16, 384, 1975.

[88] D. Khoshnevishan, P. Salminen, and M. Yor. *A note on a.s. finiteness of perpetual integral functionals of diffusions*. Elect. Comm. in Probab. 11, 2006, 108-117.

[89] J. J. Kohn. *Pseudo-differential operators and hypoellipticity*. Proc. Symp. Pure Math. 23, 61-69, 1973.

[90] A. N. Kolmogorov. *Zufallige Bewegungen*. Annals of Mathematics, 35, No. 1, 116-117, 1934.

[91] A. Kristály. *Poincaré's lemma on some non-Euclidean structures*. Chinese Annals of Mathematics, Series B, Vol. 39, iss. 2, 297-314, 2018.

[92] P. H. Labordére. *Analysis, Geometry, and Modeling in Finance: Advanced Methods in Option Pricing.* Chapman and Hall/CRC Financial Mathematics Series Book 13, 2009.

[93] J. Lamperti. *Probability.* W. A. Benjamin, Inc., 1966.

[94] P. Langevin. *Sur la theorie du mouvement brownien.* C. R. Acad. Sci. (Paris) 146, 530-533, 1908.

[95] L. Lao and E. Orsingher. *Hyperbolic and fractional hyperbolic Brownian motion.* Stochastics, 79, No. 6, 505–522, 2007.

[96] R. Léandre. *Minoration en temps petit de la densité d'une diffusion dégénérée.* J. Funct. Anal. 74, No. 2, 399-414, 1987.

[97] R. Léandre. *Majoration en temps petit de la densité d'une diffusion dégénérée.* Probability Theory and Related Fields 74(2), 289-294, June 1987.

[98] N.N. Lebedev. *Special Functions and their Applications.* Dover, New York, 1972.

[99] P. Lévy. *Le mouvement Brownien plan.* Amer. J. Math. 62, 487-550, 1940.

[100] P. Lévy. *Processus stochastiques et mouvement Brownien.* Gauthier-Villars, Paris, 1948.

[101] P. Lévy. *Calcul des probabilités-fonctions aléatoires Laplaciennes.* C. R. Acad. Sci. Paris Ser. A-B 229, 1057-1058, 1949.

[102] P. Lévy. *Calcul des probabilités-sur l'aire comprise entre un arc de la courbe du mouvement Brownien plan et sa corde.* C.R. Acad. Sci. Paris Ser. A-B 230, 432-434; Errata 689, 1950.

[103] P. Lévy. *Wiener's random functions, and other Laplacian random functions.* Proceedings 2nd Berkeley Symp., 1951.

[104] P. Malliavin. *Stochastic calculus of variations and hypoelliptic operators.* In: Proc. Inter, Symp. on Stoch. Diff. Equations, Kyoto. Wiley, 195-263, 1976.

[105] E. Mathieu. *Mémoire sur Le Mouvement Vibratoire d'une Membrane de forme Elliptique.* J. Math. Pures Appl. 13, 137–203, 1868.

[106] H. Matsumoto and M. Yor. *Exponential functionals of Brownian motion I. Probability laws at fixed time.* Probab. Surv. 2, 312–347, 2005.

[107] H. P. McKean. *Stochastic Integrals.* Academic Press, 1969.

[108] H.P. McKean. *An upper bound to the spectrum of Δ on a manifold of negative curvature.* J. Diff. Geometry, 4, 359-366, 1970.

[109] P. A. Meyer. *Probability and Potentials.* Blaisdell, 1966.

[110] R. S. Millman and G. D. Parker. *Elements of Differential Geometry.* Prentice-Hall Inc., Upper Saddle River, New Jersey, 1977.

[111] S. A. Molchanov. *Diffusion Processes and Riemannian Geometry.* Russian Math. Surveys 30: 1-63, 1975.

[112] D. Mumford. *Elastica and computer vision.* Algebraic Geometry and Its Applications. Springer- Verlag, 491-506, MR1272050 (95a:92026), 1994.

[113] D. Nualart. *The Malliavin calculus and related topics.* Springer, Berlin Heidelberg New-York, 1995.

[114] B. Øksendal. *Stochastic Differential Equations, An Introduction with Applications, 6th ed.* Springer-Verlag Berlin Heidelberg New-York, 2003.

[115] O. A. Oleinic and E. V. Radkevic. *Second order equations with non negative characteristic form.* A.M.S. and Plenum Press, 1973.

[116] G. Peccati and M. Reitzner (editors). *Stochastic Analysis for Poisson Point Processes: Malliavin Calculus, Wiener-Itô Chaos Expansions and Stochastic Geometry.* Springer, Bocconi University Press, Vol. 7, 2016.

[117] S. Pigola, M. Rigoli, K. Furutani, and A. G. Setti. *Maximum principles at infinity of Riemannian manifolds: an overview.* Matemática Contemporanea, Vol 31, 81-128, 2006.

[118] J. Pitman and M. Yor. *A guide to Brownian motion and related stochastic processes.* arXiv: Probability, 2018.

[119] P. K. Rashevskii. *About connecting two points of complete nonholonomic space by admissible curve (in Russian).* Uch. Zapiski Ped. Inst. Libknexta (2): 83–94, 1938.

[120] D. Revuz and M. Yor. *Continuous Martingales and Brownian Motion, 3th ed.* Springer-Verlag, Berlin, 2005.

[121] L. C. G. Rogers and J. W. Pitman. *Markov Functions.* The Ann. of Probability, Vol. 9, No. 4, 573-582, 1981.

[122] L.C.J. Rogers and D. Williams. *Diffusions, Markov Processes, and Martingales, Volume 2: Ito Calculus, 2nd edition,.* Cambridge University Press, Cambridge, 2000.

[123] P. Salminen and M. Yor. *Properties of perpetual integral functionals of Brownian motion with drift.* Ann. I. H. Poincar/'e, PR 41, 2005, 335-347.

[124] R. Schneider and W. Weil. *Stochastic and Integral Geometry.* Springer-Verlag, Probability and Its Applications, Berlin, 2008.

[125] L. S. Schulman. *Techniques and Applications of Path Integration.* Dover Publications, Inc. New York, 2005.

[126] L. Schwartz. *Théorie des distributions.* Publications de l'Institut de Mathématiquede l'Université de Strasbourg, No. IX-X. Nouvelle édition, entiérement corrigée,refondue et augmentée. Hermann, Paris, 1966.

[127] D. Stoyan, W. S. Kendall, S. N. Chiu, and J. Mecke. *Stochastic Geometry and Its Applications.* Wiley, Wiley Series in Probability and Statistics, 2013.

[128] D. W. Stroock. *On the growth of stochastic integrals.* Z. Wahr. verw. Geb. 18, 340–344, 1971.

[129] H. Sussmann. *On the gap between deterministic and stochastic ordinary differential equations.* Ann. Probab. 6, 19-41, 1978.

[130] H. J. Sussmann. *Orbits of families of vector fields and integrability of distributions.* Trans. Amer. Math. Soc. 180, 171-188, 1973.

[131] R. Urban. *An Explicit Formula for the Heat Kernel of a Left Invariant Operator.* ALEA, Lat. Am. J. Probab. Math. Stat. 11 (1), 299-306, 2014.

[132] S. R. S. Varadhan. *Diffusion processes in a small time interval.* Comm. Pure Appl. Math. 20, 659-685, MR 36, No. 970, 1967.

[133] S. R. S. Varadhan. *On the behaviour of the fundamental solution of the heat equation with variable coefficients.* Comm. Pure Appl. Math. 20, 431-455, MR 34, No. 8001, 1967.

[134] A. Voros. *The return of the quartic oscillator. The complex WKB method.* Annales de l'I.H.P. Physique théorique, Tome 39, No. 3, 211-338., 1983.

[135] G. Vranceanu. *Sur les espaces non holonomes.* C. R. Acad. Sci. Paris, 183, 852-854, 1926.

[136] G. N. Watson. *A Treatise on the Theory of Bessel Functions.* Cambridge University Press, second ed., 1996.

[137] T. A. Werne. *Constrained Lagrangian Circuits: Final Report for CDS 205.* Physics, 2009.

[138] D. V. Widder. *The Heat Equation.* Academic Press, London, 1975.

[139] S. T. Yau. *On the heat kernel of a complete Riemannian manifold.* J. Math. Pures Appl., ser. 9, 57, 191-201, 1978.

[140] M. Yor. *Loi de l'indice du lacet brownien et distribution de Hartman-Watson.* Z, Wahrscheinlichkeitsth, 53, 71-95, 1980.

[141] M. Yor. *Remarques sur une formule de Paul Lévy.* Séminaire de Probabilités XIV Lect. Notes in Maths. 784, Springer-Verlag, Berlin, 1980.

[142] M. Yor. *On some exponential functionals of Brownian motion.* Adv. Appl. Prob., 24, 509-531, 1992.

[143] M. Yor. *Exponential functionals of Brownian motion and related processes.* Springer-Verlag, Berlin. ISBN 3-540-65943-9. With an introductory chapter by H´elyette Geman, Chapters 1, 3, 4, 8 translated from the French by Stephen S. Wilson. MR1854494, 2001.

[144] Y. Zhou and G. S. Chirikjian. *Probabilistic models of dead-reckoning error in nonholonomic mobile robots,.* Robotics and Automation, Proceedings ICRA '03. IEEE International Conference, 1594-1599, 2003.

Index

www.ingramcontent.com/pod-product-compliance
Lightning Source LLC
LaVergne TN
LVHW010340220926
848484LV00004BB/940